W0080441

Fractals and Dynamic Systems in Geoscience

Edited by
Diego Perugini
Jörn H. Kruhl

Previously published in *Pure and Applied Geophysics* (PAGEOPH), Volume 172, No. 7, 2015

 Birkhäuser

Editors
Diego Perugini
University of Perugia
Department of Physics and
Geology
Piazza Università
06100 Perugia
Italy

Jörn H. Kruhl
TU München
Applied Crystallography and
Materials Sciences
Department for Earth and
Environmental Sciences
Theresienstr. 41
80333 München
Germany

ISBN 978-3-0348-0935-1   ISBN 978-3-0348-0936-8 (eBook)
DOI 10.1007/978-3-0348-0936-8

Library of Congress Control Number: 2015945232

Springer Basel Heidelberg New York Dordrecht London
© Springer Basel 2016

This work is subject to copyright. All rights are reserved, whether the whole or part of the material is concerned, specifically the rights of translation, reprinting, re-use of illustrations, recitation, broadcasting, reproduction on microfilms or in other ways, and storage in data banks. For any kind of use, permission of the copyright owner must be obtained.

Cover illustration: Fractal structure arising during forced injection of qz-monzonitic magma into a granitoid magma in the Aztec Wash Pluton (Nevada, USA). Photo: D. Perugini

Cover design: deblik, Berlin.

Printed on acid-free paper

Springer Basel AG is part of Springer Science+Business Media

www.springer.com

Contents

Pure Appl. Geophys. 172 (2015), 1781–1785
© 2015 Springer Basel
DOI 10.1007/s00024-015-1047-4

Pure and Applied Geophysics

Introduction to the Topical Volume "Fractals and Dynamic Systems in Geoscience"

Diego Perugini[1] and Jörn H. Kruhl[2]

1. Introduction

Chaotic dynamic systems and nonlinear processes, together with the resulting fractals and multifractals, are becoming increasingly fundamental for analyzing data and understanding processes in the Earth and environmental sciences. Many processes and phenomena, poorly known only a few years ago, can now be studied and understood with the help of conceptual models from the fields of fractals and dynamic systems. This represents a bold step towards the aim of understanding how the planet Earth works.

The focal point of this volume is the use of fractals and dynamic systems in analyzing and studying a variety of geological and geophysical processes. The aim is to provide the reader with a collection of papers highlighting the usefulness of concepts and methods belonging to fractals and chaotic dynamic systems in understanding geological processes that would be otherwise very hard or even impossible to study using only classical conceptual approaches.

This volume contains some contributions to the 6th International Conference on Fractals and Dynamic Systems in Geoscience, held in Perugia (Italy), September 30–October 2, 2013.

This forum was dedicated to recent developments in the application of fractals and dynamic systems to Earth systems, with emphases on predictability of geological risks, natural resources, and climate change.

The 21 papers presented in this volume cover well the wide range of applications of fractals and dynamic systems to the Earth and planetary sciences. As a whole, they represent the leading-edge research dedicated to the understanding of the mechanisms and processes responsible for the nonlinear dynamics triggering and characterizing geological processes.

The papers are divided into five main groups, including petrology/volcanology, seismology, geomorphology, space science, and other topics. The many fields covered by the works presented in this volume testify to the unique multi- and transdisciplinary character of the conceptual models belonging to fractal geometry and dynamic systems, spanning virtually the entire constellation of geological disciplines and beyond.

2. Petrology/Volcanology

Soesoo and Bons provide measurements of migmatitic leucosomes and granitic veins in drill cores. The authors show that the cumulative width distribution of the studied leucosomes/granitic veins follows a power law with exponents between 0.7 and 1.8. The authors also focus on fractal statistics of granitoid pluton sizes: the cumulative sizes follow power-law distributions with exponents between 0.6 and 0.8. These results support the model that the crust develops a self-organized critical state during magma generation. In this state, magma batches accumulate in a noncontinuous, stepwise manner to form ever larger accumulations. The authors suggest that there is no characteristic length or time scale in partial melting or its products and that the smallest melt segregations and km-scale plutons form the end-members of a continuous chain of mergers of magma batches.

[1] Department of Physics and Geology, University of Perugia, Perugia, Italy. E-mail: diego.perugini@unipg.it

[2] Faculty of Civil, Geo and Environmental Engineering, Technical University Munich, Munich, Germany.

The study by Albert et al. is focused on examination of magma mingling and mixing, expressed as basanitic enclaves within a phonolitic host lava. The authors report that the morphology of enclaves varies from rounded to complex finger-like structures that can be quantified by fractal geometry methods. It is shown that the logarithm of the viscosity ratio between the phonolitic magma and the basanitic enclaves ranges between 0.39 and 0.81, with most values clustering around 0.49. This allowed the authors to constrain the water content of interacting magmas. The usefulness of fractal analyses to determine preeruptive conditions of magmas is highlighted in this study.

De Campos focuses on the mixing between basalt and granite in a deep plutonic environment. The description of flow patterns and measurements of their fractal dimensions, and the evaluation of geochemical data from a Cambro-Ordovician granitic pluton are presented and discussed. Measurements of fractal dimension of flow patterns are performed at different scales. The author suggests that the compositional variability and the flow patterns of the studied pluton are greatly controlled by the onset of nonlinear dynamics that triggered chaotic mixing between a granitic and a basaltic end-member.

El Omari et al. examine numerically the cooling of a basaltic melt undergoing chaotic advection in a two-dimensional (2D) cavity with moving boundary. It is shown that different cooling rates can be obtained during thermal mixing of a single basaltic magmatic batch and that this process can induce complex temperature patterns inside the magma chamber. Results indicate that the emergence of chaotic dynamics strongly modulates temperature fields over time and greatly increases the cooling rates. This mechanism has implications for the thermal lifetime of the magmatic body and may favor the appearance of chemical heterogeneities in the igneous system as a result of different crystallization rates. The authors suggest that results from this study can provide explanations for some natural features that, to date, have received unsatisfactory explanations such as the production of magmatic enclaves, compositional zoning in mineral phases, and the generation of large-scale compositional zoning observed in many plutons worldwide.

The work of Buccianti focuses on the study of the multivariate nature of geochemical data. The author proposes an approach that can be used to investigate the shape of the frequency distribution of multivariate geochemical indices. The approach is based on the use of the transformations of the log-ratio family. The methodology proposed by the author shifts attention to the modeling of the frequency distribution of more complex indices, linking all the terms of the composition to better represent the dynamics of geochemical processes. The application to the chemistry of 616 ocean-floor basaltic glasses from the Abyssal Volcanic Glass Data File is presented.

3. Seismology

Şuţeanu performed analysis on multiple sets of earthquake networks created for the Hawaii volcanic system. The author indicates that the scale-free behavior of the connectivity distribution along the spectrum of the minimum weight values can be used to discern the interrelated earthquakes from the rest of the data set. It is highlighted that the patterns in the distributions of temporal and spatial intervals between earthquakes are similar from large to small networks. Moreover, similarities are found between the variation of the network clustering coefficient, and the variation of the exponents of the connectivity distribution and of the weight distribution. The author suggests that the synchronous variation over successive temporal windows can be related to changes in seismicity and in the life of the volcanic system.

The work by Nelson et al. focuses on understanding how a layered basalt sequence affects the propagation of a seismic wave. The authors construct detailed realistic models of basalt sequences, using fractal properties derived directly from outcrop analogues including roughness of basaltic lava flows, the latter captured using terrestrial laser scanning and satellite remote sensing. Synthetic lava flow surfaces were reconstructed using a von Karman power spectrum. P-wave velocity data were then added, and the resulting model was used to generate synthetic seismic data. The resulting stacked section shows that the ability to resolve the internal structure of the lava flows is quickly lost due to scattering and attenuation by the basalt pile. A further result from generating wide-angle data is that the appearance of a lower-velocity layer below the basalt sequence may be caused by destructive interference within the basalt itself.

The work by Matcharashvili et al. investigates the dynamical features of seismic activity in Central Asia, where strong electromagnetic (EM) soundings were performed, revealing the impact of strong electromagnetic discharges on the microseismic activity of the investigated area. Using a variety of time series analysis techniques, such as wavelet transformation, Hilbert–Huang transformation, detrended fluctuation analysis, and recurrence quantification analysis, the authors show that manmade high-energy EM irradiation essentially affects the dynamics of the seismic process in the investigated area in its temporal and spatial domains.

Michas et al. use a multifractal approach to study the time dynamics of the recent earthquake activity in the Corinth rift. The results indicate a heterogeneous clustering degree and correlations acting at all time scales, suggesting a strongly non-Poissonian behavior. In addition, the authors show that the multifractal analysis in different time periods indicates that the degree of multifractality exhibits strong variations with time, which are associated with the dynamic evolution of the earthquake activity in the rift and the transition between periods of high and low seismicity.

The work of Papadakis et al. presents analysis of the earthquake magnitude distribution for the period 1990–1998, in a broad area surrounding the epicenter of the 1995 Kobe earthquake. The authors perform a frequency–magnitude distribution analysis in the context of nonextensive statistical physics. The nonextensive parameter q_M, which is related to the frequency–magnitude distribution, is found to highlight the presence of long-range correlations and is used as an index of the physical state of the studied area. The authors report a significant increase of q_M some months before the strong earthquake (on April 9, 1994), indicating the start of a preparation phase towards the Kobe earthquake.

In order to disclose the inner time properties of complex seismograms, Telesca et al. analyze the time dynamics of P-waves of seismograms of tsunamigenic earthquakes and nontsunamigenic events, using the Fisher–Shannon method and multifractal detrended fluctuation analysis. Using jointly these two methods, the authors define a classifier, whose performance was tested by means of the receiver-operating characteristic curve that plots the true positive rate versus the false positive rate. It is shown that the classifier shows a discrimination power that can be useful for early warning of tsunami events.

Polyakov et al. explore the possibility of earthquake prediction, proposing an approach based on analysis of common short-term candidate precursors with subsequent processing of brain activity signals generated in rats. They report that the candidate precursors are observed as synchronized peaks in the nonstationarity factors, introduced within the flicker-noise spectroscopy framework for signal processing, for the high-frequency component of time series. Moreover, the authors hypothesize that these peaks correspond to the local reorganizations of the underlying geophysical system, which are believed to precede strong earthquakes. They suggest that the rodent brain activity signals could be selected as potential "immediate" (up to 2 weeks) deterministic precursors for earthquakes.

4. Geomorphology

Liucci et al. focus on the fractal analyses of landslides with an attempt to identify a possible structure in their spatial pattern. The authors apply the box-counting algorithm to maps of landslide triggering points and landslide areas and identify a scale-invariant structure. They recognize two distinct types of fractal behavior, separated by a scale value of 1 km and characterized by capacity dimensions of 1.35–1.76, in the ranges of 25 m to 1 km and 1–16 km, respectively. The authors show that the higher capacity dimension describes the spatial distribution of landslides, whereas the lower one contains information about their geometries. They also suggest that the contribution of each causal factor (i.e., predisposing and triggering factors) for the occurrence of landslide events and their spatial development could be different in the two ranges of scales identified, depending on its spatial variability at the local and the regional scale.

The work by Donadio et al. is based on an estimate of the fractal dimension of the drainage network of three large watercourses having different geologic context and tectonic styles. Their aim is to better understand the morpho-evolutionary processes of these fluvial morphotypes, to classify hydrographic patterns, and finally to compare fractal degree with some geomorphic quantitative indexes. It is suggested that, according to the geological setting and geomorphic indexes of these basins, the lower fractal dimension indicates a prevailing tectonic activity, whereas the higher fractal dimension appears to be related to stronger erosion processes on inherited landscapes.

Paliaga examines by fractal analysis the triangular facets that arise in different geomorphic systems characterizing three river basins. It is shown that the spatial dispersion of erosional facets in the subbasins, corresponding to equally distributed closure sections along the main riverbed, follows power-law distributions. The author suggests that the number of facets is related to the hydrographical network structure and reflects the way erosion cycles, induced by base level changes, behaved in the catchment.

5. Space Science

Mancinelli et al. use the data gathered during the Mariner10 and MESSENGER missions on Mercury to classify impact craters into four classes related to four main geochronological units. They show that, for each crater class, size frequency distributions show a fractal behavior. The fractal dimension is different for each crater distribution, and this allows the authors to advance a new hypothesis for the crustal structuring of Mercury due to a complex interplay between endogenic and exogenic processes.

Based on fractal geometry methods, Hossain and Kruhl studied the shock-induced fragmentation structures of basement rocks and their limestone cover in and around the Ries impact crater (Germany). Quantification was performed by automated procedures and in areas of square centimeters to square decameters. In 2D and on all scales, the fragmentation structures form complex, statistically self-similar patterns highlighting that three different

power-law relationships exist, which might reflect the effect of three fragmentation processes. Moreover, the authors show that fracture patterns are anisotropic and inhomogeneous over large areas. It is suggested that the approach proposed in their work could shed new light on impact-related fragmentation processes.

Chang et al. review a recent method, rank-ordered multifractal analysis (ROMA), explicitly constructed to analyze the intricate details of the distribution and scaling of intermittent structures. The authors apply this method to the analyses of selected examples related to the dynamical plasmas of the cusp region of the Earth's magnetosphere, velocity fluctuations of classical hydrodynamic turbulence, and the distribution of the structures of the cosmic gas obtained through large scale, moving mesh simulations. It is suggested that the ROMA method can be successfully applied to study the similar multifractal processes in extreme environments of near-Earth surroundings.

6. Other Topics

Based on fractal statistics, Blenkinsop examines lode gold deposits, geothermal wells and volcanoes, and conventional and unconventional gas wells. The author shows that mass dimensions and scaling exponents generally increase from the lode gold through geothermal wells to gas data sets, reflecting decreasing degrees of clustering. The author argues that, as all the natural resources in his study are formed by fluid fluxes in the crust, the percolation theory is an appropriate unifying framework to understand their significance. Following this idea, it is shown that none of the percolation networks that formed the studied deposits reached the percolation threshold.

The work by Şuţeanu is focused on the analysis of daily minimum and maximum surface air temperature time series from 15 high-latitude Arctic stations from Canada, Norway, and the Russian Federation. The author applies a range of analysis methods. Statistical L-moments were determined for temporal windows of different lengths. L-skewness was found to change towards more positive values, reflecting an enhancement of warm spells. In addition, Haar wavelet analysis was applied both to the entire time series and to running windows.

Persistence diagrams were generated, based on running windows advancing through time and on local slopes of Haar analysis graphs. It is suggested that this approach offers a more nuanced view on variability by reflecting its change over time on a range of temporal scales. In the light of the obtained results, the author suggests that explanations for the discrepancy between variability perception and results of pattern analysis need to be explored using an integrative approach to weather variables such as air temperature, cloud cover, precipitation, and wind.

The work of Czechowski focuses on the microscopic mechanisms that can be responsible for elongation of tails of cluster size distributions, using cellular automata. The author shows that only the appropriate forms of rebound function can lead to inverse-power tails if densities of the grid are small or moderate. For larger densities, correlations between clusters become significant and lead to elongation of tails and flattening of the distribution to a straight line on log–log scale. The author suggests that the microscopic mechanism, given by the rebound function, included in simple one-dimensional (1D) random domino automaton (RDA), can be projected on the geometric mechanism which favors larger clusters in 2D RDA.

As guest editors, we would like to thank all the authors for their efforts and patience regarding this topical volume. We also wish to acknowledge the input of these and all the other participants in making the 6th International Conference on Fractals and Dynamic Systems in Geoscience in Perugia a success. We express our appreciation for the work of the reviewers, and Editor-in-Chief for Topical Issues, Dr. Renata Dmowska, who greatly helped in managing this volume, and Mrs. Priyanka Ganesh for her patience and help with the several problems we encountered in managing submitted manuscripts with the online submission system.

(Received January 28, 2015, accepted January 28, 2015, Published online February 12, 2015)

Pure Appl. Geophys. 172 (2015), 1787–1801
© 2014 Springer Basel
DOI 10.1007/s00024-014-0995-4

| Pure and Applied Geophysics

From Migmatites to Plutons: Power Law Relationships in the Evolution of Magmatic Bodies

Alvar Soesoo[1] and Paul D. Bons[2]

Abstract—Magma is generated by partial melting from micrometre-scale droplets at the source and may accumulate to form >100 km-scale plutons. Magma accumulation thus spans well over ten orders of magnitude in scale. Here we provide measurements of migmatitic leucosomes and granitic veins in drill cores from the Estonian Proterozoic basement and outcrops at Masku in SW Finland and Montemor-o-Novo, central Portugal. Despite the differences in size and number of measured leucosomes and magmatic veins, differences in host rock types and metamorphic grades, the cumulative width distribution of the studied magmatic leucosomes/veins follows a power law with exponents usually between 0.7 and 1.8. Published maps of the SE Australian Lachlan Fold Belt were used to investigate the distribution of granitoid pluton sizes. The granites occupy ca. 22 % of the 2.6×10^5 km^2 area. The cumulative pluton area distributions show good power law distributions with exponents between 0.6 and 0.8 depending on pluton area group. Using the self-affine nature of pluton shapes, it is possible to estimate the total volume of magma that was expelled from the source in the 2.6×10^5 km^2 map area, giving an estimated 0.8 km^3 of magma per km^2. It has been suggested in the literature that magma batches in the source merge to form ever-bigger batches in a self-organized way. This leads to a power law for the cumulative distribution of magma volumes, with an exponent m_V between 1 for inefficient melt extraction, and 2/3 for maximum accumulation efficiency as most of the volume resides in the largest batches that can escape from the source. If $m_V \geq 1$, the mass of the magma is dominated by small batches; in case $m = 2/3$, about 50 % of all magma in the system is placed in a single largest batch. Our observations support the model that the crust develops a self-organized critical state during magma generation. In this state, magma batches accumulate in a non-continuous, stepwise manner to form ever-larger accumulations. There is no characteristic length or time scale in the partial melting process or its products. Smallest melt segregations and >km-scale plotuns form the end members of a continuous chain of mergers of magma batches.

Key words: Partial melting, migmatites, leucosome, pluton size distribution, power law, fractal, self-organized criticality.

1. Introduction

The formation of felsic to intermediate magmatic intrusions is commonly viewed as a three-step process which involves (1) segregation and accumulation of melt in a partially molten source, (2) magma/melt[1] ascent, and (3) emplacement of magma (Sawyer 1994; Brown 1994; Clemens 1997, 1998; Petford et al. 2000). Melt is generated in the crust or mantle by partial melting of rocks on the μm- to mm-scale and is followed by accumulation and ascent to form >km-scale volumes in the form of plutons, batholiths, and volcanic formations. Thus, the whole range of magmatic process may involve more than 10 orders of magnitude in length scale and deals with a variety of physical–chemical processes on different scales and different levels within the crust and/or mantle. Time scales also vary over many orders of magnitude, from seconds to hours for the propagation of dykes (Emerman and Marrett 1990; Lister and Kerr 1991; Clemens and Mawer 1992) to several millions of years for a thermal event that causes partial melting (Brown et al. 1999; Petford et al. 2000).

Partial melting experiments suggest that the initial melt resides at grain junctions in isolated microscopic melt pockets or forms a thin film of liquid along grain boundaries (Jurewicz and Watson 1984; Walte et al. 2003, 2007; Sawyer 2014). The classical view is that melt segregation or drainage from the solid rock matrix starts when a large number of such grain-scale melt domains become connected to allow the melt to percolate through the rock (Sawyer 2001; Brown and

[1] Institute of Geology, Tallinn University of Technology, Ehitajate tee 5, Tallinn 19086, Estonia. E-mail: alvar.soesoo@ttu.ee

[2] Department of Geosciences, Eberhard Karls University Tübingen, Wilhelmstr. 56, 72074 Tübingen, Germany. E-mail: paul.bons@uni-tuebingen.de

[1] "Melt" is pure molten (liquid) rock, while "magma" refers to liquid melt that may contain floating solid crystals or entrained pieces of solid rock. In this paper we use the term "magma" throughout.

Solar 1999; Wark et al. 2003; Jackson et al. 2003; Hasalova et al. 2008).

Migmatites are the link between initial melt accumulation from microscopic melt films and droplets, and bigger magma volumes. The term migmatite was introduced by the Finnish petrologist, J. Sederholm, who derived this term from the Greek word for mixture. Migmatites are composite rocks, which display both metamorphic and magmatic components. The magmatic component is usually found as patches or veins, so called leucosomes, of frozen magma. Several ideas have been proposed for the formation of migmatites, such as partial melting (e.g. Winkler 1961), injection of foreign magmas (Sederholm 1907), metamorphic differentiation (Ashworth and McLellan 1985; Lindh and Wahlgren 1985), and metasomatism (Misch 1968; Olsen 1984). In recent years, partial melting is considered to be the only dominant migmatite forming process. Thus, the migmatitic leucosomes represent the first step of accumulation and segregation of magma from its local source.

Two end-member models are currently considered for the segregation and accumulation of magma from migmatites to its final emplacement in plutons: (1) Melt flow is governed by the classical flow through connected channels, such as pores, veins, and dykes; and (2) step-wise merging of magma batches.

The first model requires the existence of a connected network of channels through which the magma can flow; the melt connectivity threshold must be overcome (Vigneresse et al. 1996). This threshold fraction, which depends on the wetting angle in porous aggregates, was originally estimated to be a few tens of percent (van der Molen and Paterson 1979), but is now generally regarded to be only a few percent (Rushmer 1995; Laporte and Watson 1995; Vigneresse et al. 1996).

At larger scales, the magma is mostly residing in leucosomes and veins, which, when connected, are envisaged to drain the magma from the source (e.g. Weinberg and Searle 1998; Nicolas and Jackson 1982; Brown 1994; Weinberg 1999; Olson et al. 2004; Hobbs and Ord 2010). This idea is that of "rivulets that feed rivers" or a "rooted vein network"

(Brown and Solar 1998; Petford and Koenders 1998; Weinberg 1999), where the smallest channels feed into ever-bigger channels, finally into the largest dykes that transect the crust and feed plutons (Clemens and Mawer 1992).

Bons et al. (2004, 2010) argued that neither a connected melt network, nor reaching any threshold is required to accomplish magma segregation and magma transport and extraction can take place at very low magma fractions. One problem with the connected channel network model is that flow only occurs after a full (self-organized) network has developed. Local connectivity, however, will in reality already lead to a transfer of magma and (partial) destruction of the network. It is thus questionable if large-scale connectivity is ever achieved. According to the second model, flow is, therefore, discontinuous and magma accumulates in steps in increasingly larger veins or hydrofractures (Maaløe 1987; Takada 1990; Bons et al. 2001; Bons and van Milligen 2001; Bons et al. 2004). A "hydrofracture" is a brittle fracture that opens and propagates mainly because of the internal pressure from the contained liquid (magma or fluid) and not by applied tectonic stresses (Weertman 1971). The second model is based on the observations that melt- or fluid-filled hydrofractures become unstable when they exceed a certain length (Weertman 1971). The instability arises from gradients in effective normal stress that may act on a hydrofracture along its length. These can result from the increase in lithostatic pressure with depth, which differs from the increase in pressure inside a steep hydrofracture if the density of the melt is different from that of the host rock. This limits the vertical length of a stable magma-filled hydrofracture to several tens to hundreds of metres (Secor and Pollard 1975). Once instability is reached, hydrofractures may start to propagate at one end and simultaneously close at the other end (Bons and van Milligen 2001). Batches of magma can thus move together with their containing hydrofractures, which is the crucial difference with the percolation flow model, where magma moves through a stationary network of fractures.

Experiments by Bons and van Milligen (2001), Urtson and Soesoo (2007) and numerical modelling by Bons et al. (2004) indicated that the step-wise merger of batches leads to power law distributions of batch volumes:

$$N_{\geq V} = k_V V^{-m_V}. \qquad (1)$$

Here $N_{\geq V}$ is the number of batches larger than volume (V), k_V the number larger than unit volume and m_V the distribution coefficient. In this paper we present measurements of size distributions of magma batches, from thin leucosomes/veins to plutons, in the geological record. Volumes of leucosomes, veins, or plutons cannot usually be determined directly in the geological record, where observation is normally restricted to 2D outcrop or 1D scan lines or drill cores. We will use the subscript A (area) for size measurements in 2D and H (width or thickness) for measurements in 1D. We will show that the observed size distributions are indeed power law, which is consistent with the second model of step-wise segregation and accumulation of magma, from its source in migmatites through to the final emplacement of magma in >km-scale plutons.

2. Data

2.1. Self-Similarity and Scale-Invariance in Migmatites

Both the spatial distribution and width distribution of leucosomes, veins, and dykes were subjects of several studies (e.g. Tanner 1999; Bons et al. 2004; Soesoo et al. 2004a; Brown 2007; Urtson and Soesoo 2009, Bons et al. 2010; Bonamici and Duebendorfer 2010). The results are somewhat diverse—both power law (fractal) (Soesoo et al. 2004a; Bons et al. 2004; Bonamici and Duebendorfer 2010) and non-fractal (e.g. Marchildon and Brown 2003) leucosome-width statistics have been reported, as well as fractal distributions of leucosome spacing (Tanner 1999; Soesoo et al. 2004a; Soesoo and Bons 2013).

Here we provide measurements of migmatitic leucosomes and granitic veins in drill cores and outcrops. Their width and spacing distributions (Soesoo et al. 2004a) were recorded in a number of drill cores from the Estonian Proterozoic basement (Soesoo et al. 2004b), in outcrops at Masku, southwestern Finland, and at Montemor-o-Novo, central Portugal

(Urtson and Soesoo 2009). Migmatites at these localities represent rocks with a variety of chemical compositions and metamorphic assemblages.

The Estonian drill cores F-265 and F-266 penetrate granulite facies rocks of the Jõhvi structural zone, NE Estonia (Soesoo et al. 2004b, 2006), drill core F-122 amphibolite facies metavolcanites and metasediments of the Tallinn zone, northern Estonia, and drill cores F-156 and F-268 amphibolite facies metasediments of Alutaguse zone, eastern Estonia (Soesoo et al. 2004b). The migmatites at Masku (60°32.52′N; 22°08.00′E), southwestern Finland, formed during Proterozoic granulite facies metamorphism (Mengel et al. 2001). The migmatitic rocks at the Montemor-o-Novo outcrop (38°38.12′N 8°12.54′E) in Portugal belong to the Evora massif and are produced by high-grade metamorphism of the Seria Negra Group metasediments, probably of Variscan or Cadomian age (Pereira and Silva 2002).

The thickness of leucosomes and their spacing were measured along the axis of drill cores or along line scans on outcrops using a measuring tape (Table 1). Generally, the resolution of measurements was limited to 2–3 mm. Leucosomes with thicknesses below this value, close to the size of the individual mineral grains, were not counted. When plotted on a bi-logarithmic graph with measured thickness on the x-axis and the number of leucosomes thicker than that value on the y-axis, the data follow a power law which is defined by a straight line with the distribution exponent being equal to its slope (Figs. 1, 2, 3). Here we use n_H for the observed exponent, to emphasize that this exponent is affected by sectioning effects of the 1D scan line through 3D space.

An alternative way to present data is the density distribution, which represents the number of objects belonging to an interval, divided by the interval length. The density distribution is determined with the box-counting method (similar to Bonamici and Duebendorfer 2010; Tanner 1999). In theory, the exponent of the density distribution is increased by 1 compared to the exponent of cumulative distribution $n_{density} = n_{cumulative} + 1$ (e.g. Bonnet et al. 2001). The density distribution has the advantage of being free of the unfavorable curvature (censoring) effect of the cumulative distribution at the larger scale where the number of objects approaches 1. Also, the

Table 1

Statistics of measured migmatitic leucosomes, veins and intrusion's surface sizes

Location	Coordinates	Type	Length/Area	Number of measurements	Magma fraction %	Exponent, cumulative	Exponent, box-counting
Drill core						n_H	
Estonia	59°34.43′N, 25°31.83′E	F-122 core	27 m	102	23.5	0.89	0.78
Estonia	59°28.91′N, 26°16.88′E	F-265 core	95 m	550	28.0	0.83	0.79
Estonia	59°28.19′N, 26°15.37′E	F-266 core	56 m	580	19.3	1.41	0.77
Estonia	59°26.56′N, 26°13.30′E	F-268 core	51 m	248	44.4	0.54 1.03	0.88
Estonia	59°24.46′N, 26°24.61′E	F-156 core, all measurements	40 m	450	24.0	1.19	0.77
Estonia	59°24.46′N, 26°24.61′E	F-156, larger veins only		295		1.05	0.73
Estonia	59°24.46′N, 26°24.61′E	F-156, only leucosomes,		155		1.79	0.59
Outcrop						n_H	
Masku, Finland	60°32.52′N 22°08.00′E	Scan line	5 m	178	39.6	0.93	0.84
Montemor-o-Novo, Portugal	38°38.12′N, 08°12.54′E	Scan line	43 m	713	71.0	1.14	0.92
Intrusions						n_A	
SE Australia	Fig. 1 in Chappell *et al.* (2012)	Map	2.6×10^5 km²	532	22	0.77; >40 km² 0.60; 7 40 km² 0.34; 2–7 km²	

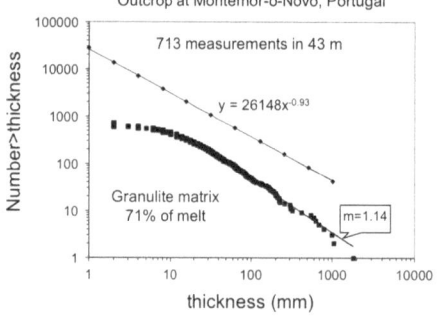
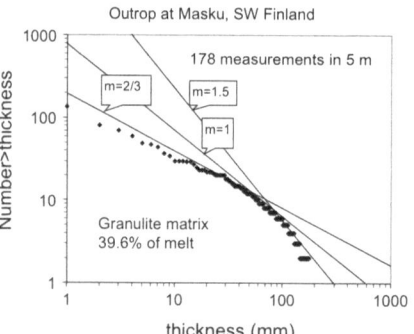

Figure 1

Distribution of leucosome and vein thicknesses at Masku (Finland) and Montemor-o-Novo (Portugal) outcrops. Note that the smaller leucosomes have power law exponent (*m* denotes cumulative distribution of n_H) close to 2/3, while larger leucosomes have an exponent of 1.5. At Montemor-o-Novo, most leucosomes follow an exponent of 1.14 (except smaller leucosomes). The results of the box-counting method are for reference

deviation of the data from the power law trend at the small scale of the distribution due to insufficient sampling of the smallest objects (truncation effect) is exaggerated by the density distribution which helps to define the range of the validity of the power law and thus estimate the correct distribution exponents. The size of the bins where objects are distributed

according to their size is critical as it determines the smoothness of the data (BONNET *et al.* 2001). In this study, logarithmic binning was used for plotting the density distribution data.

In addition to leucosome width data, the amount of melt during the late stage of melting can be estimated by integration of leucosome thickness

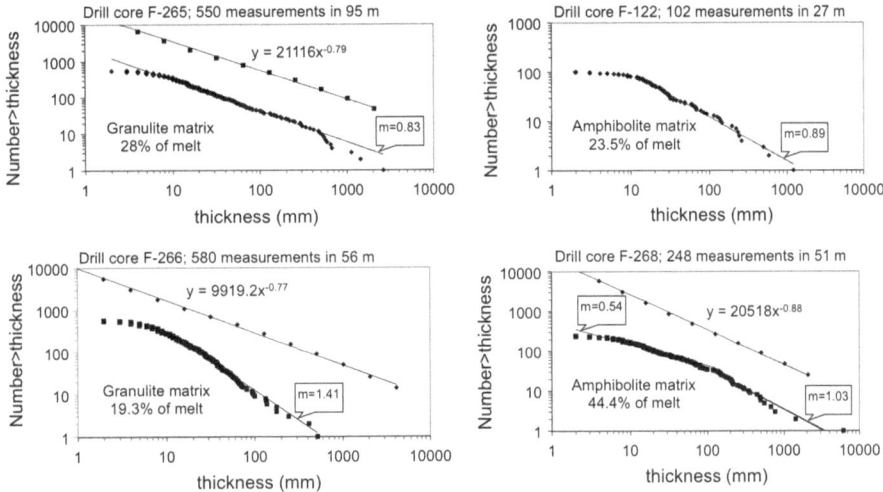

Figure 2
Distribution of leucosome and vein thicknesses in drill cores from the Estonian Precambrian granulite and amphibolite facies rocks with a range of total melt (from 19 to 44 %). The smaller leucosomes tend to deviate from the main trend. For drill core F-268, two trends are shown with exponents (m denotes cumulative distribution of n_H) of 0.54 and 1.03. The results of the box-counting method are for reference

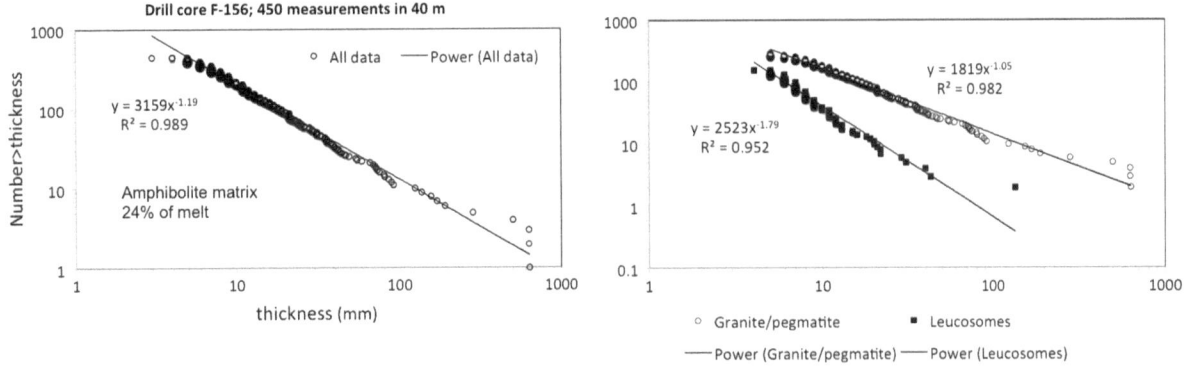

Figure 3
Distribution of leucosome and vein thicknesses in drill core F-156 from the Estonian Precambrian amphibolite rocks. While total measurements of 450 leucosomes and granitic veins give an exponent (m denotes cumulative distribution of n_H) of 1.19; the leucosomes and granitic/pegmatitic veins show distinctive trends of 1.79 and 1.05, respectively

along the measurement line. However, the calculated melt amount describes the minimal melt amount as quite large melt fractions may reside in the smallest leucosomes remaining below the resolution limit. The observed migmatite outcrops and drill core samples yielded minimal apparent melt fractions from 19 to 71 % (Figs. 1, 2, 3). The high grade rocks at Masku (Finland) and Montemor-o-Novo (Portugal) outcrops show relatively large percentages of melt varying from 39.6 to 71 % (Fig. 1).

Leucosomes and magmatic veins are abundant within the Estonian Proterozoic crystalline basement.

They vary from large (up to over 1 m wide) granitic veins and lenses to millimetre-scale thin lenses, veins, and patches. Drill cores F-265 and F-266 consist of granulite facies rocks. In these cores mostly granitic veins dominate in the mafic gneiss. Several large leucosomes are also observed. About 550 granitic veins were measured in a section of 95 m in F-265 and 580 veins in F-266 (Fig. 2). The veins form ca. 28 % of the whole rock section in F-265, while in a similar drill core (F-266) the melt percentage is around 20 %. The power law exponents (fractal dimension in a loose sense) range from 0.8 to

1.4. Drill core F-122 penetrates amphibolite facies rocks and shows about 23 % of melt. The leucosome width distribution exponent is 0.89, which is similar to the granulitic drill core F-265. The amphibolite facies rock of the drill core F-268 shows extensive melting of 44 %. The power law of distribution of leucosomes and veins is less pronounced; two trends can be seen (Fig. 3), with exponents of 0.54 and 1.03. Drill core F-156 penetrates volcanic-sedimentary biotite gneisses of amphibolite facies origin. About 450 leucosomes and veins were measured in a section of 40 m (Fig. 3). The percentage of melted material (leucosomes and granitic veins) is about 24 % (9.7 m from 40 m). This drill core presents a wide size variety of leucosomes and veins (up to coarse-grained pegmatitic veins). Except for very small leucosomes and large pegmatitic veins, the power law is well pronounced giving an exponent of 1.19. When dividing small and large leucosomes and veins, two trends are evident with exponents of 1.05 (granitic veins) and 1.79 (leucosomes, see Fig. 3).

2.2. Distribution of Intrusion Sizes

Having established that power law volume distributions are common on the small scale in the source regions of magma, we now address the size distribution of intrusions that result from the ascent of melt through the crust. For this purpose, we take the Lachlan Fold Belt (LFB) in southeastern Australia as a case study. The Palaeozoic Lachlan Fold Belt occupies the southeastern corner of the Australian continent and it has a total area of close to 300,000 km^2. There was very extensive igneous activity in the LFB in Silurian and Devonian times, and during the Carboniferous in the northeastern corner of the belt. Massive quantities of granitic magma were produced, and currently about 875 lithological units of granite are recognized (CHAPPELL et al 1991, 2012). Most of the granites were emplaced into low-grade flysch sediments of Ordovician age, or else into older granites, or volcanic rocks of the same general magmatic episode. Presently, the area exposes the Paleozoic upper mid-crustal section with a large number of granitoid intrusions (WHITE and CHAPPELL 1983; CHAPPELL 1984; SOESOO and NICHOLLS 1999; SOESOO 2000). Because of the lack of later

sedimentary cover, the intrusions are exposed and well investigated (WHITE and CHAPPELL 1983; SOESOO and NICHOLLS 1999). To our knowledge, the LFB provides the most extensive and complete exposure of a large number of granitoid intrusions, relatively undisturbed by later tectonic or sedimentary events.

To investigate the distribution of granitic intrusion sizes, we used published maps of intrusions that occupy 22 % of the 2.6×10^5 km^2 area. Figure 4 shows the masks of all intrusive rocks shown in Fig. 1 of CHAPPELL et al. (2012) and the much coarser map of LI et al. (2009). We will refer to these maps as Map A and B, respectively. Areas of all plutons with areas larger than 1 km^2, excluding those touching the edge of the map, but not those touching the coast line, are plotted in a cumulative log–log plot of number ($N_{\geq A}$) of exposures larger than a certain area (A in km^2) against that area (Fig. 4c). Map A has a much higher resolution than Map B, with 570 individual exposures of intrusions in Map A and only 72 in Map B. Despite this, the cumulative area distributions for both maps are quite similar for the largest ca. 50 intrusions.

The 109 largest intrusions, with areas >40 km^2, in Map A follow a power-law distribution with an exponent $n_A = 0.766 \pm 0.017$ (95 % confidence; Fig. 5). The next largest 192 intrusions, with areas between 7 and 40 km^2, also follow a power law, but with a smaller exponent of $n_A = 0.596 \pm 0.005$. The next 160 intrusions ≥ 2 km^2 have an even smaller exponent of $n_A = 0.340 \pm 0.004$.

3. Interpretation

BONS et al. (2004) suggested that melt batches in the source merge to form ever-bigger batches in a self-organized way. This leads to a power law distribution of melt volumes (Eq. 1) with the exponent m_V between 1 for inefficient melt extraction and 2/3 for maximum extraction efficiency as most of the volume resides in the largest batches that can escape from the source (BONS et al. 2004; SOESOO et al. 2004a). The question to address is whether the power law distributions of leucosome widths in the migmatites and of intrusion areas in the map view can be related to the volume distributions, as

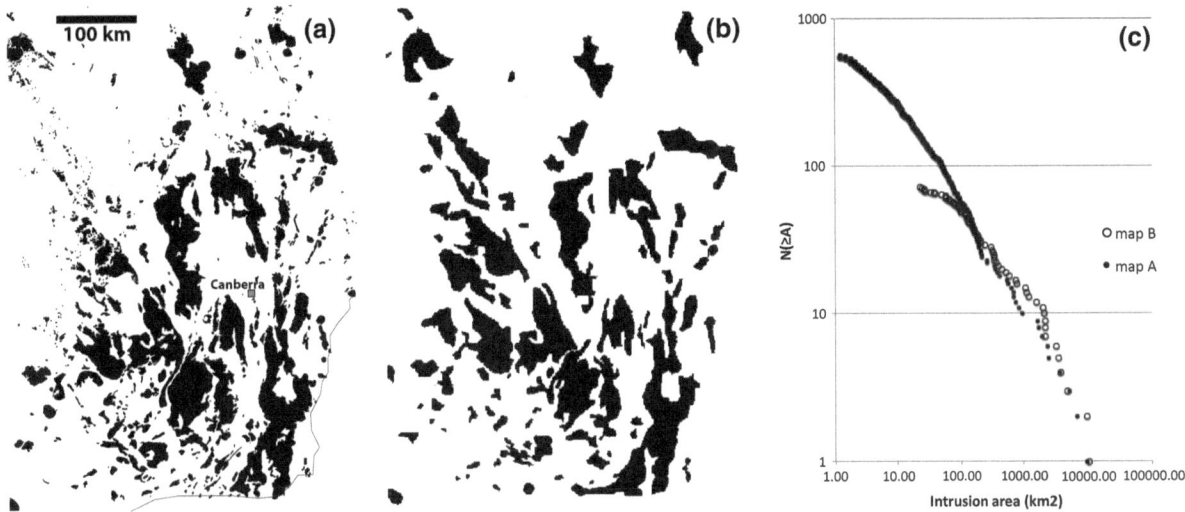

Figure 4
Intrusions in the Lachlan Fold Belt in southeastern Australia based on **a** Fig. 1 in Chappell *et al.* (2012) and **b** Li *et al.* (2009). Map B was adjusted to match the scale and projection of Map A. **c** Logarithmic graph of number of exposed intrusion ($N_{\geq A}$) larger than area (*A* in km^2) plotted against that area. Note that despite great differences in detail in the two maps, the cumulative area distributions for the largest tens of intrusions are very similar

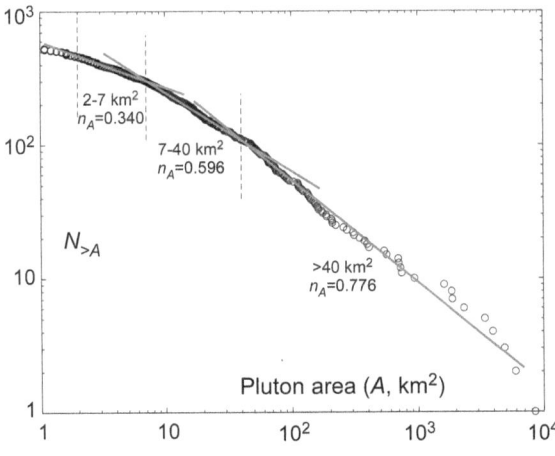

Figure 5
Log-log plot of area of granitoid plutons in Map A of the Lachlan Fold Belt. Three power law regimes can be defined

predicted by Bons *et al.* (2004). To determine how the observed exponents n_H and n_A relate to the volume exponent m_V, the shape of the magma bodies (leucosomes, veins, plutons) needs to be considered, as well as the chance that they are observed in the map or scan line.

Large magma bodies are in general flatter than small ones. Leucosomes tend to have width (*H*) and length (*L*) ratios in the order of tens or more. It is generally assumed that cracks are approximately penny-shaped and can be described with the following relationship between width (*H*) and length (*L*) (Olson 2003; Philipp 2012):

$$H = bL^a. \tag{2}$$

Although the exponent *a* is usually assumed to be unity, for example, as observed by Philipp (2012) for calcite-filled veins, Olson (2003) argued that a sub-linear relationship with $a = 0.5$ is also possible. We are not aware of published *H* versus *L* measurements for leucosomes and, therefore, assume that *a* can vary from 0.5 to 1. On a much larger scale, individual plutons are on average tabular bodies that are self-affine in their shape (McCaffrey and Petford 1997; Cruden 2005). For both *a* and *b*, Cruden (2005) obtained values of about 0.6 when the length unit is kilometer. Larger plutons are thus on average flatter than small ones. It should be noted that Eq. (2) holds for single plutons. Many of the intrusions mapped and analyzed in Figs. 4 and 5 are composite bodies, consisting of numerous individual intrusions. When two individual plutons impinge on each other, they will be counted as a single intrusion. Their *H/L* ratio will be smaller than predicted by Eq. (2).

3.1. Volume Distribution of Leucosomes in Migmatites

Thin leucosomes are shorter than thick ones, and therefore, the chance (P) of them being intersected by the drill core or scan line is smaller: P is proportional to L. The frequency of observed (f_{obs}) leucosomes is thus their frequency in a volume of migmatite multiplied by P:

$$f_{(V)obs} = f_{(V)} P_{(V)} \qquad (3)$$

The (negative) exponent of the volume frequency distribution (the derivative of Eq. 1) is $m_V + 1$ and the volume of a leucosome is proportional to its width and the square of its length, which gives:

$$
\begin{aligned}
f_{(V)obs} &\propto V^{-m_V-1} L = V^{-m_V-1} V^{1/(2+a)} \\
&= V^{1/(2+a)-m_V-1} \Leftrightarrow N_{(\geq V)obs} \propto V^{1/(2+a)-m_V}
\end{aligned}
$$
$$(4)$$

This can be converted to the cumulative width distribution that is observed in a scan line or drill core to obtain:

$$
\begin{aligned}
N_{(\geq H)obs} &\propto \left(H^{1+2/a} \right)^{1/(2+a)-m_V} \Leftrightarrow n_H \\
&= (1+2/a)(1/(2+a)-m_V) \qquad (5)
\end{aligned}
$$

Figure 6 shows the relationship between the volume distribution coefficient (m_V) and the observed width distribution coefficient in a scan line (n_H). Most values of n_H measured in this study scatter around unity, suggesting that m_H is at the low range, close to 2/3, which is the value for maximum concentration efficiency.

3.2. Volume Distribution of Plutons

Intrusions form within a certain depth range (ΔZ), and the current land surface in SE Australia is apparently within this range as many intrusions are exposed. The larger an intrusion, the taller it is and, therefore, the larger the chance that it is exposed at the current land surface. When $H \geq \Delta Z$, the chance (P) of exposure is 100 %. Smaller intrusions have an exposure chance of less than 100 %. The chance of exposure decreases with decreasing height and, therefore, decreasing area. Some small intrusions are missing from the data set, as these are either still

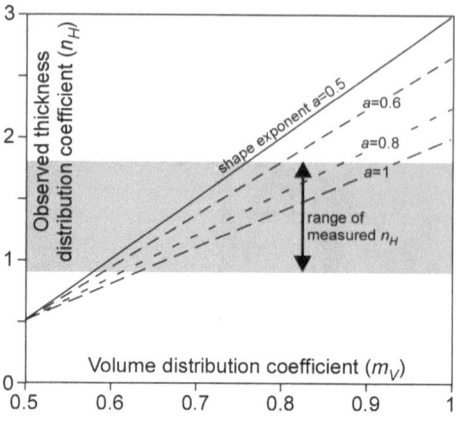

Figure 6
The expected observed power law exponent of width distributions (n_H) as a function of the volume distribution exponent (m_V), for different shape exponents a

fully buried or, alternatively, completely eroded away. This causes a flattening of the cumulative area frequency trend for those areas with $H < \Delta Z$. We suggest that the change in slope at $A \approx 40$ km^2 is caused by this effect and that smaller areas are under represented. As the trend for $A > 40$ km^2 is not disturbed by exposure effects, we can use this trend to estimate the total number of intrusions with $A \geq 1$ km^2 to be 1,845 (1,694–2,004 95 % confidence interval). This implies that 70 % of all these intrusions are not exposed.

The chance of exposure (P) of an intrusion depends on its height (H) and the intrusion depth range (ΔZ):

$$P_{(H)} = \frac{H}{\Delta Z} \text{ (for } H \leq \Delta Z) \qquad (6)$$

In analogy to Eq. (2) we relate the height of an intrusion to its area with:

$$H = \beta A^\alpha \qquad (7)$$

Considering that A is proportional to L^2, the exponent α would be $a/2$ (i.e. ≈ 0.3) for a single pluton (CRUDEN 2005) and β depends on the actual shape of the intrusion. Considering that many of the intrusions under consideration here are composite intrusions, which would, therefore, on average be flatter than single plutons, we expect $\alpha < 0.3$. The <100 % exposure chance of intrusions <40 km^2 reduced their frequency (f) by a factor P:

$$f_{(A)}^{exp} = P_{(A)}f_{(A)} = -P_{(A)}\frac{dN_{\geq A}}{dA} = \frac{\beta A^{\alpha}}{\Delta Z}n_A k_A A^{-(n_A+1)}$$
$$= \frac{\beta n_A k_A}{\Delta Z}A^{\alpha-(n_A+1)} \quad (H_{(A)} \leq \Delta Z)$$

$$(8)$$

Here f stands for the true frequency of exposures and f^{exp} for the frequency of exposed intrusions of a certain area. The slope of the cumulative frequency of exposed intrusions is that of their frequency distribution plus one:

$$N_{\geq A}^{exp} \propto A^{\alpha-n} \quad (9)$$

For the fit range between 7 and 40 km^2 we find $\alpha = 0.17$ and for areas <7 km^2, $\alpha = 0.43$. α in the intermediate area range (7–40 km^2) is smaller than 0.3, which confirms that the composite plutons have a smaller H/L ratio than single plutons, as was expected. The smallest plutons have a high value of α, which would imply a high H/L ratio, but factors other than the shape may play a role in the distribution of such small intrusions, such as mapping resolution.

We have now derived an estimate of the total number of intrusions larger than a certain area, and we know their area frequency distribution and their shape exponent. It would be of interest also to know the volume distribution to estimate the total amount of magma generated in the area. For this, the intrusion depth range ΔZ needs to be known or estimated. The frequency (f_A) distribution is the slope of the cumulative distribution ($N_{\geq A}$). Although the parameters k_A and n were fitted to data in three ranges, the transition from one distribution to another needs to be calculated as the area where the frequency of one distribution equals that of the other. We then find:

$$f_1 = f_2 \Leftrightarrow n_1 k_1 A^{-n_1-1} = n_2 k_2 A^{-n_2-1} \Leftrightarrow A$$
$$= \left(\frac{n_2 k_2}{n_1 k_1}\right)^{\frac{1}{n_2-n_1}} \quad (10)$$

We find that intrusions with $A \geq 178$ km^2 have 100 % exposure chance ($P = 1$) and that the transition from the smallest to the medium range is at 67 km^2 where $P = 0.85$. If we assume that the smallest intrusions are vertical cylinders that follow Eq. (3), we obtain using

$$A = \pi\left(\frac{1}{2}L\right)^2 \quad P = \frac{H}{\Delta Z} = \frac{b}{\Delta Z}\left(\frac{4A}{\pi}\right)^{a/2} \Leftrightarrow \Delta Z$$
$$= \frac{b}{P}\left(\frac{4A}{\pi}\right)^{a/2}$$

$$(11)$$

From this we obtain $\Delta Z = 2.7$ km. With this we can determine β using $H_{(67\ km2)} = 0.85$:

$$H = \beta A^{\alpha} \Leftrightarrow \beta = \frac{H}{A^{\alpha}}, \text{ giving } \beta = 1.11. \quad (12)$$

Assuming a cylindrical shape, the largest intrusion would have $L \approx 100$ km and $H \approx 10$ km. Using the values of α and β obtained here, this intrusion would have a predicted thickness of $H \approx 5$ km. Clearly, this is less than predicted by CRUDEN (2005), but a smaller H is to be expected as the largest intrusion is certainly a composite intrusion.

We can now proceed to estimate the total volume of intrusions in Map A, using:

$$V = HA = \beta A^{1+\alpha} \quad (13)$$

$$N_{\geq V} = k_A \beta^{\frac{n}{1+\alpha}}V^{\frac{-n}{1+\alpha}} \quad k_V = k_A \beta^{\frac{n}{1+\alpha}} = 1980 \text{ and}$$
$$m = \frac{n}{1+\alpha} = 0.65 \quad (14)$$

It is of interest to note that $m = 0.65$ is very close to the expected minimum value of 2/3 for m, which corresponds to maximum extraction efficiency (BONS et al. 2004).

From the cumulative volume distribution, the total volume of all intrusions can be derived by integrating the product of frequency and volume:

$$V_{tot} = \int_{V_{min}}^{V_{max}} V f_V dV = \int_{V_{min}}^{V_{max}} m k_V V^{-m} dV$$
$$= \frac{m k_V}{1-m}\left|(V_{max})^{1-m} - (V_{min})^{1-m}\right|$$
$$\approx \frac{m k_V}{1-m}(V_{max})^{1-m}, \quad (15)$$

where V_{min} and V_{max} are the smallest and largest volume under consideration. Because $(1 - m) < 1$, only V_{max} needs to be considered if $V_{max} \gg V_{min}$. Remains to determine the value of V_{max}, which follows from the fact that only one volume is equal or larger than the maximum volume:

$$N_{\geq V_{max}} = 1 = k_V V_{max}^{-m} \Leftrightarrow V_{max} = k_V^{1/m}$$
$$= 1.1 \cdot 10^5 \text{km}^3. \qquad (16)$$

Finally, the estimated total volume of intrusions, including those currently not exposed or shown on the map is 2.1×10^5 km^3. As the total land area of map A is 2.6×10^5 km^2, 0.8 km^3 of magma would have been expelled from the source per km^2. This value is close to the 1 km^3 estimated by ZEN (1992) for the same area.

4. Discussion

Magma bodies from small (>2–3 mm thick) to very large (>10,000 km^2 in area) follow power law distributions in their size measurements (width and height, respectively). As the shape of magma volumes, from small, magma-filled fractures to large plutons, is generally assumed to be self-similar or self-affine (Eq. 2), it can be assumed that volume distributions also follow a power law. This suggests that the process of segregation and accumulation of magma from its source to its final emplacement as plutons produces power law distributions over the full range of scales under consideration here, from small leucosomes to large plutons. There is, however, a significant gap between the two length scales considered above. As can be seen in Figs. 1, 2, and 3, a power law is observed in most migmatites over at least two orders of magnitude of length scale (about 1 cm–1 m). This is a good indication that small leucosomes, a few cm-wide granitic veins, and larger (>5 cm) granitic veins are related and represent a common magma transport (pathway) system. The range of widths is limited for several reasons.

At the small end the data are biased by the difficulty to clearly recognize the thinnest veins when their width approaches the grain size in the rock (mm-scale). As long as the true volume distribution (m_V) is smaller than unity, this under sampling is not a major issue, as these veins actually contain a small proportion of the magma. Continuous scan lines and drill cores are limited to a few tens of meters, where the thickest vein is about 1 m thick in the migmatites under consideration. To increase the range to ca. 10 m wide veins, about ten times longer scan lines or

drill cores would be needed, which are not available. However, BROWN (2005) measured granitic veins thicker than a few cm in outcrop and found a power law relationship (with an exponent $n_H \approx 1.1$) up to a width of ca. 5 m. An upper limit of the power law regime as observed in the migmatites is not known. An estimate can be derived by taking $n_H \approx 1$ and using the observation that in a scan line of about 50 m the maximum vein width is about 1 m. Taking 10 km as the maximum height of the migmatite zone, the maximum vein thickness would then be on the order of 200 m. However, this estimate relies on the unproven assumption that the same exponent extends to this scale. Alternatively, the distribution may be truncated at some scale, if there is a threshold melt volume where melt batches can escape the source region (BONS et al. 2004; BONAMICI and DUEBENDORFER 2010).

Power law distributions of leucosomes and veins have been attributed to some form of self-organization of the melt distribution. Most authors favour melt distribution to be in a self-organized network of connected melt-bearing leucosomes and veins (e.g. BROWN and SOLAR 1998; PETFORD and KOENDERS 1998; WEINBERG and PODLADCHIKOV 1994; WEINBERG 1999; HALL and KISTERS 2012). This model was criticized by BONS et al. (2010), who argued that such a network is not sustainable, as local and transient transport of melt would cause the collapse of connectivity. The alternative model is that step-wise accumulation of batches of melt would lead to power law distributions of melt volumes. This model is supported by analogue experiments (BONS and VAN MILLIGEN 2001; URTSON and SOESOO 2007) and numerical modelling (BONS et al. 2004). Contrary to classical models of magma percolating through a network of ever-bigger veins (WEINBERG 1999), veins are rarely connected, as their connection is only transient. This is consistent with the observation in outcrops that leucosomes tend to be parallel.

As long as the ambient temperature is high enough, magma in the leucosomes and veins remains liquid. However, plutons are formed by magma that ascended through cooler crust. Only large magma batches can ascend to the emplacement level, as smaller ones would freeze along the way. These large magma batches have a power law size distribution and merge to form plutons at the emplacement level at shallower crustal levels. Depending on the temporal spacing of ascent events, the size of ascending magma batches and size of the growing pluton, two end members scenarios can be envisaged:

1. Previous magma batches are completely frozen when a new one arrives. The previous magma batches now form part of the host rock of the new intrusion and the magmas cannot mix or mingle. Detailed mapping of structures and geochemical and petrological characteristics may reveal this and the pluton is regarded as a "composite pluton".
2. Previous magma batches are not yet (completely) frozen when a new one arrives. The newly arrived magma can mix and mingle with the previous magma batches and a single pluton develops with a potentially homogenized magma.

Indications for the growth of plutons by multiple magma batches are found in several granitic plutons and volcanic rocks (e.g. Cambray *et al.* 1995; Slater et al. 2001; Glazner *et al.* 2004), although mixing in a growing pluton may obscure the primary evidence.

The formation of plutons by amalgamation of individual magma batches is effectively the same as the formation of larger leucosomes from the merger of smaller ones. We observe in the migmatites that this leads to power law size distributions and plutons, therefore, also show these distributions. The formation of plutons is thus not a distinct process from magma segregation and accumulation, but instead the end of a single chain of transport and merger steps from the smallest ($<$mm^3) to the largest ($\gg$km^3) scale.

The question arises whether the remaining melt in leucosomes would actually represent the melt distribution during melt generation and formation of plutons. In case of melt percolation through a fracture network, fractures would adapt their aperture according to the melt pressure (Hobbs and Ord 2010). As melt drains away, fractures would close. In that case, a post-mortem analysis of the geological record would not reveal the original melt-volume distribution during melt flow, and it is questionable whether fractal distributions are to be expected. On the contrary, a power law or fractal spatial distribution of remaining melt is predicted by the alternative model

of step-wise merging melt batches. The numerical model of Bons *et al.* (2004) and analogue experiments by Soesoo (Urtson and Soesoo 2007) indicate that the removal of large melt batches does not affect the distribution of remaining batches.

4.1. Self-Organization

The scale invariant nature of the distribution of melt in former magmatic systems as observed in migmatites indicates that magma segregation, transport, and accumulation can be described as a criticality system. Such a system exhibits rate-independent dissipation in inhomogeneous environments; in this case, melt transport in rocks. The scaling in a partial melt–magma accumulation system emerges from the interplay between order and disorder, external dynamics and quasistatic driving forces (Pérez-Reche *et al.* 2008). There are two end-member models of the critical behaviour—the random field Ising (RFIM) and pinning-depinning (PD) approaches, which differ mainly by the amount of disorder in the system, the latter having disorder as an irrelevant parameter (Chauve *et al.* 2001). The two approaches are fundamentally different. The first model describes regimes that are dominated by nucleation, while the second deals exclusively with propagation. It has been shown that criticality in the first class of models is classical, as in second order phase transitions, while in the second class it is self-tuning in the sense that infinitely slow driving brings the system automatically to a critical state (Dahmen and Ben-Zion 2009). In the PD model, this self-turning is interpreted as self-organised criticality (SOC). The SOC state is achieved if the system is driven through minor adjustments, which can be modified by feedback mechanisms. In the SOC state (e.g. Bak *et al.* 1988), a system (rock + magmatic liquid) adjusts itself to accommodate transport through self-tuning and feedback adjustments.

A typical feature of an SOC system is that there is a strong dynamic balance between input and output and that any small perturbation can (but not necessarily must) lead to a chain reaction or avalanche (large-scale merging of melt batches and emplacement of large melt volumes). From one point of view, the partially molten rock system is comparable to the sand-pile model of BAK *et al.* (1988). When sand is sprinkled on a pile, it develops a semi-stable SOC state (from time to time having a proper SOC state), where sand leaves the pile in avalanches. The continuous input is thus balanced by intermittent bursts of output. The size distribution of the avalanches follows a power law, with many small avalanches and rare bigger ones. The latter distribution was also replicated numerically by BONS *et al.* (2004) and experimentally by BONS and VAN MILLIGEN (2001), and URTSON and SOESOO (2007).

As Pérez-Reche *et al.* (2008) recognize, SOC is an idealised state that is marginally accessible, but that most systems jump from popping (POP) behaviour, which is characterized by a large number of small avalanches to snapping (SNAP) behaviour characterized by very large avalanches. In partial melting and melt accumulation, the system behaviour is likely dominated by POP-type avalanches, while SNAP behaviour is an extreme case resulting in large avalanches and local system collapse. Different fractal dimensions (or ranges) are to be expected for the POP and SNAP regimes. This may potentially be the reason why some small and larger leucosomes/magmatic veins show different fractal dimension (e.g. Fig. 3, leucosomes vs. granite and pegmatite veins), but altogether still follow power law distributions.

The analogy with the sand-pile model is far from perfect. One difference is that the model does not preserve any power law features inside the system. Once sprinkling is stopped, the only indication of the SOC state is the critical taper of the pile. This is different in the partial-melt system under consideration here, where the geological record does preserve frozen melt batches, which represent the products of "avalanches". Only the biggest of these escaped from the system and are evidently difficult to observe, except for some indicative collapse structures (BONS *et al.* 2008).

One similarity with the ideal sand-pile model, which we suggest is of importance here, is that the size distribution of avalanches is not controlled by the sprinkling rate. The sprinkling rate only determines the intervals between avalanches. The different migmatites analyzed in this study may very well have experienced different melt production rates, yet the spatial distribution of remaining melt is remarkably similar. The age range (>100 million years) of the plutons in the LFB by far exceeds the formation time of individual plutons, which is up to ~10 million years (GLAZNER *et al.* 2004). Yet the size distribution of the plutons spread over a large area and of different ages appears to follow a single power law. This indicates that the underlying process that controls this distribution is independent of geographical position and age.

5. Conclusions

Magma is generated by partial melting from μm-scale in its source rocks and may accumulate to form ≫km-scale volumes. We measured thicknesses of leucosomes and veins in migmatites along scan lines at various localities, as well as pluton areas on the map scale. In all cases we found power law relationships between size and (cumulative) frequency. These results show that during melt segregation and transport the magma system is controlled by self-organised criticality that governs the topology of magmatic bodies from migmatitic leucosomes to plutons and batholiths.

Theoretical and field observations indicate that initial melts accumulate in a non-continuous, stepwise manner to form larger accumulations. There is no characteristic length or time scale in the partial melting process or its products. There is no fundamental difference between smallest magma-filled veins in migmatites and large plutons, which are merely end members of a chain of mergers of magma batches that create ever-larger volumes. The volume distributions inferred from the measurements in line scans and maps indicate that the cumulative volume distribution coefficient, m_V, is close to 2/3, which the value of maximum concentration efficiency. The distribution appears independent of local geological or petrological factors as is probably mostly dictated by the self-organized critical state.

Acknowledgments

This study was supported by the Estonian Ministry of Education and Research target research project no. SF0140016s09 and by grant no. 8963 (ESF) to AS.

REFERENCES

ASHWORTH, J., and McLELLAN, E., Textures, In Migmatites (ed. Ashworth. J.) (Blackie, Glasgow 1985) pp. 180–203.

BAK, P., TANG, C., and WIESENFELD, K. (1988), *Self-organized criticality*, Phys. Rev. A. *38*, 364–374.

BONAMICI, C.E., and DUEBENDORFER, E.M., (2010), *Scale-invariance and self-organized criticality in migmatites of the southern Hualapai Mountains, Arizona*, J. Struct. Geol. *32*, 1114–1124.

BONNET, E., BOUR, O., ODLING, N. E., DAVY, P., MAIN, I., COWIE, P., and BERKOWITZ, B. (2001), *Scaling of fracture systems in geological media*, Rev. Geophys. *39*, 347–383.

BONS, P.D., BECKER, J.K., ELBURG, M.A., URTSON, K. (2010), *Granite formation: stepwise accumulation or connected networks?* Earth Environ. Sci. Trans. Roy. Soc. Edin. *100*, 105–115.

BONS, P.D., DRUGUET, E., CASTAÑO, L.M., ELBURG, M.A. (2008), *Finding what is not there anymore: recognizing missing fluid and magma volumes*, Geology 36, 851–854.

BONS, P.D., ARNOLD, J., ELBURG, M.A., KALDA, J., SOESOO, A., van MILLIGEN, B.P. (2004), *Melt extraction and accumulation from partially molten rocks*, Lithos 78, 25–42.

BONS, P.D., DOUGHERTY-PAGE, J., and ELBURG, M.A. (2001), *Stepwise accumulation and ascent of magmas*, J. Met. Geol. *19*, 627–633.

BONS, P.D., and van MILLIGEN, B.P. (2001), *A new experiment to model self-organized critical transport and accumulation of melt and hydrocarbons from their source rocks*, Geology 29, 919–922.

BROWN, M. (2007), *Crustal melting and melt extraction, ascent and emplacement in orogens: mechanisms and consequences*, J. Geol. Soc. London *164*, 709–730.

BROWN, M.A., Synergistic effects of melting and deformation: an example from the Variscan belt, western France, In deformation mechanisms, rheology and tectonics: from minerals to the lithosphere (eds. Gapais, D., Brun, J. P., and Cobbold, P. R.) (J. Geol. Soc. London, Spec. Pub. 243, 2005a) pp. 205–26.

BROWN, M. A., BROWN, M., CARLSON, W. D., and DENISON, C. (1999), *Topology of syntectonic melt-flow networks in the deep crust: inferences from three-dimensional images of leucosome geometry in migmatites*, Am. Miner. *84*, 1793–1818.

BROWN, M., SOLAR, G.S. (1999), *The mechanism of ascent and emplacement of granite magma during transpression: a syntectonic granite paradigm*, Tectonophysics *312*, 1–33.

BROWN, M., and SOLAR, G. S. (1998), *Shear-zone systems and melts: feedback and self-organization in orogenic belts*. J. Struct. Geol. *20*, 211–27.

BROWN, M. (1994), *The generation, segregation, ascent and emplacement of granite magma: the migmatite-to-crustally-derived granite connection in thickened orogens*, Earth-Sci. Rev. *36*, 83–130.

CAMBRAY, F.W., VOGEL, T.A., and MILLS, J.G. (1995), *Origin of compositional heterogeneities in tuffs of the Timber Mountain Group: the relationship between magma batches and magma transfer and emplacement in an extensional environment*, Geophys. Res. *100*, 15793–15805.

CHAPPELL, B. W., BRYNT, C. J., WYBORN, D. (2012), *Peraluminous I-type granites*, Lithos *153*, 142–153.

CHAPPELL, B. W., ENGLISH, P. M., KING, P.L., WHITE, A. J., WYBORN, D. (1991), *Granites and related rocks of the Lachlan Fold Blet (1:1 250 000 scale map)*, Bureau of Mineral Resources, Geology and Geophysics, Australia.

CHAPPELL, B.W. (1984), *Source rocks of I- and S-type granites in the Lachlan Fold Belt, southeastern Australia*, Philosoph. Transac. Roy. Soc. London A. *310*, 693–707.

CHAUVE, P., LE DOUSSAL, P., and WIESE. K. J. (2001), *Renormalization of pinned elastic systems: how does it work beyond one loop ?* Phys. Rev. Letts. *86*, 1785.

CLEMENS, J. D. (1998), *Observations on the origin and ascent mechanism of granitic magma*, J. Geol. Soc. London *155*, 843–851.

CLEMENS, J.D., DROOP, G.T.R., and STEVENS, G. (1997), *High-grade metamorphism, dehydration and crustal melting: a reinvestigation based on new experiments in the silica-saturated portion of the system KAlO3-MgO-SiO2-CO2 at P < 1.GPa*, Contrib. Mineral. Petrol. *129*, 308–325.

CLEMENS, J.D., and MAWER, C.K. (1992), *Granitic magma transport by fracture propagation*, Tectonophysics *204*, 339–360.

CRUDEN, A.R., Emplacement and growth of plutons: implications for rates of melting and mass transfer in continental crust, In Evolution and differentiation of the continental crust (eds. Brown, M., and Rushmer, T.) (Cambridge University Press, New York 2005) pp. 455–519.

DAHMEN, K.A., and BEN-ZION, Y., The physics of jerky motion in slowly driven magnetic and earthquake fault systems, in encyclopedia of complexity and systems science Vol.5, (Springer, New York 2009) pp. 5021–5037.

EMERMAN, S.H., and MARRETT, R. (1990), *Why dikes?* Geology *18*, 231–233.

GLAZNER, A. F., BARTLEY, J. M., COLEMAN, D. S., GRAY, W., and TAYLOR, R. Z. (2004), *Are plutons assembled over millions of years by amalgamation from small magma chambers?* GSA Today *14*, 4–11.

HALL, D., and KISTERS, A. (2012), *The stabilization of self-organised leucogranite networks—Implications for melt segregation and far-field melt transfer in the continental crust*, Earth Planet Sci. Letts. *355–356*, 1–12.

HASALOVA, P., SCHULMANN, K., LEXA, O., STIPSKA, P., HROUDA, F., ULRICH, S., HALODA, J., and TYCOVA, P. (2008), *Origin of migmatites by deformation-enhanced melt infiltration of orthogneiss: a new model based on quantitative microstructural analysis*, J. Met. Geol. *26*, 29–53.

HOBBS, B.E., and ORD, A. (2010), *The mechanics of granitoid systems and maximum entropy production rates*, Phil. Trans. R. Soc. A. *368*, 53–93.

JACKSON, M. D., CHEADLE, M. J., and ATHERTON, M. P. (2003), *Quantitative modeling of granitic melt generation and segregation in the continental crust*, J. Geophys. Res. *108*, 2332–2353.

JUREWICZ, S. R., and WATSON, E. B. (1984*), Distribution of partial melt in a felsic system: the importance of surface energy*, Contrib. Mineral. Petrol. *85*, 25–29.

LAPORTE, D., and WATSON, E. B. (1995*), Experimental and theoretical constraints on melt distribution in crustal sources: the effect of crystalline anisotropy on melt interconnectivity*, Chem. Geol. *124*, 161–184.

LI, W., JACKSON, S.E., PEARSON, N.J., ALARD, O., and CHAPPELL, B.W. (2009), *The Cu isotopic signature of granites from the Lachlan Fold Belt, SE Australia*. Chem. Geol. *258*, 38–49.

LINDH, A., and WAHLGREN, C. (1985), *Migmatite formation at subsolidus conditions—an alternative to anatexis*, J. Met. Geol. *3*, 1–12.

LISTER, J.R., and KERR, R.C. (1991), *Fluid-mechanical models of crack propagation and their application to magma transport in dykes*, J. Geophys. Res. *96*, 10049–10077.

MAALØE, S. (1987), *The generation and shape of feeder dykes from mantle sources*, Contrib. Mineral. Petrol. *96*, 47–55.

MARCHILDON, N., and BROWN, M. (2003), *Spatial distribution of melt-bearing structures in anatectic rocks from Southern Brittany, France: implications for melt transfer at grain- to orogen-scale*, Tectonophysics *364*, 215–235.

MCCAFFREY, K.J.W., and PETFORD, N. (1997), *Are granitic intrusions scale invariant?* J. Geol. Soc. London *154*, 1–4.

MISCH, P. (1968), *Plagioclase compositions and non-anatectic origin of migmatitic gneisses in N. Cascade Mountains of Washington State*, Contrib. Mineral. Petrol. *17*, 1–70.

MENGEL, K., RICHTER, M., and JOHANNES, W. (2001), *Leucosome-forming small-scale geochemical processes in the metapelitic migmatites of the Turku area, Finland*, Lithos *56*, 47–73.

NICOLAS, A., and JACKSON, M. (1982), *High temperature dikes in peridotites: origin by hydraulic fracturing*. J. Petrol. *23*, 568–82.

OLSEN, S. N. (1984), *Mass-balance and mass-transfer in migmatites from the Front Range, Colorado*, Contrib. Mineral. Petrol. *85*, 30–44.

OLSON, S. N., MARSH, B. D., and BAUMGARTNER, L. P. (2004), *Modelling mid-crustal migmatite terrains as feeder zones for granite plutons: the competing dynamics of melt transfer by bulk versus porous flow*. Transac. Roy. Soc. Edin. Earth Sci. *95*, 49–58.

OLSON, J.E. (2003), *Sublinear scaling of fracture aperture versus length: an exception or the rule?* J. Geophys. Res. Solid Earth *108*(B9), 2413.

PEREIRA, M. F., and SILVA, J. B. (2002), *The geometry and kinematics of enclaves in sheared migmatites from the Evora Massif, Ossa-Morena Zone (Portugal)*, Geogaceta *31*, 199–202.

PÉREZ-RECHE F-JOSÉ, TRUSKINOVSKY, L., and ZANZOTTO, G. (2008), *Driving-induced crossover: from classical criticality to self-organized criticality*, Phys. Rev. Letts. *101*, 230601.

PETFORD, N., CRUDEN, A.R., MCCAFFREY, K.J.W., and VIGNERESSE, J.-L. (2000), *Granite magma formation, transport and emplacement in the Earth's crust*, Nature *408*, 669–673.

PETFORD, N., and KOENDERS, M. A. (1998), *Self-organisation and fracture connectivity in rapidly heated continental crust*, J. Struct. Geol. *20*, 1425–34.

PHILIPP, S.L. (2012), *Fluid overpressure estimates from the aspect ratios of mineral veins*, Tectonophysics *581*, 35–47.

RUSHMER, T. (1995), *An experimental deformation study of partially molten amphibolite: application to low-melt fraction segregation*, J. Geophys. Res. *100*, 15681–15695.

SAWYER, E.W. (2014), *The inception and growth of leucosomes: microstructure at the start of melt segregation in migmatites*, J. Met. Geol. *32*, 695–712.

SAWYER, E.W. (2001), *Melt segregation in the continental crust: distribution and movement of melt in anatectic rocks*, J. Met. Geol. *19*, 291–309.

SAWYER, E. W. (1994), *Melt segregation in the continental crust*, Geology *22*, 1019–1022.

SECOR, D.T., and POLLARD, D.D. (1975), *On the stability of open hydraulic fractures in the Earth's crust*, Geophys. Res. Lett. *2*, 510–513.

SEDERHOLM, J. (1907), *On granite and gneiss*, Bull. Comm. Geol. Finl. *23*, 1–110.

SLATER, L., MCKENZIE, D., GRÖNVOLD, K., and SHIMIZU, N. (2001), *Melt generation and movement beneath Theistareykir, NE Iceland*, J. Petrol. *42*, 321–354.

SOESOO, A., and BONS, P., Partial melting of Earth' rocks: fractals and analogue modelling approach, In (Ed. Perugini, D.) (VI International Conference on fractals and dynamic systems in geosciences, Perugia, Italy 2013) pp. 71–72.

SOESOO, A., KOSLER, J., and KULDKEPP, R. (2006), *Age and geochemical constraints for partial melting of granulites in Estonia*, Mineral. Petrol. *86*, 277–300.

SOESOO, A., KALDA, J., BONS, P.D., URTSON, K., and KALM, V. (2004a), *Fractality in geology: a possible use of fractals in the studies of partial melting processes*, Proc. Eston. Acad. Sci., Geol. *53*, 13–27.

SOESOO, A., PUURA, V., KIRS, J., PETERSELL, V., NIIN, M., and ALL, T. (2004b), *Outlines of the Precambrian basement of Estonia*, Proc. Eston. Acad. Sci., Geol. *53*, 149–164.

SOESOO, A. (2000), *Fractional crystallisation of mantle-derived melts as a mechanism for some I-type granite petrogenesis: an example from Lachlan Fold Belt, Australia*, J. Geol. Soc. London *157*, 135–150.

SOESOO, A., and NICHOLLS, I.A. (1999), *Mafic rocks spatially associated with Devonian felsic intrusions of the Lachlan Fold Belt: a possible mantle contribution to crustal evolution processes*, Austr. J. Earth Sci. *46*, 725–734.

TAKADA, A. (1990), *Experimental study on propagation of liquid-filled crack in gelatin: shape and velocity in hydrostatic stress condition*, J. Geophys. Res. *95*, 8471–8481.

TANNER, D. C. (1999), *The scale-invariant nature of migmatite from the Oberpfalz, NE Bavaria and its significance for melt transport*, Tectonophysics *302*, 297–305.

URTSON, K., and SOESOO, A. (2009), *Stepwise magma migration and accumulation processes and their effect on extracted melt chemistry*, Est. J. Earth Sci. *58*, 246–258.

URTSON, K., and SOESOO, A. (2007), *An analogue model of melt segregation and accumulation processes in the Earth's crust*, Est. J. Earth Sci. *56*, 3–10.

VAN DER MOLEN, I., and PATERSON, M.S. (1979), *Experimental deformation of partially melted granite*, Contrib. Mineral. Petrol. *70*, 299–318.

VIGNERESSE, J.L., BARBEY, P., and CUNEY, M. (1996), *Rheological transitions during partial melting and crystallization with application to felsic magma segregation and transfer*, J. Petrol. *37*, 1579–1600.

WALTE, N. P., BONS, P. D., PASSCHIER, C. W., and KOEHN, D. (2003), *Disequilibrium melt distribution during static recrystallization*, Geology *31*, 1009–1012.

WALTE, N.P., BECKER, J.K., BONS, P.D., RUBIE, D.C., and FROST, D.J. (2007), *Liquid distribution and attainment of textural equilibrium in a partially-molten crystalline system with a high-dihedral-angle liquid phase*, Earth Planet. Sci. Lett. *262*, 517–53.

WARK,, D.A., WILLIAMS, C.A., WATSON, E.B., and PRICE, J.D. (2003), *Reassessment of pore shapes in microstructurally equilibrated rocks, with implications for permeability of the upper mantle*, J. Geophys. Res. *108*, 2050.

WEERTMAN, J. (1971), *Theory of water-filled crevasses in glaciers applied to vertical magma transport beneath ocean ridges*, J. Geophys. Res. *76*, 1171–1183.

WEINBERG, R.F. (1999), *Mesoscale pervasive felsic magma migration: alternatives to dyking*, Lithos *46*, 393–410.

WEINBERG, R.F., and PODLADCHIKOV, Y.Y. (1994), *Diapiric ascent of magmas through power law crust and mantle*, J. Geophys. Res. *99*, 9543–9559.

WEINBERG, R.F., and SEARLE, M.P. (1998), *The Pangong injection complex, Indian Karakoram: a case of pervasive granite flow through hot viscous crust*, J. Geol. Soc. London *155*, 883–891.

WHITE, A.J.R., and CHAPPELL, B.W. (1983*), Granitoid types and their distribution in the Lachlan Fold Belt, southeastern Australia*, Geol. Soc. Am. Mem. *159*, 21–34.

WINKLER, H. (1961), *Die Genese von Graniten und Migmatiten auf Grund neuer Experimente*, Geol. Rundsch. *61*, 347–364.

ZEN, E. (1992), *Using granite to image the thermal state of the source terrain*, Trans. R. Soc. Edinb. Earth Sci. *83*, 107–114.

(Received April 16, 2014, revised September 5, 2014, accepted November 19, 2014, Published online December 3, 2014)

Pure Appl. Geophys. 172 (2015), 1803–1814
© 2014 Springer Basel
DOI 10.1007/s00024-014-0917-5

Pure and Applied Geophysics

Fractal Analysis of Enclaves as a New Tool for Estimating Rheological Properties of Magmas During Mixing: The Case of Montaña Reventada (Tenerife, Canary Islands)

HELENA ALBERT,[1] DIEGO PERUGINI,[2] and JOAN MARTÍ[3]

Abstract—The volcanic unit of Montaña Reventada on the island of Tenerife (Canary Islands, Spain) is an example of magma mingling and mixing in which the eruptive process was triggered by an intrusion of basanite into a phonolite magma chamber. The eruption started with emplacement of a basanitic scoria deposit followed by emplacement of a phonolitic lava flow characterized by the presence of mafic enclaves. These enclaves represent approximately 1 % of the outcrop and are basanitic, phono-tephritic and tephri-phonolitic in composition. The morphology of each enclave is different, varying from rounded to complex finger-like structures usually with cuspate terminations. In this study we quantified textural heterogeneity related to the enclaves generated by the mixing process and thus provided a new perspective on the 1100 AD Montaña Reventada eruption. The textural study was performed by use of fractal geometry methods and the results show that the logarithm of the viscosity ratio between the phonolitic magma and the enclaves ranges between 0.39 and 0.81, with a mode at 0.49. This enables us to infer the water content is 2–2.5 wt% for the phonolitic magma and 1.5–2 wt% for the basanitic magma and the enclaves.

Key words: Magma mixing, Fractal analysis, Tenerife, Viscosity, Water content, Basanite.

1. Introduction

Fractal geometry has recently been used in several studies to study mixing processes. In these studies magma mixing is described as a chaotic process (FLINDERS and CLEMENS, 1996; PERUGINI *et al.*, 2002, 2003a, 2006; PERUGINI and POLI, 2000; POLI and PERUGINI, 2002), which implies that study of magma mixing can be kinematically constrained to the study of the stretching and folding of magmas and the diffusion processes that originate between them (PERUGINI and POLI, 2012). These studies have provided a new perspective for calculating properties and conditions related to magma chamber dynamics, for example the viscosity ratio between magmas and the proportions of the magmas involved in the mixing process.

The physical mixing (mingling) of two magmas with contrasting physical properties tends to result in the formation of enclaves (PERUGINI *et al.*, 2007) and/or the presence of flow banding, which has been described by some authors as the result of stretching and folding of magma filaments (PERUGINI *et al.*, 2003b, 2004). When the two magmas equilibrate thermally, chemical diffusion also occurs, thereby generating compositions that are intermediate between the two initial compositions (PETRELLI *et al.*, 2006; PERUGINI *et al.*, 2012, 2013; MORGAVI *et al.*, 2013). Chemical diffusion is facilitated by the stretching and folding processes, because of the increase of contact area between the two magmas.

During magma mixing two kinds of region, coherent and active, are generated (PERUGINI *et al.*, 2003b). Coherent regions remain generally unaffected by diffusion and, therefore, the composition of the original magmas is preserved. In contrast with coherent regions, active regions are strongly affected by stretching and folding, and chemical diffusion produces a composition that is intermediate between the two magmas involved in the mixing process. This means that magma mixing should not be interpreted as a linear process in which different hybrid compositions correspond to different amounts of intruded magma. Typically,

[1] Central Geophysical Observatory, Spanish Geographic Institute (IGN), Madrid, Spain. E-mail: hln.albert@gmail.com

[2] Department of Earth Sciences, University of Perugia, Perugia, Italy.

[3] Institute of Earth Sciences Jaume Almera, CSIC, Barcelona, Spain.

Reprinted from the journal

when a mafic magma intrudes into a felsic magma chamber and mixing starts, magmas of different hybrid composition will be produced, because of the generation of coherent and active regions in which the proportion of the two magmas and the degrees of chemical diffusion vary (PERUGINI et al., 2003b).

Magma mixing has been observed in the volcanic deposits of Tenerife (Canary Islands) in pyroclastic rocks (WOLFF, 1985; MARTÍ et al., 1990; EDGAR et al., 2002, 2007) and lava flows (ARAÑA, 1985; ARAÑA et al., 1989, 1994). The products of some of these eruptions are excellent examples of magma mixing processes resulting from the presence of enclaves and/or stretching and folding structures. One of the best known cases of magma mixing on Tenerife is the 1100 AD (CARRACEDO et al., 2007) Montaña Reventada eruption (ARAÑA et al., 1994; WIESMAIER et al., 2011). The eruption center is located in the island's northwestern rift zone, on the southwestern flank of the Teide-PicoViejo (TPV) active central volcanic complex (ABLAY and MARTÍ, 2000). The eruption, in which a batch of mafic magma from deep in the rift zone probably intruded into the phonolitic chamber of the TPV, generated basaltic scoria followed by emplacement of a thick phonolite lava flow characterized by the presence of multiple enclaves. Previous studies (ARAÑA et al., 1994; WIESMAIER et al., 2011) focusing on the geochemistry and mineralogy of the eruption products show that these enclaves are the result of mixing between basanitic and phonolitic magmas. Although several studies have been performed on phonolitic magma storage conditions in Tenerife (ANDÚJAR et al., 2010, 2013; ANDÚJAR and SCAILLET, 2012a, b), to the best of our knowledge there is no information about basanitic magma.

The purpose of the study reported in this paper was to provide a new perspective of the 1100 AD Montaña Reventada eruption by quantifying the textural heterogeneities generated by the mixing process. We show that a textural study performed by use of fractal geometry methods can be useful for calculating the water content and the viscosity of the enclaves and the basanitic magma, providing new and useful data about this magma.

2. Geological Setting

TPV started to grow approximately 180–190 ka ago in the interior of the caldera of Las Cañadas (ABLAY and MARTÍ, 2000; MARTÍ et al., 2008) (Fig. 1). This volcanic depression originated as a result of several vertical collapses of the former Tenerife central volcanic edifice (Las Cañadas edifice) caused by the explosive emptying of high-level magma chambers. Occasional large-scale, lateral collapses of the volcano flanks also occurred and modified the resulting caldera depression (MARTÍ et al., 1994, 1997; MARTÍ and GUDMUNDSSON, 2000). Construction of the present central volcanic complex on Tenerife involved the formation of these twin stratovolcanoes, which are derived from the interaction between two different shallow magma systems that have evolved simultaneously and have given rise to a complete magma series from basalt to phonolite (ABLAY et al., 1998; MARTÍ et al., 2008).

TPV mostly consists of mafic-to-intermediate products, in which felsic materials are volumetrically subordinate overall (MARTÍ et al., 2008). Felsic products, however, predominate in the most recent eruptions that have occurred from the central vents and a multitude of vents distributed on the edifice's flanks (Fig. 1). Both mafic and phonolitic magmas have erupted from these vents. The Santiago del Teide and Dorsal rift axes (Fig. 1), the two main tectonic lineations currently active on Tenerife, probably join beneath the TPV complex (CARRACEDO, 1994; ABLAY and MARTÍ, 2000). Some flank vents on the western side of Pico Viejo are located on eruption fissures that are sub-parallel to fissures located further down the Santiago del Teide rift and define the main rift axis. This is the case of the Montaña Reventada eruption (Fig. 1) studied in this paper. On the eastern side of Teide some flank vents define eruption fissures that run parallel to the Dorsal rift.

The eruptive history of the TPV comprises a main phase of eruption of mafic-to-intermediate lavas that form the core of the volcanoes and also infill most of the Las Cañadas depression and the adjacent La Orotava and Icod valleys. Approximately 35 ka ago the first phonolites appeared and

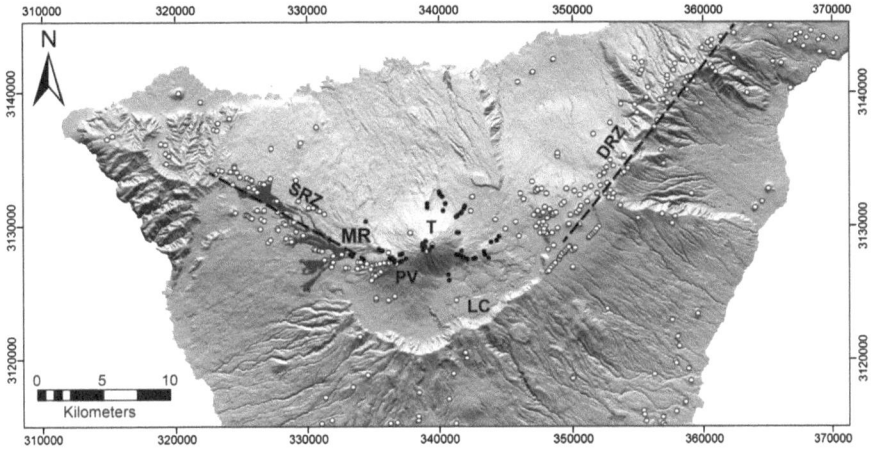

Figure 1

Location of the Montaña Reventada lavas (*MR*) and illustration of rift zones and visible vents. *Black dots* correspond to felsic vents and *white dots* to mafic and intermediate vents. *SRZ* Santiago rift zone, *DRZ* Dorsal rift zone, *LC* Las Cañadas, *T* Teide, *PV* Pico Viejo (projection UTM 28 N). Modified from MARTÍ et al., 2011

since then have become the predominant material in the TPV eruptions. Basaltic eruptions have also continued and are mostly associated with the two main rift zones. Available petrological data suggest that the interaction between a deep basaltic and a shallow phonolitic magmatic system beneath central Tenerife controls the eruption dynamics of TPV (MARTÍ *et al.*, 2008). Most of the phonolitic eruptions from TPV show signs of magma mixing, suggesting that eruptions are induced by intrusion of deep basaltic magmas into shallow phonolitic reservoirs.

The 1100 AD Montaña Reventada eruption is a clear case of a mafic eruption in the NW rift zones in which basanitic magma interacted with phonolitic magma from the TPV system. The result was a Strombolian eruption that generated a welded scoria deposit of basanitic composition and a phonolitic lava flow with clear evidence of mixing. The basanitic scoria deposit has a maximum thickness of 2 m proximal to the vent. The lava flow was mainly emplaced a few kilometers from the vent and has an average thickness of 12 m. The total volume (DRE) of the deposit estimated from geological mapping by MARTÍ *et al.*, (2008) is 0.054 km^3.

3. Methodology

3.1. Fractal Dimension

The enclaves contained in the phonolitic lava flow can be classified according to their shapes, which vary from highly irregular to almost round (Fig. 2). A few have angular profiles. Previous studies have suggested that angular enclaves correspond to more mafic compositions (ARAÑA *et al.*, 1994; WIESMAIER *et al.*, 2011), because of fragmentation of the contact surface between the intruding mafic magma and the cooler phonolitic magma (WIESMAIER *et al.*, 2011).

Photographs of 67 samples were taken normal to the surface of the enclaves to delineate the contact between the enclaves and the host rock. The images were processed by use of NIH (National Institute of Health) software (ImageJ) to generate binary images in which enclaves and host rock were replaced by black and white pixels, respectively (Fig. 2). The contact tracing operation was repeated several times to estimate the error, which was found to be approximately 2–3 %.

Enclaves with hybrid compositions can be studied by using fractal geometry methods to analyze the morphology of their complex margins. The

Figure 2
Examples of enclaves and their binary images. From **a** to **c** the morphology becomes less complex. Enclaves **a** and **c** have cuspate termination

complexity of the morphology of the enclaves was quantified by the fractal dimension (D_{box}). To compute this value the box-counting method was used; this consists in placing square mesh of different sizes (r) over the image then counting the number of boxes (N) that contain part of the image (Fig. 3).

3.2. Viscosity

PERUGINI and POLI (2005) proposed a method for establishing the relationship between the complexity of the morphology of the interface between two fluids and the their viscosity ratio (V_R), which is defined as the ratio of the viscosity of the host fluid to that of the driving fluid. After several fluid-mechanical experiments they derived the following empirical relationship:

$$\log(V_R) = 0.013 \times e^{3.34 D_{box}} \qquad (1)$$

which shows that the complexity of the interface increases with the viscosity contrast. This empirical relationship can be applied to natural cases to

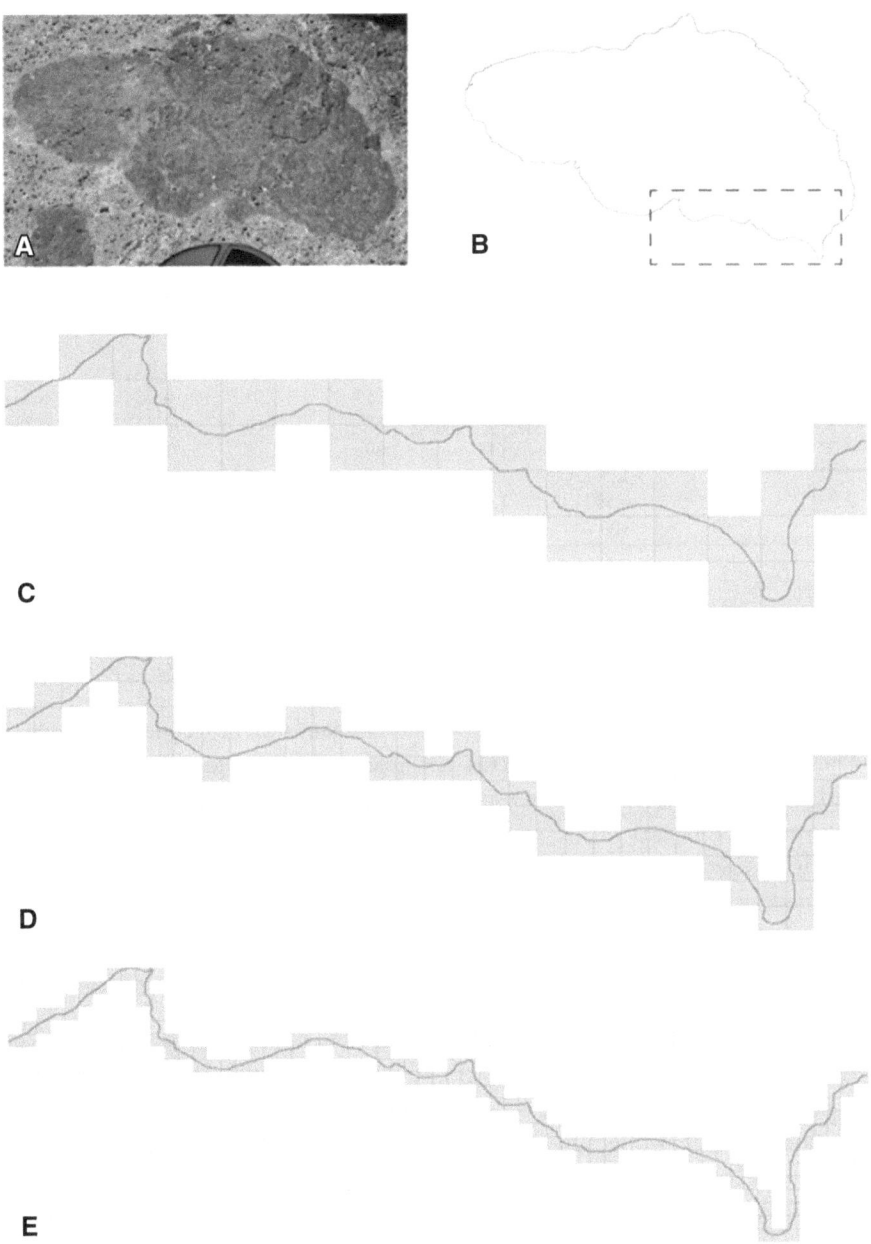

Figure 3

a Original image of the enclave. **b** Thresholded contact between the enclave and the host rock. **c–e** Magnified area indicated by the *dashed line* in **a** showing the procedure used to measure the fractal dimension (D_{box}) by use of the box-counting method. Square mesh of different sizes (r) is laid over the contact area between the enclaves and the host rock. The number of boxes (N) containing *black pixels* is counted

estimate the viscosity ratio between two magmas coming into contact during a mixing process. The relationship is valid only while the two magmas can be regarded as fluids, therefore before a significant amount of crystallization has occurred. In this study the viscosity ratio was calculated by using the fractal dimension of the morphology of the enclaves. Because this was a bi-dimensional study D_{box} varied between 1 and 2.

To estimate the viscosity of the magmas as a function of whole rock composition and temperature we used the model produced by GIORDANO *et al.*,

Table 1

Major element average from WIESMAIER et al., (2011) and ARAÑA et al., (1994) recalculated to 100 after adding H_2O

	Phonolite			Basanite					Enclaves	
	2 wt% H_2O	2.5 wt% H_2O	3 wt% H_2O	0.5 wt% H_2O	1 wt% H_2O	1.5 wt% H_2O	2 wt% H_2O	2.5 wt% H_2O	1.5 wt% H_2O	2 wt% H_2O
SiO_2	58.05	57.76	57.46	46.79	46.55	46.32	46.08	45.85	50.24	49.98
TiO_2	1.15	1.14	1.13	3.35	3.34	3.32	3.30	3.29	2.56	2.55
Al_2O_3	18.39	18.30	18.20	17.26	17.17	17.08	16.99	16.91	17.61	17.52
FeO_{tot}	4.25	4.23	4.21	10.11	10.06	10.01	9.95	9.91	7.93	7.89
MnO	0.16	0.16	0.16	0.18	0.18	0.18	0.18	0.18	0.17	0.17
MgO	1.16	1.16	1.15	4.54	4.52	4.49	4.47	4.45	3.25	3.24
CaO	2.21	2.19	2.18	9.15	9.10	9.05	9.01	8.96	6.83	6.79
Na_2O	7.70	7.66	7.62	4.93	4.91	4.88	4.86	4.84	6.23	6.19
K_2O	4.61	4.59	4.57	1.90	1.90	1.89	1.88	1.87	2.70	2.69
P_2O_5	0.32	0.32	0.32	1.29	1.29	1.28	1.27	1.27	0.97	0.96
H_2O	2.00	2.50	3.00	0.50	1.00	1.50	2.00	2.50	1.50	2.00
Sum	100	100	100	100	100	100	100	100	100	100

(2008). Whole rock analyses were taken from WIESMA-IER et al., (2011) and ARAÑA et al., (1994). Average compositions were recalculated to 100 after adding H_2O and recalculating all Fe to FeO_{tot} (Table 1). This procedure was used because the calculated viscosity ratios correspond to magmas located in the magma chamber. Therefore, it was necessary to calculate the viscosity of magmas under plausible conditions at that depth. Previous experimental work has shown that phonolitic magma erupting from the TPV flank vents was stored at temperatures of ≈ 900 °C, at pressures of ≈ 50 MPa and with $\approx 2.5 \pm 0.5$ wt% dissolved H_2O (ANDÚJAR et al., 2010, 2013; ANDÚJAR and SCAILLET, 2012a, b). Consequently, we considered 2, 2.5 and 3 wt% H_2O for the phonolitic magma. Because there is no information for the water content of basanite, a range of 1–2.5 wt% H_2O was taken into account. An average composition of enclaves with 65–70 % basanite and 35–30 % phonolite was contemplated, because in the Harker diagrams shown by WIESMAIER et al., (2011), which also include the analysis conducted by ARAÑA et al., (1994), data cluster at approximately 65–70 % of mafic magma. We propose this amount as representative of the mafic magma present in the system and the other values as a result of different degrees of mixing, because of the active and coherent regions. The average composition was recalculated to 100 after adding 1.5 and 2 wt% H_2O.

4. Results

When the box-counting method is applied to fractal patterns the following relationship is satisfied (MANDELBROT, 1982):

$$N = r^{-D_{box}} \qquad (2)$$

Equation 1 can be also written as:

$$\log(N) = -D_{box} \times \log(r) \qquad (3)$$

The slope of the linear interpolation of the $\log(r)$ vs. $\log(N)$ graph is equal to $-D_{box}$ (Fig. 4). Figure 4a–c illustrate the application of Eqs. 2 and 3 to the three enclaves given in Fig. 2a–c. From a to c the D_{box} value decreases with the complexity of the morphology of the interface. The D_{box} of the 67 images of the enclaves ranges between 1.01 and 1.23 (Fig. 5) and has a mode at $D_{box} = 1.09$. A histogram with the regression coefficients of the linear interpolation is given in Fig. 6.

The calculation of the logarithm of the viscosity ratio from D_{box} according to Eq. 1 is shown in Fig. 7. Figure 7b is a detail of Fig. 7a that focuses on the variation in $\log V_R$. The $\log V_R$ range between 0.39 and 0.81 and the distribution of the values can be seen in Fig. 8. The class with $V_R = 0.49$ has the highest frequency.

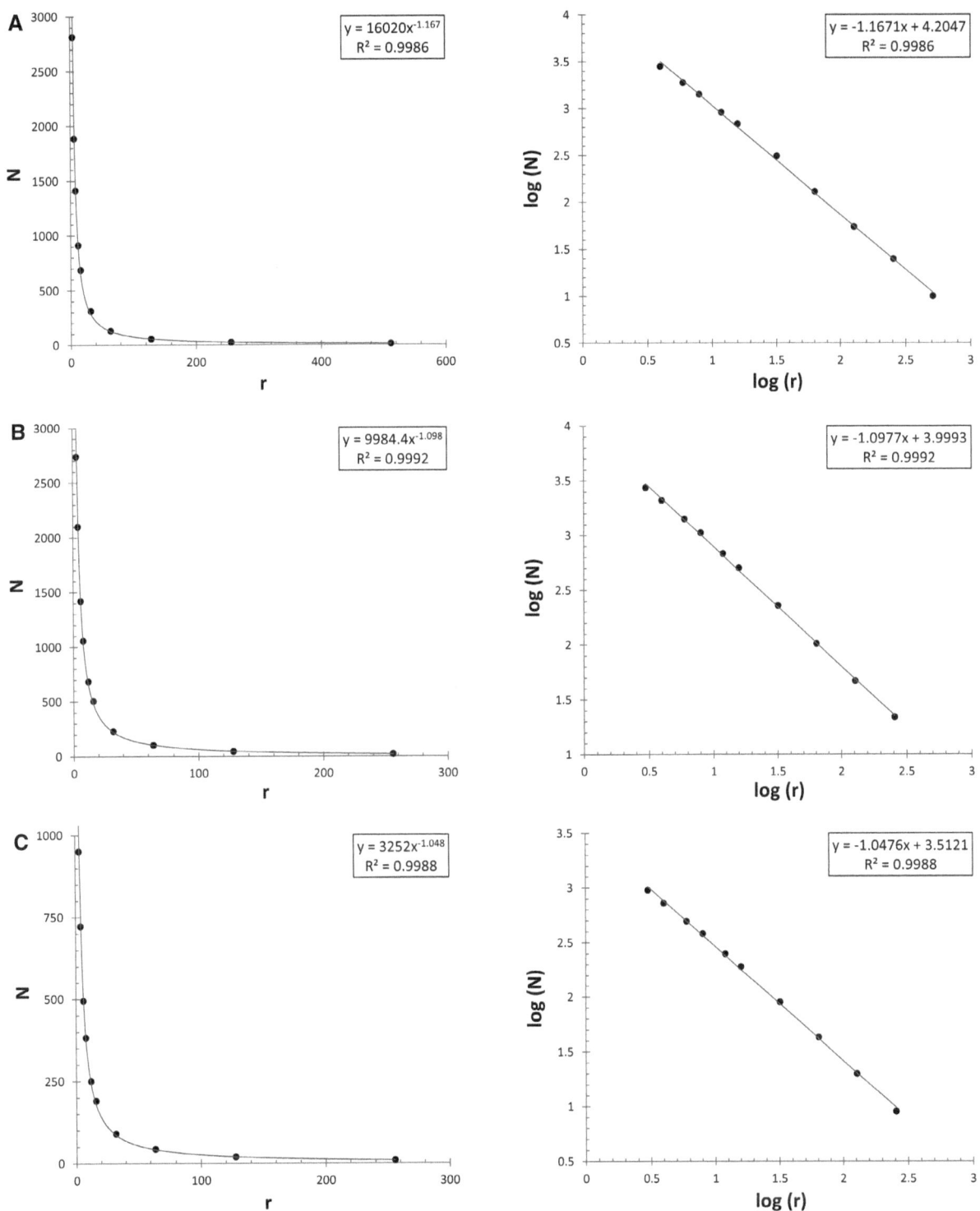

Figure 4
Calculation of fractal dimension (D_{box}). The slope of the linear interpolation of $\log(r)$ vs. $\log(N)$ is equal to $-D_{box}$

Reprinted from the journal

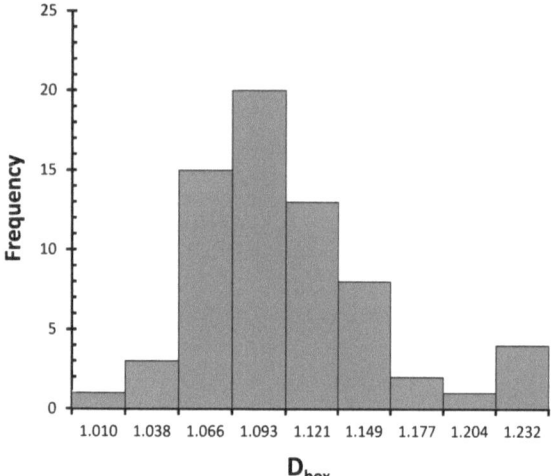

Figure 5

Frequency histogram displaying the distribution of values of the fractal-dimensions (D_{box}) of the enclaves in Montaña Reventada

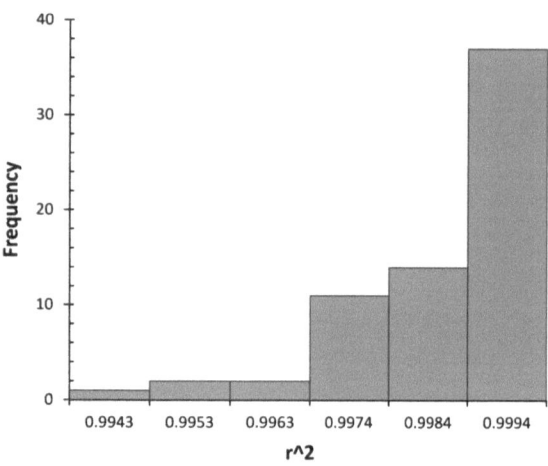

Figure 6

Frequency histogram displaying the distribution of values of the regression coefficients for the linear interpolation of $\log(r)$ vs. $\log(N)$. Values are always greater than 0.99, indicating the excellent linear fitting of the data

Knowing the viscosity of the phonolite (Table 2) and the viscosity ratio between the phonolitic magma and the enclaves, the average viscosity of the enclaves can be calculated as follows:

$$V_R = \frac{\mu_{phonolite}}{\mu_{enclave}} \quad (4)$$

$$\mu_{enclave} = \frac{\mu_{phonolite}}{V_R} \quad (5)$$

For the viscosity of the phonolite, computed values with 2.5 ± 0.5 wt% of dissolved H_2O were

considered. For V_R the minimum and maximum values were used (0.39 and 0.81) and also the value with the highest frequency (0.49). The values obtained are listed in Table 3.

5. Discussion

PERUGINI and POLI (2005) state that different D_{box} values correspond to different V_R. Accordingly, we focus here on the relationship between changes in V_R and changes in magma composition because of different degrees of mixing. More precisely, we propose the existence of a mixing area in which the two magmas interacted and produced magma of intermediate composition. Some of the enclaves considered have cuspate terminations (Fig. 2c), which have been used as evidence (PERUGINI et al., 2007) of the detachment of the enclaves from the mafic magma and their move toward more felsic magma. As WIESMAIER et al., (2011) showed, some enclaves still have mingling structures. Enclaves with cuspate terminations, mingling structures, and variable composition—and hence variable D_{box} and V_R—must have originated in this mixing zone throughout the whole process.

In the first stage, when the basanite ($\approx 1,200$ °C) reached the phonolite magma chamber (≈ 900 °C), the more mafic enclaves, characterized by quenching and angulate shapes, were generated by disruption of the layer formed as a consequence of the thermic contrast between the two magmas. According to FOLCH and MARTÍ (1998) and SNYDER (2000), during this first stage of mixing the temperature of the basanite started to fall and the temperature of the phonolite started to increase, hence the V_R decreased, thereby facilitating the mixing process between the basanite and the phonolite. The cooling and consequent crystallization of the mafic magma caused accumulation of gas bubbles at the interface between the two magmas that led, in some cases, to production of vesiculated blobs of mafic magma inside the felsic magma (EICHELBERGER, 1980; THOMAS and TAIT, 1997). This is consistent with the fact that the vesiculated enclaves of Montaña Reventada have higher D_{box} (Figs. 2a, b, 4a, b) and are therefore closer in composition to the mafic end-member. While the basanite continued to ascend to the surface,

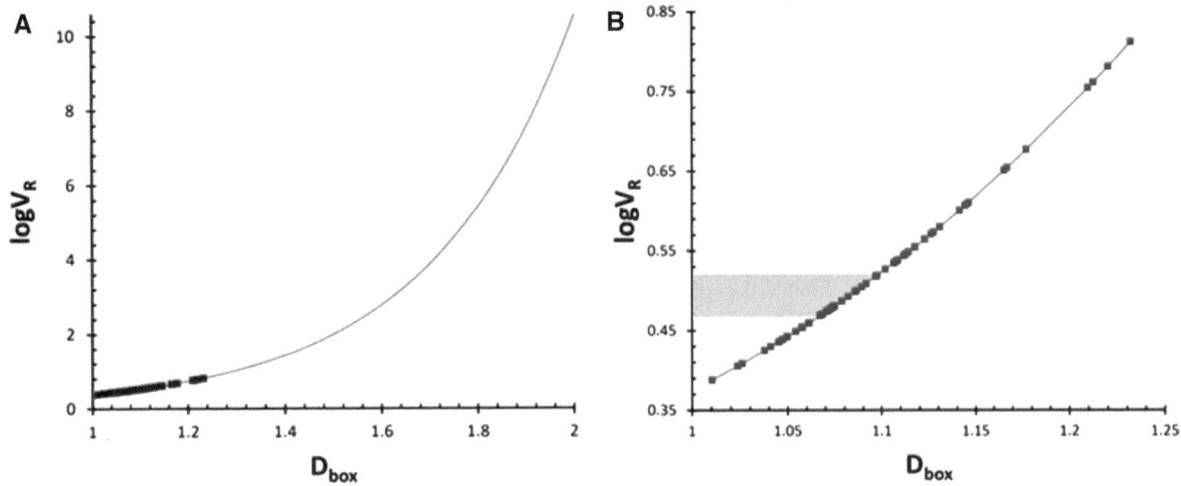

Figure 7
Variation of D_{box} vs. $\log(V_R)$ for the studied enclaves. **a** The curve shows the exponential fit of Eq. 3 considering $1 < D_{box} < 2$. **b** Magnification of the range of variation of D_{box} vs. $\log(V_R)$ for the hybrid enclaves. The *shaded area* indicates the highest frequency range

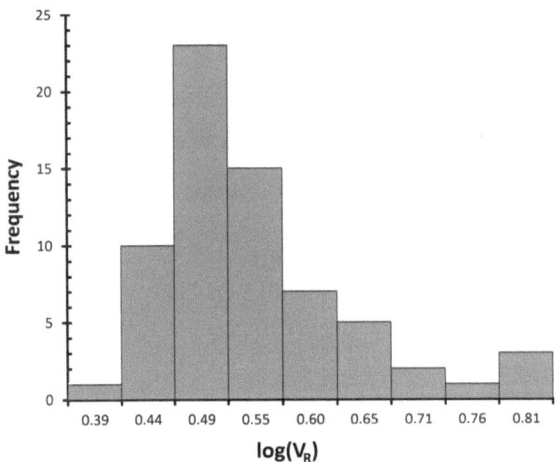

Figure 8
Frequency histogram displaying the distribution of values of $\log(V_R)$ calculated by use of Eq. 3. Class 0.49 corresponds to the shaded area in Fig. 7b

mingling structures were generated in the contact area. These structures were captured in the blobs of magma, which detached from this zone through the felsic magma and generated enclaves. The presence of both coherent and active regions yielded enclaves with different amounts of mafic and felsic magma. Chemical diffusion inside the enclaves produced different hybrid composition and hence enclaves with interface morphology characterized by different D_{box}. The morphology of some enclaves could correspond to different D_{box} values over their contours. This is consistent with the existence of coherent and active regions within the same enclave. Throughout the process the mixing zone becomes more homogenous, because of mingling and diffusion and because the V_R continued to decrease and thus to generate enclaves with lower D_{box}.

Table 2

Logarithm of the viscosity of the phonolite, the basanite, and the enclaves with 65–70 % mafic magma

wt% H_2O: T (°C)	Phonolite			Basanite					Enclaves	
	2	2.5	3	0.5	1	1.5	2	2.5	1.5	2
	$\log\mu$ (Pa·s)									
900	4.11	3.83	3.57	4.28	3.75	3.43	3.19	3.00	3.77	3.49
1,000	3.31	3.06	2.84	3.13	2.72	2.47	2.28	2.13	2.84	2.62
1,100	2.64	2.42	2.23	2.24	1.92	1.71	1.56	1.43	2.11	1.92
1,200	2.07	1.88	1.71	1.54	1.28	1.10	0.97	0.87	1.50	1.34

31

On the basis of the V_R between the phonolitic magma and the enclaves, it is possible to estimate the range of viscosities of the enclaves. Because the viscosity is closely related to the water content this enables us to estimate a plausible range of dissolved water content in the enclaves. The enclaves were generated at a temperature lower than the basanite and higher than the phonolite and are the result of the mixing of these two magmas with different viscosities. Hence the viscosity range of the enclaves must be between those of the basanite and the phonolite. This constraint reduces the water content of the phonolite, the basanite, and the enclaves to only two possible combinations of the values. As shown in Fig. 9 if the water content of the phonolite is 2 or 2.5 wt% the water content of the basanite must be 1.5 or 2 wt% respectively. Because

$V_R = 0.49$ is a mode value it can be regarded as being related to the percentage of mafic magma present in the system and the other values because of different degrees of mixing. The viscosity of the considered enclaves (Table 2) overlap with the curve of $V_R = 0.49$ (Table 3) for a water content of 1.5 or 2 wt%.

Another important question concerning the Montaña Reventada eruption is whether all the basanite that intruded into the phonolitic magma was erupted or not. The enclaves represent just ≈ 1 % of the outcrop (ARAÑA et al., 1994) and, according to estimates from the stratigraphic sections in ARAÑA et al., (1994) and WIESMAIER et al., (2011), the amount of basanite erupted corresponds to up to 15 % of the total erupted products generated during the eruption. This suggests that residual basanite magma was still stored in the magma chamber, which may have either evolved into more differentiated compositions or crystallized to form a denser body. Data from subsequent eruptions on Tenerife are unable to shed any further light on this question.

Table 3

Logarithm of the viscosity of the enclaves calculated accordingly with the V_R values and the viscosity of the phonolite

Phonolite:	2 wt% H_2O			2.5 wt% H_2O		
V_R:	0.39	0.49	0.81	0.39	0.49	0.81
T (°C)	Enclaves $\log\mu$ (Pa·s)					
900	3.72	3.62	3.30	3.44	3.34	3.02
1,000	2.92	2.82	2.50	2.67	2.57	2.25
1,100	2.25	2.15	1.83	2.03	1.93	1.61
1,200	1.68	1.58	1.26	1.49	1.39	1.07

6. Conclusions

The results of the fractal study conducted on Montaña Reventada show that enclaves with different D_{box}—and hence different composition—represent

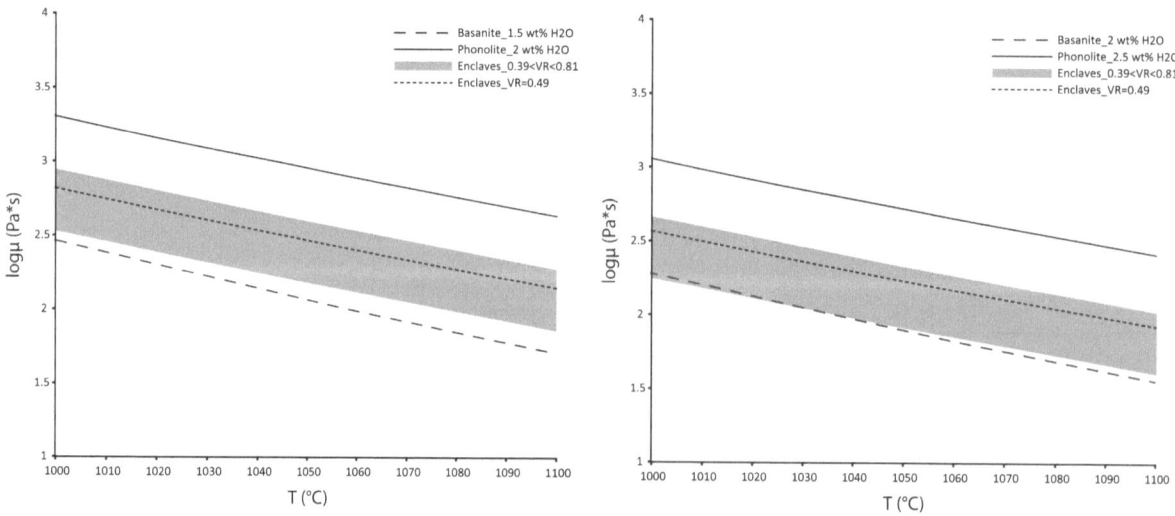

Figure 9
Variation of logarithm of viscosity with temperature

different degrees of mixing in the system, even though all could have been generated at the same time and with the same percentage of mafic magma.

Fractal analysis of enclaves of Montaña Reventada offers clues to constraining the dissolved water content and the viscosities of the enclaves and the basanitic magma. The initial estimate, based on previous work, of the water content of the phonolite of 2–3 wt% can be reduced in this case to 2–2.5 wt%. Acceptable values for the basanite and the enclaves range between 1.5 and 2 wt%.

Acknowledgments

H. Albert was funded by the Spanish Geographic Institute (IGN). This research was partially funded by the IGN and the European Commission (FP7 630 Theme: ENV.2011.1.3.3-1; Grant 282759: VUE-LCO). We are grateful for the help and support provided by C. López, H. Lamolda, and A. Felpeto. We thank the two anonymous reviewers and the editor for their constructive comments that helped to improve the manuscript. We also thank the Teide National Park for their permission to undertake this research. The English text was corrected by Michael Lockwood.

REFERENCES

ABLAY, G.J., and MARTÍ, J. (2000), *Stratigraphy, structure, and volcanic evolution of the Pico Teide–Pico Viejo formation, Tenerife, Canary Islands*, Journal of Volcanology and Geothermal Research, *103* (1), 175–208.

ABLAY, G.J., CARROLL, M.R., PALMER, M.R., MARTÍ, J., and SPARKS, R.S.J. (1998), *Basanite–phonolite lineages of the Teide–Pico Viejo volcanic complex, Tenerife, Canary Islands*, Journal of Petrology, *39* (5), 905–936.

ANDÚJAR, J., and SCAILLET, B. (2012a), *Experimental constraints on parameters controlling the difference in the eruptive dynamic of phonolitic magmas: the case from Tenerife (Canary Islands)*, Journal of Petrology, *53* (9), 1777–1806.

ANDÚJAR, J., and SCAILLET, B. (2012b), *Relationships between pre-eruptive conditions and eruptive styles of phonolite–trachyte magmas*, Lithos, *152*, 122–131.

ANDÚJAR, J., COSTA, F., and MARTÍ, J. (2010), *Magma storage conditions of the last eruption of Teide volcano (Canary Islands, Spain)*, Bulletin of Volcanology, *72*, 381–395.

ANDÚJAR, J., COSTA, F., and SCAILLET, B. (2013), *Storage conditions and eruptive dynamics of central versus flank eruptions in volcanic islands: the case of Tenerife (Canary Islands, Spain)*, Journal of Volcanology and Geothermal Research, *260*, 62–79.

ARAÑA, V., APARICIO, A., GARCIA CACHO, L., and GARCIA GARCIA, R. (1989), Mezcla de magmas en la región central de Tenerife. Los volcanes y la caldera del Parque Nacional del Teide (Tenerife, Islas Canarias), Vol. Ministerio de Agricultura Pesca y Alimentación, 269–298.

ARAÑA, V., Evolución y mezcla de magmas en Las Cañadas del Teide, In Mecanismos eruptivos y estructuras profundas de volcanes españoles e italiano (eds Araña V. and Coello J.) (Reun. Cient. CSIC-CNR, Islas Canarias 1985) pp. 38–42.

ARAÑA, V., MARTÍ, J., APARICIO, A., GARCÍA-CACHO, L., and GARCÍA-GARCÍA, R. (1994), *Magma mixing in alkaline magmas: an example from Tenerife, Canary Islands*, Lithos, *32* (1), 1–19.

CARRACEDO, J.C. (1994), *The Canary Islands: an example of structural control on the growth of large oceanic-island volcanoes*, Journal of Volcanology and Geothermal Research, *60* (3), 225–241.

CARRACEDO, J.C., RODRÍGUEZ BADIOLA, E., GUILLOU, H., PATERNE, M., SCAILLET, S., PÉREZ TORRADO, F.J., PARIS, R., FRA-PALEO, U., and HANSEN, A. (2007), *Eruptive and structural history of Teide Volcano and rift zones of Tenerife, Canary Islands*, Geological Society of America Bulletin, *119* (9), 1027–1051.

EDGAR, C.J., WOLFF, J.A., NICHOLS, H.J., CAS, R.A.F., and MARTÏ, J. (2002), *A complex Quaternary ignimbrite-forming phonolitic eruption: the Poris member of the Diego Hernández Formation (Tenerife, Canary Islands)*, Journal of Volcanology and Geothermal Research, *118* (1), 99–130.

EDGAR, C.J., WOLFF, J.A., OLIN, P.H., NICHOLS, H.J., PITTARI, A., CAS, R.A.F., and MARTÍ, J. (2007), *The late Quaternary Diego Hernandez Formation, Tenerife: Volcanology of a complex cycle of voluminous explosive phonolitic eruptions*, Journal of Volcanology and Geothermal Research, *160* (1), 59–85.

EICHELBERGER, J.C. (1980), *Vesiculation of mafic magma during replenishment of silicic magma reservoirs*, Nature, *288*, 446–450.

FLINDERS, J., and CLEMENS, J.D. (1996), *Non-linear dynamics, chaos, complexity and enclaves in granitoid magmas*, Transactions of the Royal Society of Edinburgh-Earth Sciences, *87* (1), 217–224.

FOLCH, A., and MARTÍ, J. (1998), *The generation of overpressure in felsic magma chambers by replenishment*, Earth and Planetary Science Letters, *163*, 301–314.

GIORDANO, D., RUSSELL, J.K., and DINGWELL, D.B. (2008), *Viscosity of magmatic liquids: a model*, Earth and Planetary Science Letters, *271* (1), 123–134.

MANDELBROT, B.B., The fractal geometry of nature (San Francisco, CA, 1982).

MARTÍ, J., AND GUDMUNDSSON, A. (2000), *The Las Cañadas caldera (Tenerife, Canary Islands): an overlapping collapse caldera generated by magma-chamber migration*, Journal of Volcanology and Geothermal Research, *103* (1), 161–173.

MARTÍ, J., GEYER, A., ANDUJAR, J., TEIXIDÓ, F., and COSTA, F. (2008), *Assessing the potential for future explosive activity from Teide–Pico Viejo stratovolcanoes (Tenerife, Canary Islands)*, Journal of Volcanology and Geothermal Research, *178* (3), 529–542.

MARTÍ, J., HURLIMANN, M., ABLAY, G.J., and GUDMUNDSSON, A. (1997), *Vertical and lateral collapses on Tenerife (Canary Islands) and other volcanic ocean islands*, Geology, *25* (10), 879–882.

MARTÍ, J., MITJAVILA, J., and ARAÑA, V. (1994), *Stratigraphy, structure and geochronology of the Las Cañadas caldera (Tenerife, Canary Islands)*, Geological Magazine, *131* (6), 715–727.

MARTÍ, J., MITJAVILA, J., and VILLA, I. M. (1990), *Stratigraphy and K-Ar ages of the Diego Hernández wall and their significance on the Las Cañadas Caldera formation (Tenerife, Canary Islands)*, Terra Nova, *2* (2), 148–153.

MARTÍ, J., SOBRADELO, R., FELPETO, A., and GRACÍA, O. (2011), *Eruptive scenarios of phonolitic volcanism at Teide–Pico Viejo volcanic complex (Tenerife, Canary Islands)*, Bulletin of Volcanology, *74* (3), 767–782.

MORGAVI, D., PERUGINI, D., DE CAMPOS, C.P., ERTEL-INGRISCH, W., and DINGWELL, D.B. (2013), *Morphochemistry of patterns produced by mixing of rhyolitic and basaltic melts*, Journal of Volcanology and Geothermal Research, *253*, 87–96.

PERUGINI, D., and POLI, G. (2000), *Chaotic dynamics and fractals in magmatic interaction processes: a different approach to the interpretation of mafic microgranular enclaves*, Earth and Planetary Science Letters, *175* (1), 93–103.

PERUGINI, D., and POLI, G. (2005), *Viscous fingering during replenishment of felsic magma chambers by continuous inputs of mafic magmas: field evidence and fluid-mechanics experiments*, Geology, *33* (1), 5–8.

PERUGINI, D., and POLI, G. (2012), *The mixing of magmas in plutonic and volcanic environments: analogies and differences*, Lithos, *153*, 261–277.

PERUGINI, D., BUSÀ, T., POLI, G., and NAZZARENI S. (2003a), *The role of chaotic dynamics and flow fields in the development of disequilibrium textures in volcanic rocks*, Journal of Petrology, *44* (4), 733–756.

PERUGINI, D., DE CAMPOS, C.P., DINGWELL, D.B., and DORFMAN, A. (2013), *Relaxation of concentration variance: a new tool to measure chemical element mobility during mixing of magmas*, Chemical Geology, *335*, 8–23.

PERUGINI, D., DE CAMPOS, C.P., ERTEL-INGRISCH, W., and DINGWELL, D.B. (2012), *The space and time complexity of chaotic mixing of silicate melts: implications for igneous petrology*, Lithos, *155*, 326–340.

PERUGINI, D., PETRELLI, M., and POLI, G. (2006), *Diffusive fractionation of trace elements by chaotic mixing of magmas*, Earth and Planetary Science Letters, *243* (3), 669–680.

PERUGINI, D., POLI, G., and GATTA, G.D. (2002), *Analysis and simulation of magma mixing processes in 3D*, Lithos, *65* (3), 313–330.

PERUGINI, D., POLI, G., and MAZZUOLI, R. (2003b), *Chaotic advection, fractals and diffusion during mixing of magmas: evidence from lava flows*, Journal of Volcanology and Geothermal Research, *124* (3), 255–279.

PERUGINI, D., VALENTINI, L., and POLI, G. (2007), *Insights into magma chamber processes from the analysis of size distribution of enclaves in lava flows: a case study from Vulcano Island (Southern Italy)*, Journal of Volcanology and Geothermal Research, *166* (3), 193–203.

PERUGINI, D., VENTURA, G., PETRELLI, M., and POLI, G. (2004), *Kinematic significance of morphological structures generated by mixing of magmas: a case study from Salina Island (Southern Italy)*, Earth and Planetary Science Letters, *222* (3), 1051–1066.

PETRELLI, M., PERUGINI, D., and POLI, G. (2006), *Time-scales of hybridisation of magmatic enclaves in regular and chaotic flow fields: petrologic and volcanologic implications*, Bulletin of Volcanology, *68* (3), 285–293.

POLI, G., and PERUGINI, D. (2002), *Strange attractors in magmas: evidence from lava flows*, Lithos, *65* (3), 287–297.

SNYDER D. (2000), *Thermal effects of the intrusion of basaltic magma into a more silicic magma chamber and implications for eruption triggering*, Earth and Planetary Science Letters, *175*, 257–273.

THOMAS N., and TAIT S.R. (1997), *The dimensions of magmatic inclusions as a constraint on the physical mechanism of mixing*, Journal of Volcanology and Geothermal Research, *75*, 167–178.

WIESMAIER, S., DEEGAN, F.M., TROLL, V.R., CARRACEDO, J.C., CHADWICK, J.P., and CHEW, D.M. (2011), *Magma mixing in the 1100 AD Montaña Reventada composite lava flow, Tenerife, Canary Islands: interaction between rift zone and central volcano plumbing systems*, Contributions to Mineralogy and Petrology, *162* (3), 651–669.

WOLFF, J.A. (1985), *Zonation, mixing and eruption of silica-undersaturated alkaline magma: a case study from Tenerife, Canary Islands*, Geological Magazine, *122* (6), 623–640.

(Received July 3, 2014, revised July 22, 2014, accepted July 24, 2014, Published online August 15, 2014)

Pure Appl. Geophys. 172 (2015), 1815–1833
© 2014 Springer Basel
DOI 10.1007/s00024-014-0940-6

Chaotic Flow Patterns from a Deep Plutonic Environment: a Case Study on Natural Magma Mixing

Cristina P. De Campos[1]

Abstract—This work focuses on the mixing between basalt and granite in a deep plutonic environment. The description of mixing patterns and measurements of fractal dimensions, and the evaluation of geochemical data from a Cambro–Ordovician granitic pluton are summarized and discussed. Different morphologic domains within the pluton reveal concentric fragmented and/or folded layers of granite in a gabbro/granite mixed matrix. This stands in contrast to two predominantly regular gabbroic regions. These regular regions are separated by tightly stretched filament areas, in which mixing is enhanced. Sharp and gradational contacts between granitic and gabbroic domains depict the interplay among frozen flows (mingling) and convection-enhanced diffusion processes (mixing). Measurements of fractal dimensions at different scales and analysis of normalized concentration variance for major elements point towards magma mixing: the compositional variability and flow patterns of the studied pluton have been greatly controlled by a natural chaotic mixing process between a granitic and a basaltic end-member. During the mixing process, coeval fractional crystallization no doubt contributed to increasing the complexity of the system. However, since flow, and therefore mixing, stops with temperature decrease, flow patterns must have retained the predominant morphology and composition of the moment at which both contrasting magmas came together and froze. Flow patterns have been preserved. With further temperature decrease, fractional crystallization took over and hybrid rocks were generated from the fractionation of magmas previously mixed in different proportions.

Key words: Magma mixing versus magma mingling, Mixing patterns, Pluton, Fractals, Concentration variance.

1. Introduction

Magma mixing has been widely described in the literature for both volcanic and plutonic environments during the whole evolutionary history of our planet (e.g., Sparks *et al.* 1977; Eichelberger 1980; Anderson 1982; Bacon 1986; Didier and Barbarin 1991; Abe 1997; Blundy and Sparks 1992; Wiebe 1994; Perugini *et al.* 2004; De Campos *et al.* 2004a, b; Kratzman *et al.* 2009). When volcanic and plutonic environments are compared, remarkable differences between the quality and quantity of information should be taken into account, with the differences in timescales being the most important (e.g., Perugini and Poli 2012). As a consequence, during mixing processes, each of these magmatic systems follows different cooling paths. In the plutonic environment, the magmatic system remains at relatively high temperature and pressure conditions for a longer period of time. In a large pluton, additional reintrusion and remelting under long cooling times may lead to different crystallization episodes, followed by mineralogical reequilibrium and a late recrystallization (e.g., Bateman 1995; Hibbard 1995; Bergantz 2000). In such an environment, magmatic enclaves—so-called mafic microgranular enclaves (MMEs; e.g., Didier and Barbarin 1991)—are considered to be the major markers for interaction processes. The presence of xenocrysts and disequilibrium textures and contrasting isotopic ratios, for both granites and their enclaves, is a common feature of these systems and has been widely discussed in the literature (e.g., Hibbard 1981, 1995; Perugini *et al.* 2005; Slaby *et al.* 2010, 2011).

This work focuses on the challenges of applying new approaches in magma mixing, especially chaos theory, to a plutonic environment. The structure of Santa Angélica, a complexly zoned granitic pluton, has been chosen because it shows: (1) two mappable gabbroic nuclei, (2) irrefutable signs of interaction between two contrasting granitic and basaltic/gabbroic magmas, and (3) enclave swarms that can also be mapped as a separate lithologic unit.

[1] Department of Earth and Environmental Sciences, Ludwig-Maximilian-University (LMU), Theresienstrasse 41/III, 80333 Munich, Germany. E-mail: decampos@lmu.de

Reprinted from the journal

In this work, new concepts and tools which emerged from chaos theory, in both numerical modeling and experiments on magma mixing, have been applied to study and interpret data from natural outcrops in a plutonic environment. We quantified the morphology of mixing patterns at different scales by measuring their fractal dimensions. The observed mixing patterns and measured fractal dimensions were then compared with those obtained from mixing experiments from the literature.

New whole-rock geochemical data from a sampling campaign along a profile crossing homogeneous regions, where compositions are closer to the liquidus before fractionation, have been analyzed and quantified. Different homogenization degrees for different chemical elements and their relationships are presented, discussed, and also compared with experimental data from the literature. Petrologic implications of the presented results are discussed.

2. Magma Mixing versus Mingling: the State of the Art

Given its ubiquity in nature, insight into the mechanisms underlying mixing of magmas is of great scientific relevance. The two main parameters which control the mixing process are convection and diffusion (e.g., OTTINO 1989 and references therein). From a kinematic point of view, mixing is governed by the interplay between stretching and folding. Chaotic advection has been defined as a way to generate small-scale structures in the spatial distribution of flow fields, through the stretching and folding property of chaotic flows (e.g., AREF et al. 2014 and references therein). This chaotic dynamics is thought to quickly evolve from smooth initial conditions into a complex pattern of filaments or sheets, tending exponentially fast to complex geometric patterns. Chaotic mixing processes in any fluid system, including magmatic ones, play a key role in generating scale-invariant mixing patterns, i.e., fractals (OTT and ANTONSEN 1988; WADA 1995; FLINDERS and CLEMENS 1996; DE ROSA et al. 2002; POLI and PERUGINI 2002; PERUGINI et al. 2003, 2006).

Mineralogical and geochemical studies on mixed rocks and their melted products have highlighted the importance of chaotic mixing dynamics in producing heterogeneities (e.g., HIBBARD 1995; WALLACE and BERGANTZ 2002; COSTA and CHAKRABORTY 2004; PERUGINI et al. 2005, 2008; DE CAMPOS et al. 2008, 2010; SLABY et al. 2010, 2011; MORGAVI et al. 2013a). Additional studies with natural and analogue materials pointed out the controlling role played by chaotic mixing dynamics in generating the complexity of morphological patterns found in rocks (FLINDERS and CLEMENS 1996; POLI and PERUGINI 2002; PERUGINI et al. 2002, 2005, 2006, 2012; DE CAMPOS et al. 2011; PERUGINI and POLI 2012; MORGAVI et al. 2013b).

At high temperature, convection will bring together contrasting melt compositions as layers or filaments. This interaction is expected to produce large contact interfaces between contrasting melts. Along interfaces, chemical exchanges will be strongly enhanced. In a magmatic system (volcanic or plutonic) following the same dynamic and rheological laws, as for any other fluid, this process can theoretically propagate from the kilometer/meter to the micrometer length scale. Experiments and numerical modeling show that this process will lead to variable degrees of hybridization, which will be controlled by the different mobility of chemical elements (e.g., PERUGINI et al. 2006, 2008, 2012; DE CAMPOS et al. 2004a, 2008, 2010, 2011; PERUGINI and POLI 2012; MORGAVI et al. 2013a, b).

In the petrologic literature, the term "magma mingling" has been applied to indicate physical dispersion of magmas, with no chemical interaction being involved. On the other hand, magma mixing means that not only convection, but also diffusion, and therefore chemical exchanges, are operating in the system, generating hybrid compositions. Unless there is clear evidence for the absence of chemical exchange in the system, the term "magma mixing" should therefore be more appropriate (e.g., PERUGINI and POLI 2012).

In this work, although we preferentially describe the general process as magma mixing, the designation magma mingling, for structures interpreted as such in the literature, is still used. A discussion on the most appropriate designation is presented in the Conclusions.

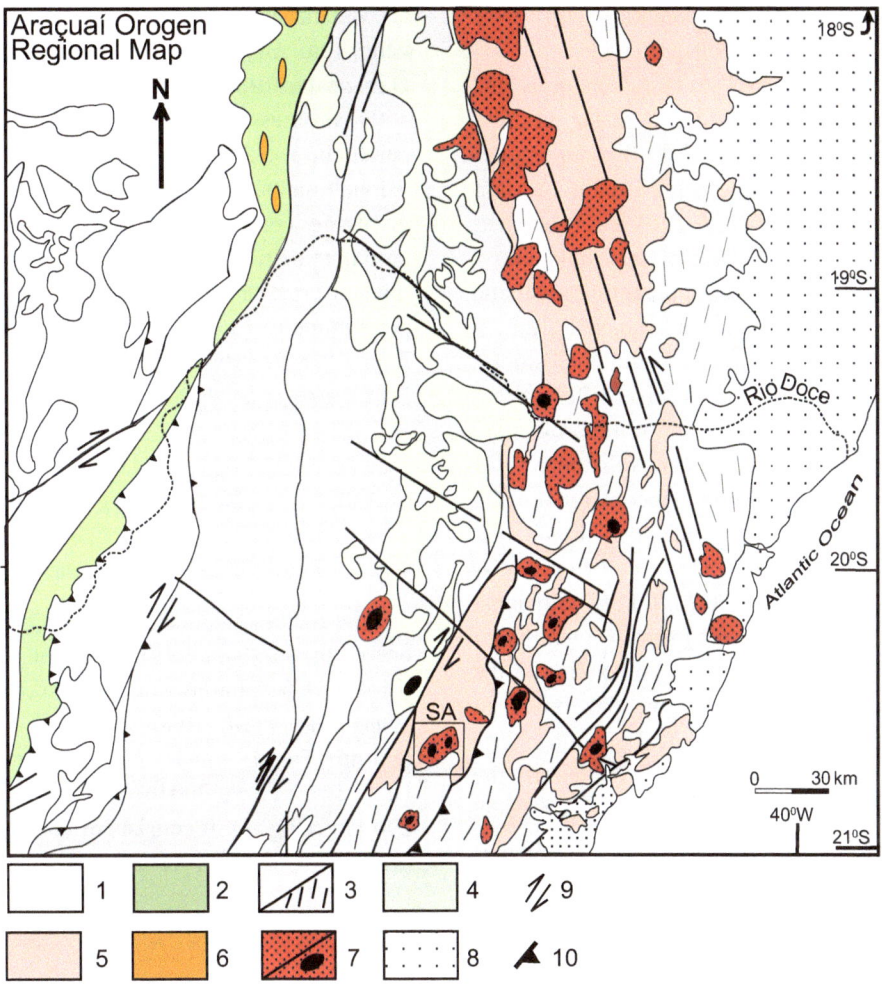

Figure 1

The Late Neoproterozoic–Cambrian Araçuaí orogen along the Atlantic continental margin, north of latitude 21°S, in eastern Brazil. The Santa Angélica Intrusive Complex (SAIC) is framed with a square and marked "SA" (modified after DE CAMPOS *et al.* 2004b; PEDROSA-SOARES *et al.* 2011). *1* Pre-Neoproterozoic successions (Archean to Paleoproterozoic); *2* Neoproterozoic rift, passive margins, and oceanic successions (c. 875 Ma); *3* Neoproterozoic basins and paragneissic complexes; *4* G1-supersuite (630–585 Ma); *5* G2- (585–560 Ma) and G3-supersuites (545–525 Ma); *6* G4-supersuite (530–500 Ma); *7* G5-supersuite (520–480 Ma); *8* Phanerozoic covers; *9* oblique to strike slip faults or ductile shear zones; *10* thrust and detachment faults. Location of the SAIC is framed and highlighted with "SA" in the southern part of the map. Additional information in Sect. 3

3. The Neoproterozoic–Cambrian Araçuaí Orogen: Regional Geology and the Granitic Magmatism

The structure of Santa Angélica is a relatively small (c. 200 km^2) granitic pluton from the Late Neoproterozoic–Cambrian Araçuaí orogen, which encompasses the entire region between the São Francisco Craton and the Atlantic continental margin, north of latitude 21°S, in eastern Brazil (Fig. 1).

The most remarkable feature of this crustal segment is the large amount of different plutonic igneous rocks of Late Neoproterozoic up to Cambro–Ordovician ages, depicting a long-lasting (c. 630–480 Ma) succession of granite production events. Granitic rocks cover one-third of the orogenic region, exposing shallow to deep crustal levels over an area of more than 350,000 km^2. Granites record the whole evolutionary history of the Araçuaí orogen, from the

37

subduction-controlled precollisional stage, up to the postcollisional gravitational collapse.

Based on field relations, structural features, and geochemical and geochronological data, granites from this orogen were grouped into five supersuites (G1–G5; DE CAMPOS et al. 2004b; PEDROSA-SOARES et al. 2008, 2011; GRADIM et al. 2014 and references therein). The grouping of diverse rock units into a supersuite has been strictly based on petrological and geochemical similarities, and constrained by zircon U–Pb ages. Supersuites include batholiths, stocks, and other minor intrusive bodies. Five different tectonic episodes during the evolution of the orogen are related to granites, namely precollisional (G1 supersuite, c. 630–585 Ma), syncollisional (G2 supersuite, c. 585–560 Ma), late collisional (G3 supersuite, c. 560–530 Ma), and postcollisional (G4, c. 530–500 Ma; G5, c.520–480 Ma). Late collisional refers to the transitional stage from the waning of convergent forces to the extensional relaxation of the orogen, generally accompanied by the onset of delamination and convective removal of lithospheric mantle. The postcollisional stage is related to the climax of the gravitational collapse of the orogen, which is coeval to asthenosphere ascent and includes plutons that cut and disturb the regional tectonic trend (PEDROSA-SOARES and WIEDEMANN-LEONARDOS 2000; DE CAMPOS et al. 2004b; PEDROSA-SOARES et al. 2001, 2008; ALKMIM et al. 2006; GRADIM et al. 2014).

The different crustal levels exposed along the Araçuaí orogen expose distinct parts, sizes, and characteristics of G5 bodies. In general, crustal depth increases from north to south and from west to east, so that small G5 bodies crop out preferentially in the southern and eastern regions of the orogen (DE CAMPOS et al. 2004b; PEDROSA-SOARES et al. 2011; Fig. 1). The most outstanding features revealed are the roots of funnel-shaped intrusions with inverse zoning, displaying interfingering of mafic to intermediate rocks in the core and syenomonzonitic to granitic borders, together with widespread evidence of magma mingling and mixing (e.g., BAYER et al. 1987; SCHMIDT-THOMÉ and WEBER-DIEFENBACH 1987; WIEDEMANN 1993; MENDES et al. 1997, 1999, 2005; MEDEIROS et al. 2001, 2003; WIEDEMANN et al. 2002; DE CAMPOS et al. 2004b; MENDES and DE CAMPOS 2012).

Metaluminous to peraluminous, high-K calc-alkalic, I- to A-type granitoids (e.g., HORN and WEBER-DIEFENBACH 1987; WIEDEMANN 1993; MENDES et al. 1999) progressively evolve into markedly alkalic to peralkalic rocks (LUDKA et al. 1998). These postcollisional melts originated from contrasting sources, involving contributions from an enriched mantle, partial remelting from a mainly metaluminous continental crust and dehydration melting from slightly peraluminous rocks (e.g., WIEDEMANN et al. 1995; LUDKA et al. 1998; MENDES et al. 1997, 1999; DE CAMPOS et al. 2004b; GRADIM et al. 2014).

4. The Santa Angélica Pluton

4.1. Structural Geology: Previous Mapping

The Santa Angélica Intrusive Complex (SAIC) is one of the most interesting examples of the postcollisional G5 plutons in this mobile belt (e.g., WIEDEMANN et al. 1986; BAYER et al. 1986, 1987; SCHMIDT-THOMÉ and WEBER-DIEFENBACH 1987; DE CAMPOS et al. 2004b). The SAIC covers about 200 km^2 (Fig. 2). It consists of an elliptically shaped intrusion composed of several roughly concentric

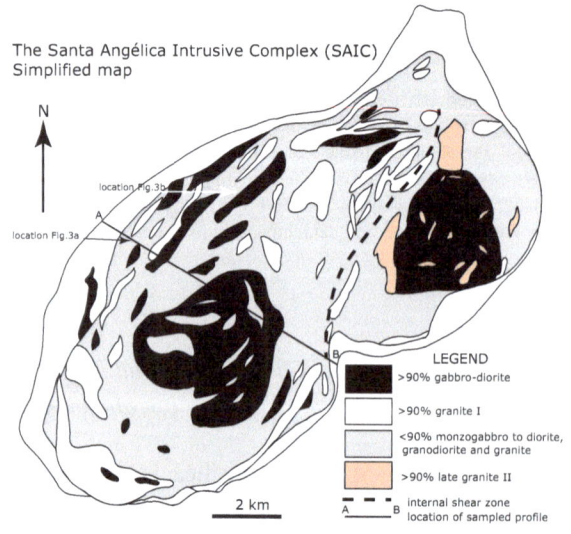

Figure 2

Simplified map of the Santa Angélica Intrusive Complex, a postcollisional G5 pluton in the Araçuaí mobile belt (mod. after BAYER et al. 1987) with location of sampled profile and outcrop from Fig. 3a, b

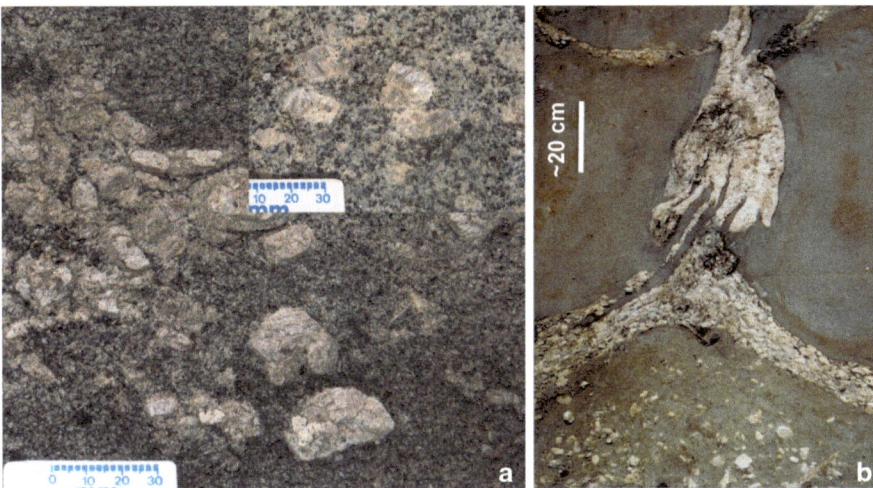

Figure 3
a Detail of the hybrid zone located on Fig. 2, showing mantled feldspars and different hybridization degrees of the monzodioritic to granodioritic matrix. The picture at the *upper right* was taken ca. ½ m from the major one; **b** Stretched and folded structures between microgabbro and granite originated from clear interaction between contrasting magmas. Pirineu Ridge, north of the SAIC, location on Fig. 2

lens-shaped granitic layers, elongated lenses of gabbroic to dioritic rocks, and tightly packed heterogeneous enclave swarms (Fig. 3a, b). The complex intrudes an antiformal structure with a northwest-striking hinge, in contrast to the hinge lines of major structures that strike northeast. Country rocks are high-grade metamorphosed biotite-garnet-sillimanite gneisses and biotite-hornblende granodioritic to tonalitic gneiss (G1/G2 pre- to syncollisional orthogneisses). These are locally migmatized, showing subvertical foliation with parallel contacts to the borders of the intrusive complex, which dip inwards towards the intrusion. Regarding the enclosing gneisses, widespread migmatization is reported as older than 541 Ma (U–Pb zircon ages for the Muniz Freire orthogneiss and enclosing rocks, from PEDROSA-SOARES *et al.* 2008; GRADIM *et al.* 2014). Additional U–Pb zircon dating for amphibolite facies metamorphism and migmatization in the enclosing rocks is 573 + 12 Ma (Celina-metatexite, from BAYER *et al.* 1987). Since then, no further signs of superimposed orogenic deformation have been documented for this region.

However, the intensity of migmatization is stronger near to the pluton contact, where the enclosing gneisses exhibit nebulitic fabrics consistent with partial melting. The foliation inside the intrusive complex is normally subvertical. Type II granite (Fig. 2) is fine-grained and intruded in a late brittle phase. Common signs of mixing between gabbro-diorite and type I granite can be observed throughout the intrusion as large mingled/mixed zones formed by (1) filaments, (2) sheets, or (3) enclave swarms at different hybridization degrees with the type I granite (Fig. 3a, b). A main NE–SW active shear zone caused more intensive mixing, producing a fine-banded rock consisting of monzodioritic and granitic layers. This NE–SW shear zone separates the two gabbro-dioritic nuclei (Fig. 2) with no signs of continuation outside the structure (WIEDEMANN *et al.* 1986; BAYER *et al.* 1986, 1987; DE CAMPOS *et al.* 2004b).

4.2. Geochemical and Mineralogical Data from the SAIC: the State of the Art

Mineralogical and whole-rock geochemical data point towards a high-K calc-alkalic to slightly alkalic expanded suite (Fig. 4a, b). Rock types grade from biotite-hypersthene-augite gabbro (monzogabbro), biotite-diorite (monzodiorite), to allanite-biotite granitoids (SCHMIDT-THOMÉ and WEBER-DIEFENBACH 1987). Granitic to gabbroic rocks are metaluminous and both enriched in incompatible elements (HORN and WEBER-DIEFENBACH 1987). This is mostly characteristic for the mafic to intermediary rocks.

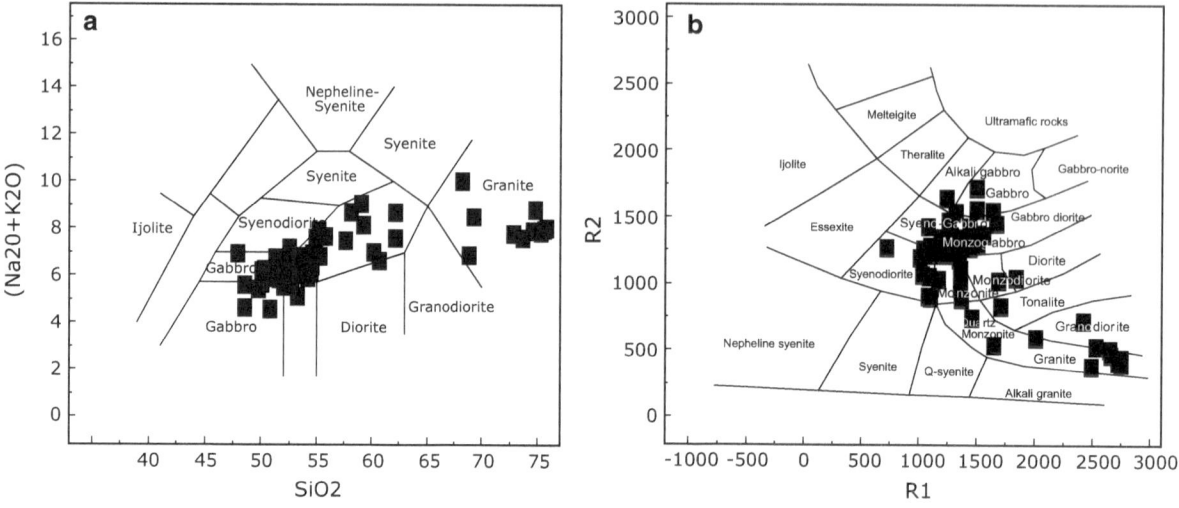

Figure 4

TAS diagram (total alkalis versus SiO$_2$, after Cox *et al.* 1979) with composition of whole rocks ranging from gabbro to granite; **b** R_1–R_2 diagram (after DE LA ROCHE *et al.* 1980) showing an even more expanded sequence of hybrid rocks ranging from gabbro to granite. $R_1 = 4Si - 11(Na + K) - 2(Fe + Ti)$; $R_2 = 6Ca + 2$ Mg $+$ Al. This classification is closer to the petrographic classification of SAIC rocks from the literature

Anomalous K$_2$O, Ba, Sr, and light rare earth element (LREE) contents, as well as high contents in high-field strength (HFS) elements such as Ti, Y, Nb, P, and Zr are also typical (e.g., DE CAMPOS *et al.* 2004b; PEDROSA-SOARES *et al.* 2008 and references therein).

Magma mingling and mixing, together with fractional crystallization, were recognized as main differentiation processes during the evolution of these igneous sequences (WIEDEMANN *et al.* 1986, 2002; SCHMIDT-THOMÉ and WEBER-DIEFENBACH 1987; HORN and WEBER-DIEFENBACH 1987; MENDES *et al.* 1997; MEDEIROS *et al.* 2001).

In comparison with average values of gabbros and diorites (NOCKOLDS *et al.* 1978), rocks from the SAIC have higher Na$_2$O, K$_2$O, TiO$_2$, and P$_2$O$_5$ and relatively standard Cr, Ni, and Co values.

Sr isotopic ratios ($_{87}$Sr/$_{86}$Sr values) ranging from 0.709 to 0.707 (SÖLLNER *et al.* 1991, 2000), together with K$_2$O values for basic and intermediate rocks, possibly reflect heterogeneous mixing with the granites suggesting a hybrid origin for the basic to intermediate rocks. Additionally, very low $\varepsilon_{Nd(t)}$ values are found for basic rocks, e.g., $\varepsilon_{Nd(500Ma)} = -19.76$ for a gabbrodiorite from the core region, sampled close to sample 39b (Table 2), and $\varepsilon_{Nd(500Ma)} = -20.74$ (DE CAMPOS *et al.* 2004b),

for an allanite granite from the border region, sampled close to sample 50 (Table 2).

High Ba and Sr values, over 1,500 and 1,000 ppm, respectively, have been correlated to an abnormal mantle enrichment episode (WIEDEMANN *et al.* 1995; LUDKA *et al.* 1998). In contrast, almost identical Sr–Nd isotopic ratios for both gabbrodiorite and granite (SÖLLNER *et al.* 1991; DE CAMPOS *et al.* 2004b) point towards a highly homogenized mixed system, which is not consistent with the homogenization degree of the whole magmatic body at outcrop scale. A granitic melt mixed with a mantle-derived basaltic melt to produce a hybrid monzogabbro with the same isotopic signature as the granite (DE CAMPOS *et al.* 2004b).

In comparison with other intrusions for the same time–space episode, the smaller spreading of data points between granite and monzogabbro from the SAIC is consistent with field and petrographic evidence of effective mixing between gabbro and granite in this intrusion (e.g., HORN and WEBER-DIEFENBACH 1987; DE CAMPOS *et al.* 2004b; MENDES and DE CAMPOS 2012). An alternative could be that mixing with a granitic magma may have occurred at different stage, depth, and temperature, before and after main fractional crystallization took over (DE CAMPOS *et al.* 2004b).

At the end of the orogen, the progressive crustal relaxation greatly favored the generation of A-type magmas. The rising of mantelic magmas along preexisting ductile shear zones, together with the coeval reheating of the crust, induced the new episode of partial melting and mixing between the contrasting magmas (e.g., MENDES *et al.* 1997, 1999; LUDKA *et al.* 1998; MEDEIROS *et al.* 2001; WIEDEMANN *et al.* 2002, DE CAMPOS *et al.* 2004b). While granitic rocks have been recognized as originated from the partial melting of the continental crust, an enriched mantle has been interpreted as source for the basic magmas.

U–Pb zircon ages constrain the magmatic crystallization of both monzogabbro and granite from Santa Angélica to 513 ± 8 Ma (SÖLLNER *et al.* 2000). Reintrusion episodes such as dykes and later plugs have not been evaluated together with the contact flow lines between the coeval main granite I and gabbro/monzogabbro intrusion. A later granite generation episode has been mapped and dated in the northeast region of the pluton: granite II dated at 492 ± 15 Ma (SÖLLNER *et al.* 2000; WIEDEMANN *et al.* 2002). Therefore, all contact lines with granite II have been removed for the calculations in this work. From 480 Ma up to about 135 Ma, there is no more evidence of tectonomagmatic activity in the region (e.g., PEDROSA-SOARES *et al.* 2008 and references therein). From 135 Ma on, increasing lithospheric thinning is reported with vertical block movements associated with intrusion of basaltic dykes, culminating with the breaking up of Gondwana between Africa and Brazil further to the east (ALMEIDA *et al.* 2013).

Hence, all previous data and observations point towards the issue that stretched and folded structures described and measured in this work (e.g., Fig. 3a, b) are not related to reintrusion episodes or migmatites but result from frozen-in interactive flow structures from the anorogenic coeval intrusion of contrasting magmas in the lower crust. No signs of compressional deformation due to regional folding have been reported or observed for this intrusion so far (this work, WIEDEMANN *et al.* 1986; BAYER *et al.* 1987).

Therefore, structures depicted on the SAIC map are related to the coeval intrusion of contrasting magmas and their hybridization process.

5. Fractal Analysis of Flow Patterns

One of the most striking characteristics in geology is the self-similarity shown by many natural objects, i.e., their scale invariance. For analyzing outcrops and enclave geometry, numerical modeling indicates a scale-invariant, i.e., fractal (bilogarithmic), distribution (e.g., DE ROSA *et al.* 2002; PERUGINI *et al.* 2003). For the analysis of magma mixing structures, especially due to the expected development in time of a chaotic process (e.g., ALLÈGRE and TURCOTTE 1986; OTTINO 1989), numerical modeling associated with experiments also evidences a scale-invariant, i.e., fractal (bilogarithmic), distribution (e.g., WADA 1995; PERUGINI *et al.* 2003; DE CAMPOS *et al.* 2011; MORGAVI *et al.* 2013b).

The time evolution of magma mixing processes therefore generates fractal morphologies, which are controlled by the stretching and folding of the fluids (in our case, magmas) involved in the process (e.g., OTTINO 1989; FLINDERS and CLEMENS 1996; PERUGINI *et al.* 2006). This is the reason why we used fractal statistics to quantify the complexity of mixing patterns.

The simplest and most widely used technique to measure the fractal dimension on images is the "box-counting" method. This technique has been used before to quantify mixing patterns in both plutonic and volcanic rocks (e.g., WADA 1995; PERUGINI *et al.* 2003; PERUGINI and POLI 2005). The same method has also been used to quantify mixing patterns for experiments with analogue and natural basaltic and rhyolitic melts (DE CAMPOS *et al.* 2011; MORGAVI *et al.* 2013b).

For the purpose of evaluating the fractal dimension, we started by extracting the image boundary. For the first evaluation, the original image from Fig. 2 was converted to a binary (black and white) image, so that the main interfaces between mafic gabbro and felsic granitic magmas are replaced by a black line (Fig. 5a–c). The interface between the two melts was then used to measure the fractal dimension of the mixing patterns using ImageJ software (ABRAMOFF *et al.* 2004).

For the box-counting technique, a square mesh of size r is laid over the image and the number of boxes N_r containing black pixels and belonging to the interface between the two melts is counted (e.g., MANDELBROT 1982).

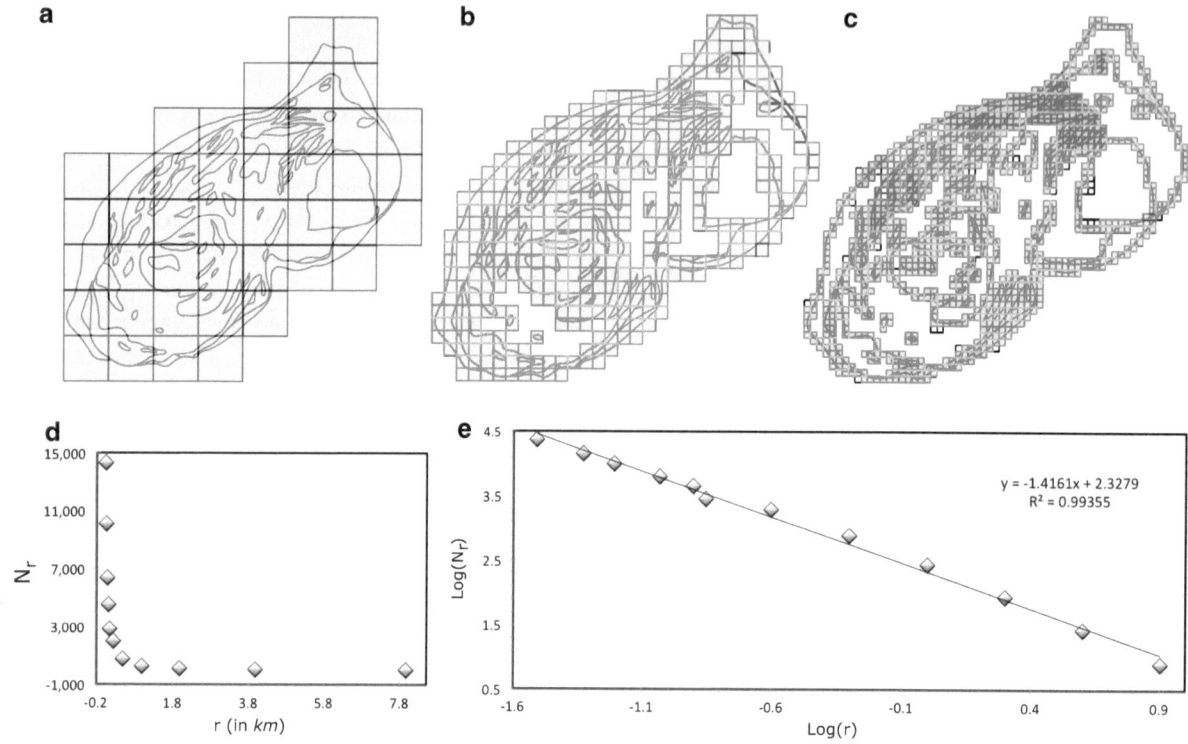

Figure 5

a–c Graphical representation of the applied box-counting method on mixing patterns from the simplified geological map of the Santa Angélica structure; for this counting only the contact lines between coeval rock units were considered. Contacts with granite II (see Fig. 2) were not taken into consideration. A mesh of different sizes (r), coherent with the structure scale, overlays the original image. The number of boxes (N_r) containing part of the contact interface is then counted; **d** plot of r against N_r, for the repeated box-counting procedure on images (**a–c**); **e** log–log plot of r against N_r for the geological map of SAIC, showing the linear relationship between these two parameters and, therefore, the fractal nature of the geological contacts. The equation of the regression line gives a slope of 1.416, corresponding to the fractal dimension (D_{box}). See text for details

For fractal patterns, the following relationship is satisfied:

$$N_r = r^{-D_{box}}. \qquad (1)$$

Using logarithms, Eq. (1) can also be written as

$$\log(N_r) = -D_{box} \cdot \log(r) \qquad (2)$$

A structure is considered as fractal if the obtained data lie on a straight line in a log–log plot. The fractal dimension (D_{box}) is estimated from the slope resulting from the linear interpolation of the $\log(r)$ versus $\log(N_r)$ graph.

In our case, the routine of the used software counts the number of boxes of given size needed to cover a one-pixel-wide, binary (black on white) border. The procedure is then repeated for boxes that are 2 to 1,024 pixels wide. Pixels have been

recalculated as millimeters and rescaled for the different images (Table 1).

Figure 5d shows the variation of the number of boxes (N_r) containing black pixels belonging to the interface between the two melts against the box size (r), whereas the graph in Fig. 5e depicts the corresponding log–log plot. It is shown that data points

Table 1

Summary of the different parameters for the rescaling of images. Values of fractal dimension (D_{box})

Outcrop image		Rescaling		Map		Rescaling	
Pixels	1,243	cm	329.39	Pixels	2,874	km	761.61
Pixels	993	cm	263.14	Pixels	2,721	km	721.06
FD	1.32				1.42		

FD fractal dimension

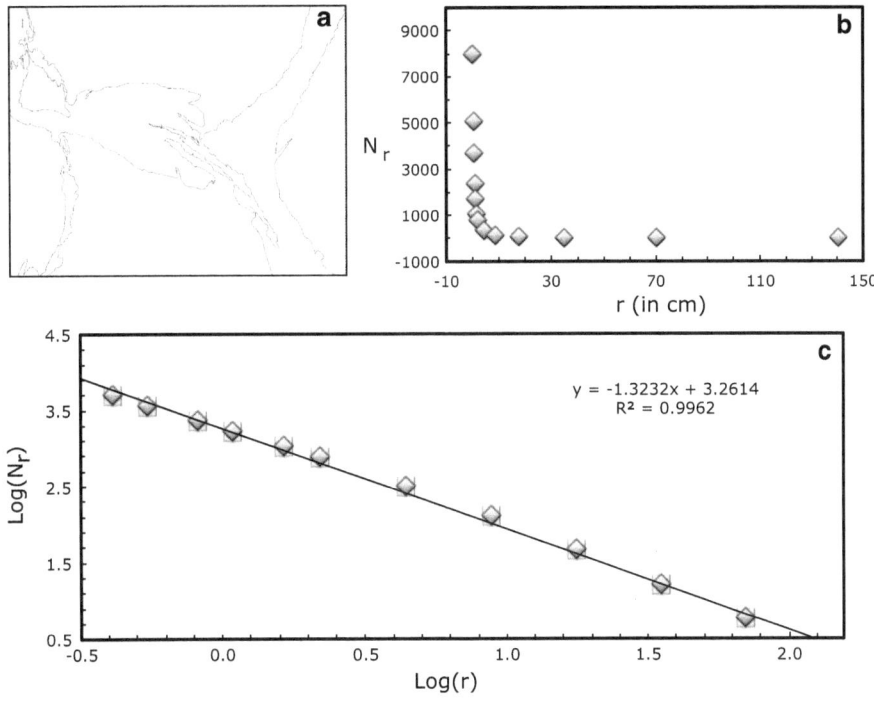

Figure 6

a Contact interfaces between microgabbro and granite traced from the binary images of Fig. 3b. **b** Plot of r against N_r, for the repeated box-counting procedure on image (**a**); **c** log–log plot of r against N_r for the contact interfaces of image (**a**), showing a linear relationship between these two parameters and hence the fractal nature of the interaction contacts. The equation of the regression line gives a slope of 1.416, corresponding to the fractal dimension (D_{box}). See text for details

follow a linear distribution, indicating that interfaces are fractal.

The evaluation of the image from Fig. 3b followed the same procedure described above for the geological map in Fig. 2. The image boundary from Fig. 3b was extracted and converted to the binary (black and white) image shown in Fig. 6a. The main interfaces between mafic microgabbro and felsic granitic magmas have been replaced by a black line (Fig. 6a). The interface between the two melts was used to measure the fractal dimension of the mixing patterns again using ImageJ software (ABRAMOFF et al. 2004).

Figure 6b shows the variation of the number of boxes (N_r) containing black pixels belonging to the interface between the two melts against the box size (r), whereas the graph in Fig. 6c depicts the corresponding log–log plot. It is shown that data points once more follow a linear distribution, indicating that interfaces are fractal.

Values of the fractal dimension of the mixing patterns were measured in the range of threshold values appropriate for each case, for instance, where D_{box} shows little variation. In particular, the values of the fractal dimension for the geological map and the outcrop were 1.42 and 1.32, respectively.

5.1. Comparison with Experimental Results from the Literature: Some Implications

Recently, the morphology of mixing patterns from chaotic mixing experiments with analogue silicate melts (haplobasalt and haplogranite from DE CAMPOS et al. 2011) and with natural melts (basalt/gabbro and granite/rhyolite from MORGAVI et al. 2013b) was quantified by measuring the fractal dimension. A linear relationship was derived between mixing time and morphological complexity. The main purpose and interest of such experiments was to obtain a fully chaotic mixing system. Therefore, the chosen fluid-dynamics protocol contained all fundamental building blocks for enhancing mixing, which are: combined action of stretching and folding dynamics, and chemical diffusion between interacting fluids. It is noteworthy that, before these experiments, mixing processes were only simulated numerically because of the difficulty of setting up suitable experimental systems for high-temperature and high-viscosity melts. This is the reason why the experimental temperature was kept superliquidus (1,300 °C) to allow reasonable experimental times for the extreme viscosity ratios. These experiments therefore replicate magma mixing processes, and despite constraints related to the too short mixing times, they represent an important improvement towards a tight link between experiments and nature.

In our natural case, several parameters are unknown, such as the mixing time, the proportion and exact composition of mixed end-members, and the temperature. Here, we compare our results with those from experiments with similar systems in order to shed light on open questions.

For patterns obtained from controlled chaotic experiments with haplobasalt (anorthite-diopside) and haplogranite, a value of 0.91 for the fractal dimension was measured using the box-counting method (DE CAMPOS et al. 2011). For patterns obtained using the same experimental setup but with natural melts at different times of 53, 106, and 212 min, values for the fractal dimension, measured using the same box-counting method, were 1.15, 1.47, and 1.68, respectively (MORGAVI et al. 2013b). Our values of fractal dimension for the geological map and the outcrop, rescaled for km and cm, are 1.42 and 1.32, respectively. These values are closer to those from short-time patterns measured for shorter mixing times at higher temperature but with natural melts.

When leaving controlled laboratory conditions to apply the insights obtained from experiments to a natural case, additional complexities and uncertainties emerge. However, when experiments are performed with natural melts (MORGAVI et al. 2013b), with viscosity ratios close to the real world, the natural environment can be simulated in a more accurate way, so that flow patterns can be compared with each other. As a partial conclusion, the obtained fractal dimensions point towards chaotic flows generated by mixing two contrasting magmas (basalt and rhyolite). The difference between the fractal dimension from the map (1.42 for a km scale) and the outcrop (1.32 for a cm to m scale) could be caused by different local mixing times and/or local differences in the viscosity contrast (Table 1). In such a large natural magma chamber such as the SAIC, complexities are therefore to be expected.

Theoretical and experimental studies on mixing highlight the importance of stretching and stirring of fluid particles under chaotic advection, enhancing mixing. In particular, a flow can be partitioned into regions of qualitatively different fluid motion. Separatrices divide the fluid domain into different regions: the inner core regions, the inner and outer recirculation regions, and the outer flow (e.g., RIZZI and CORTELEZZI 2011; AREF et al. 2014 and references therein).

Geometric patterns from experiments with natural silicate melts (basalt and rhyolite) at high temperatures are far more complex. Those obtained under a journal bearing system (MORGAVI et al. 2013b), which should only generate chaotic filament regions, produce complex microenclave patterns far from the expected morphology of regular chaotic flows obtained by mixing experiments with Fe-free haplobasalt and haplogranite from DE CAMPOS et al. (2011) with the same geometry.

Regions of qualitatively different fluid motion have been recognized. In the structure of Santa Angélica, one main shear zone (Fig. 2) could be a suitable candidate for dividing the fluid domains into two separate vortex regions, where separate inner stagnant core regions were mapped. The restriction of the internal shear zone to the magmatic structure suggests that it was generated by the flow and not superimposed by later deformation. It has no connection to regional structures which envelop the large magma bulge.

Inner and outer recirculation regions are filament rich and under diverse hybridization degrees (Fig. 3a, b). Regarding a possible candidate for the outer flow (e.g., RIZZI and CORTELEZZI 2011 and references therein) in the SAIC, the border region consists of over 95 % granite (Fig. 2).

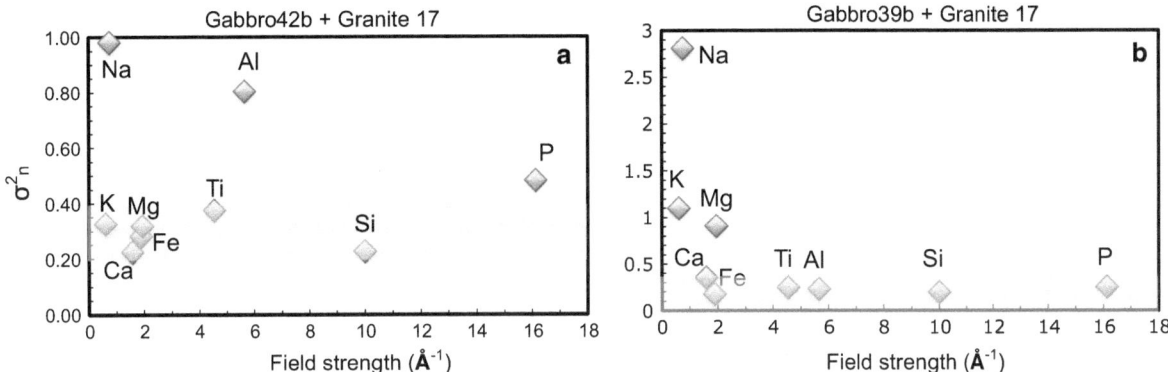

Figure 7

a, b Variation of the normalized concentration variance (σ_n^2) of major element oxides as a function of field strength (calculated as nominal charge Z of the element divided by ionic radius r, after Shannon 1976) for all analyzed samples; Normalization was calculated for samples 42b (**a**) and 39b (**b**) as least evolved and sample 17 as a most evolved end-member. To fulfill the assumption, the normalized concentration variance should plot along a similar (σ_n^2) range. For the case of the fine-grained basalt 42b (**a**), five major elements fulfill this condition. For the coarse-grained gabbro 39a, five major elements fulfill this condition, but Na and K plot far away from the expected range (**b**). See text for details

6. Geochemical Data: This Study

For the evaluation of the compositional range in the SAIC, we sampled over 50 outcrops according to the following criteria: (1) all sampled outcrops are located along a straight NW–SE profile with maximum offset of ±50 m (Fig. 7); (2) to avoid mixing or hybridization through sampling, sampled outcrops correspond to the most homogeneous isotropic texture, with <10 % pheno- or xenocrysts, so that the whole-rock composition is expected to be close to the matrix composition and, therefore, to the original magma generating the sampled plutonic rock; (3) sample sizes were calculated in accordance with the grain size, with mean sample sizes of at least c. 2 kg for fine-grained rocks up to c. 40 kg for coarse-grained granites; (4) analyzed samples were fresh with no visible signs of weathering after petrographic control; (5) during sampling, special attention was paid to avoid sites with mineral segregation, which could cause additional complexities generated by visible crystal settling from fractional crystallization.

6.1. Sample Preparation and Results

Around 500 kg of samples were prepared and original volumes reduced according to the previous criteria. Forty-six samples as rock powder fulfilling all criteria were further prepared for X-ray fluorescence, inductively coupled plasma (ICP)-mass spectroscopy (MS) laser ablation, and isotopic analysis. In this work we only treat results obtained from whole-rock analysis.

Major elements were measured at the X-ray Fluorescence Lab of the Department of Geology, Federal University of Rio de Janeiro, using a Philips PW2400 spectrometer (Rh tube). The loss on ignition was obtained by the weight of the sample before and after heating at 950 °C for 0.5 h. Major elements were detected after melting 1.2 g of powdered material with lithium tetraborate. Light elements were measured under the following conditions: flow detector, pentaerythritol (PET) crystal analyzer, Ge, and a 40 kV/ 70 mA power tube. Heavy elements were analyzed with a sealed detector, LIF200/LIF220 crystal analyzers, and a 50 kV/50 mA power tube. Based on analysis of standards, the relative analytical error was estimated as <1 % for Si and Al, 1–2 % for Fe, Mg, and Ca, and 3–5 % for Ti, Na, and K. Calibration curves were obtained from the following international standards: NIM-P, 521-84n, GBW07112, GIT-IWG, ANRT, BE-N GIT, PM-S GIT, CRPG BR, AN-G GIT, GBW07104, GBW07110, GBW07111, AC-E, GS-N, N-MA, and CRPG GH.

All dry normalized data are presented in Table 2. Chemical classification of all analyzed samples is depicted in Fig. 4a, b, showing a wide compositional

Table 2

X-ray fluorescence data for major elements in the basaltic and rhyolitic end-member glasses used in the mixing experiments

s	SiO_2	TiO_2	Al_2O_3	FeO_t	MnO	MgO	CaO	Na_2O	K_2O	P_2O_5
1	54.482	2.108	15.731	8.710	0.133	4.447	7.137	3.295	2.783	1.174
2	52.788	2.797	14.931	9.531	0.130	4.893	7.743	3.055	2.483	1.649
3	54.243	2.048	16.017	8.287	0.126	4.781	7.489	3.231	2.652	1.125
4a	53.305	2.630	15.040	9.792	0.140	4.364	7.333	3.299	2.698	1.399
4b	51.477	2.937	14.816	10.580	0.156	4.563	7.834	3.249	2.667	1.720
5b	52.451	3.225	14.767	9.710	0.134	4.400	7.592	3.074	2.827	1.820
6	59.378	1.849	16.089	7.471	0.136	2.193	4.149	3.500	4.623	0.613
8	53.325	2.106	15.445	9.157	0.144	5.756	7.971	3.152	1.925	1.020
9	62.209	1.545	16.153	6.106	0.099	1.695	3.097	3.665	5.010	0.422
10	68.880	0.503	16.538	2.940	0.065	0.853	3.259	3.886	2.951	0.125
12	54.616	2.427	15.143	8.938	0.135	4.198	7.023	3.344	2.974	1.200
13	74.778	0.356	13.148	1.622	0.045	0.204	0.997	3.087	5.707	0.057
14a	52.083	1.942	16.160	8.269	0.114	6.443	8.516	3.231	2.280	0.961
14b	55.271	2.209	16.341	8.983	0.147	2.922	5.428	3.685	4.173	0.841
15	51.309	2.627	16.732	10.218	0.151	4.535	6.448	3.665	3.061	1.254
17	75.825	0.222	13.375	1.275	0.029	0.200	1.074	3.188	4.772	0.041
20	73.660	0.311	14.061	1.982	0.064	0.449	1.845	3.494	4.058	0.075
21a	75.289	0.205	13.596	1.367	0.047	0.252	1.404	3.227	4.562	0.051
21b	54.592	2.254	16.419	8.451	0.124	3.793	6.472	3.735	3.139	1.022
22	74.666	0.229	13.761	1.521	0.035	0.291	1.581	3.436	4.426	0.052
23a	72.854	0.339	14.641	1.883	0.060	0.405	1.990	3.678	4.076	0.074
23b	54.288	2.642	15.047	10.255	0.138	3.395	6.657	3.340	3.074	1.165
25	53.252	2.627	16.041	9.477	0.136	3.872	6.517	3.467	3.277	1.334
27	62.240	1.457	15.680	6.440	0.090	2.161	3.794	3.730	3.811	0.598
28	58.218	1.666	17.452	6.946	0.118	2.147	4.132	4.384	4.292	0.644
31	57.679	2.101	16.058	8.639	0.145	2.293	4.991	3.627	3.833	0.634
32	75.691	0.132	13.862	1.006	0.040	0.142	1.185	3.580	4.329	0.032
33	50.370	3.858	14.899	11.119	0.151	3.927	7.361	3.169	3.047	2.097
34	52.653	1.760	18.105	8.048	0.136	4.977	7.674	3.558	2.246	0.843
35	50.149	2.588	16.118	10.704	0.159	4.775	8.096	3.683	2.488	1.241
36	60.215	0.861	19.231	5.547	0.158	1.611	5.224	4.297	2.639	0.217
37	60.773	0.862	19.409	5.091	0.146	1.506	5.418	4.122	2.460	0.214
38a	69.307	0.398	15.773	3.012	0.059	0.620	2.281	4.413	4.030	0.108
39a	48.555	1.510	15.886	9.793	0.146	10.739	8.113	2.825	1.780	0.653
39b	47.952	2.916	17.313	12.527	0.174	3.740	6.841	3.535	3.370	1.631
40	54.909	2.163	16.776	8.918	0.150	3.109	5.343	4.528	3.018	1.085
41	48.629	2.286	14.892	10.802	0.168	7.758	8.905	2.823	2.709	1.028
42a	50.880	1.771	14.323	10.398	0.189	8.388	7.863	2.911	1.609	1.668
42b	49.819	2.396	16.367	10.158	0.156	6.129	8.410	3.195	2.200	1.170
43	59.153	1.498	17.458	6.062	0.118	2.032	4.087	4.287	4.732	0.573
44	53.180	1.961	15.674	9.204	0.197	5.379	6.807	3.871	2.849	0.878
45	50.098	2.979	15.520	10.594	0.161	4.908	8.512	3.197	2.437	1.594
46	50.580	2.560	15.966	11.406	0.164	5.566	6.927	3.412	2.427	0.991
47	51.660	2.152	16.258	10.543	0.153	5.855	6.727	3.473	2.305	0.874
48	55.822	2.299	16.365	8.858	0.126	2.702	5.298	3.790	3.815	0.925
49	55.339	2.400	15.573	9.687	0.131	3.123	5.983	3.366	3.415	0.983
50	68.222	0.726	15.426	2.961	0.058	0.698	1.770	3.735	6.225	0.180

Concentrations given in weight percentage. For analytical conditions see Sect. 6.1

range from granite to gabbros with a clear compositional gap between 62 and 70 % SiO_2.

Viscosity contrasts between the most contrasting gabbro (39b) and granite (17) were calculated after GIORDANO *et al.* (2008) for the temperature range of 900–1,300 °C. Calculated values are compared with those from end-members (basalt and rhyolite) from mixing experiments (MORGAVI *et al.* 2013b) in Table 3.

Table 3

Values of viscosity (η) of two end-members (39b and 17) for different temperatures, compared with those from end-members from the literature; calculations after GIORDANO et al. (2008)

T (°C)	T (K)	Log η, Gb (Pa s)	Log η, Gr (Pa s)	Obs.
900	1,173	3.82	6.32	Viscosity of samples 39b
1,000	1,273	2.81	5.35	and 17 (Table 1)
1,100	1,373	2.02	4.54	calculated after
1,200	1,473	1.38	3.85	GIORDANO et al. (2008)
1,300	1,573	0.86	3.26	
1,350	1,623	0.86	4.75	MORGAVI et al. (2013a, b, c); see reference for compositions

Additional geochemical results from the literature have already been discussed in Sect. 4.2

6.2. Quantification of Mixing Efficiency and Element Mobility: Concentration Variance

Recently, the coeval mobility of all elements present in a magmatic system during mixing was calculated from mixing experiments within rhyolite/basalt and analogue melts (PERUGINI et al. 2013; MORGAVI et al. 2013c). According to these works, the main factors controlling element mobility are (1) compositional and rheological variations within system, (2) the appropriate thermodynamics for contrasting multicomponent melts, and (3) the complete strain history. In a natural scenario, these are unknown variables. For a given outcrop scenario, PERUGINI et al. (2013) proposed an alternative way to normalize the effects of absolute diffusive and convective histories and apply a quantity commonly used in the fluid dynamics literature (e.g., ROTHSTEIN et al. 1999; LIU and HALLER 2004) to evaluate the degree of homogenization of fluid mixtures: the concentration variance (σ^2).

The variance of concentration for a given chemical element (C_i) from a set of n likely values can be written as

$$\sigma^2(C_i) = \frac{\sum_{i=1}^{N}(C_i - \mu_i)^2}{N}, \qquad (3)$$

where N is the number of samples (i.e., 46, the number of analyzed outcrops), C_i is the concentration of element i, and γ is the mean composition for this

element i. This measure decreases with mixing time due to the progressive increase in homogeneity.

For evaluation of data from mixing experiments, the range of element concentration in the initial experimental samples is known (PERUGINI et al. 2012, 2013; MORGAVI et al. 2013c). In those cases, variance values were normalized to the initial variance of each element in the end-members for comparative purposes. This comparative value is the concentration variance, or simply variance, and is calculated from the following relationship:

$$\sigma_n^2 = \frac{\sigma^2(C_i)_t}{\sigma^2(C_i)_{t=0}}, \qquad (4)$$

where $\sigma^2(C_i)_t$ and $\sigma^2(C_i)_{t=0}$ are the concentration variance of a given chemical element (C_i) at the end of the experiment (at time t) and at the beginning before mixing starts (i.e., at time $t = 0$, the initial variance), respectively.

In this work we apply this concept of variance for the study of natural plutonic rocks, which have been analyzed for major element contents: SiO_2, TiO_2, Al_2O_3, MnO, FeO_t, MgO, CaO, Na_2O, K_2O, and P_2O_5. In our natural case we consider the following premises: (1) our population is a restricted population of the studied system; (2) nevertheless, each sample represents a composition close to the initial local hybrid liquid from which the sampled rock was generated through a later fractional crystallization; (3) initial end-members have compositions close to the most contrasting rocks of the system; (4) in the studied natural scenario, the initial most contrasting compositions (SiO_2-poorer gabbro and SiO_2-richer granite) are not necessarily the end-members before mixing; e.g., the chosen most contrasting compositions may be a result of a previous mixing and/or contamination at depth. In a natural scenario the exact composition of the end-members before mixing is therefore a missing parameter.

Here, we also calculate the concentration variance of a population of large, homogeneous samples and compare the results with those from experiments. Different end-member candidates for the least evolved end-member will be tested and results compared with the experiments with similar materials.

The concentration variance (σ_n^2) was calculated for the transect depicted in Fig. 2 for all analyzed

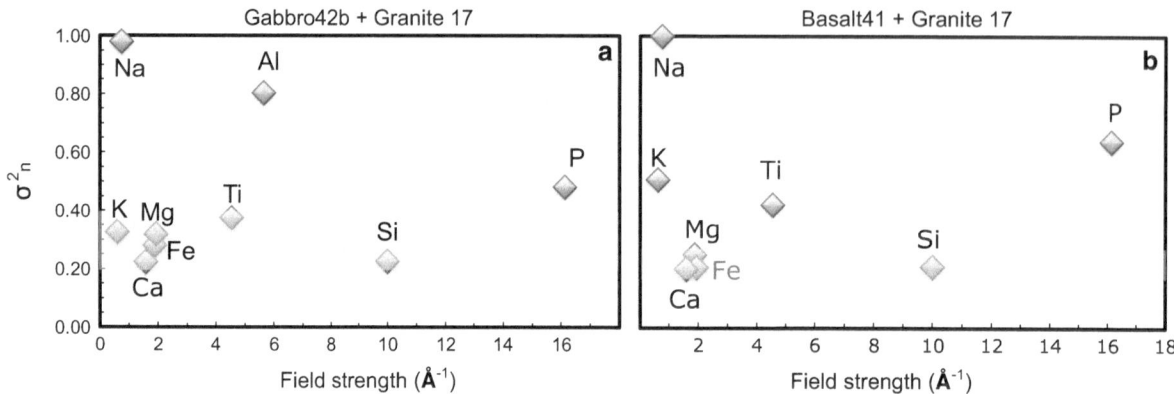

Figure 8

(**a**, **b**) Variation of the normalized concentration variance (σ_n^2) of major element oxides as a function of field strength (calculated as the nominal charge Z of the element divided by its ionic radius r, after SHANNON 1976) for all analyzed samples; Normalization is compared between samples 42b (**a**) and 41 (**b**) as a least evolved and sample 17 as a most evolved end-member. To fulfill the assumption, the normalized concentration variance should plot along a similar (σ_n^2) range. For the fine-grained basalt 42b (**a**), five major elements fulfill this condition, while for the medium-grained gabbro 41, only four elements plot along a similar (σ_n^2) range (**b**). Discussion in the text

chemical elements; results are shown in Figs. 7a, b and 8a, b. Three different gabbroic compositions were chosen and plotted against a most evolved granite sample.

For comparison with results from PERUGINI et al. (2012), the concentration variance (σ_n^2) for each chemical element is plotted against the field strength Z/r, calculated as the nominal charge Z divided by the ionic radius r (after SHANNON 1976). Results from the calculations of the concentration variance from experiments with analogue materials (anorthite-diopside + haplogranite from PERUGINI et al. 2012) reveal that different element mobilities (σ_n^2) are relatively similar and of the same order of magnitude. K_2O shows the highest mobility (i.e., the lowest concentration variance, σ_n^2), followed by CaO, MgO, Na_2O, Al_2O_3, and SiO_2.

Results from mixing experiments with natural samples (MORGAVI et al. 2013b, c) confirm different homogenization degrees of a given element at a given space–time. The concentration variance decreases progressively in time, as expected for the evolution of a mixing process. When compared with the results from mixing experiments with anorthite-diopside and haplogranite (PERUGINI et al. 2012), the σ_n^2 values further confirm that the variation in the concentration variance with time is different, although similar. As a consequence, despite a more complex chemical system, the rate of

σ_n^2 decay (relaxation of concentration variance) increases in a different sequence from that measured during experiments with analogue materials: SiO_2, TiO_2, Al_2O_3, MgO, CaO, FeO_{tot}, K_2O, and Na_2O, indicating that the rate of homogenization of chemical elements increases when moving from SiO_2 to Na_2O (MORGAVI et al. 2013b).

When end-members are well known, as in the case of experiments, the normalized concentration variance of all elements should only range from 1 ($\sigma_n^2 = 1.0$), at time $t = 0$ (i.e., before mixing starts) to 0, when the normalized concentration variance achieves the ideal hybrid composition at time t. In a mixed system, the correlation between the drop of concentration variance with mixing time is therefore: 1 at time $t = 0$ (i.e., before mixing starts) and 0 at time t. However, experiments have shown that, especially for mobile elements such as K, Na, and Al (BINDEMAN and DAVIS 1999; DE CAMPOS et al. 2011), uphill diffusion (e.g., WATSON and JUREWICZ 1984) is thought to be responsible for changes in the expected mathematical model.

In a natural scenario, the exact composition of the end-members before mixing is an additional missing parameter. Therefore, we expect that the normalized concentration variance for natural samples may plot outside the 1–0 range.

In our case, three different gabbroic compositions were chosen as tentative candidates for the least

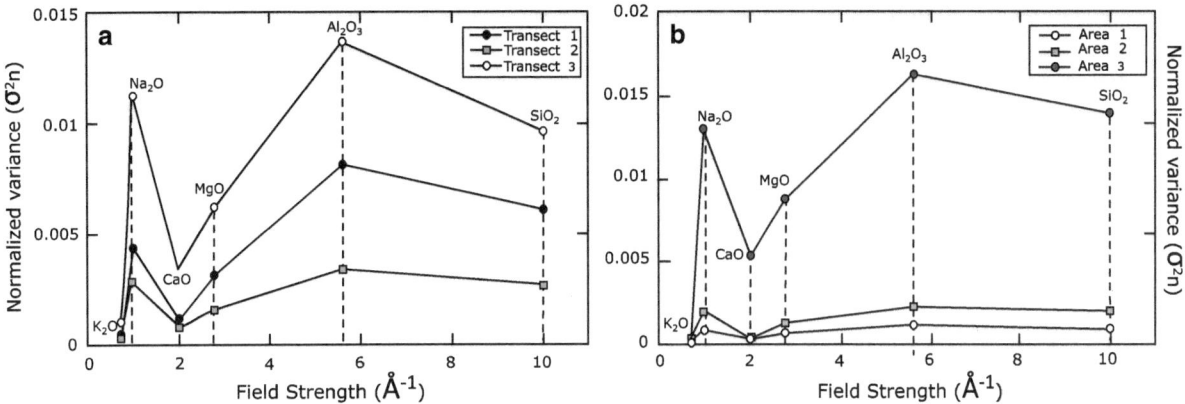

Figure 9

From PERUGINI *et al.* (2012): Variation of the normalized concentration variance (σ_n^2) of major element oxides as a function of field strength (calculated as the nominal charge Z of the element divided by the ionic radius r, after SHANNON 1976) for samples from a chaotic mixing experiment with silicate melts (haplobasalt and haplogranite). **a** Values from three different transects. **b** Values calculated for three different areas. Discussion in the text

evolved end-member and plotted against a most evolved granite sample 17 (compositions in Table 2). Tested end-member candidates were (1) a fine-grained gabbro 39a from the gabbroic core region, (2) a medium-grained gabbro 41, and (3) a fine-grained basalt 42b.

The σ_n^2 values for all elements from the fine-grained basalt 42b (Fig. 7a) plot below the 1–0 expected range. At least five major elements plot along a similar (σ_n^2) range.

Results for Na and K for the coarse-grained gabbro 39a plot outside the expected range (Fig. 7b). Results for most major elements for the medium-grained gabbro 41 show even worse results, with values falling in a broader (σ_n^2) range (Fig. 8a, b).

The concentration variance (σ_n^2) values for major chemical elements from homogeneous outcrops point towards a complex chemical hybridization process that could precede crystallization. However, for our tested end-members, the concentration variance of at least five major elements plot along the same range.

The mobility of Na and K is known from experiments to be higher in this system (BAKER 1990, 1991; Fig. 9—from PERUGINI *et al.* 2012). Values outside the claimed range are therefore expected, if the least evolved candidate is a product of previous mixing. Other explanations for higher σ_n^2 values are (1) onset of crystallization and local flow-enhanced differential concentration of high-temperature minerals, such as apatite, pyroxene, biotite, and Fe–Ti oxides, all present is these rocks; (2) autometasomatic reactions, mostly with Na and K, with local recrystallization while cooling; and (3) faster elements such as Na and K have been previously hybridized in the magma, as pointed out before.

7. Conclusions

In this work we aim to answer three main questions regarding a complexly zoned plutonic intrusion, the SAIC:

(a) Are the rocks from the SAIC a product of mixing of end-members?

(b) If yes, are the claimed end-members of the system present among sampled outcrops?

(c) Can both the measurement of fractal dimensions at different scales and the calculation of concentration variances of major elements from homogeneous outcrops help solve these questions?

To answer these questions, we made the assumption that no significant influence of the long-plutonic cooling history would have "blended together" mixing and postmixing events. Since original quenched textures are reported to be preserved in the original magmatic fabrics (e.g., WIEDEMANN *et al.* 1986; SCHMIDT-THOMÉ and WEBER-DIEFENBACH 1987; BAYER *et al.* 1987; DE CAMPOS *et al.* 2004b), these have not been modified by subsequent local recrystallization in a significant way. Local recrystallization and late metassomatic replacements are restricted to the rock matrix and did not influence the contact contour lines, nor change the fractal dimension at kilometer or meter scales in a significant way.

As previously discussed, the term "magma mingling" has been applied to indicate the physical dispersion of magmas, when no chemical interaction is involved. Magma mixing means that chemical exchanges are present in the system with the formation of hybrid compositions. In the case of the SAIC, there is clear evidence of chemical exchange in the system, with the production of hybrid homogenized rocks ranging in composition from biotite-hypersthene-augite-gabbro-norite to allanite-biotite-granite.

Fractal dimensions measured for the geological map and an outcrop point towards chaotic flows generated by mixing two contrasting magmas (basalt and rhyolite). The difference between the fractal dimension from the map (1.42 for a km scale) and the outcrop (1.32 for a cm to m scale) could be caused by different local mixing time and/or local differences in viscosity contrast. In a relatively large magma chamber such as the SAIC, complexities are to be expected.

Values of the normalized concentration variance for different end-member candidates plot along the same range for at least four major elements: Ca, Mg, Si, and Fe. For the coarser-grained gabbro 39a, results for Na and K outside the expected range may be explained by the higher mobility of Na and K, or could reflect that the claimed least evolved end-member is a product of a previous mixing. Concentration variance (σ_n^2) values for major chemical elements point towards a complex chemical hybridization process, which could precede main crystallization.

Results from the measurement of fractal dimension and normalized variance of major element contents confirm isotopic, petrographic, and geochemical conclusions from the literature: magma mixing, not only mingling, is the major differentiation process at the SAIC.

During chaotic mixing, exponential drops in values of variance point towards a very efficient process leading to a quick homogenization of initial contrasting compositions. The Santa Angélica Intrusive Complex (SAIC) is far from homogenization for most major elements. It is therefore a natural snapshot in this process at depth.

Regions of qualitatively different fluid motion comparable to theoretical and experimental results from chaotic fluid motion studies, from the literature, have been recognized in the SAIC. A main shear zone divides flow domains into two separate vortex regions, where separate inner stagnant core regions were mapped. The restriction of the internal shear zone to the magmatic structure suggests that it was generated by the flow and not superimposed by a later deformation.

Summing up all data and observations:

- From the evaluation of contact lines solely between granite I and gabbro/monzogabbro, we recognize a major mixing episode, which we preliminarily call ME 1 (mixing episode 1);
- From the additional numerical modeling we conclude that the original end-members could not be recognized, despite three decades of intensive field and geochemical work in the area and new resampling (this work): granite I and gabbro-diorite are not the original end-members any more. This implies that mixing episode 1 (ME 1) has been relatively short but efficient. The uncontaminated end-members, especially the basalt, probably originate from even deeper depths;
- A newer restricted mixing/mingling episode has been observed between the reintrusion of granite II and monzogabbro. This is named ME 2 but has not been evaluated in this work.

Therefore, we conclude that the main rock units from the SAIC (granite I, gabbro-diorite, and hybrid monzogabbro to diorite) have been generated by chaotic mixing between a basaltic and a granitic magma. However, at least two different mixing episodes have been recorded. When reintrusion stopped and temperature decreased, magmatic interactions froze in and plutonic rocks crystallized in situ with minor signs of local recrystallization and subsolidus mineralogical replacements. A thorough review on the petrology and mineralogy of the SAIC is in preparation.

All conclusions presented here point towards the issue that most of the enclaves, the regular core regions, the hybrid unit, the stretched and folded lenses, and layers cropping out in this pluton today result altogether from chaotic interactions between anorogenic magmas (granite I and gabbro/basalt) intruded in the lower crust during a single intrusive episode.

Acknowledgments

This work brings together information collected by the author and collaborators during decades of field and laboratory work at both the UFRJ, Rio de Janeiro, and the LMU, Munich. I am grateful to J. C. Mendes and S. R. de Medeiros for discussions and for the use of the preparation and analytical facilities of the UFRJ (X-ray fluorescence laboratory). Thanks are due to D. B. Dingwell and the GeoCenter of the LMU for support over the last decade. Discussions with A. Goretkin, D. Morgavi, and J. Kruhl are acknowledged. I am also very grateful to Ilya Bindeman and an unknown reviewer for their suggestions on the first version of this work.

REFERENCES

ABE, Y. (1997), *Thermal and chemical evolution of the terrestrial magma ocean.* Phys. Earth Planet. Inter. *100*, 27–39.

ABRAMOFF, M.D., MAGALHAES, P.J., and RAM, S.J. (2004), *Image processing with ImageJ.* Biophotonics Int. *11*, 36–42.

ALLÈGRE, C.J., and TURCOTTE, D.L. (1986), *Implications of a two-component marble-cake mantle.* Nature *323*, 123–127.

ALKMIM, F.F., MARSHAK, S., PEDROSA-SOARES, A.C., PERES, G.G., CRUZ, S.C.P., and WHITTINTON, A. (2006), *Kinematic evolution of the Araçuaí–West Congo orogen in Brazil and Africa: nutcracker tectonics during the Neoproterozoic assembly of Gondwana.* Precambrian Res. *149*, 43–64.

ALMEIDA, J., DIOS, F., MOHRIAK, U.W., VALERIANO, C.M., HEILBRON. M., EIRADO, and L.G., TOMAZZOLI, E. (2013), *Pre-rift tectonic scenario of the Eo-Cretaceous Gondwana break-up along SE Brazil–SW Africa: insights from tholeiitic mafic dyke swarms.* In: MOHRIAK, W.U., DANFORTH, A., POST, P.J., BROWN, D.E., TARI, G.C., NEMČOK, M., SINHA, S.T. (eds) Conjugate Divergent Margins. Geological Society, London, Special Publications, 369, The Geological Society of London. doi:10.1144/SP369.24.

ANDERSON, D.L. (1982), *Isotopic evolution of the mantle: the role of magma mixing.* Earth Planet. Sci. Lett. *57*, 1–12.

AREF, H., BLAKE, J.R., BUDIŠIĆ, M., CARTWRIGHT, J.H.E., CLERCX, H.J.H., FEUDEL, U., GOLESTANIAN, R., GOUILLART, E., LE GUER, Y., van HEIJST, G.F., KRASNOPOLSKAYA, T.S., MACKAY, R.S., MELESHKO, V.V., METCALFE, G., MEZIĆ, I., MOURA, A.P.S., EL OMARI, K., PIRO, O., SPEETJENS, M.F.M., STURMAN, R., THIFFAULT, J.L. and

TUVAL, I. (2014), *Frontiers of chaotic advection.* Journal-ref: EPL 105, 30003: 61 pp (in print).

BACON, C.R. (1986), *Magmatic inclusions in silicic and intermediate volcanic rocks.* J. Geophys. Res. *91*, 6091–6112.

BAKER, D. (1990), *Chemical interdiffusion of dacite and rhyolite— anhydrous measurements at 1 Atm and 10 Kbar, application of transition-state theory, and diffusion in zoned magma chambers.* Contrib. Mineral. Petrol. *104*, 407–423.

BAKER, D. (1991), *Interdiffusion of hydrous dacitic and rhyolitic melts and the efficacy of rhyolite contamination of dacitic enclaves.* Contrib. Mineral. Petrol. *106*, 462–473.

BATEMAN, R. (1995), *The interplay between crystallization, replenishment and hybridisation in large felsic magma chambers.* Earth Sci. Rev. *39*, 91–106.

BAYER, P., HORN, H.A., LAMMERER, B., SCHMIDT-THOMÉ, R., WEBER-DIEFENBACH, K. and WIEDEMANN, C. (1986), *The Brasiliano Mobile Belt in Southern Espírito Santo (Brazil) and its igneous intrusions.* Zentralblatt für Geologie und Paleontologie I, *9/10*, 1429–1439.

BAYER, P., HORN, H.A., SCHMIDT-THOMÉ, R. and WEBER-DIEFENBACH, K. (1987), *Complex concentric granitoid intrusions in the Coastal Mobile Belt, Espírito Santo, Brazil: the Santa Angélica pluton-an example.* Geologische Rundschau 76, 357-371.

BERGANTZ, G.W. (2000), *On the dynamics of magma mixing by reintrusion: implications for pluton assembly processes.* J. Struct. Geol. *22*, 1297–1309.

BINDEMAN, I.N. and DAVIS, A.M. (1999), *Convection and redistribution of alkalis and trace elements during the mingling of basaltic and rhyolitic melts.* Petrology 7, 91–101.

BLUNDY, J.D., and SPARKS, R.S.J. (1992), *Petrogenesis of mafic inclusions in granitoids of the Adamello massif, Italy.* J. Petrol. *33*, 1039–1104.

COSTA, F. and CHAKRABORTY, S. (2004), *Decadal time gaps between mafic intrusion and silicic eruption obtained by chemical zoning patterns in olivine.* Earth Planet. Sci. Lett. *227*, 517–530.

COX, K.G., BELL, J.D., and PANKHURST R.J. (1979), *The interpretation of igneous rocks.* Allen and Unwin, London, 450.

DE CAMPOS, C.P., DINGWELL, D.B., and FEHR, K.T. (2004a), *Decoupled convection cells from mixing experiments with alkaline melts from Phlegrean Fields.* Chem. Geol. *213*, 227–251.

DE CAMPOS, C.P., MENDES, J.C., LUDKA, I.P., DE MEDEIROS, S., COSTA-DE-MOURA, J., and WALLFASS, C. (2004b), *A review of the Brasiliano magmatism in southern Espírito Santo, Brazil, with emphasis on post-collisional magmatism.* In: WEINBERG, R., TROUW, R.A., FUCK, R., AND HACKSPACHER, P. (eds). The 750-550 Ma Brasiliano event of South America. Journal of the Virtual Explorer, Electronic Edition ISSN 1441-8142, 17, Paper 1.

DE CAMPOS, C.P., DINGWELL D.B., PERUGINI D., CIVETTA L., and FEHR, T.K. (2008), *Heterogeneities in magma chambers: insight from the behaviour of major and minor elements during mixing experiments with natural alkaline melts.* Chem. Geol. 256, 131–145.

DE CAMPOS, C.P., ERTEL-INGRISCH, W., PERUGINI, D., DINGWELL, D.B. and POLI, G. (2010), *Chaotic Mixing in the System Earth: Mixing Granitic and Basaltic Liquids.* In: SKIADAS, C.H. and DIMOTILAKIS, I., Eds. Chaotic Systems Theory and Applications. International Publication Company, 51–58.

DE CAMPOS, C.P., PERUGINI, D., ERTEL-INGRISCH, W., DINGWELL, D.B., and POLI, G.(2011), *Enhancement of magma mixing efficiency by chaotic dynamics: an experimental study.* Contrib. Mineral. Petrol. *161*, 863–881.

DE LA ROCHE, H., LETERRIER, J., GRAND CLAUDE, P. and MARCHAL, M. (1980), *A classification of volcanic and plutonic rocks using R1–R2 diagrams and major element analyses—its relationships with current nomenclature*. Chem. Geol. *29*, 183–210.

DE ROSA, R., DONATO, P., and VENTURA, G. (2002), *Fractal analysis of mingled/mixed magmas: an example from the Upper Pollara eruption (Salina Island, Southern Tyrrhenian Sea, Italy)*. Lithos *65*, 299–311.

DIDIER, J. and BARBARIN, B. (1991), *Enclaves and Granite Petrology*. Developments in Petrology, 13, Elsevier, Amsterdam.

EICHELBERGER, J.C. (1980), *Vesiculation of mafic magma during replenishment of silicic magma reservoirs*. Nature *288*, 446–450.

FLINDERS, J., and CLEMENS, J.D., (1996), *Non-linear dynamics, chaos, complexity and enclaves in granitoid magmas*. Trans. R. Soc. Edinb. Earth Sci. *87*, 225–232.

GIORDANO D., RUSSEL J.K., and DINGWELL, D.B. (2008), *Viscosity of magmatic liquids: a model*. Earth Planet. Sci. Lett. *271*, 123–134.

GRADIM, C., RONCATO, J., PEDROSA-SOARES, A.C., CORDANI, U., DUSSIN, I., ALKMIN, F.F., QUEIROGA, G., JACOBSON, T., DA SILVA, L.C., and BABINSKI, M. (2014), *The hot back-arc zone of the Araçuaí orogen, Eastern Brazil: from sedimentation to granite generation*. Braz. J. Geol. *44*, 1, 155–180. doi:10.5327/Z2317-4889201400010012.

HIBBARD, M.J. (1981), *The magma mixing origin of mantled feldspar*. Contrib. Mineral. Petrol. *76*, 158–170.

HIBBARD, M.J. (1995), *Petrography to petrogenesis*. Prentice Hall.

HORN, H.A., and WEBER-DIEFENBACH, K. (1987), *Geochemical and genetic studies of three inverse zoned intrusive bodies of both alkaline and subalkaline composition in the Araçuaí-Ribeira Mobile Belt (Espírito Santo, Brazil)*. Revista Brasileira de Geociências *17*, 4, 488–497.

KRATZMAN, D.J., CAREY, S., SCASSO, R., and NARANJO, J.A. (2009), *Compositional variations and mixing in the 1991 eruptions of Hudson volcano*. Chil. Bull. Volcanol. *71*, 419–439.

LIU, W., and HALLER, G. (2004), *Strange eigenmodes and decay of variance in the mixing of diffusive tracers*. Physica D *188*, 1–39.

LUDKA, I.P., WIEDEMANN, C.M., and TÖPFNER, C. (1998), *On origin of incompatible elements in the Venda Nova pluton, state of Espírito Santo, southeast Brazil*. J. S. Am. Earth Sci. *11*, 473–486.

MANDELBROT, B.B. (1982), *The Fractal Geometry of Nature*. W.H. Freeman, New York.

MEDEIROS, S.R. DE, WIEDEMANN, C.M., and VRIEND, S. (2001), *Evidence of Mingling between contrasting magmas in a deep plutonic environment. The example of Várzea Alegre in the Panafrican/Brasiliano Belt in Brazil*. Anais da Academia Brasileira de Ciências, *73*, 1, 99.119.

MEDEIROS, S.R., DE, MENDES, J.C., MCREATH, I., and WIEDEMANN, C.M. (2003), *U-Pb and Rb-Sr dating and isotopic signature of the charnockitic rocks from Varzea Alegre intrusive complex, Espírito Santo, Brazil*. In: IV South American Symposium on Isotope Geology, Short papers, II, 609–612.

MENDES, J.C., MCREATH, I., WIEDEMANN, C.M., and FIGUEIREDO, M.C.H. (1997), *Charnoquitóides do maciço de Várzea Alegre: um exemplo de magmatismo cálcio-alcalino de alto K no arco magmático do Espírito Santo*. Revista Brasileira de Geociências *27*, 1, 13–24.

MENDES, J.C., WIEDEMANN, C.M., and MCREATH, I. (1999), *Charnockitic Magmatic Rocks from the Várzea Alegre Massif, Espírito Santo, Southeast Brazil: Conditions of Formation*. Revista Brasileira de Geociências *29*, 1, 47–54.

MENDES, J.C., MEDEIROS, S.R. DE, MCREATH, I., and DE CAMPOS, C.P. (2005), *Cambro-Ordovician Magmatism in SE Brazil: U-Pb and Rb-Sr Ages, combined with Sr and Nd Isotopic Data of Charnockitic Rocks from the Várzea Alegre Complex*. Gondwana Res. *8*, 3, 1–9.

MENDES, J.C, and DE CAMPOS, C.P. (2012), *Norite and charnockites from the Venda Nova pluton, SE Brazil: Intensive parameters and some petrogenetic constraints*. Geosci. Front. *3*, 6, 789–800. doi:10.1016/j.gsf.2012.05.009.

MORGAVI, D, PERUGINI, D, DE CAMPOS, C.P., ERTEL-INGRISCH, W., LAVALLÉE, Y., MORGAN, L., and DINGWELL, D.B. (2013a), *Interactions between rhyolitic and basaltic melts unraveled by chaotic magma mixing experiments*. Chem. Geol. *346*, 199–212.

MORGAVI, D, PERUGINI, D, DE CAMPOS, C.P., ERTEL-INGRISCH, W., and DINGWELL, D.B. (2013b), *Morphochemistry of patterns produced by mixing of rhyolitic and basaltic melts*. J. Volcanol. Geotherm. Res. *253*, 87–96.

MORGAVI, D, PERUGINI, D, DE CAMPOS, C.P., ERTEL-INGRISCH, W., and DINGWELL, D.B. (2013c), *Time evolution of chemical exchanges during mixing between rhyolitic and basaltic natural melts*. Contrib. Mineral. Petrol. *166*, 615–638.

NOCKOLDS, S.R., KNOX, R.W.O.B., and CHINNER, G.A. (1978), *Petrology for Students*. Cambridge Univ. Press, Cambridge.

OTT, E., and ANTONSEN, T.M. (1988), *Chaotic fluid convection and the fractal nature of passive scalar gradients*. Phys. Rev. Lett. *61*, 2839–2842.

OTTINO, J.M. (1989), *The kinematics of mixing: stretching, chaos and transport*. Cambridge University Press, Cambridge.

PEDROSA-SOARES, A.C., and WIEDEMANN-LEONARDOS, C.M. (2000), *Evolution of the Araçuaí Belt and its connection to the Ribeira Belt, Eastern Brazil*. In: Tectonic Evolution of South America. Eds: CORDANI, U., MILANI, E.J., THOMAZ, A. and CAMPOS, D.A. 1st edition., I, 265–285.

PEDROSA-SOARES, A.C., NOCE, C.M., WIEDEMANN, C.M., and PINTO, C.P. (2001), *The Araçuaí-West Congo Orogen in Brazil: an overview of a confined orogen formed during Gondwanaland assembly*. Precambrian Res. *110*, 307–323.

PEDROSA-SOARES, A.C., ALKMIM, F.F., TACK, L., NOCE, C.M., BABINSKI, M., SILVA, L.C., and MARTINS-NETO, M.A. (2008), *Similarities and differences between the Brazilian and African counterparts of the Neoproterozoic Araçuaí–West-Congo orogen*. Geological Society of London, Special Publications *294*, 153–172.

PEDROSA-SOARES, A.C., DE CAMPOS, C.P., NOCE, C., SILVA, L.C., NOVO, T., RONCATO, J., MEDEIROS, S., CASTAÑEDA, C., QUEIROGA, G., DANTAS, E., DUSSIN, I., and ALKMIM, F.F. (2011), *Late Neoproterozoic-Cambrian granitic magmatism in the Araçuaí Orogen (Brazil), the Eastern Brazilian Pegmatite Province and related mineral resources*. In: SIAL, A.N., BETTENCOURT, J., DE CAMPOS, C.P., and FERREIRA, V.P. (eds.) Granite-Related Ore Deposits. Geological Society of London, Special Publications *350*, 25–51. doi:10.1144/SP350.3.

PERUGINI, D., POLI, G., and GATTA, G. (2002), *Analysis and simulation of magma mixing processes in 3D*. Lithos *65*, 313–330.

PERUGINI, D., POLI, G., and MAZZUOLI, R. (2003), *Chaotic advection, fractals and diffusion during mixing of magmas: evidence from lava flows*. J. Volcanol. Geotherm. Res. *124*, 255–279.

PERUGINI, D., VENTURA, G., and POLI, G. (2004), *Kinematic significance of morphological structures generated by mixing of magmas: a case study from Salina Island (Southern Italy)*. Earth Planet. Sci. Lett. *222*, 1051–1066.

PERUGINI, D., POLI, G., and VALENTINI, L. (2005), *Strange attractors in plagioclase oscillatory zoning: petrological implications.* Contrib. Mineral. Petrol. *149*, 482–497.

PERUGINI, D. and POLI, G. (2005), *Viscous fingering during replenishment of felsic magma chambers by continuous inputs of mafic magmas: field evidence and fluid- mechanics experiments.* Geology 33, 5–8.

PERUGINI, D., PETRELLI, M., and POLI, G. (2006), *Diffusive fractionation of trace elements by chaotic mixing of magmas.* Earth Planet. Sci. Lett. *243*, 669–80.

PERUGINI, D., DE CAMPOS, C.P., DINGWELL, D.B., PETRELLI, M., and POLI, G. (2008), *Trace element mobility during magma mixing: preliminary experimental results.* Chem. Geol. *256*, 146–157.

PERUGINI, D. and POLI, G. (2012), *The mixing of magmas in plutonic and volcanic environments: analogies and differences.* Lithos *153*, 261–277.

PERUGINI, D., DE CAMPOS, C.P., ERTEL-INGRISCH, W., and DINGWELL, D.B. (2012), *The space and time complexity of chaotic mixing of silicate melts: implications for igneous petrology.* Lithos *155*, 326–340.

PERUGINI, D., DE CAMPOS, C.P., DINGWELL, D.B., and DORFMAN, A. (2013), *Relaxation of concentration variance: a new tool to measure chemical element mobility during mixing of magmas.* Chem. Geol. *335*, 8–23.

POLI, G. and PERUGINI, D. (2002), *Strange attractors in magmas: evidence from lava flows.* Lithos *65*, 287–297.

RIZZI, F. and CORTELEZZI, L. (2011), *Stirring, stretching and transport generated by pairs of like-signed vortices* J. Fluid Mech. *674*, 244–280.

ROTHSTEIN, D., HENRY, E., and GOLLUB, J.P. (1999), *Persistent patterns in transient chaotic fluid mixing.* Nature *401*, 770–772.

SHANNON, R.D. (1976), *Revised effective ionic radii and systematic studies of interatomic distances in halides and chalcogenides.* Acta Crystallogr. A *32*, 751–767.

SCHMIDT-THOMÉ, R. and WEBER-DIEFENBACH, K. (1987), *Evidence for froze-in magma mixing in Brasiliano calc-alkaline intrusions The Santa Angélica pluton, southern Espírito Santo, Brazil.* Revista Brasileira de Geociências *17*, 498–506.

SLABY, E., GOTZE, J., WORNER, G., SIMON, K., WRZALIK, R., and SMIGIELSKI, M. (2010), *K-feldspar phenocrysts in microgranular magmatic enclaves: a cathodoluminescence and geochemical study of crystal growth as a marker of magma mingling dynamics.* Lithos *105*, 85–97.

SLABY, E., SMIGIELSKI, M., SMIGIELSKI, T., DOMONIK, A., SIMON, K., and KRONZ, A. (2011), *Chaotic three-dimensional distribution of Ba, Rb, and Sr in feldspar megacrysts grown in an open magmatic system.* Contrib. Mineral. Petrol. *162*, 909–927.

SÖLLNER, F., LAMMERER, B., and WEBER-DIEFENBACH, K. (1991), *Die Krustenentwicklung in der Küstenregion nördlich von Rio de Janeiro/Brasilien.* Münchener Geowissenschaftliche Hefte 11 München, Friedrich Pfeil Verlag. 4.

SÖLLNER, F., LAMMERER, B., and WIEDEMANN-LEONARDOS, C.M. (2000), *Dating the Ribeira Mobile Belt.* Zeitschrift für Angwandte Geologie SH *1*, 245–255.

SPARKS, S.R.J., SIGURDSSON, H., and WILSON, L. (1977), *Magma mixing: a mechanism for triggering acid explosive eruptions.* Nature *267*, 315–318.

WADA, K. (1995), *Fractal structure of heterogeneous ejecta from the Meakan volcano, eastern Hokkaido, Japan: implications for mixing mechanism in a volcanic conduit.* J. Volcanol. Geotherm. Res. *66*, 1/4, 69–79.

WATSON, E.B. and JUREWICZ, S.R. (1984), *Behavior of alkalis during diffusive interaction of granitic xenoliths with basaltic magma.* J. Geol. *92*, 121–131.

WALLACE, G. and BERGANTZ, G. (2002), *Wavelet-based correlation (WBC) of crystal populations and magma mixing.* Earth Planet. Sci. Lett. *202*, 133–145.

WIEBE, R.A. (1994), *Silicic magma chambers as traps for basaltic magma: the Cadillac Mountain intrusive complex, Mount Desert Island, Maine.* J. Geol. *102*, 423–437.

WIEDEMANN, C.M., BAYER, P., HORN, H., LAMMERER, B., LUDKA, I.P., SCHMIDT-THOMÉ, R., and WEBER-DIEFENBACH, K. (1986), *Maciços Intrusivos do Sul do Espírito Santo e seu Contexto Regional.* Revista Brasileira de Geociências *16*, 1, 24–37.

WIEDEMANN, C.M. (1993), *The evolution of the early Paleozoic, late to post-collisional magmatic arc of the Coastal Mobile Belt, in the State of Espírito Santo, eastern Brazil.* Anais da Academia Brasileira de Ciências *65*, 1, 163–181.

WIEDEMANN, C.M., MENDES, J.C., and LUDKA, I.P. (1995), *Contamination of Mantle Magmas by Crustal Contributions - Evidence from the Brasiliano Mobile Belt in the State of Espírito Santo, Brazil.* Anais da Academia Brasileira de Ciências *67*, 2, 279–292.

WIEDEMANN, C.M., MEDEIROS, S.R. DE, MENDES, J.C., LUDKA, I.P., and COSTA-DE-MOURA, J. (2002), *Architecture of Late Orogenic Plutons in the Araċuaí-Ribeira Folded Belt, Southeast Brazil.* Gondwana Res. *19*, 381–399.

(Received April 15, 2014, revised September 12, 2014, accepted September 22, 2014, Published online October 19, 2014)

Reprinted from the journal

Pure Appl. Geophys. 172 (2015), 1835–1849
© 2015 Springer Basel
DOI 10.1007/s00024-014-1029-y

Pure and Applied Geophysics

Cooling of a Magmatic System Under Thermal Chaotic Mixing

Kamal El Omari,[1] Yves Le Guer,[1] Diego Perugini,[2] and Maurizio Petrelli[2]

Abstract—The cooling of a basaltic melt undergoing chaotic advection is studied numerically for a magma with a temperature-dependent viscosity in a two-dimensional (2D) cavity with moving boundary. Different statistical mixing and energy indicators are used to characterize the efficiency of cooling by thermal chaotic mixing. We show that different cooling rates can be obtained during the thermal mixing of a single basaltic magmatic batch undergoing chaotic advection. This process can induce complex temperature patterns inside the magma chamber. The emergence of chaotic dynamics strongly modulates the temperature fields over time and greatly increases the cooling rates. This mechanism has implications for the thermal lifetime of the magmatic body and may favor the appearance of chemical heterogeneities in the igneous system as a result of different crystallization rates. Results from this study also highlight that even a single magma batch can develop, under chaotic thermal advection, complex thermal and therefore compositional patterns resulting from different cooling rates, which can account for some natural features that, to date, have received unsatisfactory explanations, including the production of magmatic enclaves showing completely different cooling histories compared with the host magma, compositional zoning in mineral phases, and the generation of large-scale compositional zoning observed in many plutons worldwide.

Key words: Magma cooling, chaotic advection, thermal lifetime of magma chambers, temperature-dependent viscosity, crystallinity, thermal eigenmodes, numerical simulation.

List of symbols

A Area (m^2)
C_p Heat capacity (J/kgK)
k Thermal conductivity (W/mK)
L Wall characteristic length (m)

Dimensionless numbers

Nu   Nusselt number

Φ Crystal fraction
Pr   Prandtl number
Re   Reynolds number
T^* Dimensionless temperature
X Rescaled dimensionless temperature

Greek symbols

ρ Fluid density (kg/m^3)
σ Standard deviation
τ Period of modulation (s)

Subscript

m   Mean

1. Introduction

Thermal equilibrium is of great interest in both petrology and volcanology because the possible evolution of magmatic systems strongly depends on the development of thermal and compositional heterogeneities which, in turn, can influence the capability of the magmatic mass to differentiate and/or erupt. It has recently been shown that magmatic systems can exhibit wide compositional heterogeneity in both space and time (e.g., Perugini and Poli 2012; Perugini *et al.* 2012). This heterogeneity is the result of complex processes developing in magmatic masses and is mostly considered to be a result of chaotic dynamics (i.e., magma mixing; De Campos 2011; Morgavi *et al.* 2013; Perugini *et al.* 2012). The presence of chaotic behaviors in igneous systems is widely reported in literature (e.g., Bergantz 2000; Petrelli *et al.* 2011; Perugini and Poli 2012; Perugini *et al.* 2012), but surprisingly, there are few contributions addressing the thermal behavior of a magmatic body experiencing chaotic dynamics. Since the late 1990s, different works on thermal advection of

Electronic supplementary material The online version of this article (doi:10.1007/s00024-014-1029-y) contains supplementary material, which is available to authorized users.

[1] Laboratoire SIAME, Fédération CNRS IPRA, Université de Pau et des Pays de l'Adour (UPPA), 64000 Pau, France. E-mail: kamal.elomari@univ-pau.fr; yves.leguer@univ-pau.fr
[2] Department of Physics and Geology, University of Perugia, Perugia, Italy. E-mail: diego.perugini@unipg.it; maurizio.petrelli@unipg.it

high-viscosity fluids have shown that temperature fields can be strongly modulated by the onset of chaotic dynamics (SAATDJIAN and LEPREVOST 1998; LEFEVRE *et al.* 2003; MOTA *et al.* 2007; EL OMARI and LE GUER 2010a; LE GUER and EL OMARI 2012), but they are mostly confined to industrial processes. Therefore, study of the thermal evolution of a magmatic body in a chaotic environment could be of great interest and would start to fill the gap in previous and recent literature. The development of different thermal domains in space and time may, in fact, strongly influence the cooling history of the magmatic mass and therefore crystallization, leading to the formation of volumes of melts with strong rheological and compositional differences. As an example, crystal-size distributions have long been known to be influenced by the cooling kinetics of the magma (WINKLER 1949). Different cooling histories may also induce differential rates of growth for crystals. In this work, we focus on the cooling kinetics of a batch of magmatic melt undergoing very low advection. In particular, we present numerical simulations of chaotic thermal advection aimed at understanding the space and time modulation of the temperature field during cooling of a mafic melt. The temperature dependence of the magma viscosity is taken into account. Results are discussed in the light of the timescales of cooling for the magmatic body and the impact of this process on the evolution of the magmatic mass. The production of different thermal domains in which magma crystallization may proceed with differential efficiency is also discussed.

2. *The Physical Problem*

2.1. *Conceptual Model and Properties of the Mafic Magma*

The study of several natural magmatic systems has repeatedly led to the inference that the magma dynamics are governed by chaotic dynamics (e.g., FLINDERS and CLEMENS 1996; DE ROSA *et al.* 2002; PERUGINI *et al.* 2003, 2006). The fact that magma dynamics is chaotic means that its investigation can be reduced, as a first approximation, to the study of

stretching and folding of the silicate melt. This approach has enabled the investigation of the interplay between flow fields and the modulation of geochemical composition in the magmatic system (e.g., PERUGINI *et al.* 2003, 2006). Despite the simplicity of such an approach, it is important to note that it is capable of generating structures and compositional patterns that mimic those observed in natural rocks (e.g., PERUGINI *et al.* 2003; PERUGINI and POLI 2012). This observation confirms that a system exhibiting chaotic advection contains much that is necessary for replicating the fluid-dynamic evolution of a magma body. Thus, irrespective of the specific processes responsible for advection (e.g., convection, flow in conduits, etc.), chaotic dynamics of magmas is a very powerful conceptual tool for addressing the complexity of this petrogenetic and volcanological process. Guided by this conceptual model, we consider here the thermal chaotic mixing in a 2D rectangular cavity filled with a Newtonian mafic magma (Fig. 1). The numerical system contains all essential ingredients and fundamental building blocks to replicate the basic fluid dynamics of a magma body, consisting of stretching and folding processes. We consider that the cavity is not open to mass fluxes, implying that, once in place, the mafic batch behaves as a closed system. The thermophysical properties of the mafic magma are given in Table 1. The initial temperature T_i of the molten magma is chosen as 1,200 °C (liquidus temperature), and the

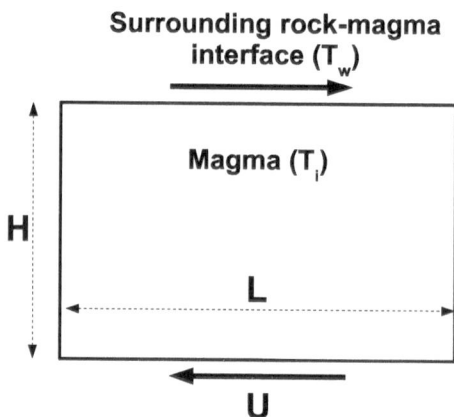

Figure 1
Sketch of the two-dimensional magma chamber (aspect ratio 0.6)

Table 1

Thermophysical properties of the melted mafic magma at the initial temperature T_i

Mafic magma property	Value	Unit
Density, ρ	2,750	kg m^{-3}
Thermal conductivity, k	2,2	W m^{-1} K^{-1}
Thermal diffusivity, α	8×10^{-7}	m^2 s^{-1}
Specific heat, C_p	1,000	J kg^{-1} K^{-1}
Dynamic viscosity, μ ($T_i = 1,200\ °C$)	100	Pa s

temperature of the surrounding rock–magma interface T_w is 600 °C. All the thermophysical properties are considered independent of temperature except the viscosity, as explained in the following section. The corresponding Prandtl number at the initial temperature of the magma is Pr = 45,450. For this high Prandtl number, the magma flow is characterized by a flow momentum much higher than the heat diffusion. Classically, for a steady laminar flow, the wall effects will be felt further inside the magma chamber for the velocity field than for the thermal field, thus the magma cooling will be governed by what happens in the vicinity of the walls (EL OMARI and LE GUER 2010). We will see that this mechanism will differ when chaotic advection enters into action. We limit the study to magmas at the early stages of crystallization, and we assume that no release of latent heat of crystallization occurs during the formation of crystals (see Sect. 2.2). The crystallization is envisaged as a phenomenon directly linked to the increases in the viscosity of the melt during cooling. Although these assumptions might appear quite severe, as reported in the introductory section, the aim of this study is to investigate the sole effect of chaotic advection on the thermal field to assess the kinetics of cooling of a mafic magma. As another assumption, we consider that the density remains constant during the cooling of the magma. Thus, natural convection due to buoyancy is not possible inside the magma chamber (i.e., convective thermal plumes BRANDEIS and JAUPART 1986). Assuming a constant density within the magma chamber is in keeping with the idea that we study natural magmas at the early stages of the differentiation process, as stated above.

2.2. Magma Viscosity and Viscosity–Crystallinity Relation

Magma viscosity is primarily linked to its silica content. The higher the amount of silica in a magma, the more viscous the magma will be due to the strong silicon–oxygen bonds which produce silica chains. Since we are investigating a mafic magma at the early stages of crystallization (i.e., low chemical variations), we neglect the link of viscosity to chemical composition. The gas content also affects the magma viscosity. We assume that the system contains a fixed amount of gas phases (closed system) and that their content is below the saturation value (i.e., no gaseous exsolution occurs). Other factors that strongly influence magma viscosity are temperature and crystal content. The viscosity increases with decreasing temperature and with increasing crystal content. In this study we consider a pseudo-Newtonian fluid (with no yield stress and a viscosity not dependent on strain rate) for which the temperature dependence of the viscosity is modeled by an exponential law that can simulate the rapid increase in viscosity when tiny crystals form during cooling (MC BIRNEY and MURASE 1984; SPERA 2000; COSTA and MACEDONIO 2003; GIORDANO *et al.* 2008):

$$\mu = \mu_0 \exp(B(1 - T^*)) \tag{1}$$

with μ_0 the viscosity at the initial temperature ($T^* = 1$), here considered equal to 100 Pa s. B is the Pearson number that takes into account the increase of viscosity with cooling (LE GUER and EL OMARI 2012). In this study, three cases are considered: the non-temperature-dependent, reference case ($B = 0$), and cases with moderate ($B = 5$) and greater temperature dependence ($B = 10$). The parameter B determines how fast the viscosity increases as the temperature of the magma is lowered. The viscosity range is from 100 Pa s for the mafic melt at the initial temperature, near its liquidus temperature, to about 10^6–10^7 Pa s for the same magma now containing a certain amount of crystals. The assumption that the magma is a Newtonian fluid is valid when we consider temperatures near the liquidus. This fact was confirmed by the measurements of SATO (2005) for the 1707 basalt from Mount Fuji volcano, for which the viscosity is almost constant against shear rate at 1,210 °C, corresponding to a melt with very low crystal content. As the temperature is lowered (subliquidus temperatures), a non-Newtonian shear thinning might appear depending on several parameters such as the quantity and morphology of crystals (CIMARELLI et al. 2011) and the applied strain rate (CARICCHI et al. 2007). MADER et al. (2013) extensively reviewed the rheology of two-phase magmas, reporting that the rheology of a crystal-bearing igneous system mostly depends on the crystal fraction Φ, the critical crystal fraction at which particles cannot move past one another (i.e., the maximum packing fraction Φ_{cr}), and the flow index n. The latter is a function of the Φ/Φ_{cr} ratio and of the crystal aspect ratio (MADER et al. 2013). As reported by MADER et al. (2013), a magmatic system is always Newtonian for $\Phi/\Phi_{cr} < 0.5$. For higher values of Φ/Φ_{cr} (i.e., $0.5 < \Phi/\Phi_{cr} < 0.8$), the Newtonian behavior still persists for values of the flow index n equal to or greater than 0.9. The values of Φ_{cr} are not easy to unravel in natural magmatic systems (MADER et al. 2013). Using a geometric approach, Φ_{cr} is about 0.74, if crystals are subspherical. The value of Φ_{cr} is lower for disordered systems ($\Phi_{cr} \approx 0.64$; MADER et al. 2013). Assuming $\Phi_{cr} = 0.656$ (smooth particles; MADER et al. 2013), the behavior of the system can be safety considered to be Newtonian for crystal content Φ up to 0.32. Studying a natural system, MARSH (1981) states that the limit of the phenocrystal content observed in basaltic lavas is $\Phi_{cr} = 0.55$; above this critical point the viscosity of the magma is so high that it cannot erupt as lava. The maximum packing density corresponds to a minimum of liquid content in the melt or equivalently to a minimum of porosity, which is highly dependent on the morphology of the crystals and their polydispersity. The viscosity of the magma also depends intimately on the shape of the crystals and their polydispersity. Moreover, the nucleation and growth rates of these crystals are closely linked to the local cooling rate encountered inside the magma. Magma cooling rates are related to the thermal fields, which are intimately linked to the flow kinematics. To summarize the behavior described by MADER et al. (2013), a mafic magma with low crystal content ($\Phi < 30\%$) can be reasonably considered as a Newtonian fluid if the crystal aspect ratio is not larger than 3.2. For higher content of crystals and/or larger

values of the crystal aspect ratio, a rheological transition characterized by a rapid increase of the apparent viscosity occurs. A complete review of the transition from Newtonian to non-Newtonian behavior lies beyond the scope of this work, and to be conservative, we limit the discussion of results to magmatic systems with crystal content below 30% (i.e., $\Phi = 0.3$). Classically, the model given in the literature to link the apparent viscosity to the degree of crystallinity is the Krieger–Dougherty model (KRIEGER and DOUGHERTY 1959) derived from the Einstein–Roscoe model (ROSCOE 1952):

$$\frac{\mu}{\mu_0} = \left(1 - \frac{\Phi}{\Phi_{cr}}\right)^{-\beta}, \qquad (2)$$

where Φ_{cr} is the critical volume fraction and the exponent β is a fitting parameter correlated with Φ_{cr} to account for particle shape (CIMARELLI et al. 2011). This model is appropriate to fit the data mainly for low crystal concentrations.

2.3. Chaotic Advection Flow

The chaotic advection phenomenon is now well known to enhance fluid mixing, reactive mixing or heat transfer in many industrial processes (AREF 1984; OTTINO 1989). It is also recognized to be of particular importance for natural phenomena in various earth domains such as volcanology (METCALFE et al. 1995; PERUGINI and POLI 2012; RENJITH et al. 2013), atmospheric sciences or oceanic dispersion (pollutants, black tides or plankton blooms; LOPEZ et al. 2001). Chaotic advection becomes particularly interesting for applications where the viscous effects are large compared with inertial effects (i.e., very low Reynolds numbers). This can be encountered for problems which involve small physical dimensions (typically for microfluidic applications), for very low velocities or for very large viscosities. The flow of a magma couples two of these elements (i.e., large viscosity and low velocity of the chamber wall, if present). For two-dimensional chaotic advective flows, a simple unsteady velocity field is able to generate very complex tracer patterns (concentration or temperature fields) (AREF 2002). The global mixing mechanism comprises a stirring phase related to the stretching and folding of fluid elements (as blobs and filaments) and a mixing phase. By chaotic advection, we mean that nearby fluid elements separate from each other exponentially in time in particular domains of the flow. That is why mixing is greatly enhanced due to the efficient generation of interfaces between scalars, which then facilitates the diffusion of the scalar through these interfaces (the mixing phase itself). It is not necessary for the velocity field itself to be chaotic as in turbulence to obtain chaotic trajectories of particles. A necessary constraint to produce chaotic advection in two-dimensional flow is to break the time invariability of the streamlines, as obtained by considering an unsteady flow with moving magma chamber walls in the present study. The 2D rectangular cavity we have chosen for the study of the magma flow is a classical geometry for chaotic advection (CHIEN et al. 1986; JANA et al. 1994; LIU et al. 1994). It has a height/length ratio ($H/L=$) of 0.6. As a stirring protocol, a constant velocity is imposed alternately on the side walls during each half period (Fig. 2). Thus, a nondimensional period τ controls the efficiency of the thermal mixing. In this study, a highly laminar magmatic flow is chosen with a reference Reynolds number equal to 1 for the simulations. This reference Reynolds number is based on the viscosity of the initial hot magma μ_0 (at $T_i = 1{,}200\ °C$). As the viscosity at the initial temperature of the magma is fixed, the product $U \cdot L$ is constant and the couple (U, L) has to be chosen. Thus, U and L are dependent variables for this problem. For the temperature-dependent viscosity cases, when the magma viscosity increases, the Reynolds number becomes smaller. Thus, a simple shear flow is applied to the magma chamber and the stirring protocols correspond to alternating movement of the two side walls at a given constant velocity, originating chaotic dynamics. As explained above, this protocol has been chosen to trigger chaotic advection in the magmatic system according to natural evidence and previous studies where magma mixing processes have been widely documented (e.g., PERUGINI and POLI 2012; PERUGINI et al. 2006, 2007).

The chosen period is one allowing the chamber side walls to be swept alternately five times each ($\tau = 10$). In order to show the time scales involved in the chaotic mixing mechanism for the magmatic

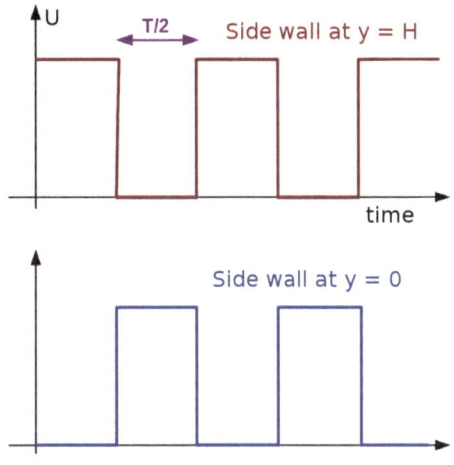

Figure 2
Temporal modulation of the magma chamber walls (stirring protocol)

system, in Table 2, for Re $= 1$, the time unit L/U and the real time of a period of stirring are given for three sizes L of the magma chamber (the associated velocities U are fixed as explained above). As an indication, the operating shear rate in the magmatic body is also given.

In a recent work, EL OMARI and LE GUER (2010a) showed that the global thermal chaotic mixing is very sensitive to the wall kinematics and that the stagnation parabolic points at the wall play a key role. Our objective in this study is also to verify this point for the thermal mixing occurring during the cooling of the magmatic mass.

3. Mathematical Formulation

3.1. Conservation Equations and Flow Parameters

The governing equations (Navier–Stokes, energy, and continuity) are solved for the 2D cavity described above. Some hypotheses are considered: gravity is considered normal to the cavity plane, thus no natural convective flow can develop inside the cavity as the density is not dependent on the temperature as indicated in the Sect. 1; additionally no viscous heating effect is considered. This last assumption is justified even with the consideration of the exponential increase of the viscosity with temperature because the velocities considered are very low; the Brinkman number (Br) which characterizes the

Table 2

Time unit and real time associated to a period of stirring for Reynolds number equal to 1 and different sizes of magma chamber

Re = 1	Time unit L/U	Real time for a period $10 L/U$	Shear rate
$L = 100$ m $U = 3.63 \times 10^{-4}$ m s^{-1}	3.19 days	$\approx$32 days	$\approx 10^{-5}$ s^{-1}
$L = 1{,}000$ m $U = 3.63 \times 10^{-5}$ m s^{-1}	0.87 years	$\approx$8.72 years	$\approx 10^{-7}$ s^{-1}
$L = 10{,}000$ m $U = 3.63 \times 10^{-6}$ m s^{-1}	87.2 years	$\approx$872 years	$\approx 10^{-9}$ s^{-1}

The corresponding initial shear rate is also given

relative importance of the viscous dissipation in the energy equation is very low:

$$\text{Br} = \frac{\mu_0 \cdot U^2}{k \cdot (T_i - T_w)} \ll 1. \tag{3}$$

Thus, the viscous heating term is neglected in the energy equation (LE GUER and EL OMARI 2012). Additional nonlinearities are introduced into the problem via the viscosity law, which is taken to depend on temperature, and also through the inclusion of inertia. Considering the parameter values given in Table 2, the Strouhal number is around 0.1, which is not too small. As a consequence, the flow does not satisfy the quasisteady hypothesis and the acceleration and deceleration phases imposed on the walls will influence the mixing inside the whole cavity.

The unsteady convective heat transfer cooling problem is governed by the nondimensional conservation equations for mass, momentum, and energy:

$$\nabla \cdot \mathbf{v} = 0, \tag{4}$$

$$\partial_t \mathbf{v} + (\mathbf{v} \cdot \nabla)\mathbf{v} = -\nabla \mathbf{p} + \frac{1}{\text{Re}} \nabla \cdot \left(\frac{\mu}{\mu_0} \left(\nabla \mathbf{v} + (\nabla \mathbf{v})^{\text{T}} \right) \right), \tag{5}$$

$$\partial_t T^* + \mathbf{v} \cdot \nabla T^* = \frac{1}{\text{RePr}} \nabla^2 T^*, \tag{6}$$

where

$$\text{Re} = \frac{\rho U L}{\mu_0}, \ \text{Pr} = \frac{\mu_0 C_p}{k}, \ \text{and } T^* = \frac{T - T_w}{T_i - T_w}$$

The characteristic scales considered for this nondimensional problem are the cavity length L, the

velocity of the wall U, the time U/L, and the pressure ρU^2. With the above definition of the dimensionless temperature, the maximum temperature difference between the walls and the fluid is always 1. $T^* = 1$ corresponds to the beginning of the thermal mixing (all the magma is at 1,200 °C), and $T^* = 0$ to the magma at the rock temperature of 600 °C. Thus, the 2D unsteady convective heat transfer cooling problem is governed by only two nondimensional numbers (Re and Pr). Another point has to be mentioned, concerning the latent heat of crystallization that is released into the mafic magma during its cooling. For the mafic magma with 30 % of mass fraction crystallized (corresponding to the high viscosity encountered for the low temperatures), we have estimated the Stefan number (characterizing the ratio of sensible heat to latent heat) to be above 20. This high ratio indicates that, as a first approximation, the latent heat of crystallization can be neglected, despite the role it may play in localized areas near solidification fronts.

3.2. Numerical Method

The continuity and Navier–Stokes equations were solved, as well as the energy conservation equation, by means of an in-house code (Tamaris) based on the unstructured finite-volume method. Spatial schemes approximating convective and diffusive fluxes are second-order accurate. The convective fluxes are approximated by the high-resolution nonlinear CU-BISTA scheme in order to reduce the numerical diffusion. This is crucial to avoid overshoot of the thermal diffusion in the energy equation. Time advancement is ensured by the implicit, second-order-accurate, three-time-step Gear scheme, while the pressure and velocity fields are coupled by the SIMPLE algorithm. A parallel algebraic multigrid (AMG) solver is used to resolve the obtained linear systems, since a sufficiently fine mesh of 54,000 computational cells was used to capture possible striations arising in the temperature field. This mesh was chosen after a thorough study of the dependence of the results on mesh size. The MCIA supercomputer at Bordeaux was used for the parallel calculations. More details about the numerical methods used and code validations are given elsewhere

(EL OMARI and LE GUER 2009, 2010a, b, 2012; BAMMOU *et al.* 2013).

3.3. Thermal Mixing Indicators

We mainly use three statistical mixing indicators to characterize the efficiency of heat transfer by chaotic thermal mixing: the mean temperature, the variance, and the Nusselt number. Their evolutions are followed over time during the magma cooling. The mean temperature of the magma T_m^*, which represents the energy extracted from the fluid across the walls (EL OMARI and LE GUER 2010), is given by

$$T_m^* = \frac{1}{\sum_c A_c} \left(\sum_c A_c T_c^* \right), \tag{7}$$

where the summation is made over all 2D computational cells of area A_c (the subscript c is for cell). Indeed, the mean temperature evolution can be seen as an indicator of the ratio of the total energy supplied to the fluid from the initial time considered to the time t:

$$E(t) = \rho C_p V_{chamber} (T_m(t) - T_W). \tag{8}$$

Hence, the mean temperature is asymptotically bounded by the fixed temperature imposed on the walls (i.e., T_w or $T_m^* = 0$ for nondimensional temperature). The second indicator is the variance σ^2 of the fluid temperature, which represents the level of homogenization of the scalar temperature inside the 2D magma chamber (STREMLER 2008):

$$\sigma^2 = \frac{1}{\sum_c A_c} \sum_c \left(A_c (T_c^* - T_m^*)^2 \right). \tag{9}$$

For a thermal mixing problem without any source of scalar, the advection–diffusion equation indicates that the temperature fluctuations evolve towards a uniform state ($\sigma^2 = 0$) at a decay rate given by the product of twice the thermal diffusivity α and the temperature gradients induced by the flow inside the magma chamber (THIFFEAULT 2012). Thus, for our case, as the thermal diffusivity is given and the flow field chosen, the speed at which the magma will be cooled will be studied by following the decay rate of σ^2 over time for different temperature-dependent viscosity cases. The last indicator used is the mean

Nusselt number $\overline{Nu}$, which characterizes the strength of heat transfer across a wall. It represents the dimensionless temperature gradient in a direction normal to the wall:

$$\overline{Nu} = \frac{1}{S_W} \int_{S_W} \nabla T^* \cdot \mathbf{n} \, d\mathbf{S}. \tag{10}$$

Since we consider here that the wall temperature (the temperature of the surrounding rock–magma interface) is kept constant during the cooling, the parietal heat flux exhibits a nonuniform distribution along the boundaries of the chamber (i.e., a nonuniform temperature gradient distribution). This distribution is closely related to the complex fluid flow kinematics inside the magma chamber.

4. Results and Discussion

4.1. Flow Streamlines, Temperature, and Crystallinity Field Patterns

In Fig. 3, one can observe the streamlines in the cavity flow system corresponding to the separate motion of each wall for the non-temperature-dependent viscosity case. Despite their simplicity, the combination of these two very simple Eulerian velocity fields produces very complex irregular Lagrangian trajectories that consequently give complex temperature patterns, as shown in Fig. 4, which presents several snapshots of the temperature field taken at different periodic times for two values of the Pearson number **B**. We observe highly elongated and folded striations produced by chaotic advection near the rock walls of the magma chamber, with cold magma tongues penetrating inside the chamber; these structures lead to the existence of flow regions with very different temperature levels. For the case $B = 0$, one can notice the existence of a large unmixed thermal zone in the central part of the domain with two filaments emanating from it which move together in the magmatic chamber during chaotic advection. This island of hot magma is transported across the magma chamber without shape distortion, undergoing weaker cooling compared with parietal areas (BRES-LER 1997). After a transient stage, the spatial

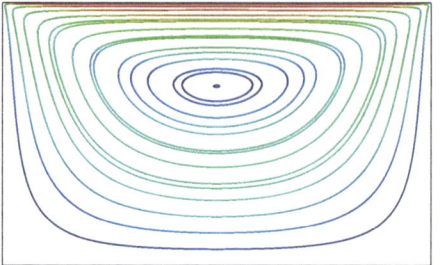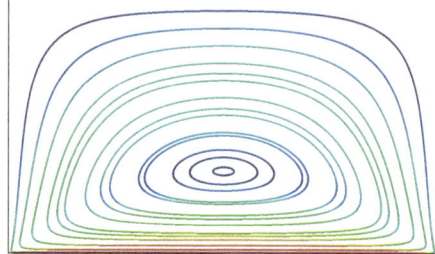

Figure 3
Flow streamlines induced by the wall movements during the first (*left*) and second (*right*) half-period of the stirring protocol

distribution of the temperature patterns takes the same form at each periodic time but the amplitude of the dimensionless temperature differences tends towards 0; these self-similar structures are called thermal strange eigenmodes (EL OMARI and LE GUER 2010; LIU and HALLER 2004) and are the signature of an underlying fractal structure in the flow. The spatial structure of the temperature field is smooth; this is due to the relatively high value of the thermal diffusivity and is also the reason why only relatively large temperature striations are observed. By comparison, the spatial structures of the concentration patterns would have presented more lamellar structures due to the lower value of the molecular diffusivity. The thermal diffusion blurs the fine-scale structure of the representative thermal strange eigenmode. The direct consequence of the existence of this thermal strange eigenmode is that the crystallization front in such a chaotic flow will not last long parallel to the walls, as classically considered for the cooling of a magmatic chamber (HUPPERT 2000). The result could be the existence of a magmatic enclave, which would appear after complete solidification of the magmatic mass (e.g., DIDIER and BARBARIN 1999; PERUGINI and POLI 2000). This process could also account for the widespread occurrence of compositional zoning in mineral phases (e.g., GINIBRE et al. 2002; PERUGINI et al. 2005). In fact, as minerals are transferred among the different dynamic regions, they can undergo multiple episodes of resorption and growth according to the thermal and compositional features of the different regions existing within the magmatic system. At larger length scales, this could also generate compositional zoning in the magmatic mass, as observed in several plutons worldwide (e.g.,

MAHOOD and FRIDRICH 1982; HECHT and VIGNERESSE 1999). The effect of temperature-dependent viscosity is noticeable, since these hot zones disappear for $B = 5$. This is due to the strong increase of the magma viscosity in the vicinity of the cold walls, which promotes the carriage of magmatic liquid by the moving walls and thus improves the stirring. The period that corresponds to the establishment of a persistent temperature pattern seems to be much longer than before for the case $B = 0$. The self-similar patterns also seem to appear much later. This fact is confirmed below by observations of the probability distribution functions (PDFs) of temperature. The case with greater temperature dependence ($B = 10$) is illustrated in Fig. 5. The strong effect of the temperature-dependent viscosity on the velocity field is clearly shown in the evolution of the streamlines. At $t = 5\tau$, for example, the streamlines are deformed and a small new vortex appears in a corner of the magma chamber. The temperature patterns are correlatively modified: they do not align with the streamlines and are, in some regions of the chamber, perpendicular to them. A complex spatial distribution of the temperature gradients over the whole domain is thus obtained.

In Fig. 6, the crystallinity fields are plotted for the case $B = 5$ using the Krieger–Dougherty model. We recall that this model is appropriate for low crystal fractions up to the critical value of $\Phi_{cr} = 0.4$ (MADER et al. 2013). Starting with zero crystallinity at the first instant (all the magma is at the liquidus temperature), the crystallinity rapidly reaches values around 0.3 in some convoluted zones of the cavity. These zones correspond exactly to the cold zones in the temperature patterns of Fig. 4 (for $B = 5$). High degrees of

Non temperature-dependent viscosity case (B=0)

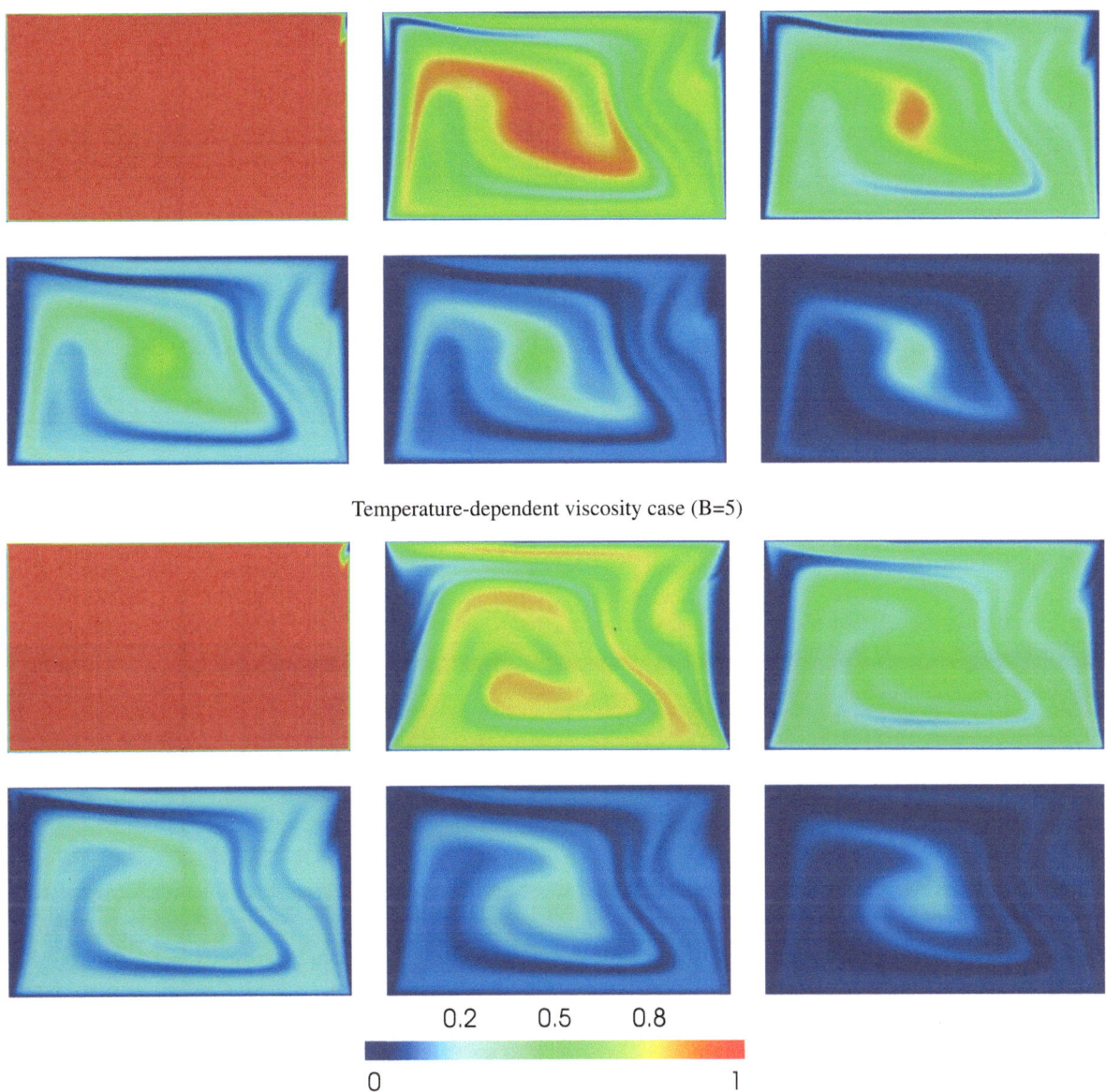

Temperature-dependent viscosity case (B=5)

0.2   0.5   0.8

0   1

Figure 4
Temperature fields at different periodic instants $t = 0$, 5τ, 10τ, 15τ, 20τ, and 25τ (from *left* to *right* and *top* to *bottom*), for $B = 0$ and 5

crystallinity are thus observable deep inside the magma chamber (for $t = 5\tau$ and 10τ), allowing the formation of solid enclaves that could travel inside the cavity. Thus, the solidification fronts will not develop parallel to the walls. From $t = 20\tau$, the crystallinity becomes almost homogeneous in the cavity around the value 0.3, close to the critical crystallinity for Newtonian behavior.

4.2. Statistical Indicator Evolutions

The impact of the temperature dependence of viscosity on the mixing of magma can be appreciated in the above-mentioned global statistical indicators. In Fig. 7, one can observe the evolution of the mean temperature T_m^* inside the magmatic chamber over

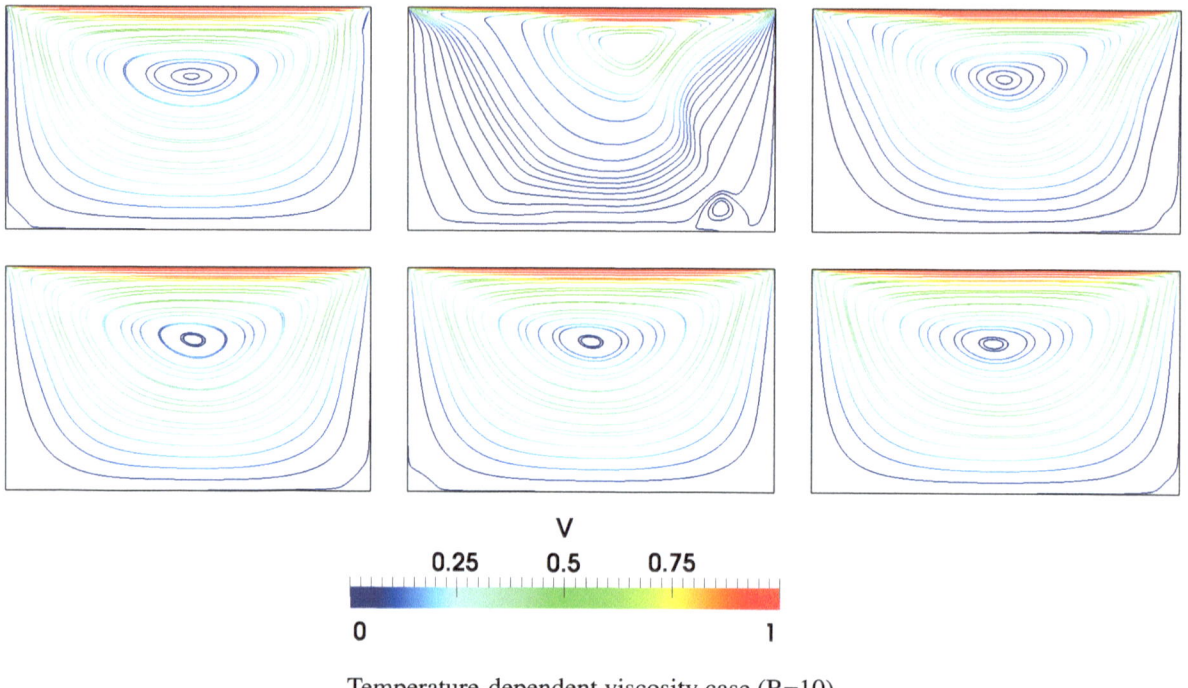

V

0.25   0.5   0.75

0   1

Temperature-dependent viscosity case (B=10)

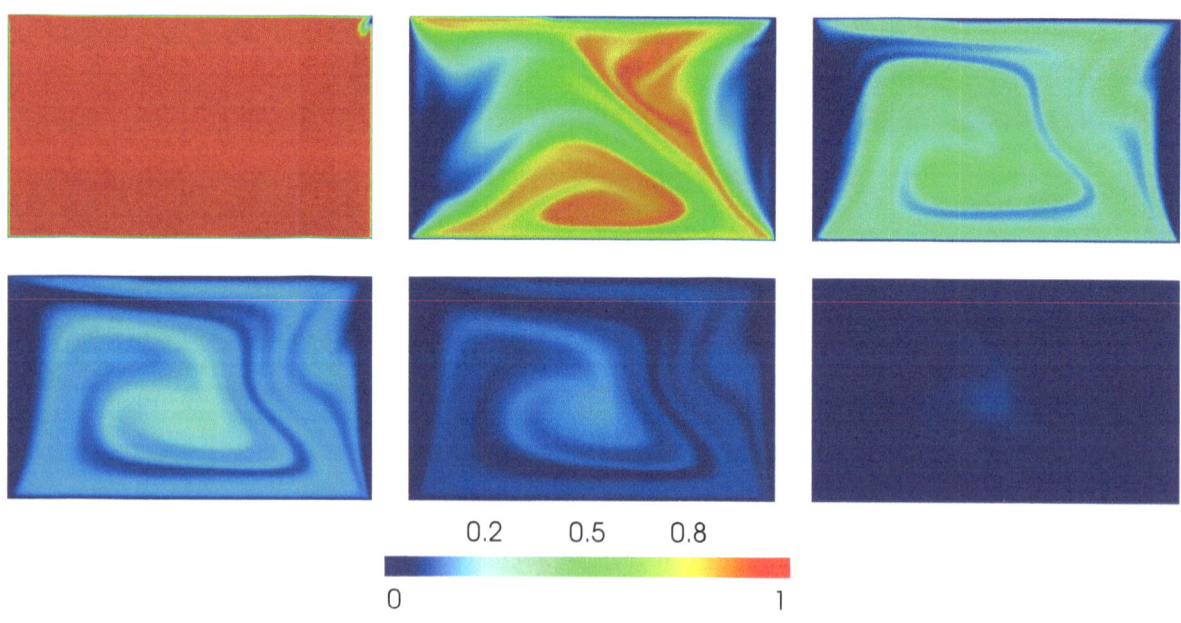

0.2   0.5   0.8

0   1

Figure 5

Flow streamlines and temperature fields at different periodic instants $t = 0, 5\tau, 10\tau, 15\tau, 20\tau$, and 25τ (from *left* to *right* and *top* to *bottom*), for $B = 10$

time (number of periods) for the three different mixing cases ($B = 0$, $B = 5$, and $B = 10$). This represents the amount of energy extracted from the magma across the walls from time 0 to time t. This

figure shows that the mean temperature decreases exponentially towards the wall temperature $T^* = 0$ when chaotic mixing is present ($B = 0$ to $B = 10$ cases), and much more rapidly when the viscosity is

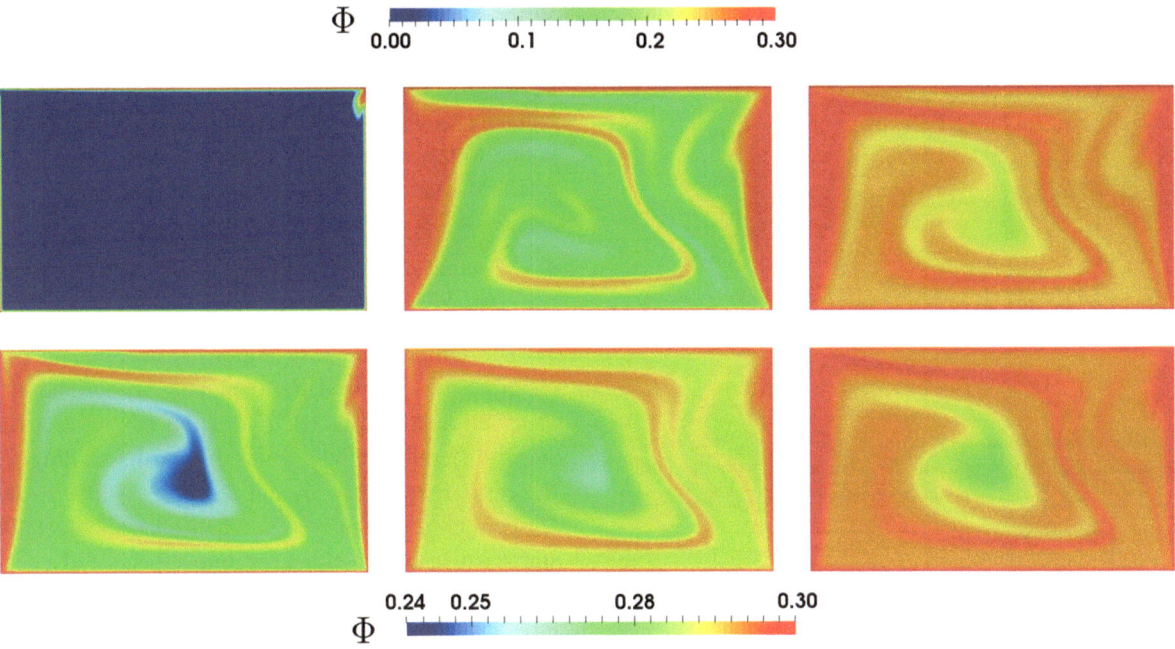

Φ 0.00   0.1   0.2   0.30

0.24   0.25   0.28   0.30
Φ

Figure 6
Crystallinity fields at different periodic instants $t =0$, 5τ, 10τ, 15τ, 20τ, and 25τ (from *left* to *right* and *top* to *bottom*), for $B = 5$ (in relation with the temperature fields of Fig. 4). The *color scale* is different for *top* and *bottom* rows

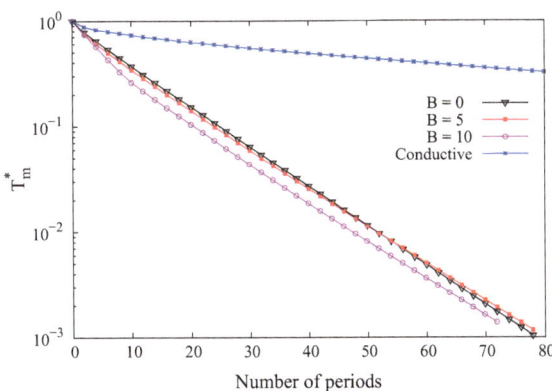

Figure 7
Evolution of the mean temperature for different values of the temperature-dependent viscosity coefficient **B**. Comparison with the purely conductive case

high (higher **B** value). However, the final decay rates are identical. The purely conductive case (fixed walls) is also given for reference; the slow nature of the thermal mixing is evident, since it operates only by thermal diffusion. This case is also characterized by a temperature field with isotherms parallel to the magma chamber walls.

The standard deviation evolutions are given in Fig. 8 for the same cases. We recall that the variance σ^2 represents the level of homogenization of the temperature inside the magma chamber. We observe first of all a phase which corresponds to the creation of the temperature gradients. This phase is much longer (up to five periods) and the gradient more pronounced when **B** is higher. After this transient time, the smoothing of the temperature gradients is observed, being most important for the case $B = 10$. For longer times (after 50 periods), the evolutions are quite similar, with the same decay rate for $B = 0$ and $B = 10$. For the purely conductive case, the temperature homogenization across the magma chamber is very low. This once more confirms the cooling efficiency during chaotic mixing.

Figure 9 illustrates the parietal heat transfer along one of the moving walls. It is characterized by the mean Nusselt number evolution with time, which is found to be globally exponential. The large oscillations of the Nusselt values observed for the three cases are due to the variation of the temperature gradients along the wall resulting from the periodic modulation of the wall movement. For $B = 10$, the

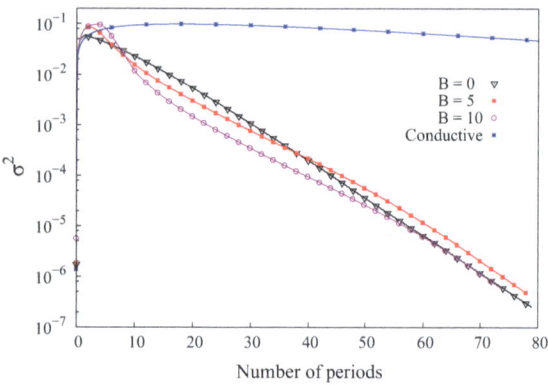

Figure 8

Evolution of the variance for different values of the temperature-dependent viscosity coefficient **B**. Comparison with the purely conductive case

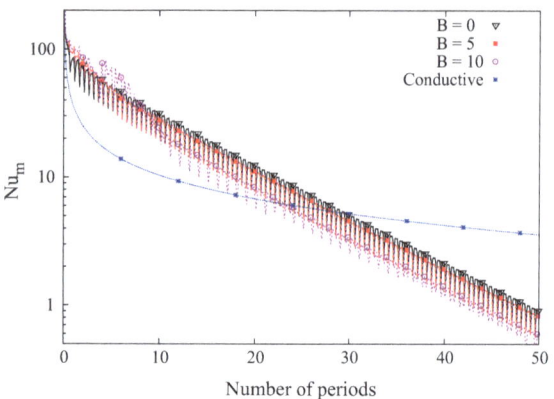

Figure 9

Evolution of the mean Nusselt number along one side wall for different values of the temperature-dependent coefficient B and for the purely conductive case

strong nonlinearity introduced by the temperature-dependent viscosity results in significant irregular fluctuations in the evolution of the Nusselt number.

Without chaotic advection inside the magma chamber (the conductive case), the heat transfer displays a significant slowdown over time. The asymptotical behavior of the Nusselt number evolution in this case is due to the very low penetration of the temperature gradient zone inside the magma chamber.

4.3. Temperature Probability Distribution Functions and Recurrent Patterns

The temperature distributions inside the magma chamber are analyzed by displaying the probability distribution functions (PDFs) of the temperatures at seven different periodic times during the cooling for the cases $B = 0$, $B = 5$, and $B = 10$. At first all the temperatures were $T^* = 1$. When chaotic advection starts, a distribution of temperature appears and a peak is observed, revealing the most represented temperature. This peak moves in time to the left while its height increases. The shift of the PDF towards the left illustrates that the hot temperatures disappear during the thermal mixing process. A narrower distribution is obtained when the mean temperature approaches the wall temperature $T^* = 0$. This effect is much more pronounced for the case $B = 10$, for which the right tail of the distribution is shorter.

An important feature of this thermal chaotic flow is revealed by the PDFs of the rescaled dimensionless temperature X, defined as

$$X = \frac{T^* - T_m^*}{\sigma}. \quad (11)$$

When the dimensionless temperature difference is rescaled by the standard deviation σ, we observe, for $B = 0$, that the distribution for the last periods (from

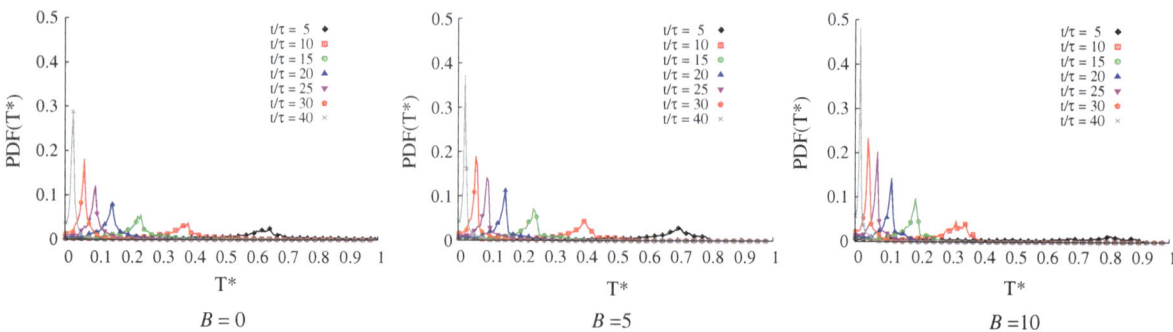

Figure 10

PDFs of the temperature T^* at seven different periodic times for $B = 0$, $B = 5$, and $B = 10$

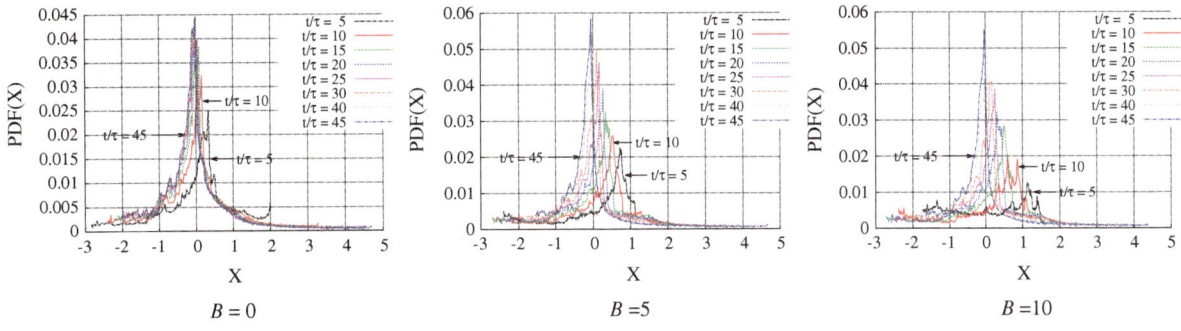

Figure 11
PDFs of the rescaled dimensionless temperature X at seven different periodic times for $B = 0$, $B = 5$, and $B = 10$

$t/\tau = 20$ to $t/\tau = 45$) superimpose (Fig. 11). The superimposed PDFs, plotted for different times during the mixing process but at the same phase of the period, characterize the self-similarity of the dissipation mechanism and are the signature of the existence of a thermal strange eigenmode in the periodic magmatic flow. This is characterized by the production of persistent patterns in the magmatic flow that develop after a transient time (see Fig. 4). These patterns arise from a subtle combination of stretching, folding, and thermal diffusion as the flow is periodically exactly the same (ROTHSTEIN *et al.* 1999; PIKOVSKY and POPOVYCH 2003; LIU and HALLER 2004; EL OMARI and LE GUER 2010). The convergence to the thermal strange eigenmode is also associated with the exponential decay of the dimensionless temperature variance (see Fig. 8). With the increase in the temperature dependence and the nonlinearity due to the stronger link between the velocity field (which is not that of Fig. 3) and the temperature field ($B = 5$ and $B = 10$ cases), the temperature patterns lose their self-similarity and the flow becomes more complex.

5. Conclusions

In this study we investigated the effects of chaotic advection for thermal mixing during the cooling of a mafic magma with temperature-dependent viscosity. The implications for magmatic systems are important. One of the most significant is related to the thermal lifetime of magma chambers. We have shown that, when the thermal field of a magma chamber is governed by chaotic dynamics, it develops complex structures that are strongly modulated by advective flow fields. This generates different thermal domains due to different cooling efficiencies depending on the ability of chaotic thermal mixing to homogenize the thermal fields; in some cases poorly mixed regions may remain. These correspond to volumes of melt in which heat is transferred with strongly different efficiencies to the surrounding volumes of melt. In strongly stretching regions, heat transfer is fast because of the formation of large contact interfaces. Due to higher heat transfer rates, these regions cool faster than weakly stretching regions, allowing the preservation of chemical structures. This is particularly true for closed systems (i.e., magma chambers that are not refilled by new magma), where the total energy of the system is equal to the initial energy of the system. On the contrary, open systems (e.g., magma chambers continuously refilled by new hot magma) will be able to mix more deeply as a result of the prolonged energy input to the system. In weakly stretching regions, heat dissipation is slower than in strongly stretching areas, because it occurs mainly via diffusion. This allows for the formation of volumes of melt that can potentially preserve the initial temperature for a longer time. Here, chemical homogenization is also very slow due to lower values of chemical diffusion coefficients with respect to thermal diffusion coefficients. As a result, after the beginning of chaotic advection, the magmatic system quickly evolves towards a configuration in which different thermal domains exist. The development of chaotic thermal mixing also contributes to an increase in the space and time complexity of the magmatic system. Moreover, we have shown that the temperature-dependent viscosity effect is manifested by the introduction of additional complexity in the flow; it also destroys the self-similar character of the temperature fields. This first study was made with the consideration of a certain number of assumptions about the physical phenomena involved. To get closer to a more appropriate magma behavior, consideration of further physical processes could be investigated by numerical simulations for the magma flow, for example, the buoyancy effect if the shear has a component parallel to the gravity field (differentiation), the viscous heating generated by viscous friction near the walls, the non-Newtonian shear-thinning rheological behavior of the magma with or without the existence of a yield stress, and the crystallization kinetics during cooling (including a physical model of magma solidification).

Acknowledgments

We wish to acknowledge the European Research Council for Consolidator Grant CHRONOS (No. 612776) and PRIN no. 2010TT22SC_004 (D.P.). The authors were granted access to the HPC resources of MCIA Bordeaux (France).

Appendix: Supplementary Material

Supplementary material (videos) related to this article can be found online at:

http://siame.univ-pau.fr/live/transport-multiphasique/Galerie/magma-mixing.

REFERENCES

AREF, H. (1984), *Stirring by chaotic advection*, J. Fluid Mech. *143*, 1–21.

AREF, H. (2002), *The development of chaotic advection*. Phys. Fluids 2002, *14*, 1315.

BAMMOU, L., EL OMARI, K., BLANCHER, S., LE GUER, Y., BENHAMOU, B., MEDIOUNI, T. (2013), *A numerical study of the longitudinal thermoconvective rolls in a mixed convection flow in a horizontal channel with a free surface*, Int. J. Heat Fluid Flow, *42*, 265–277.

BERGANTZ, G.W. (2000), *On the dynamics of magma mixing by reintrusion: Implications for pluton assembly processes*, Journal of Structural Geology, *22*(9), 1297–1309.

BRANDEIS, G. and JAUPART, C. (1986), *On the interaction between convection and crystallization in cooling magma chambers*, Earth and Planetary Science Letters, 345–361.

BRESLER, L., SHINBROT, T., METCALFE, G., OTTINO, J.M. (1997), *Isolated mixing regions: origin, robustness and control*, Chem. Eng. Sci. *52*, 1623–1636.

CARICCHI, L., BURLINI, L., ULMER, P., GERYA, T., VASSALLI, M., PAPALE, P. (2007), *Non-Newtonian rheology of crystal-bearing magmas and implications for magma ascent dynamics*, Earth and Planetary Science Letters, *264*(3–4), 402–419.

CHIEN, W.L., RISING, H., and OTTINO, J.M. (1986), *Laminar mixing and chaotic mixing in several cavity flows*, J. Fluid Mech., *170*, 355, 1986.

CIMARELLI, C., COSTA, A., MUELLER, S., MADER, H.M. (2011), *Rheology of magmas with bimodal crystal size and shape distributions: Insights from analog experiments*, Geochemistry, Geophysics, Geosystems, *12*(7), Q07024.

COSTA, A. and MACEDONIO, G. (2003), *Viscous heating in fluids with temperature-dependent viscosity: implications for magma flows*, Nonlin. Processes Geophys., *10*, 545–555.

DE CAMPOS C.P., PERUGINI D., ERTEL-INGRISCH, W., DINGWELL, D.B. and POLI, G. (2011), *Enhancement of magma mixing efficiency by chaotic dynamics: an experimental study*, Contributions to Mineralogy and Petrology, *161*, 863–881.

DE ROSA, R., DONATO, P. and VENTURA, G. (2002), *Fractal analysis of mingled/mixed magmas: an example from the Upper Pollara eruption (Salina Island, southern Tyrrhenian Sea, Italy)*, Lithos, 65, 299–311.

DIDIER, J., and BARBARIN, B. (1999), Enclaves and Granite Petrology, Developments in Petrology, *13*, (Elsevier, Amsterdam), 625 pp.

EL OMARI, K. and LE GUER, Y. (2009), *A numerical study of thermal chaotic mixing in a two rod rotating mixer*, Computational Thermal Sciences, *1*, 55–73.

EL OMARI, K. and LE GUER, Y. (2010a), *Alternate rotating walls for thermal chaotic mixing*. Int. J. of Heat and Mass Transfer, *53*, 123–134.

EL OMARI, K. and LE GUER, Y. (2010b), *Thermal chaotic mixing of power law fluids in an alternate rotating walls mixer*, Int. J. Non-Newtonian Fluid Mech., *165*, 11–12, 641–651.

EL OMARI, K. and LE GUER, Y. (2012), *Laminar mixing and heat transfer for constant heat flux boundary condition*, Heat and Mass Transfer, *48*(8), 1285–1296.

FLINDERS, J. and CLEMENS, J.D. (1996), *Non-linear dynamics, chaos, complexity and enclaves in granitoid magmas*, Trans. R. Soc. Edinburgh Earth Sci. 87, 225–232.

GINIBRE, C., KRONZ, A. and WORNER, G. (2002), *High-resolution quantitative imaging of plagioclase composition using accumulated back-scattered electron images: New constraints on oscillatory zoning*, Contributions to Mineralogy and Petrology, *142*, 436–448.

GIORDANO D., RUSSEL J.K. and DINGWELL D.B. (2008), *Viscosity of magmatic liquids: a model*, Earth and Planetary Sciences, *271*, 123–134.

HECHT, L. and VIGNERESSE, J.L. (1999), A multidisciplinary approach combining geochemical, gravity and structural data: implications for pluton emplacement and zonation, Geological Society, London, Special Publications, *168*, 95–110.

HUPPERT, H.E. (2000), Geological fluid mechanics, Perspectives in Fluid Dynamics. A collective introduction to current research. G.K. Batchelor, H.K. Moffat and M.G. Worster Ed. (Cambridge University Press), 447–506.

JANA, S.C., METCALFE, G., OTTINO, J.M. (1994), *Experimental and computational studies of mixing in complex Stokes flows: the vortex mixing flow and multicellular cavity flows*, J. Fluid Mech., *269*, 199–246.

KRIEGER, I. and DOUGHERTY, T. (1959), *A mechanism for non-Newtonian flow in suspension of rigid spheres*, Trans. Soc. Rheol., *3*, 137–152.

LE GUER, Y., EL OMARI K. (2012), Chaotic advection for thermal mixing, Advances in Applied Mechanics. In: E. van der Giessen and H. Aref Series Ed. Guest Ed. H.J.H. Clercx and M.F.M. Speetjens (San Diego: Academic), 45, 189–237.

LEFEVRE, A. and MOTA, J.P.B. and RODRIGO, A.J.S. and SAATDJIAN, E. (2003), *Chaotic advection and heat transfer enhancement in Stokes flows*, International Journal of Heat and Fluid Flow, *24*(3), 310–321.

LIU, W., HALLER, G. (2004), *Strange eigenmodes and decay of variance in the mixing of diffusive tracers*, Physica D, *188*, 1–39.

LIU, M., PESKIN, R.L., MUZZIO, F.J. and LEONG, C.W. (1994), *Structure of the stretching field in chaotic cavity flows*, AIChE Journal, *40*(8), 1273–1286.

LOPEZ, C., NEUFELD, Z., HERNANDEZ-GARCIA, E. and AND HAYNES, P.H. (2001), *Chaotic advection of reacting substances: Plankton dynamics on a meandering jet*, Physics and Chemistry of the Earth, Part B: Hydrology, Oceans and Atmosphere, *26*, 4, 313–317.

MADER, H.M., LLEWELLIN E.W., MUELLER, S.P. (2013), *The rheology of two-phase magmas: A review and analysis*, Journal of Volcanology and Geothermal Research, *257*, 135–158.

MAHOOD, G., and FRIDRICH, C., 1982, *Differentiation in waxing and waning magma chambers (abs.)*, Geological Society of America Abstracts with Programs, *14*, 7, 553–554.

MARSH, B.D. (1981), *On the crystallinity, probability of occurrence and rheology of lava and magma*, Contrib. Mineral. Petrol., *78*, 85–98.

MC BIRNEY, A.R. and MURASE, T. (1984), *Rheological properties of magmas*, Ann. Rev. Earth Planet. Sci. *12*, 337–357.

METCALFE, G., BINA, C.R. and OTTINO J.M. (1995), *Kinematic considerations for mantle mixing*, Geophys. Res. Lett. *22*(7), 743–746.

MORGAVI D., PERUGINI, D., DE CAMPOS, C.P., ERTEL-INGRISCH, W., LAVALLÉE, Y., MORGAN, L. and DINGWELL, D.B. (2013), *Interactions between rhyolitic and basaltic melts unraveled by chaotic mixing experiments*, Chemical Geology, *346*, 199–212.

MOTA, J.P.B., RODRIGO, A.J.S., SAATDJIAN, E. (2007), *Optimization of heat-transfer rate into time-periodic two-dimensional Stokes flows*, International Journal for Numerical Methods in Fluids, *53*(6), 915–931.

OTTINO, J.M., The Kinematics of Mixing: Stretching, Chaos and Transport (Cambridge University Press, Cambridge 1989).

PERUGINI D., DE CAMPOS, C.P., ERTEL-INGRISCH, W. and DINGWELL, D.B. (2012), *The space and time complexity of chaotic mixing in silicate melts: implications for igneous petrology*, Lithos, *155*, 326–340.

PERUGINI, D., DE CAMPOS, C.P., DINGWELL, D.B. and DORFMAN, A. (2012), *Relaxation of concentration variance: A new tool to measure chemical element mobility during mixing of magmas*, Chemical Geology, *335*, 8–23.

PERUGINI, D., PETRELLI, M., POLI, G. (2006), *Diffusive fractionation of trace elements by chaotic mixing of magmas*, Earth Planet. Sci. Lett., *243*, 669–680.

PERUGINI, D., POLI, G. and VALENTINI, L. (2005), *Strange attractors in plagioclase oscillatory zoning: petrological implications*, Contributions to Mineralogy and Petrology, *149*(4), 482–497.

PERUGINI, D. and POLI, G. (2012), *The mixing of magmas in plutonic and volcanic environments: Analogies and differences*, Lithos, *153*, 261–277.

PERUGINI, D. and POLI, G., (2000), *Chaotic dynamics and fractals in magmatic interaction processes: a different approach to the interpretation of mafic microgranular enclaves*, Earth Planet. Sci. Lett., *175*, 93–103.

PERUGINI, D. and VALENTINI, L. and POLI, G. (2007), *Insights into magma chamber processes from the analysis of size distribution of enclaves in lava flows: A case study from Vulcano Island (Southern Italy)*, Journal of Volcanology and Geothermal Research, *166*, 193–203.

PERUGINI D., G. POLI G., MAZZUOLI R. (2003), *Chaotic advection, fractals and diffusion during mixing of magmas: evidence from lava flows*, J. Volc. Geotherm. Res., *2615*:1–25.

PETRELLI, M., PERUGINI, D., POLI, G. (2011), *Transition to chaos and implications for time-scales of magma hybridization during mixing processes in magma chambers*, Lithos, *125*(1–2), 211–220.

PIKOVSKY, A., POPOVYCH, O. (2003), *Persistent patterns in deterministic mixing flows*, Europhys. Lett., *61*(5):625–631.

RENJITH, M.L., CHARAN, S.N., SUBBARAO, D.V., BABU, E.V.S.S.K. and RAJASHEKHAR, V.B. (2013), Grain to outcrop-scale frozen moments of dynamic magma mixing in the syenite magma chamber, Yelagiri Alkaline Complex, South India, Geoscience Frontiers, in press, 1–20.

ROSCOE, R. (1952) *The viscosity of suspensions of rigid spheres*, British Journal of Applied Physics, *3*, 267–269.

ROTHSTEIN, D., HENRY, E., and GOLLUB, J.P. (1999), *Persistent patterns in transient chaotic fluid mixing*, Nature, *401*:770–772.

SAATDJIAN, E. and LEPREVOST, J.C. (1998), *Chaotic heat transfer in a periodic two-dimensional flow*, Physics of Fluids, *10*, 8, 2102–2104.

SATO, H. (2005), *Viscosity measurement of subliquidus magmas: 1707 basalt of Fuji volcano*, Journal of Mineralogical and Petrological Sciences, *100*, 133–142.

SPERA, F.J. (2000), Physical Properties of Magma, in: Sigurdsson, H. (Ed.), Encyclopedia of Volcanoes. Academic Press, San Diego, CA, 171–189, 2000.

STREMLER, M.A. (2008), Mixing measures, Encyclopedia of Microfluidics and Nanofluidics (ed. Li D., New York, Springer), 1376–82.

THIFFEAULT, J.-L. (2012), *Using multiscale norms to quantify mixing and transport*, Nonlinearity 25, 1–44.

WINKLER, H.G.F. (1949), *Crystallization of basaltic magma as recorded by variation of crystal-size in dikes*, Mineral. Mag., *28*, 557–574.

(Received April 4, 2014, revised December 16, 2014, accepted December 23, 2014, Published online February 1, 2015)

Reprinted from the journal

Pure Appl. Geophys. 172 (2015), 1851–1863
© 2014 Springer Basel
DOI 10.1007/s00024-014-0963-z

▌Pure and Applied Geophysics

Frequency Distributions of Geochemical Data, Scaling Laws, and Properties of Compositions

ANTONELLA BUCCIANTI[1]

Abstract—Many random processes occur in geochemistry. Accurate predictions of the manner in which elements or chemical species interact with each other are needed to construct models able to treat the presence of random components. Although modelling of frequency distributions with some probabilistic models (for example Gaussian, log-normal, Pareto) has been well discussed in several fields of application, little attention has been devoted to the features of compositional data and, in particular, to their multivariate nature. In this contribution an approach coherent with the properties of compositional information is proposed and used to investigate the shape of the frequency distribution of geochemical indices obtained by robust multivariate analysis. The purpose is to understand data-generation processes from the perspective of compositional theory. The approach is based on use of transformations of the log-ratio family, each with peculiar theoretical and practical advantages, depending on the statistical methods adopted. Accordingly, because, in compositional data, all the relevant information about one term (x_i) of a D-part composition is contained in the ratios to each of the remaining parts $x_2,\ldots, x_D$, analysis of single variables is abandoned. The proposed methodology directs attention to modelling of the frequency distribution of more complex indices, linking all the terms of the composition to better represent the dynamics of geochemical processes. An example of its application is presented and discussed on the basis of consideration of the chemistry of 616 ocean floor basaltic (OFB) glasses from the abyssal volcanic glass data file (AVGDF) of the Smithsonian Institution.

Key words: Frequency distributions, compositional data, normal model, lognormal model, power laws, fractals.

1. Introduction

Geochemical variables actually observed are the consequence of several events, some of which may be poorly defined or imperfectly understood. These variables tend to change with time and space but, despite their complexity, may share specific common traits and it is possible to model them stochastically. Description of the frequency distribution of the abundances of geochemical elements has been an important target of research, attracting attention for at least 100 years, starting with CLARKE (1889, 1924), and continued by GOLDSCHMIDT (1933) and WEDEPOHL (1955). However, it was AHRENS (1954a, b) who focussed on the effect of skewness distributions, for example the log-normal distribution, regarded by him as a fundamental law of geochemistry.

The effect of normal and lognormal distribution has been also discussed by REIMANN and FILZMOSER (1999) and by MATEU-FIGUERAS and PAWLOWSKY-GLAHN (2008). It is well known that log-normal distributions are generated by processes that follow the law of proportionate effect, and the term "multiplicative process" is used to describe the underlying model. The success of this type of distribution in geochemistry is related to the wide application of the theory of successive random dilutions (SRD; OTT 1990). In general, if a succession of n dilutions occurs, the resulting concentration c_n will be the product of n individual dilution factors and the original concentration c_0:

$$c_n = \alpha_1\alpha_2\cdots\alpha_n c_0 = c_0 \prod_{i=1}^{n} \alpha_i. \qquad (1)$$

The product occurs because the final mixture of one dilution becomes the starting mixture for the next, giving us a proportionate process. The concentration in each successive state is a constant proportion of the concentration in any previous state. If the proportions are always constant, the process will be deterministic. If the entire set of n dilutions is repeated, all the dilution factors will be the same, and exactly the same concentration c_n, will result.

However, real processes in the environment seldom give the same final concentrations, even if they are repeated, because many complex factors affect

[1] Department of Earth Sciences, University of Florence, Via G. La Pira 4, 50121 Florence, Italy. E-mail: antonella.buccianti@unifi.it

the dilutions, causing random change. If the entire process is repeated several times with independent dilution factors and the resulting final concentrations are plotted on a histogram, the distribution will have a single mode and will be right-skewed. As n increases, log c_n approaches a normal distribution, in accordance with the central limit theorem (AITCHISON and BROWN 1954), and the variance of log c_n increases with n. In summary, a series of n independent successive random dilutions of an initial concentration creates a distribution that is approximately log-normal, irrespective of the distributions of the dilution factors. Several investigations have, however, provided strong support for the hypothesis that the distribution of the abundance of minor and trace elements is fractal or multifractal, because of the presence of one or more heavily skewed or heavily tailed distributions (ALLÉGRE and LEWIN 1995; MA *et al.* 2014). These are known as power law distributions, also often referred to as Pareto or Zipfian distributions. Computer-generated fractals are self-similar over an infinite range of scales and their geometrical properties can be adequately described by one variable only, the fractal dimension. For natural fractals, however, self-similarity is not valid in the strict sense but is valid statistically over a finite range of scales. Consequently, the fractal dimension may change inside different scale ranges because of the presence of non-linear interactions between different scales. Thus the entire range of scales can be divided into an infinite number of subsets and for each subset its own self-similarity law is valid for its own fractal dimension. A natural fractal may be thus regarded as a multifractal, i.e., a superposition of an infinite number of monofractals.

Because of their features, multifractal models can describe complex phenomena in geochemical distributions, thus improving our knowledge of them (GONCALVES 2001).

In this work the frequency distributions of the major and minor elements in basaltic volcanic glasses collected on the ocean floors of the world were analysed from a compositional multivariate perspective as an alternative to traditional univariate modelling. In this respect, the compositional approach proposed by FILZMOSER *et al.* (2009a, b), who criticised traditional analysis of single geochemical variables, is

limited, because investigation of these remains at the univariate level. The purpose of the work was to determine which type of probability density function (normal, lognormal, Pareto and similar) is most suitable for capturing the partition of the elements in this geological material when complex geochemical multivariate indices are investigated (LIEBOVITCH and SCHEURLE 2000; BUCCIANTI 2011).

2. Materials and Methods

2.1. The Data Base

Study of volcanic glasses is of fundamental importance because, compared with crystalline basalts, they represent the magmatic liquid compositions directly. The Smithsonian Institution has recently offered the possibility of using data for the ocean floor basaltic (OFB) glasses from the abyssal volcanic glass data file (AVGDF) (MELSON *et al.* 2002). Results from analysis by laser-ablation (LA)-ICP-MS of 60 elements in 616 samples have been discussed by JENNER and O'NEILL (2012). The samples analysed are from the Atlantic ($n = 307$), Pacific ($n = 265$), Indian ($n = 41$), and Caribbean Oceans ($n = 3$) and represent most of the major geographical regions characterized by different spreading rates and local tectonic environments. As reported by JENNER and O'NEILL (2012), the samples cover a wide range of major element chemistry from primitive to moderately evolved (9.7 to approx. 3.5 wt% MgO) and are approximately similar to the entire AVGDF. MgO % approximates a normal frequency distribution whereas K, P, and Ti have log-normal distributions with little skewness or kurtosis. An equal tail of depleted samples thus balances the number of enriched samples. This type of behaviour is comparable with that of the entire AVGDF so that, in general, a continuum in OFB is present, confirming previous results (HOFMANN 2003).

Differences among the diverse tectonic settings (spreading ridges ($n = 497$), seamounts ($n = 37$), fracture zones ($n = 71$), aseismic ridges ($n = 4$), and back-arc basins ($n = 7$)) were not revealed by investigation of the frequency distributions. However, it should be noted that the investigation was performed on single variables in the original samples and not by considering the properties of compositional data, with the exception of use of the log function.

Our proposal was to investigate the frequency distribution of major and minor oxides after having transformed them into complex geochemical indices by robust multivariate analysis, as explained in the next section.

2.2. Statistical Methods

Compositional data are vectors of proportions (PEARSON 1897), percentages, ppm, or similar, thus adding to a given constant, known as closed data (CHAYES 1960). They are multivariate and represent parts of the same whole, conveying only relative information among those parts. Consequently, analysis performed by choosing one compositional part only, which occurs when frequency distributions are investigated, cannot be fully informative.

The peculiar mathematical properties of compositional data arise from the fact that they are vectors of positive values where each value describes the contribution of one of D parts (AITCHISON 1982, 1986). Because of these features, compositional data pertain to a particular sample space, called a simplex, whose geometry is governed by a specific metric (AITCHISON 1986; PAWLOWSKY-GLAHN and EGOZCUE 2001; BUCCIANTI and MAGLI 2011). The D-part simplex, S^D, is a subset of the D-dimensional real space. For example, when $D = 2$, it can be represented as a line, for $D = 3$ as a planar triangle (the well known ternary diagram), and for $D = 4$ as a tetrahedron. The simplex sample space, when governed by Aitchison geometry, has all the properties of a $(D - 1)$ dimensional Euclidean space (BILLHEIMER et al. 2001; PAWLOWSKY-GLAHN and EGOZCUE 2001; EGOZCUE and PAWLOWSKY-GLAHN 2006). To achieve this result an inner product, a norm (the length of a vector or modulus) and a distance (a measure of difference between vectors), called the Aitchison distance, have been defined. Internal and external operations called perturbation and powering are then associated. Applications of these concepts in several fields of investigation can be found in BUCCIANTI et al. (2006) and PAWLOWSKY-GLAHN and BUCCIANTI (2011).

To move data from the simplex to real space, compositions have to be represented by coordinates that are, in fact, real vectors. A family of log-ratio transformations has been proposed (AITCHISON 1982; EGOZCUE et al. 2003) to achieve this. Among the different methods that can be used, the isometric log-ratio (ilr) conversion has theoretical advantages and practical properties, particularly when the concept of balance between groups of parts is considered (EGOZCUE and PAWLOWSKY-GLAHN 2005). A practical way of constructing balances is by using sequential binary partitions (SBP) of the composition $\mathbf{x}$. In each of the $D - 1$ partition steps of an SBP, the compositional parts are split into two non-overlapping groups. The resulting $D - 1$ ilr-coordinates represent balances between these groups in R^{D-1}:

$$\mathrm{ilr}_i(x) = \sqrt{\frac{r_{i+} \times r_{i-}}{r_{i+} + r_{i-}}} \log \frac{g(c_{i+})}{g(c_{i-})}, \ i = 1, 2, \ldots D - 1$$

(2)

where c_{i+}, c_{i-} denote the groups of parts separated in the i-th step of the SBP, r_{i+}, r_{i-} are the number of parts included in c_{i+}, c_{i-}, respectively, and $g(\cdot)$ is the geometric mean of its argument. As a natural way of representing compositions AITCHISON (1982) proposed several other types of transformation, one being the centred log-ratio (clr) conversion. The clr transformation from S^D to R^D is defined as:

$$\mathrm{clr}(\mathbf{x}) = \left[\log \frac{x_1}{g(\mathbf{x})}, \log \frac{x_2}{g(\mathbf{x})} \ldots, \log \frac{x_D}{g(\mathbf{x})} \right], \quad (3)$$

where $g(\mathbf{x}) = (x_1 x_2 \cdots x_D)^{1/D}$ is the geometric mean of the parts in $\mathbf{x}$. Not all the conversions are able to enhance the relationships among the terms of the composition. The use of log-ratios with common denominators (additive log-ratio (alr) transformation) is problematic in some instances. These log-ratios represent non-orthogonal directions in the simplex (EGOZCUE and PAWLOWSKY-GLAHN 2005). This is an important point, because the use of ratios with the same denominator is a common practice in bivariate analysis (ROLLINSON 1992, 1993).

In this work the behaviour of major and minor oxides representing the compositions of oceanic volcanic glasses has been analysed. To achieve this, robust principal-components analysis (rPCA) was performed after transformation of the data by using the log-centred (clr) conversion. The presence of normal, log-normal, or power law distributions, and the presence of fragmented processes, has been then checked by analysing the behaviour of the scores. Before this investigation the homogeneity of the data set was checked to verify:

1. the presence of anomalous samples; and
2. sources of variability because of geographical areas or tectonic settings.

To manage compositional data correctly the isometric log-ratio conversion (ilr) was adopted.

In all cases, adoption of robust methods enabled in-depth investigation of the data structure (HUBERT 2005; VERBOVEN and HUBERT 2005; DASZYKOWSKI *et al.* 2007).

2.3. *Probability Density Functions and Geochemical Processes*

In geochemistry probability density functions are usually adopted to visualize and model the behaviour of single variables. In this respect the lognormal distribution is used when processes follow the law of proportionate effect (KAPTEYN 1903). The term multiplicative process is also used to describe this underlying model, but only a small change from the lognormal generative process yields a power law distribution. AITCHISON and BROWN (1954) suggested this link for the first time, discussing whether a power law distribution or a lognormal model represented the best fit in investigation of economic income. Use of power-law modelling has been successful in exploration geochemistry for identification of threshold values compared with the background level (LI *et al.* 2003). In this respect, CHENG *et al.* (1994) noted the limits of traditional methods and suggested that the background and anomalies correspond to different power-law distributions, thus creating a multifractal appearance. ALLÈGRE and LEWIN (1995) showed that the ordinary distribution of trace elements could be normal, multimodal, fractal, or multifractal. In accordance with these authors' ideas, final element concentration distributions for materials that have had a long geological history characterised by several enrichment and/or depletion episodes have been found to be multifractal. XIE and YIN (1993) showed

there is a hierarchy of geochemical dispersion patterns from local to global and from microscopic to continental. Thus, any geochemical dispersion pattern can be explained as the combination of many sub-patterns at different hierarchical levels nested within one another. The consequence is that the spatial distributions of element concentration in the Earth's materials are clustered at many different scales. And it is well know that cluster objects in nature are often fractal (FEDERER 1988).

When different fitting methods are applied to observed distribution functions it is not easy to understand which approximation is true, power law, normal (theoretically impossible for positive variables), or log-normal. Which scaling exponents are better able to represent reality? The answer depends on the range of scales selected for the fitting (ABRAMENKO 2012). For a narrow range, observational data can be well approximated by a linear fit in a double-logarithmic plot (a power law) but for a broader range the observed distribution may be curved. A log-normal fit is capable of capturing the curvature of the observed probability density function on a broad range of scales working better than the power law fit. All the above types of fitting are only models intended to represent a multifractal system with a finite number of scalar variables. In the following sub-sections the main features of these models will be explained in brief by taking into account representations of compositional data also.

2.3.1 *Power Law Distributions*

A non-negative random variable X is said to have a power law distribution if:

$$\Pr[X \geq x] \cong cx^{\alpha} \quad \text{for} \quad x > 0 \qquad (4)$$

for constants $c > 0$ and $\alpha > 0$, so the tails fall asymptotically in accordance with the power α. For a power law distribution α usually falls in the range $0 < \alpha \leq 2$, in which case X has infinite variance. If $\alpha \leq 1$ then X also has an infinite mean. The scaling exponent α is usually called the fractal dimension. From a general perspective, the model leads to much heavier tails than other common models explaining the concentrations of minor and trace elements in geological materials (MA *et al.* 2014). An interesting feature of this distribution model is that if X has a power law distribution, then in a log–log plot of $\Pr[X \geq x]$ the pattern of points will be described by a straight line.

2.3.2 Log-Normal and Normal Distributions

A random variable X follows a lognormal probability density function if the random variable $Y = \log(X)$ has a normal (i.e., Gaussian) distribution. If the normal distribution Y is represented by the density function:

$$f(y) = \frac{1}{\sqrt{2\pi}\sigma} e^{-(y-\mu)^2/2\sigma^2}, \qquad (5)$$

with μ mean, σ^2 variance, and $-\infty < y < \infty$, then the lognormal distribution is given by:

$$f(x) = \frac{1}{\sqrt{2\pi}\sigma x} e^{-(\log x - \mu)^2/2\sigma^2}. \qquad (6)$$

The lognormal distribution has finite moments but may be very similar in shape to the power law distribution. If X is lognormal, and the variance of the corresponding normal distribution is large, then in a log–log plot the behaviour can appear to be linear for several orders of magnitude (MITZENMACHER 2004). By considering all the previous comments it is clear that the sources of variability in a data set could govern the suitability of one probability model compared with another.

2.3.3 Models for Compositional Data

The most typical distribution models for compositions are the Dirichlet and the additive logistic normal (or normal on the simplex) (AITCHISON et al. 2003; MATEU-FIGUERAS et al. 2005; MATEU-FIGUERAS and PAWLOWSKY-GLAHN 2008; MATEU-FIGUERAS et al. 2013). As an alternative, because compositional data have to be transformed from simplex space to real space, coordinates obtained by the ilr transformation or by application of the concept of balance can be modelled by use of classical methods (EGOZCUE et al. 2003; EGOZCUE and PAWLOWSKY-GLAHN 2005; FILZMOSER et al. 2009a, b).

Consider, for example, a D-part composition $\mathbf{x} = (x_1, x_2, \ldots, x_D)$. It does not matter which are the

units of the variables or whether or not the composition is almost constant. The element x_1 is taken to be a reference element whose frequency distribution has to be investigated in comparison with the rest. Thus, consider the following balance or log-contrast (l_c):

$$l_c = \sqrt{\frac{D-1}{D}} \log \frac{x_i}{(x_2 x_3 \cdots x_D)^{1/(D-1)}}, \qquad (7)$$

representing a ilr coordinate generated by the first step of an SBP where x_1 is separated from the other parts; in the successive step, the SBP process is repeated at the same level so that x_2 becomes the reference element and x_1 substitutes it in the denominator of Eq. (7). If all the terms of the composition are considered, a family of D coordinates is obtained and their frequency distribution can be investigated by considering the effect of each component in relation to the behaviour of the rest of the composition (FILZMOSER et al. 2009a, b).

The approach is advantageous, because statistical analysis can be performed in real space. It seems to be also partially able to capture the proportional effect of the element partition processes that characterize crystallization and melting phenomena in solid geological materials (ALLÈGRE and LEWIN 1995). However, the analysis again remains at the univariate level and the joint nature of the members of a composition is practically lost. The multivariate features of compositions can in fact be taken into account only by adopting multivariate methods. Thus the idea of considering an oxide or trace element isolated from its composition must be abandoned in our case.

3. Results and Discussion

The shape of the frequency distribution of the oxides of major and minor components, for example SiO_2, Al_2O_3, FeO_{tot}, MgO, CaO, Na_2O, K_2O, TiO_2, and P_2O_5, whose abundance in 616 ocean floor basaltic glasses around the world is expressed as a weight percentage, was investigated. Instead of analysing the behaviour of single components and taking into account that all the relevant information about them is contained in the ratios between the parts of the composition (EGOZCUE *et al.* 2003; EGOZCUE and PAWLOWSKY-GLAHN 2005; FILZMOSER *et al.* 2009a, b; BUCCIANTI 2013), robust multivariate analysis (rPCA) was performed (FILZMOSER *et al.* 2009a, b). In this way the variance–covariance structure of the $D = 9$ resulting clr variables was analysed, thus avoiding biasing effects on the investigation of the oxides (ALLÉGRE and LEWIN 1995). Before this study the homogeneity of the data set was checked to verify:

1. the presence of anomalous samples; and
2. sources of significant differences because of geographical areas or tectonic settings

in both cases adopting robust methods (DASZY-KOWSKI *et al.* 2007; FILZMOSER and TODOROV 2013). All the analyses were performed by using Matlab and R routines (VERBOVEN and HUBERT 2005; R DEVELOPMENT CORE TEAM 2007; EVERITT and HOTHORN 2011; WEHRENS 2011; TEMPL *et al.* 2011; VAN DEN BOOGAART 2013).

3.1. Identification of Anomalous Values

Robust methods were developed because atypical observations substantially affect classical estimates (mean and variance) of a dataset. Outliers can occur by mistake, measurement errors, or because the data set is characterised by the mixing of different populations. Robust estimators of location and scale in univariate analysis are the median, the median absolute deviation, and M-estimators. For multivariate observations robust estimates of the centre μ and the scatter matrix $\sum$ of a dataset $\mathbf{X}$ $n \times p$ (n = number of cases, p = number of variables) can be obtained by use of the minimum covariance determinant (MCD) estimator (ROUSSEEUW 1984; VERBOVEN and HUBERT 2005). In our case the procedure was applied after transformation of the compositional data into coordinates by isometric log-ratio conversion. In Fig. 1, the distance–distance plot shows the robust distance based on the MCD procedure versus the classical Mahalanobis distance for each observation. Horizontal and vertical lines are drawn at the cut-off value of the chi-squared function $\sqrt{\chi^2_{p,0.975}}$. Looking at Fig. 1, we clearly see some outlying observations on the right identified both by robust and classical analysis. They strongly differ from the other cases in the sequence. These anomalous compositions correspond to the Tuzo Wilson seamount located in the Pacific Ocean. The high K_2O and P_2O_5 content in this area have been related to the effect of metasomatism processes enriching the source region (COUSENS *et al.* 2011).

Continuing with the analysis of Fig. 1, we observe also that several cases with a large robust but a small Mahalanobis distance could not be recognised by adopting a classical approach only. These results

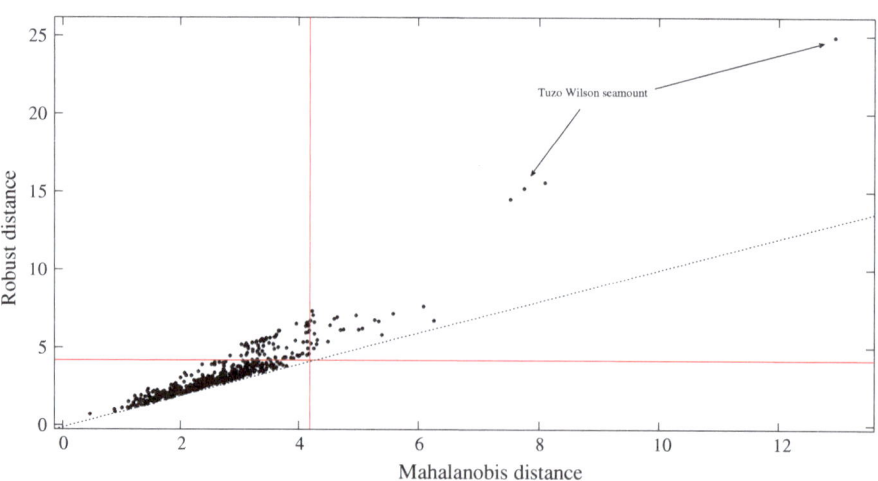

Figure 1

Distance–distance plot displaying for each observation the robust distance determined by use of the MCD estimator and the classical Mahalanobis distance. *Horizontal* and *vertical lines* are drawn at the cutoff value of $\sqrt{\chi^2_{p,0.975}}$

testify that a further check on the importance of geography and tectonic settings as potential sources of data variability is needed to interpret the behaviour of the scores obtained by the robust PCA.

3.2. Checking for Potential Sources of Variability: Geography and Tectonic Settings

The 616 analysed samples are from the Atlantic ($n = 307$), Pacific ($n = 265$), Indian ($n = 41$), and Caribbean Oceans ($n = 3$) and represent most of the major geographical regions characterized by varying spreading rates and local tectonic environments. One-way multivariate analysis of variance (MANOVA) was used to verify the presence of significant differences among geographical areas and tectonic settings. The procedure involves testing the null hypothesis of equal mean vectors across g considered groups by taking into account the intercorrelations of the independent variables. Under the classical assumptions that all groups arise from multivariate normal distributions many test statistics have been developed, as for example Wilk's lambda. However, because the effect of outliers can strongly bias the results the robust approach of TODOROV and FILZMOSER (2010) was followed (Table 1).

Results obtained by considering geography as a discriminant factor are reported in Table 2. Results with tectonic settings as grouping factor are reported in Table 3. In both cases significant chemical differences seem to characterise both belonging to a different ocean and geodynamic context, a result in contrast with the uniformity of the data base discussed by JENNER and O'NEILL (2012). This conclusion will aid interpretation of the shape of the frequency distribution of the scores obtained by the robust PCA procedure.

3.3. Joint Variables Behaviour: Robust Principal-Components Analysis

Principal-components analysis is a statistical method explaining the covariance structure of data by means of a small number of components. The components are linear combinations of the original variables and their interpretation aids understanding of different sources of variability. Classical PCA is

Table 1

Summary statistics of the ilr coordinates

Summary (x)							
ilr_1	ilr_2	ilr_3	ilr_4	ilr_5	ilr_6	ilr_7	ilr_8
Min.: 2.111	Min.: 0.554	Min.: 0.396	Min.: 9.686	Min.: −0.498	Min.: −0.691	Min.: 0.485	Min.: 0.341
1st Qu.: 4.674	1st Qu.: 0.756	1st Qu.: 0.752	1st Qu.: 0.815	1st Qu.: −0.356	1st Qu.: 0.100	1st Qu.: 1.838	1st Qu.: 1.541
Median: 5.033	Median: 0.824	Median: 0.840	Median: 0.847	Median: −0.313	Median: 0.232	Median: 2.146	Median: 1.667
Mean: 5.017	Mean: 0.836	Mean: 0.821	Mean: 0.843	Mean: −0.308	Mean: 0.249	Mean: 2.104	Mean: 1.650
3rd Qu.: 5.401	3rd Qu.: 0.900	3rd Qu.: 0.905	3rd Qu.: 0.878	3rd Qu.: −0.255	3rd Qu.: 0.375	3rd Qu.: 2.446	3rd Qu.: 1.777
Max.: 6.571	Max.: 1.503	Max.: 1.353	Max.: 0.990	Max.: −0.144	Max.: 1.078	Max.: 3.956	Max.: 2.177

Reprinted from the journal

Table 2

Results from robust MANOVA analysis using geography (ocean) as grouping factor (TODOROV and FILZMOSER 2010)

Robust one-way MANOVA (Bartlett chi^2)
Data: **x**
Wilks' lambda = 0.5836, chi^2 value = 284.486, DF = 15.473, p value <2.2e−16

Sample estimates	ilr_1	ilr_2	ilr_3	ilr_4	ilr_5	ilr_6	ilr_7	ilr_8
Atlantic	5.046	0.821	0.832	0.854	−0.304	0.252	2.114	1.655
Indian	5.453	0.787	0.818	0.837	−0.301	0.266	2.397	1.802
Pacific	4.983	0.840	0.777	0.858	−0.316	0.149	2.140	1.684

Table 3

Results from robust MANOVA analysis using tectonic settings as grouping factor (TODOROV and FILZMOSER 2010)

Robust one-way MANOVA (Bartlett chi^2)
Data: **x**
Wilks' lambda = 0.6276, chi^2 value = 262.088, DF = 16.968, p value <2.2e−16

Sample estimates	ilr_1	ilr_2	ilr_3	ilr_4	ilr_S	ilr_6	ilr_7	ilr_8
Fracture zone	4.761	0.868	0.791	0.829	−3.281	3.232	2.028	1.682
Seamount	4.982	0.824	0.845	0.851	−3.341	3.234	2.218	1.689
Spreading ridge	5.110	0.817	0.814	0.857	−3.307	3.206	2.166	1.671

highly sensitive to outlying observations, with the consequence that components are often attracted toward anomalous cases and the variability structure of regular samples is not well captured. Use of robust PCA methods enables principal components to be obtained that are not affected so much by outliers.

In our analysis the procedure described in HUBERT (2005) based on projection pursuit and MCD estimation was applied to clr-transformed variables. The cumulative percentage of variance explained by the first two components is 0.915 %, indicative of high correlation among the variables.

The outlier map in Fig. 2 visualizes the orthogonal distance from a data point to the PCA subspace versus the score distance, the latter expressing the statistical distance of a data point from the origin in the PCA sub-space. For both distances a cut-off value is determined (HUBERT 2005), and when it is exceeded the corresponding point is assigned to one of the three outlier groups: good PCA leverage points, bad PCA-leverage points, or orthogonal outliers. For the volcanic glasses database the observations up on the right correspond to the Tuzo Wilson Seamount whose anomalous behaviour has previously been identified.

Considering the cross-tabulation between oceans and outliers, results indicate that approximately 10 % of Atlantic, 41 % of Indian, and 15 % of Pacific samples are anomalous. Considering the tectonic settings, all the data from the aseismic ridges are recognised as outliers, together with the 14 % from fracture zones, 13.5 % from seamounts and approximately 14 % from spreading ridges. To summarise, a total of 14 % of anomalous data has been identified for the whole dataset.

The first component alone explains approximately 75 % of the total data variability and its investigation seems to be exhaustive of the joint behaviour of the investigated major and minor oxides. The loadings are positive for K_2O, TiO_2, and P_2O_5 and negative for the rest of the composition, discriminating elements whose geochemical behaviour is incompatible or compatible in general. LE MAITRE (1968) discusses the use of principal components based on the $D − 1$ non-zero eigenvalues and their corresponding eigenvectors. When compositional data are analysed the principal components are then log-linear contrasts of the D proportions (AITCHISON 1983):

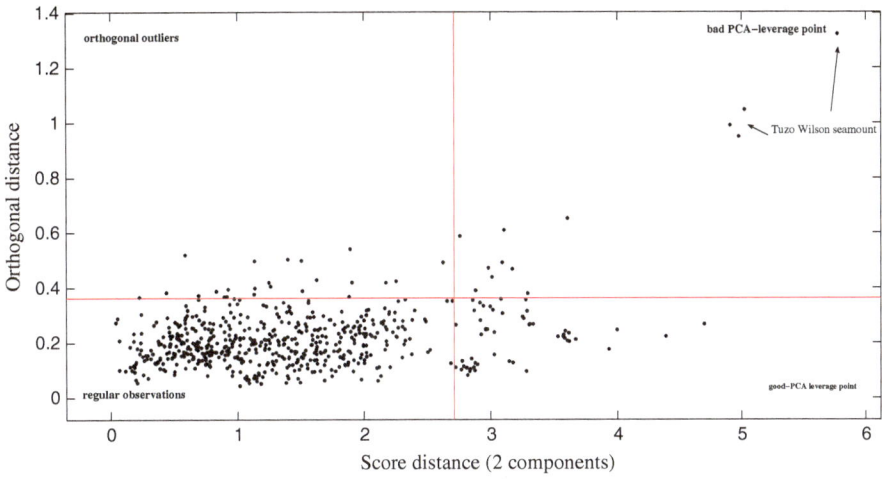

Figure 2
Robust PCA outliers map

$$\log(K_2O) + 0.38 \log(P_2O_5) + 0.15 \log(TiO_2)$$
$$- 0.38 \log(MgO) - 0.33 \log(CaO)$$
$$- 0.28 \log(Al_2O_3) - 0.26 \log(SiO_2)$$
$$- 0.19 \log(FeO_t) - 0.09 \log(Na_2O)$$
$$= k$$

Elements with positive coefficients generally show incompatible behaviour whereas elements with negative coefficients are compatible with the modal mineralogy of this type of sample (olivine, pyroxene, and plagioclase) during melting and fractionation phenomena.

By considering the frequency distribution of the first component (Fig. 3), representing the most important geochemical processes affecting the basaltic glass chemistry, it is easy to verify that standard models (normal or log-normal) are not adequate, as typically obtained for geochemical data (AGTERBERG 2007). Here the physicochemical mechanisms governing their chemistry are not self-organized and the processes of partition of the elements in space or time does not correspond to application in geochemistry of the binomial model of ALLÉGRE and LEWIN (1995).

The frequency distribution of Fig. 3 can be also displayed in diagrams in which the logarithm of the cumulative number of samples exceeding a specific component value is plotted against the component value itself. The result is reported in Fig. 4. It is apparent that the presence of different linear segments fitting the data and the curvature of some of the

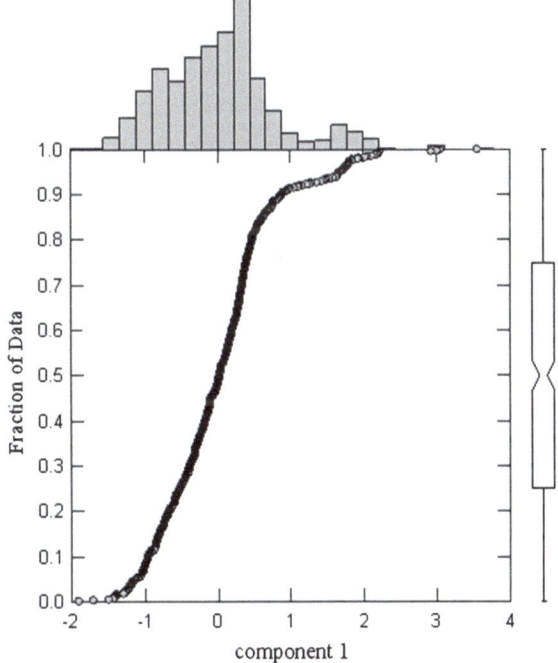

Figure 3
Quantile plot for the first component with histogram and notched box-plot on the *horizontal* and *vertical axes*, respectively

diagram emerges. Consequently, typical multifractal behaviour could describe our dataset. As already reported, the notion of multifractal behaviour is related to sets that may support self-similar measures characterized not by a single fractal dimension but a range of such dimensions. This behaviour could

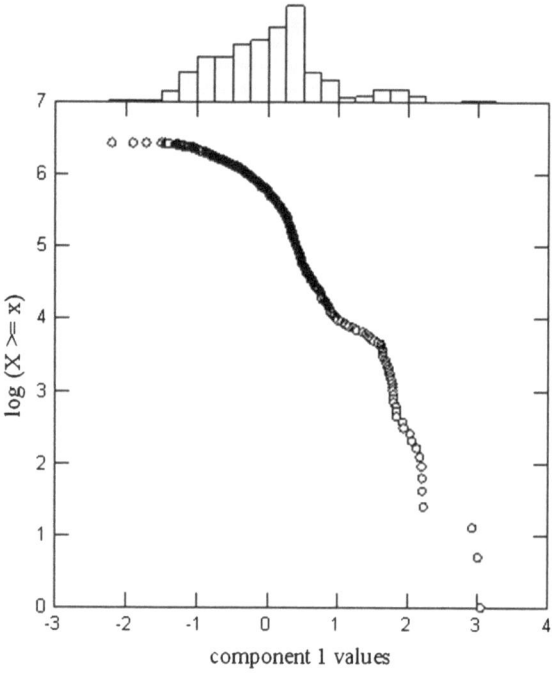

Figure 4
Cumulative number of samples exceeding a specific first component value against the component value itself on a log–log scale.
The presence of several *straight segments* indicates the multifractal
nature of the investigated phenomenon

correspond for the first component to the presence of
a mixture of several similar distributions, each
characterized by a power-law model. Each segment
represents a rule that governs the relationships
between higher and lower abundance for different
sets of values and the jointly considered variables.
This condition might correspond to the repeated
application in time or space of the chromatographic
and Rayleigh distillation processes discussed by
ALLÉGRE and LEWIN (1995) where at each step only
the most enriched material (chromatographic model)
or residual material (Rayleigh distillation) is re-
fractioned, generating a skewness in concentration
values with heavy tail distributions. The processes
could be also well described by repeated application
of the differential scaling operator, called perturbation, in the simplex geometry for compositional data
(AITCHISON 1986).

By taking into account these considerations, it is
evident that iterations of non-linear dynamic processes could have been important in defining the

partition of the elements. A crosscheck between
segments of Fig. 4 and the potential sources of
variability discussed in the Sect. 3.2 indicates that the
presence of different power laws does not perfectly
match geographical areas (oceans) and/or tectonic
settings. This probably reflects the presence of
marked breaks in the joint geochemical behaviour
of the investigated variables depending on further
variables, for example the nature of the source in the
Earths mantle. In this framework some anomalous
cases corresponding to the Tuzo Wilson seamount
located in the northeast Pacific Ocean are well
recognisable (the most separated samples in Fig. 4)
because of their highest content of K_2O and P_2O_5 in
the whole dataset. The role of metasomatism processes, which enrich the source region, is discussed in
the literature explaining the chemistry of the geodynamic context of this area of the Pacific Ocean
(COUSENS and CHASE 2011).

The multifractality of the first component indicates that the joint compositional geochemistry of
major and minor oxides has highly intermittent
character. Intermittency implies a tendency to cluster
into small-scale features of high intensity surrounded
by wide areas of less intense fluctuations in both
temporal and spatial domains. Large fluctuations in
the intermittent processes are not as rare as in
Gaussian processes and substantially affect the statistical moments thus leading to multifractality
(FRISCH 1995). In this respect the multifractality of
the first component also indicates something about its
intermittency. For an intermittent and/or multifractal
variable the probability density function is not
determined by the first and second statistical
moments and high-order moments are critical in
determination of the shape of the heavy tails. This
situation may occur for different reasons; the most
plausible, working with compositions, whose terms
are related by chemical reactions, is that the fluctuations are not because of the sum of independent
random variables but because of their product. It
happens in any fragmentation process or random
multiplicative cascade (AGTERBER 2007) and could
occur in the partition of the elements during melting
or crystallisation.

Summarising, the joint behaviour of major and
minor oxides indicates the presence of highly

fragmented geochemical processes in the spatial and temporal domain. Scores are characterised by wide intervals of low fluctuations that are intermittent with small intervals affected by extremely large fluctuations. In this respect the behaviour of the scores could represent the natural evolution of non-linear dynamic dissipative systems (the chemical reactions governing melting and crystallisation phenomena), systems far from thermodynamic equilibrium (KONDEPUDI and PRIGOGINE 1999). Further analysis of trace elements is needed to further explain these results, and the continuum of OFB discussed by HOFMANN (2003) could be investigated.

4. Conclusions

In this contribution an approach coherent with the properties of compositional information is proposed and used to investigate the shape of frequency distribution of numerical complex indices representing the joint behaviour of geochemical variables. The purpose is to propose a tool able to discover which type of processes could have governed generation the data. The approach is based on analysis of the probability density function of the first principal component obtained by robust PCA of log-transformed SiO_2, Al_2O_3, FeO_{tot}, MgO, CaO, Na_2O, K_2O, TiO_2, and P_2O_5 data, the major and minor oxides of 616 ocean floor basaltic (OFB) glasses from the abyssal volcanic glass data file (AVGDF). The first component alone explains approximately 75 % of the total data variability and seems to be indicative of the similar behaviour of the major and minor oxides investigated.

Results indicate that the frequency distribution of the first component cannot be described by use of classical normal or log-normal models, and several power laws, each with a different fractal dimension D, are thus needed. Score values are, in fact, characterised by wide intervals of low fluctuations that are intermittent with small intervals affected by extremely large fluctuations. As happens in any fragmentation process or random multiplicative cascade (AGTERBERG 2007), the fluctuations of the first component could be attributable to:

1. the product of independent random variables; and
2. the presence of non linear dynamic dissipative systems whose thermodynamical conditions were far from equilibrium (KONDEPUDI and PRIGOGINE 1999).

The dependence of this behaviour on geographical areas and tectonic settings (spatial/time domains) is significant, but not exhaustive. This probably reflects the presence of marked breaks in the abundance of compatible and incompatible elements in the analysed compositions, but also some differences in their local magmatic source in the earth's mantle.

To summarise, appreciation of the properties of compositional data and adoption of robust methods of investigation have shown the presence of significant differences for major and minor oxides of the AV-GDF not reported in literature.

Further investigation will continue by considering the joint behaviour of trace elements from the same dataset with the objective of better determination of the dynamics and profile of geochemical processes.

Acknowledgments

This work has received financial support from the University of Florence (I) through the project ex 60 %—2012 and Tuscany Region (I) and Lamma Consortium through the Geobasi project. Referees are kindly thanked for their valuable revision of the first version of this work.

REFERENCES

ABRAMENKO, V.I. (2012), *Fractal multi-scale nature of solar/stellar magnetic fields*, Proceedings of the International Astronomical Union 8(S294), 289–300.

AGTERBERG, F. (2007), *Mixtures of multiplicative cascade models in geochemistry*, Nonlinear Process Geophys 14, 201–209.

AHRENS, L. (1954a), *The lognormal distribution of the elements-II*, Geochim. Cosmochim. Acta 6, 121–131.

AHRENS, L. (1954b), *The lognormal distribution of the elements (a fundamental law of geochemistry and its subsidiary)*, Geochim. Cosmochim. Acta 5, 49–73.

AITCHISON, J. (1982), *The statistical analysis of compositional data (with discussion)*, J. R. Stat. Soc. Ser. B-Stat. Methodol. 44(2), 139–177.

AITCHISON, J. (1983), *Principal component analysis of compositional data*, Biometrika 70(1), 57–65.

Reprinted from the journal

AITCHISON, J. (1986), The Statistical Analysis of Compositional Data. Chapman and Hall, London, 463 pp. Reprint of first edition by The Blackburn Press, 2003.

AITCHISON, J. and BROWN, J. (1954), *On criteria for descriptions of income distribution*. Metroeconomica 6, 88–98.

AITCHISON, J., MATEU-FIGUERAS, G. and NG, K. (2003), *Characterization of distributional forms for compositional data and associated distributional tests*, Math Geol 35(6), 667–680.

ALLÉGRE, C. and LEWIN, E. (1995), *Scaling laws and geochemical distribution*, Earth Planet. Sci. Lett. *132*, 1–13.

BILLHEIMER, D., GUTTORP, P. and FAGAN, W. (2001), *Statistical interpretation of species composition*, J. Am. Stat. Assoc. 96(456), 1205–1214.

BUCCIANTI, A. (2011), Natural laws governing the distribution of the elements in geochemistry: The role of the log-ratio approach, in Pawlowsky-Glahn, V. and Buccianti, A. (Eds.). Compositional Data Analysis: Theory and Applications, Wiley ch 18, 255–266.

BUCCIANTI, A. (2013), *Is compositional data analysis a way to see beyond the illusion?*, Compt. and Geosc. 50, 165–173.

BUCCIANTI, A. and MAGLI, R. (2011), *Metric concepts and implications in describing compositional changes for world rivers water chemistry*, Comput. Geosci. 37(5), 670–676.

BUCCIANTI, A., MATEU-FIGUERAS, G. and PAWLOWSKY-GLAHN, V. (2006), Compositional Data Analysis in the Geosciences. From theory to practice, Geological Society, London, Special Publication 264.

CHAYES, F. (1960), *On correlation between variables of constant sum*, J. Geophys. Res. 65(12), 4185–4193.

CHENG, Q., AGTERBERG, F.P. and BALLANTYNE, S.B. (1994), *The separation of geochemical anomalies from background by fractal methods*, J. Geochem. Explor. *51*, 109–130.

CLARKE, F. (1889), *The relative abundance of the chemical elements*, Philosofical Society of Washington Bulletin *11*, 131–142.

CLARKE, F. (1924), *The composition of the earth crust*, USGS Professional paper *127*,1–117.

COUSENS, B.L., CHASE, R.L. and SCHILLING J.G. (2011). Geochemistry and origin of volcanic rocks from Tuzo Wilson and Bowie seamounts, northeast Pacific Ocean, Canadian Journal of Earth Sciences 22(11), 1609–1617.

DASZYKOWSKI, M., KACZMAREK, K., VANDER HEYDEN, Y., WALCZAK, B. (2007), *Robust statistics in data analysis—A review. Basic Concepts*, Chemom. and Intell. Lab. System. 85, 203–219.

EGOZCUE, J.J. and PAWLOWSKY-GLAHN, V. (2005), *Groups of parts and their balances in compositional data analysis*, Mathem Geol 37(7),795–828.

EGOZCUE, J.J. and PAWLOWSKY-GLAHN, V. (2006), Simplicial geometry for compositional data, in: Buccianti, A., Mateu-Figueras G. and Pawlowsky-Glahn V. (Eds.). Compositional Data Analysis in the Geosciences: from theory to practice Special Publication 264, Geological Society, London, 12–28.

EGOZCUE, J.J., PAWLOWSKY-GLAHN, V., MATEU-FIGUERAS, G. and BARCELO-VIDAL, C. (2003), *Isometric logratio transformations for compositional data analysis*, Math Geol 35(3), 279–300.

EVERITT B. and HOTHORN T. (2011), An introduction to Applied Multivariate Analysis with R, Springer, Berlin.

FEDERER, J. (1988), Fractals Plenum, New York, p. 283.

FILZMOSER, P., HRON, K. and REIMANN, C. (2009a), *Univariate statistical analysis of environmental (compositional) data: problems and possibilities*, Sci. Total Environ. 407, 6100–6108.

FILZMOSER, P., HRON, K. and REIMANN, C. (2009b), *Principal component analysis for compositional data with outliers*, Environmetrics 20(6), 621–632.

FILZMOSER, P. and TODOROV, V. (2013), *Robust tools for the imperfect world*, Information Sciences 245(1), 4–20.

FRISCH, U. (1995), Turbulence, The Legacy of A.N. Kolmogorov, Cambridge University Press.

GOLDSCHMIDT, V.M. (1933), *Grundlagen der quantitativen geochimie*, Fortschrift Mineralogie *17*(2), 112–156.

GONCALVES, M.A. (2001), *Charactersization of geochemical distributions using multifractals models*, Math. Geosc. *33*, 41–61.

HOFMANN, A. (2003), Sampling mantle eterogeneity through oceanic basalts: Isotopes, vol 2. in Treatise on Geochemistry, The mantle and Core, Carlson R.W. (Ed.), Elsevier, Amsterdam.

HUBERT, M., ROUSSEEUW P.J., and VANDEN BRANDEN K. (2005), *ROBPCA: A new approach to robust Principal Component Analysis*, Technometrics 47(1), 64–79.

JENNER F., O'NEILL H. (2012), *Analysis of 60 elements in 616 ocean floor basaltic glasses*, Geochem. Geophys. Geosyst. *13*(1), 1–11.

KAPTEYN, J. (1903), Skew frequency distributions in biology and statistics, Astronomical Laboratory, Groningen, Noordhoff.

KONDEPUDI, D. and PRIGOGINE, I. (1999), Modern thermodynamics, Wiley, Chichester, p. 486.

LE MAITRE, R.W. (1968), *Chemical variation within and between volcanic rocks series—a statistical approach*, J. Petrol. 9, 220–252.

LI, C., MA, T. and SHI, J. (2003), *Application of a fractal method relating concentrations and distances for separation of geochemical anomalies from background*, J. Geochem. Explor 77, 167–175.

LIEBOVITCH, L. and SCHEURLE, D. (2000), *Two lessons from fractals and chaos*, Complexity 5(4), 34–43.

MA, T., LI, C. and LU, Z. (2014), *Estimating the average concentration of minor and trace elements in surficial sediments using fractal methods*, J. Geochem. Explor. *139*, 207–216.

MATEU-FIGUERAS, G. and PAWLOWSKY-GLAHN, V. (2008), *A critical approach to probability laws in geochemistry*, Math Geosci. 40, 489–502.

MATEU-FIGUERAS, G., PAWLOWSKY-GLAHN, V. and BARCELO-VIDAL, C. (2005), *The additive logistic skew-normal distribution on the simplex*, Stoch. Environ. Res. Risk Assess. 19(3), 205–214.

MATEU-FIGUERAS, G., PAWLOWSKY-GLAHN, V. and EGOZCUE, J.J. (2013), *The normal distribution in some constrained sample spaces*, SORT 37, 29–56.

MELSON, W., O'HEARN, T. and JAROSEWICH, E. (2002), *A data brief on the Smithsonian abyssal volcanic glass data file*, Geochem. Geophys. Geosyst. 3(4), 2001GC000249.

MITZENMACHER, M. (2004), *A brief history of generative models fro power law and lognormal distributions*, Internet Mathematics *1*(2), 226–251.

OTT, W. (1990) *A physical explanation of the lognormality of pollutant concentrations*, Journal of Air and Waste Management Association 40(10), 1378–1383.

PAWLOWSKY-GLAHN, V. and BUCCIANTI, A. (2011), Compositional Data Analysis. Theory and Applications, Wiley, London.

PAWLOWSKY-GLAHN, V. and EGOZCUE, J. (2001), *Geometric approach to statistical analysis on the simplex*, Stoch. Environ. Res. Risk Assess. 15(5), 384–398.

PEARSON, K. (1897), *Mathematical contributions to the theory of evolution. On a form of spurious correlations which may arise when indices are used in the measurements of organs*, Proc. R. Soc. London 60, 489–498.

R DEVELOPMENT CORE TEAM (2007). R: A Language and Environment for Statistical Computing. R Foundation for Statistical Computing, Vienna, ISBN 3-900051-07-0.

REIMANN, C. and FILZMOSER, P. (1999), *Normal and lognormal data distribution in geochemistry: death of a myth. Consequences for the statistical treatment of geochemical and environmental data*, Environ. Geol. *39*(9), 1001–1014.

ROLLINSON, H.R. (1993), Using geochemical data: evaluation, presentation, interpretation, Pearson Prentice Hall, England.

ROLLINSON, H.R. (1992), *Another look at the constant sum problem in geochemistry*, Mineral. Magaz. *56*(385), 469–475.

ROUSSEEUW, P.J. (1984), *Least Median of Squares Regression*, Journal of the American Statistical Association *79*, 871–880.

TEMPL, M., HRON, K. and FILZMOSER, P. (2011), robCompositions: An R-package for robust statistical analysis of compositional data, in: Compositional Data Analysis: Theory and Applications (Pawlowsky-Glahn and Buccianti eds.), Wiley, 341–355.

TODOROV, V. and FILZMOSER, P. (2010), *Robut statistics for one-way MANOVA*, Computational Statistics and Data Analysis *54*, 37–48.

VAN DEN BOOGAART, K.G. and TOLOSANA-DELGADO R. (2013), Analyzing Compositional Data with R, Springer, Berlin.

VERBOVEN S. and HUBERT M. (2005), *LIBRA: a MATLAB library for robust analysis*, Chemom. and Intell. Labor. System. *75*, 127–136.

WEDEPOHL, H. (1955), *The composition of the Earth crust*, Geochim. Cosmochim. Acta *59*, 1217–1239.

WEHRENS, R. (2011), Chemometrics with R. Multivariate Analysis in the Natural Sciences and Life Sciences, Springer, Berlin.

XIE, X. and YIN, B. (1993), *Geochemical patterns from local to global*, J. Geochem. Explor. *47*, 109–129.

(Received April 11, 2014, revised October 20, 2014, accepted October 20, 2014, Published online November 19, 2014)

Reprinted from the journal

Pure Appl. Geophys. 172 (2015), 1865–1878
© 2014 Springer Basel
DOI 10.1007/s00024-014-0939-z

❚Pure and Applied Geophysics

Aspects of Structure in Earthquake Networks

MIRELA SUTEANU[1]

Abstract—Analysis performed on multiple sets of earthquake networks created for the Hawaii volcanic system reveals characteristics that can be associated with fundamental properties of seismicity. The scale-free behaviour of the connectivity distribution along the spectrum of the minimum weight values, which can be used to discern the interrelated earthquakes from the rest of the data set, is mirrored by a similar behaviour of the distribution of the number of linked neighbours. The patterns found in the distributions of temporal and spatial intervals between earthquakes are similar from large to small networks. Similarities are found between the variation of the network clustering coefficient, C, and the variation of the exponents of the connectivity distribution, β, and of the weight distribution, γ; their synchronous variation over successive temporal windows can be related to changes in seismicity and in the life of the volcanic system. A Zipf distribution is found for the ranked sets of magnitude values of successive network nodes. The distribution of differences between the magnitude values of successive nodes is also governed by a power law.

Key words: Seismicity, earthquakes, networks, Hawaii volcanoes, nonlinear systems, scaling properties.

1. Introduction

Current research dedicated to understanding seismicity and its ruling phenomena illustrates various aspects of the correlations found in earthquake patterns. Over the past decade, approaches based on complex networks revealed that networks built on earthquake data enjoy scaling properties (BAK *et al.* 2002; BAIESI and PACZUSKI 2004, 2005; DAVIDSEN *et al.* 2008; ZALIAPIN *et al.* 2008; ABE and SUZUKI 2012; TENENBAUM *et al.* 2012; ZALIAPIN and BEN-ZION 2013a, b; SUTEANU 2014). This article presents an analysis of earthquake networks introduced by SUTEANU (2014) and used for the study of volcanic

seismicity in Hawaii. An objective of this research is to determine whether the distributions of the spatial and temporal distances between connected nodes have a similar pattern in different networks, and, if that is the case, whether the pattern enjoys scaling properties. Since the network average clustering coefficient (C), the exponent of the connectivity distribution (β), and the exponent of the weight distribution (γ) are global parameters, which characterise the network as a whole, we investigate whether any coherent relationship can be found between them. We also inquire about whether distributions of magnitude values for successive nodes of a network enjoy specific properties, and, if that is the case, whether those properties are characteristic of all networks.

2. The Earthquake Networks

In this analysis we generate directed weighted networks of earthquakes to identify interrelated events following the method in SUTEANU (2014). The source of data is the Advanced National Seismic System (ANSS) catalog for the Big Island of Hawaii between 1 January 1989 and 31 December 2012 (64,392 earthquakes). The minimum magnitude value for catalog completeness was established in two different ways. By determining local slopes for the frequency-magnitude graph (MARAUN *et al.* 2004), a minimum magnitude value of 1.6 was found (Fig. 1b), and a b value of 0.99 ± 0.04 with 95 % confidence level was determined. By applying the maximum likelihood method (CLAUSET *et al.* 2009), the value found for the minimum magnitude was higher: 1.95; in this case, the value determined for b was 1.01 ± 0.01 with 95 % confidence level. This increase in the magnitude threshold leads to a sharp

[1] Department of Mathematics and Computing Science, Saint Mary's University, 923 Robie St., Halifax, NS B3H 3C3, Canada. E-mail: mirela.suteanu@gmail.com

Reprinted from the journal

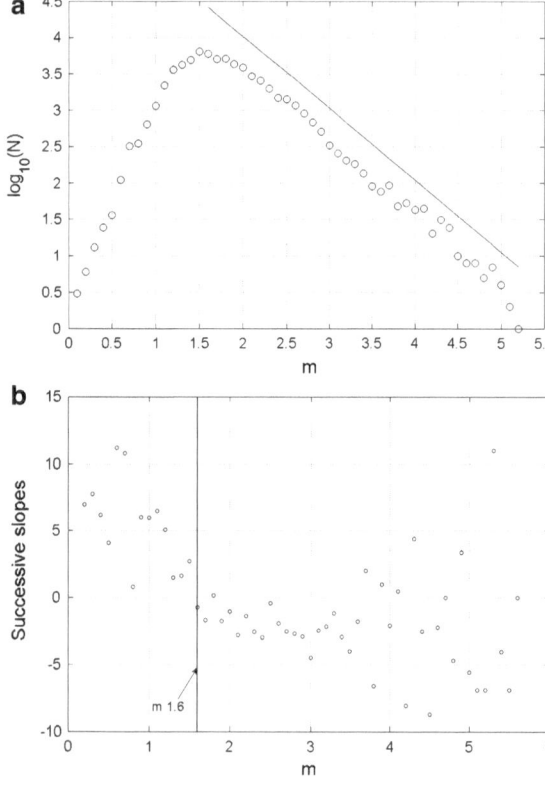

Figure 1

a Magnitude-frequency distribution for the Hawaii data set ranging from 1 January 1989 to 31 December 2012. The *line* corresponds to the *b* value obtained for magnitudes $m \geq 1.6$ ($b = 0.99 \pm 0.04$ with 95 % confidence level). **b** Local slopes determined for the magnitude-frequency distribution with a magnitude bin size $\Delta m = 0.1$. The selected threshold is marked by a *thick vertical line*

drop in the number of events included in the analysis: if the number of earthquakes is 37,451 for a minimum magnitude of 1.6, the total number of events drops to 17,917 for a minimum magnitude of 2. The network approach was applied to catalogs considered with both of the minimum value boundaries, $m \geq 1.6$ and $m \geq 2$ (Table 1). The same qualitative results were found for both thresholds.

The epicenters of the earthquakes are considered nodes of the networks and they are linked by directed weighted edges, with directions given by the temporal succession of the events. Similarly to the integrative approach introduced by BAK *et al.* (2002) and developed by BAIESI and PACZUSKI (2004, 2005), this article considers that the edges carry space-time-magnitude weights. However, there are major differences between our approach and Baiesi and Paczuski's

model, as discussed in SUTEANU (2014). The edge weight, W, is a combination of a variable in space, a variable in time, and a variable in magnitude. All three weight components are seen independently, as separate ingredients with comparable contributions to the total edge weight and a maximum value limited to 1. This approach is particularly important in the case of volcano-tectonic seismicity, where the seismic sources are diverse (tectonic stress, thermodynamic processes, dynamics of gas, fluid and solid).

Given the omnipresence of scaling relationships in earthquake distributions (OMORI 1894; UTSU 1961; KAGAN 1994; NANJO and NAGAHAMA 2000; LAPENNA *et al.* 2000; SHCHERBAKOV *et al.* 2004; CARBONE *et al.* 2005; FELZER and BRODSKY 2006; SHCHERBAKOV *et al.* 2006; LENNARTZ *et al.* 2008; BUNDE and LENNARTZ 2012; LIPPIELLO *et al.* 2012; VAROTSOS *et al.* 2012), power laws are used for the quantitative estimation of the relationships between earthquakes in space and time.

(a) Distance weight:

$$w_d = cd^r, \; r < 0, \tag{1}$$

where d is the spatial distance between the two nodes of an edge measured in km, and c is a positive constant.

(b) Time weight:

$$w_t = st^p, p < 0, \tag{2}$$

where t is the time interval between the two nodes of an edge measured in hours, and s is a positive constant.

Assuming that all earthquakes that are very close in space or in time could be related to each other, and in order to avoid singularities, a small cutoff value is used for the weights in space and time. Therefore, the method uses modified forms of Eqs. (1) and (2):

$$w_d = \begin{cases} 1, & d \leq d_{min} \\ cd^r, & d \geq d_{min}, \; r < 0 \end{cases} \tag{1'}$$

$$w_t = \begin{cases} 1, & t \leq t_{min} \\ st^p, & t \geq t_{min}, \; p < 0 \end{cases} \tag{2'}$$

The constants c and s are calculated using the boundary conditions:

Table 1

All classes

Class	T_{max} (days)	D_{max} (km)	r	p	t_{min} (h)	d_{min} (km)	H	L
B	10	30	-1.35	-1	1	1	1.00	1.01×10^{-5}
C	10	30	-1.35	-1	0.5	0.2	0.65	5.76×10^{-7}
D	30	30	-1.35	-1	1	1	1.00	3.39×10^{-6}
E	40	50	-1.35	-1	0.05	0.2	6.07	7.24×10^{-9}
F	7	10	-1.35	-1	0.05	0.1	2.72	1.44×10^{-7}
G	7	10	-1.35	-1	0.05	0.025	2.72	2.22×10^{-8}
H	8	10	-1.35	-1	0.5	0.2	0.65	3.19×10^{-6}
I	8	11	-1.35	-1	0.05	0.1	2.72	1.11×10^{-7}
J	8	10	-1.35	-1	1	1	1.00	5.60×10^{-5}
L	7	10	-1	-0.5	14	2	1.00	1.39×10^{-2}
M	7	10	-1	-0.5	1	1	1.00	1.86×10^{-3}
N	40	50	-1	-0.5	0.5	0.2	0.66	2.19×10^{-5}
O	50	50	-2	-2	1	1	1.00	6.75×10^{-11}
P	30	30	-0.5	-1.5	1	1	1.00	2.27×10^{-6}
R*	40	50	-1.6	-2	1	1	1.00	6.31×10^{-10}
S*	50	50	-2	-1.8	1	1	1.00	3.49×10^{-10}
T*	9	11	-1.37	-0.51	1	1	1.00	7.19×10^{-4}

T_{max} is the maximum time interval between events, D_{max} is the maximum distance between events, r is the exponent of the distance weight w_d (Eq. 1), p is the exponent of the time weight w_t (Eq. 2), d_{min} and t_{min} represent cutoff values, H is the highest value of the total edge weight in the class, and L is the lowest value of the total edge weight in the class

The classes tagged with * have a minimum magnitude threshold of 2, while all other classes have a minimum magnitude of 1.6

$$w_d = cd_{min}^r = 1 \qquad (1'')$$

and

$$w_t = st_{min}^p = 1 \qquad (2'')$$

(c) The magnitude weight is proportional to the magnitude of the first occurring event of the edge:

$$w_m = \frac{m}{m_{max}}, \qquad (3)$$

where m is the magnitude of the first occurring earthquake associated with the edge, and m_{max} is the maximum magnitude value in the data set.

The total weight of an edge is given by the product of the weights in space, time, and magnitude, but only the nodes that carry enough weight are selected in the network of correlated events; therefore, only edges with a total weight W higher than a minimum threshold W_{min} are chosen for this network:

$$W = \begin{cases} w_d \cdot w_t \cdot w_m, & W \geq W_{min} \\ 0, & W < W_{min} \end{cases} \qquad (4)$$

This definition allows various combinations of space-time-magnitude correlations between any two events: any node can have any number of predecessors and any number of successors, as long as its edges carry enough total weight. An example of an earthquake network is illustrated in Fig. 2.

A maximum interval of influence in time T_{max} and a maximum interval of influence in space D_{max} are assigned in order to simplify the computation, but, hypothetically, T_{max} and D_{max} may cover the whole extent of the catalog: large spatial and temporal intervals between events produce very small values of the edge weights w_d and w_t, and, implicitly, of the total edge weight W; the resulting weak links are eventually eliminated by the network definition (Eq. 4).

The network classes are sets of earthquake networks that share the same values of the parameters D_{max}, r, d_{min}, T_{max}, and p, t_{min}. A summary of all classes that have been studied is presented in Table 1. An initial network is created in every class when specific values are assigned to the above parameters, but, in general, this first network does not hold predominantly correlated earthquakes; it mainly

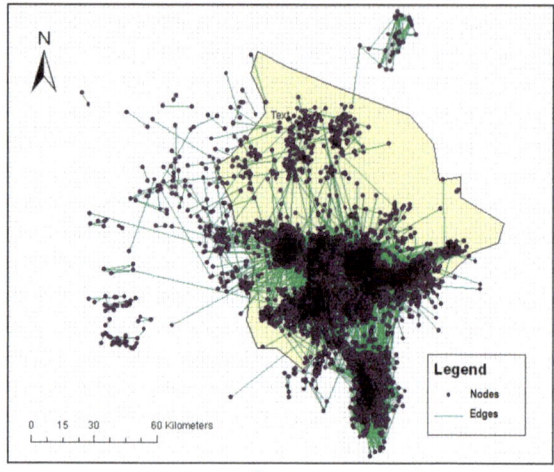

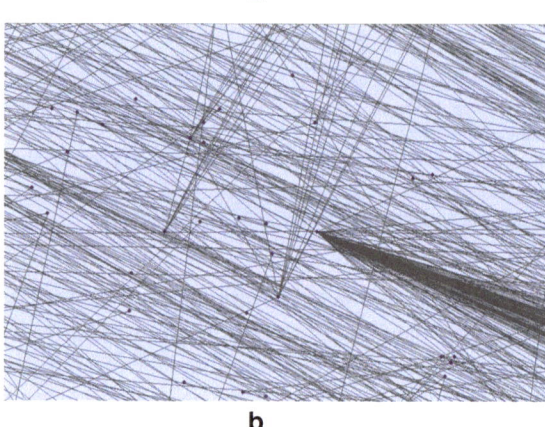

Figure 2
The earthquake network. **a** An example for the Big Island of Hawaii. **b** Zoomed-in example of *nodes* and *edges*

represents a collection of events that serves for the initiation of the method. An index 0 is used to differentiate the initial networks $B0$, $C0$, ..., $P0$ from the rest of the networks. Inside each class, a series of networks is generated by setting various threshold values for the minimum edge weight W_{min}.

In SUTEANU (2014) network parameters such as node connectivity distribution, node weight distribution (BOCCALETTI *et al.* 2006), and the exponents of these distributions, β and respectively γ (ALBERT and BARABÁSI 2002), are assessed for sets of earthquake networks in different classes. The study showed that networks having W_{min} in the middle to the upper range of the interval between the lowest edge weight value L and the highest edge weight value H in their class exhibit significant scaling properties of the node connectivity distributions, while networks with low

values of W_{min} exhibit poor or no such characteristics (Fig. 3), in agreement with the results of SUTEANU (2014). Since the higher values of weight in the strong links are associated with the earthquakes that are most likely to be related with each other, a relationship can be discerned between the major characteristics of seismicity and the structured, power law characteristics in the distributions of node connectivity.

For networks that exhibit obvious scaling properties when assessed with linear regression as in Figs. 3 and 4, a better assessment can be further performed using the maximum likelihood method (CLAUSET *et al.* 2009). An example is shown in Fig. 5a, b for the connectivity and weight distributions in network M2.

This behaviour along the spectrum of W_{min} is the same in networks with the minimum value of magnitude 1.6 and in networks with the minimum value of magnitude 2, and is robust with respect to variations of parameters r and p within the limits found in temporal and spatial distributions of earthquakes (OMORI 1894; UTSU 1961; UTSU *et al.* 1995; FELZER and BRODSKY 2006; SHCHERBAKOV *et al.* 2005, 2006; DAVIDSEN *et al.* 2008; LIPPIELLO *et al.* 2009, 2012; LENNARTZ *et al.* 2011). Since p values usually range between 0.7 and 1.8 (UTSU 1961), in this study p values were meant to cover this interval and therefore they were chosen from the larger interval [0.5, 2]. Also, since r values of 1.35–1.37 have been observed for distance distributions (FELZER AND BRODSKY 2006), or found to be equal to the earthquake's fractal distribution in the studied area (BAIESI and PACZUSKI 2004), in this article r values were chosen from the larger interval [0.5, 2] (Table 1).

The identification of correlated earthquakes may start with a certain choice of parameter values (the initial network) from the range of values exemplified previously, and, after setting increasing thresholds for the minimum weight W_{min}, end when networks with scale-free properties are found. An important aspect of this method is that, although different parameter values may be chosen initially, the same statistical population of interrelated earthquakes is identified in the end (SUTEANU 2014).

Our present analysis shows that the distributions of the spatial and temporal intervals between

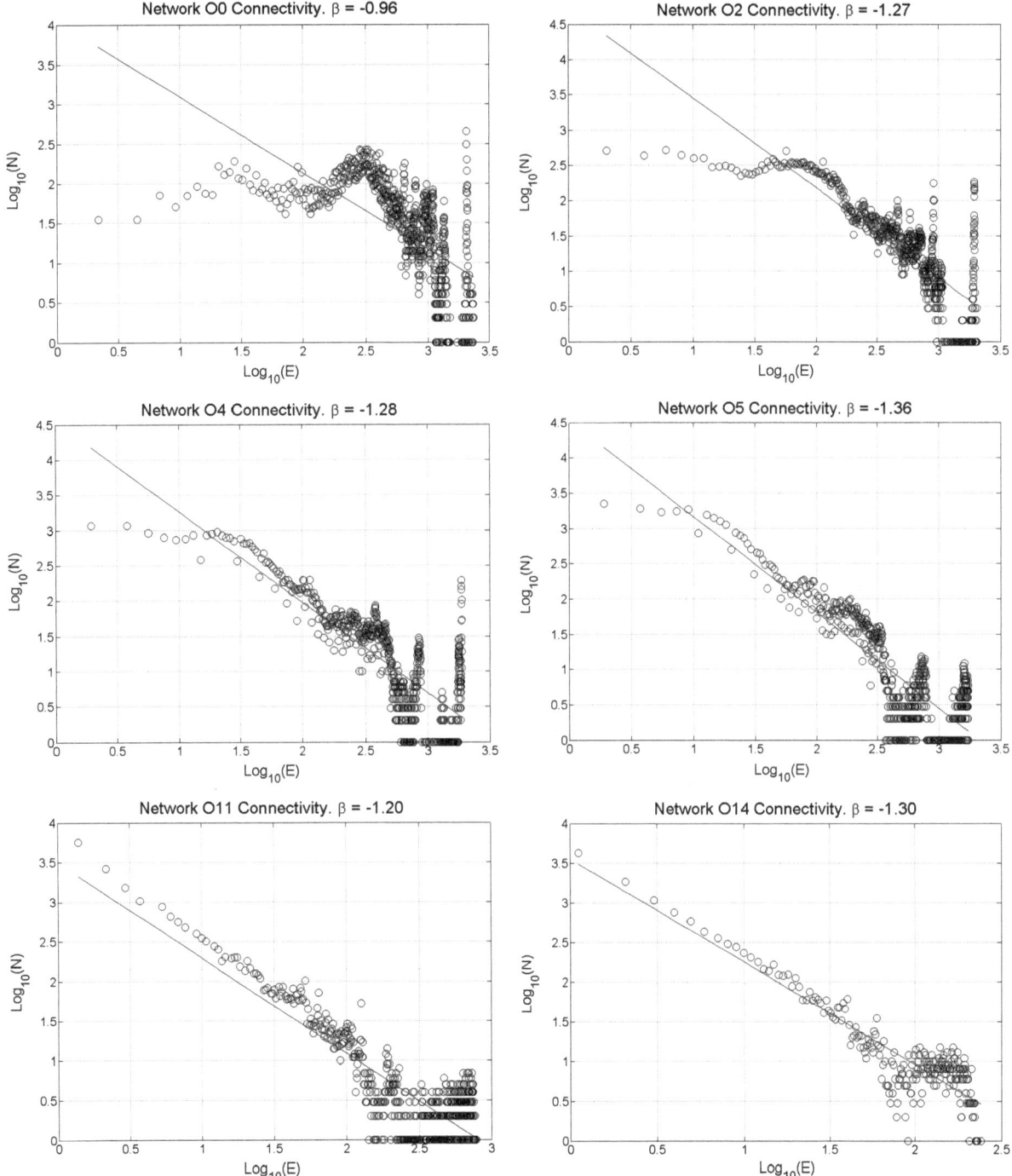

Figure 3
Connectivity distribution for networks O0, O2, O4, O5, O11, and O14. *N* (on the *Y*-axis) is the number of nodes that have E edges (on the *X*-axis)

connected nodes have a similar pattern and scaling properties in all networks, from small to large networks, from networks inside the same class to networks from different classes. An example is shown for three networks in class E (Table 2): Fig. 4 presents the connectivity distribution in the three

89

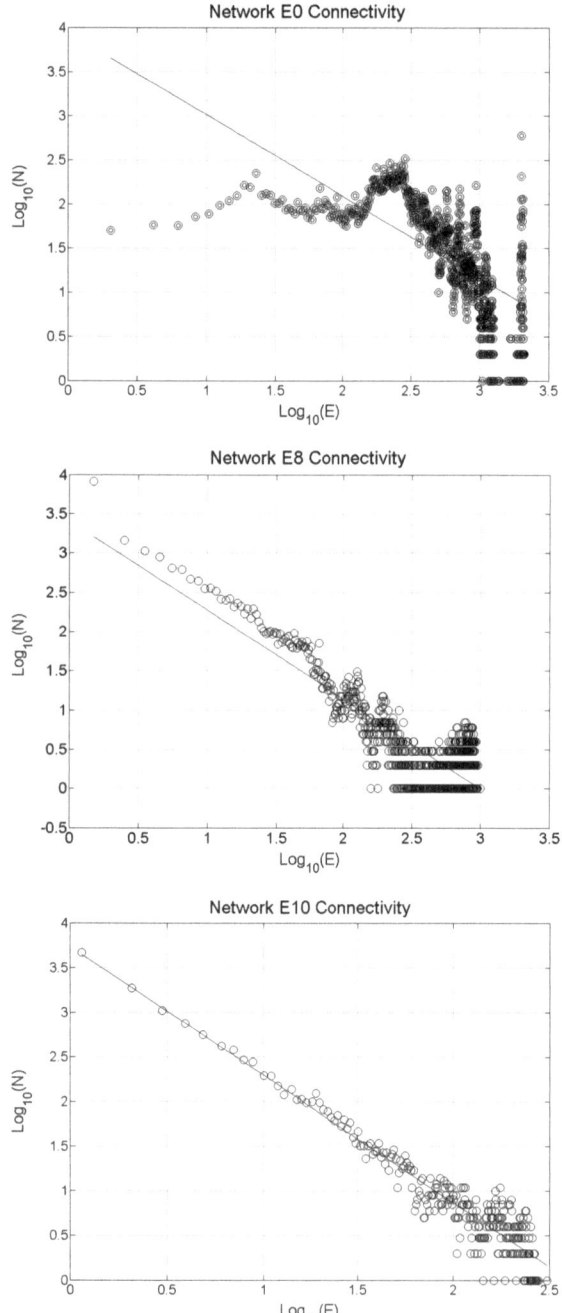

Figure 4
Connectivity distribution for networks E0, E8, and E10. N (on the Y-axis) is the number of nodes that have E edges (on the X-axis)

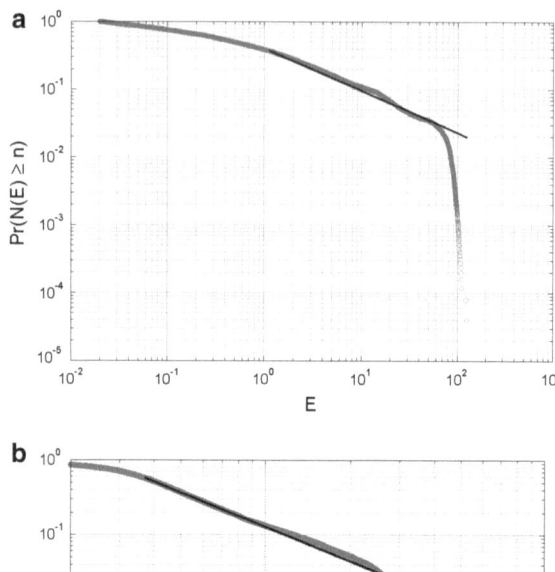

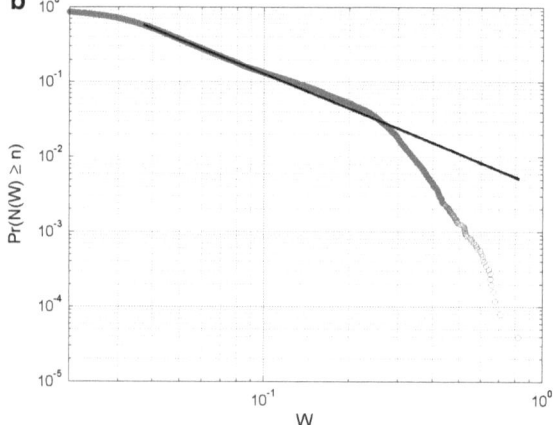

Figure 5
Maximum likelihood method applied to network M2: cumulative density function for **a** connectivity, with $\beta = -0.62 \pm 0.005$, and **b** weight, with $\gamma = -1.52 \pm 0.011$, at 95 % confidence level

networks, from the highly irregular distribution of the large initial network E0, in which most of the earthquakes are not necessarily correlated, to the scaling distributions of the networks E8 and E10. The left column in Fig. 6 shows how the distribution of distances between events changes from E0 to E8 and E10. The three distributions exhibit similarities with each other and with distributions found in other studies (DAVIDSEN *et al.* 2008; LIPPIELLO *et al.* 2009): an increase up to a maximum value, followed by a power law decrease. While the maximum value is ~ 1.2 km in E0, it shifts towards lower values in E8 (~ 0.7 km) and E10 (~ 0.3 km). The peaks at 13, 18, 23 km... in the distribution of E0 may be attributed to the spatial relations among events that are clustered around neighbouring volcanoes, distinct vents, and fracture zones (see Fig. 2a); these peaks are not present in the distributions of E8 and E10, suggesting that these networks have fewer nodes correlated over long distances.

Table 2

Class E networks

Class E definition	Range of total weight W values	Network		Number of nodes	Number of edges
		Name	W_{min}		
$T_{max} = 40$ days $D_{max} = 50$ km $r = -1.35$ $p = -1$ $d_{min} = 0.2$ km $t_{min} = 3$ min	$H = 6.07$ $L = 7.24 \times 10^{-9}$	E0	7.24×10^{-9}	37,441	8,488,767
		E8	5×10^{-5}	23,032	584,548
		E10	5×10^{-4}	14,091	117,835

E0 is the initial network that was generated using the parameter values shown in the first column. See Table 1 for the meaning of T_{max}, D_{max}, r, p, d_{min}, t_{min}, H, and L

The right column of Fig. 6 shows the change in time interval distributions from E0 to E8 and E10. Although the overall shape of the distribution is similar in the three networks and power law properties are present in all three of them, the time interval between events decreases significantly when the scaling properties of the connectivity distribution become stronger in networks E8 and E10. The increase between 7 and 15 days with a peak at 11 days from E0, which can be attributed to precursory sequences in Hawaii (CHASTIN and MAIN 2003), diminishes in E8, and completely disappears from E10, since E10 retains only the strongest links and the earthquakes that are extremely likely to be interrelated. These differences among the three networks illustrate the fact that in networks that exhibit increasingly strong scaling properties of the connectivity distribution, i.e., networks with nodes that are increasinly likely to be interrelated, the earthquakes selected for the networks are increasingly close in time and space. The same spatial and temporal patterns are observed for both thresholds of minimum magnitude of 1.6 and 2.

3. The Clustering Coefficient

A relevant measure of the extent to which the nodes in the network tend to group together is the clustering coefficient. If a node i has k_i neighbours, the local clustering coefficient C_i of the node i is defined as the actual number of links L_i between its neighbours divided by the total number of possible links between them (WATTS and STROGATZ 1998):

$$C_i = \frac{2L_i}{k_i(k_i - 1)} \qquad (5)$$

In this study, C_i is 0 for nodes with neighbours that do not have any link between them and for nodes with a node degree equal to 1.

In order to calculate the clustering coefficient, the number of links L_i between the neighbours of each node i were first calculated and assessed. The results show significant similarities between the change in the connectivity distributions towards a scaling structure and the change in the neighbours' link distributions, as illustrated in the examples from class O in Fig. 3 and 7. A description of these networks is presented in Table 3.

Figure 7 shows that the distribution of the number of links between nodes' neighbours L_i observed in the initial network O0 has a similar appearance as the connectivity distribution in O0 (Fig. 3): an approximately constant interval followed by an approximate power law tail and scattered dot patterns. Moreover, corresponding to the increasingly strong scaling properties of the connectivity distribution that appears in the networks with W_{min} in the upper range of the weight spectrum, a more prominent scaling structure seems to emerge in the L_i link distributions with the increase of the threshold W_{min}.

The network average clustering coefficient C was calculated as the average of all C_i:

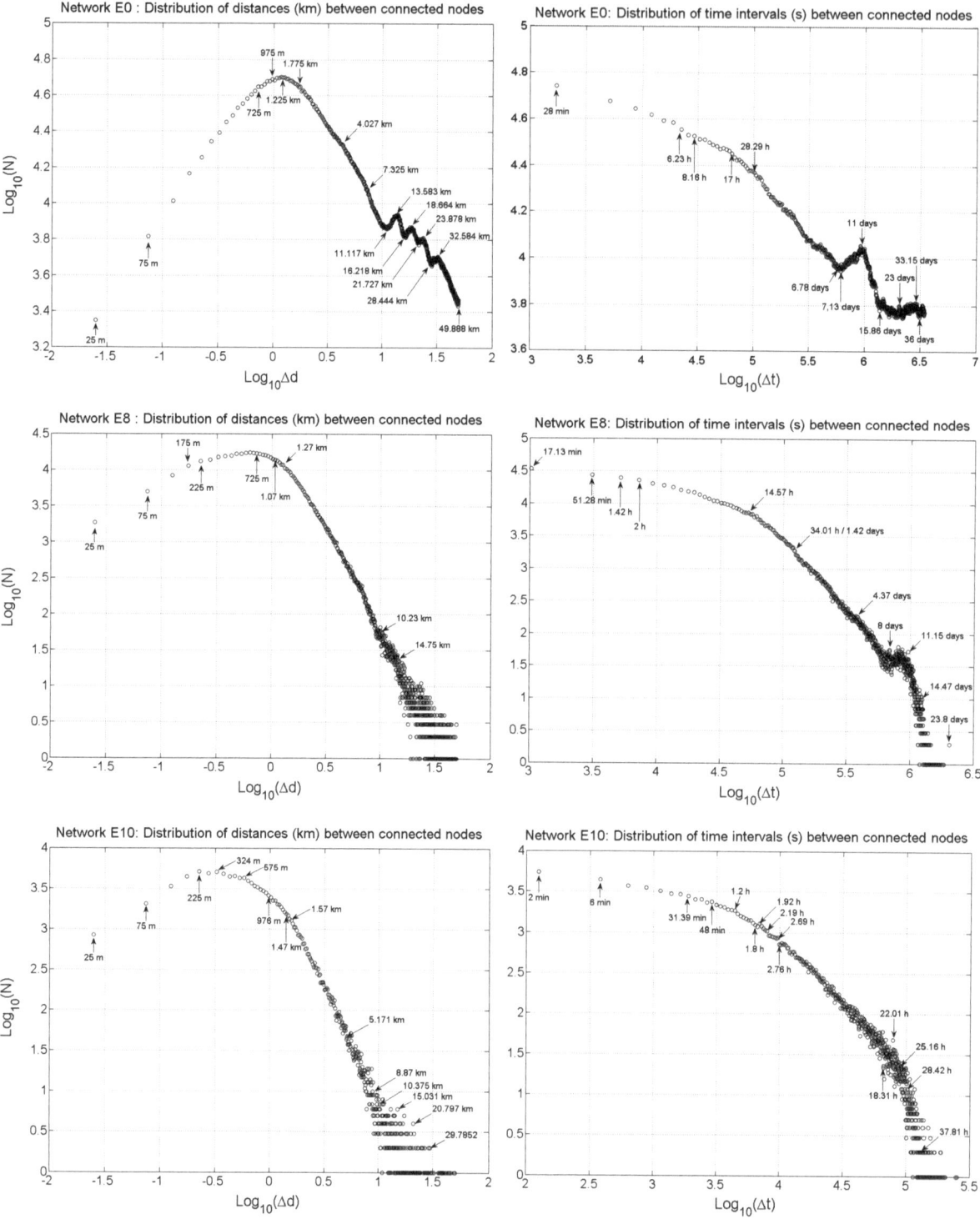

Figure 6

Left column distribution of distances (km) between any two nodes in networks E0, E8 and E10; *N* is the number of distances of Δ*d* km between any two earthquakes in each network. *Right column* distribution of time intervals (*s*) between any two nodes in the networks E0, E8, and E10; *N* is the number of time intervals of Δ*t* seconds between any two earthquakes in each network

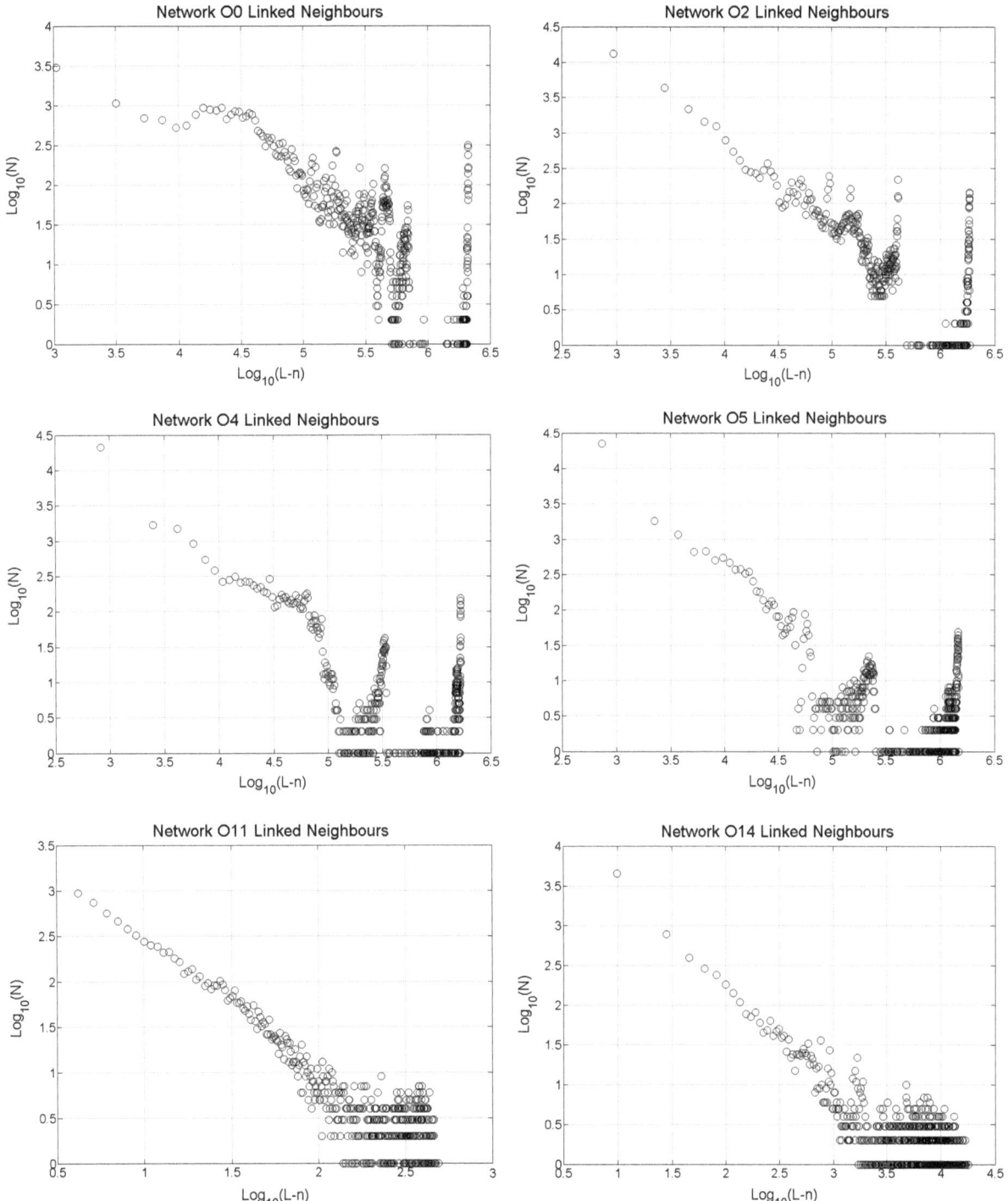

Figure 7

Distribution of actual links between the neighbours of each node for networks O0, O2, O4, O5, O11, and O14. N (on the Y-axis) is the number of nodes that have L-n ("linked-neighbour") links between their neighbours (on the X-axis)

Reprinted from the journal

Table 3

Class O networks

Class O definition	Range of total weight W values	Network		Number of nodes	Number of edges
		Name	W_{min}		
$T_{max} = 50$ days	$H = 1.00$	O0	6.75×10^{-11}	37,443	9,966,177
$D_{max} = 50$ km	$L = 6.75 \times 10^{-11}$	O2	10^{-8}	37,354	5,737,672
$r = -2$		O4	10^{-7}	37,019	3,826,589
$p = -2$		O5	5×10^{-7}	36,366	2,816,955
$d_{min} = 1$ km		O11	5×10^{-4}	21,996	444,071
$t_{min} = 1$ h		O14	10^{-2}	14,158	130,196

O0 is the initial network that was generated using the parameter values shown in the first column. See Table 1 for the meaning of T_{max}, D_{max}, r, p, d_{min}, t_{min}, H, and L

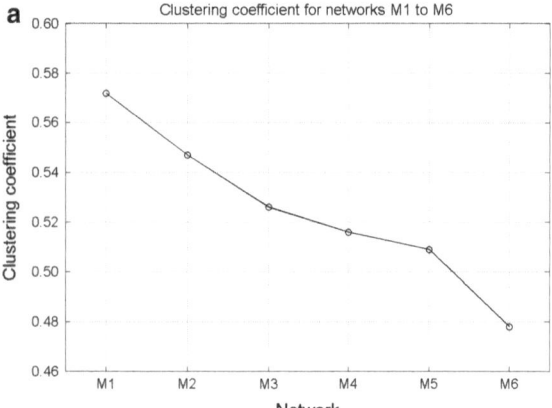

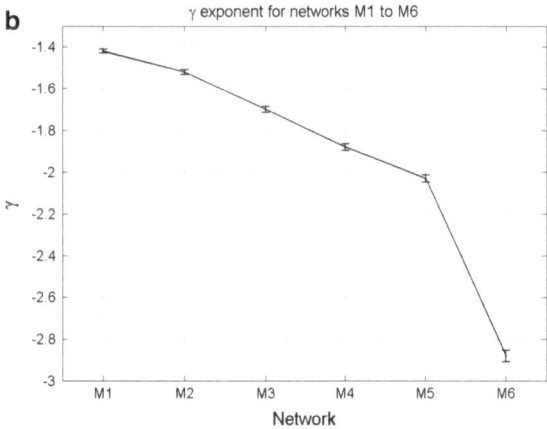

Figure 8

Networks M1 to M6: **a** the clustering coefficient and **b** the signed values of γ, the exponent of the weight distribution. The exponent γ was determined using the maximum likelihood method; *vertical bars* represent uncertainties (ranging from 0.01 to 0.028) at a 95 % confidence level

$$C = \frac{1}{N_{nodes}} \sum_{i=1}^{N_{nodes}} C_i, \qquad (6)$$

where N_{nodes} is the total number of nodes, and C_i is the clustering coefficient of node i (WATTS and STROGATZ 1998).

Compared with many other types of networks (NEWMAN 2003), rather high values of the network clustering coefficient (>0.4) are found in the earthquake networks studied in this article. High values of the network clustering coefficient have also been found by BAIESI and PACZUSKI (2005). Figure 8a shows an example of network clustering coefficient values for the networks in class M, which are described in Table 4. In this example, the corresponding values of the exponent of the weight distribution, γ, display a similar overall behaviour, as shown in Fig. 8b.

The variation of the exponents β and γ over successive event windows can be used to study the way in which relationships between earthquakes and volcanic processes change over time (SUTEANU 2014). The network is divided in successive event windows of various lengths. In the example presented in this article, the analysis was performed on network T5, in which the minimum magnitude value for catalog completeness is 2. The full description of class T is given in Table 5. Inside class T, network T5 was chosen since it exhibits scaling properties of the connectivity distribution; as previously discussed, this suggests that network T5 is populated with interrelated events. Since the number of events in

Table 4

Class M networks

Class M definition	Range of total weight W values	Network		Number of nodes	Number of edges
		Name	W_{min}		
$T_{max} = 7$ days	$H = 1$	M_0	1.86×10^{-3}	33,065	1,913,280
$D_{max} = 10$ km	$L = 1.86 \times 10^{-3}$	M1	1.50×10^{-2}	26,919	1,270,458
$r = -1$		M2	2×10^{-2}	25,396	1,088,015
$p = -0.5$		M3	3×10^{-2}	22,682	782,659
$d_{min} = 1$ km		M4	4×10^{-2}	20,440	584,686
$t_{min} = 1$ h		M5	5×10^{-2}	18,825	443,533
		M6	10^{-1}	14,006	149,235

M_0 is the initial network that was generated using the parameter values shown in the first column. See Table 1 for the meaning of T_{max}, D_{max}, r, p, d_{min}, t_{min}, H, and L

Table 5

Class T networks

Class T definition	Range of total weight W values	Network		Number of nodes	Number of edges
		Name	W_{min}		
$T_{max} = 9$ days	$H = 1$	T_0	7.19×10^{-4}	15,106	643,983
$D_{max} = 11$ km	$L = 7.19 \times 10^{-4}$	T1	10^{-3}	15,054	641,336
$r = -1.37$		T2	5×10^{-3}	13,108	468,361
$p = -0.51$		T3	7.5×10^{-3}	12,444	398,420
$d_{min} = 1$ km		T4	10^{-2}	11,976	348,196
$t_{min} = 1$ h		T5	2×10^{-2}	10,744	233,108
		T6	3×10^{-2}	9,738	167,709

T_0 is the initial network that was generated using the parameter values shown in the first column

networks with $m \geq 2$ is much smaller than in the case of networks with $m \geq 1.6$, the size chosen for successive windows in T5 was 500, as opposed to 1,000, which was used for M2 in (SUTEANU 2014). Equal-sized windows of 500 events and their corresponding networks were generated. The values of β and γ were estimated using the maximum likelihood method; uncertainties are low compared to the variation of the exponents (Fig. 9b, c; uncertainties at a confidence level of 95 % are represented as vertical bars), and the scale range is around one order of magnitude. The variation of the network clustering coefficient C in the corresponding temporal windows is shown in Fig. 9a. Since in this article we consider the signed values of β and γ, and not their absolute values, the maxima in Fig. 9b, c correspond to the minima discussed by SUTEANU (2014). The lower case letters from "a" to "h" are used to tag the maximum values of the network clustering coefficient C (Fig. 9a) and

of the exponent γ (Fig. 9b), and the corresponding areas on the cumulative number of earthquakes (see the graph in Fig. 10).

As shown in Fig. 9a–c, the overall variation of each of the three parameters over successive event windows exhibits strong similarities with the variation of the other two parameters: the majority of the local maxima of C match the local maxima of γ, and the shapes of the C and γ variations show notable similarities with the β variation.

In agreement with the results of SUTEANU (2014), the maxima in the graph of the temporal variation of the coefficient C (Fig. 9a) and of the exponent γ (Fig. 9b) can be associated with enhanced volcanic activity, major discharges of energy, changes in earthquake processes, etc. The synchronous maxima of the parameters show that, in the corresponding windows, there is an increase in the number of nodes that are highly interconnected and grouped together.

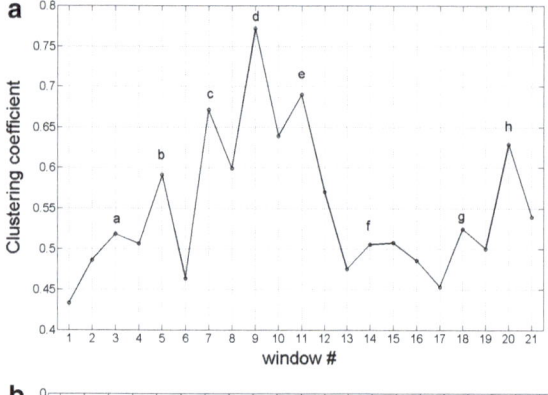

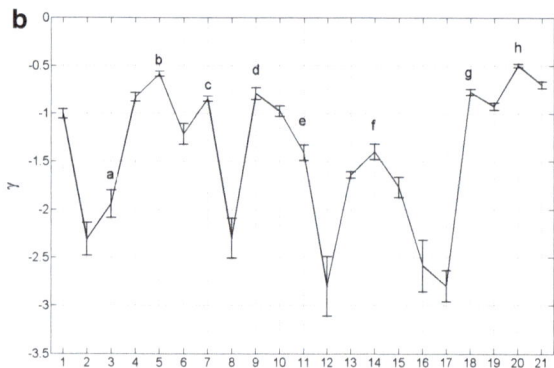

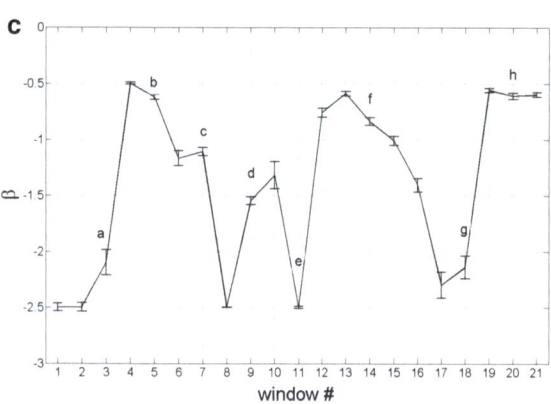

Figure 9

Successive temporal windows of network T5. **a** Variation of the network average clustering coefficient, C. **b** Variation of the weight distribution exponent, γ. **c** Variation of the connectivity distribution exponent, β. *Vertical bars* show the uncertainties at the 95 % confidence level obtained with the maximum likelihood method

We believe that the similarity in the variation of the three network parameters is another confirmation that the analysis we perform in successive temporal windows is able to reflect the change in the relationships between earthquakes over time.

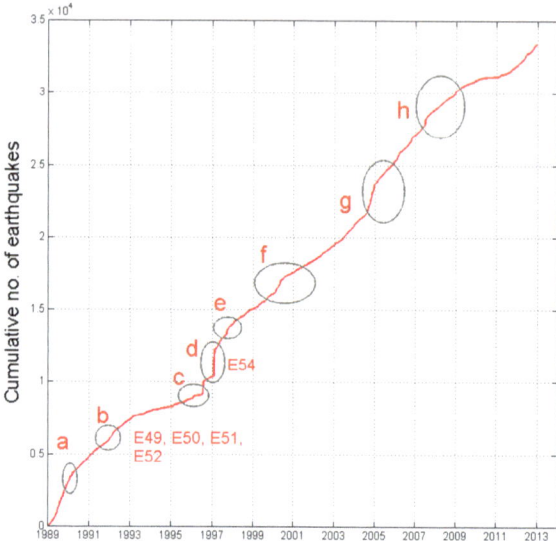

Figure 10

Cumulative number of earthquakes from January 1989 to December 2012. The *small letters* tag the areas corresponding to the maximum values of the network clustering coefficient C in successive temporal windows of network M2

4. Magnitude of Successive Nodes

All the sequences of values (m_i, m_j)—where i and j are nodes of magnitude m_i and m_j, respectively, and node i has a directed edge to node j—are ranked and assessed. This is accomplished by taking magnitude values of every two linked nodes i and j to create ordered pairs (m_i, m_j). The number of occurrences of every instance (m_i, m_j) is calculated, and the set of all occurrences is arranged in decreasing order and represented against the rank. Moreover, groups of three nodes are studied in a similar way: if the node j has a directed edge to the node k of magnitude m_k, then the sequences of values (m_i, m_j, m_k) are ranked and assessed. The results show that a Zipf law is found in both cases; the same results are found for both thresholds of minimum magnitude discussed in "The Earthquale Networks," $m \geq 1.6$ and $m \geq 2$. Examples are presented in Fig. 11: Fig. 11a shows the example of a network with strong scaling properties in the connectivity distribution, network D5, while Fig. 11b shows the example of a large initial network, network E0, which has an irregular distribution of node connectivity. This property can probably also be

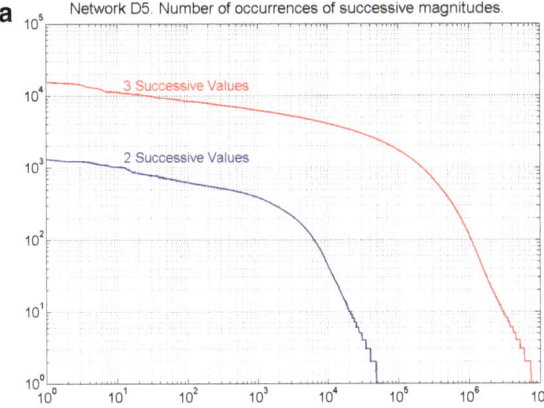

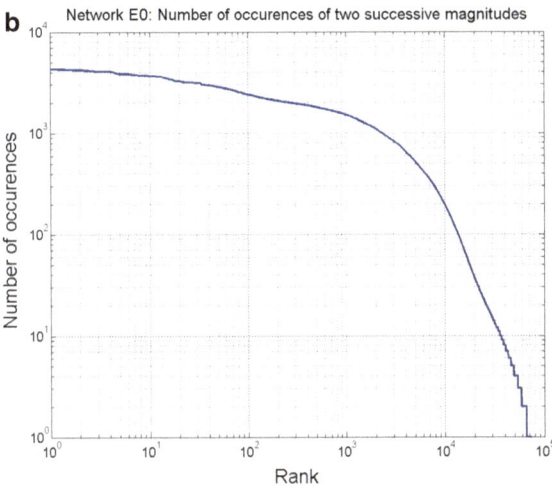

Figure 11

Magnitude values of successive nodes. **a** Magnitude of two and three successive nodes in a network that enjoys scaling properties of the connectivity distribution (network D5). **b** Pairs of magnitude values of successive nodes in a large initial network (network E0)

found for sequences of four, five, or more successive magnitude values. Since it is unlikely that the majority of nodes in E0 represent earthquakes that are correlated with each other (SUTEANU 2014), we believe that these results do not reflect characteristics of interconnected earthquakes, but rather root in the Gutenberg-Richter magnitude frequency distribution (GUTENBERG and RICHTER 1954).

5. Conclusions

The study of earthquake networks is able to reveal aspects of structure that can be reasonably associated with fundamental properties of seismicity. The scale-free behaviour of the connectivity distribution along the spectrum of W_{min}, which can be used to discern the interrelated earthquakes from the rest of the data set, is mirrored by a similar behaviour of the distribution of the number of node linked neighbours. This behaviour is robust with respect to variations in parameters within the range of values found in temporal and spatial distributions of earthquakes.

A similar pattern of increase up to a maximum followed by a power law decrease was found in the distribution of distances between earthquakes in a variety of networks: large networks, small networks, networks in the same class, or networks that belong to different classes. Similarly, an overall pattern of slow decrease followed by a power law decrease was found in the distribution of time intervals between events in various networks: networks inside each class, networks across the classes, as well as in the selective networks of interrelated events. In this latter case, the above distributions reflect the increased proximity in space and time of the correlated events.

In comparison to most of the biological, technological, or social networks, high values have been found for the network average clustering coefficient, which generally ranges between 0.4 and 0.6. A notable similarity was found between the variation of the network clustering coefficient, C, along the series of networks in a class, and the variation of the exponent of the the weight distribution, γ. Moreover, the synchronous variation of C and γ in successive temporal windows and the strong similarities with the variation of β can be related to changes in the characteristics of seismicity and in the life of the volcanic system.

The intrinsic scale-free structure of seismicity is also revealed in the Zipf distribution found for the ranked number of occurrences of groups of magnitude values of successive network nodes.

REFERENCES

ABE, S., and SUZUKI, N. (2012), *Dynamical evolution of the community structure of complex earthquake network*, EPL *99* (3) 39001, doi: 10.1209/0295-5075/99/39001.

ALBERT, R., and BARABÁSI, A.-L. (2002), *Statistical mechanics of complex networks*, Rev. Mod. Phys., *74*, 47–97.

BAIESI, M., and PACZUSKI, M. (2004), *Scale-free networks for earthquakes and aftershocks*, Phys. Rev. E, *69*, 066106-1–8.

BAIESI, M., and PACZUSKI, M. (2005), *Complex networks of earthquakes and aftershocks, Nonlinear Processes in Geophysics, 12,* 1–11.

BAK, P., CHRISTENSEN, K., DANON, L., and SCANLON, T. (2002), *Unified scaling law for earthquakes,* Phys. Rev. Lett., *88,* doi:10.1103/PhysRev Lett., 88, 178501.

BOCCALETTI, S., LATORA, V., MORENO, Y., CHAVEZ, M., and HWANG, D.-U. (2006), *Complex networks: structure and dynamics,* Phys. Reports, *424,* 175–308.

BUNDE, A., and LENNARTZ, S. (2012), *Long-term correlations in earth sciences,* Acta Geophys., *60,* 3, 562–588.

CARBONE, V., SORRISO-VALVO, L., HARABAGLIA, P., GUERRA, I. (2005), *Unified scaling law for waiting times between seismic events,* Europhys. Lett., *71,* 6, 1036–1042.

CHASTIN, S.F.M., and MAIN, I.G. (2003), *Statistical analysis of daily seismic event rate as a precursor to volcanic eruptions,* Geophys. Res. Lett., *30,* 13, 1671. doi:10.1029/2003GL016900.

CLAUSET, A., SHALIZI, C.R. and NEWMAN, M.E.J. (2009), *Power-law distributions in empirical data,* SIAM Review 51(4), 661–703.

DAVIDSEN, J., GRASSBERGER, P., and PACZUSKI, M. (2008), *Networks of recurrent events, a theory of records, and an application to finding causal signatures in seismicity,* Phys. Rev. E, *77,* 066104.

FELZER, K. R., and BRODSKY, E. E. (2006), *Decay of aftershock density with distance indicates triggering by dynamic stress,* Nature, *441,* 735–738.

GUTENBERG, B., and RICHTER, C. F., Seismicity of the Earth (Princeton University Press, Princeton 1954).

KAGAN, Y.Y. (1994), Observational evidence for earthquakes as a nonlinear dynamic process, Elsevier Physica D, Nonlinear Phenomena, *77,* 1-3, 160–192.

LAPENNA, V., MACCHIATO, M., PISCITELLI, S., and TELESCA L. (2000), *Scale invariance properties in seismicity of Southern Apennine Chain (Italy),* Pure Appl. Geophys., *157,* 4, 589–602.

LENNARTZ, S., LIVINA, V. N., BUNDE, A., and HAVLIN S. (2008), *Long-term memory in earthquakes and the distribution of interoccurrence times,* Europhys. Lett., *89,* 69001, doi:10.1209/0295-5075/81/69001.

LENNARTZ, S., BUNDE, A., and TURCOTTE, D.L. (2011), *Modelling seismic catalogues by cascade models: Do we need long-term magnitude correlations?,* Geophysical Journal International, *184,* 1214–1222.

LIPPIELLO, E., CORRAL, A., BOTTIGLIERI, M., GODANO, C., and DE ARCANGELIS, L. (2012), *Scaling behavior of the earthquake intertime distribution: Influence of large shocks and time scales in the Omori law,* Phys. Rev. E, *86,* 6–2, 066119, doi:10.1103/PhysRevE.86.066119.

LIPPIELLO, E., DE ARCANGELIS, L., and GODANO, C. (2009), *The role of static stress diffusion in the spatio-temporal organization of aftershocks,* Phys. Rev. Lett., *103,* 038501, doi:10.1103/PhysRevLett.103.038501.

MARAUN, D., RUST, H. W. and TIMMER, J. (2004), *Tempting Long-Memory – on the Interpretation of DFA Results,* Nonlinear Processes Geophys., *11,* 495–503.

NANJO, K., and NAGAHAMA, H. (2000), *Spatial distribution of aftershocks and the fractal structure of active fault systems,* Pure Appl. Geophys., *157,* 4, 575–588.

NEWMAN, M.E.J. (2003), *The Structure and Function of Complex Networks,* SIAM REVIEW, *45,* 2, 167–256.

OMORI, F. (1894), *On the aftershocks of earthquakes,* J. College of Science, Imperial University of Tokyo, *7,* 111–200.

SHCHERBAKOV, R., TURCOTTE, D.L., and RUNDLE, J.B. (2004), *A generalized Omori's law for earthquake aftershock decay,* Geophys. Res. Lett., *31,* L11613, doi:10.1029/2004GL019808.

SHCHERBAKOV, R., TURCOTTE, D.L., and RUNDLE, J.B. (2005), *Aftershock statistics,* Pure and Applied Geophysics, *162,* 1051–1076.

SHCHERBAKOV, R., TURCOTTE, D.L., and RUNDLE, J.B. (2006), *Scaling properties of the Parkfield aftershock sequence,* B. Seismol. Soc. Am., *96,* 4B, 376–S384, doi:10.1785/0120050815.

SUTEANU, M. (2014), *Scale free properties in a network-based integrated approach to earthquake pattern analysis,* Nonlinear Processes in Geophysics, *21,* 427–438.

TENENBAUM, J.N., HAVLIN, S., and STANLEY, H.E. (2012), *Earthquake networks based on similar activity patterns,* Phys. Rev. E *86,* 046107.

UTSU, T. (1961). *A statistical study of the occurrence of aftershocks,* Geophysical Magazine, *30,* 521–605.

UTSU, T., OGATA, Y., and MATSU'URA, R.S. (1995), *The Centenary of the Omori Formula for a Decay Law of Aftershock Activity,* J. Phys. Earth, *43,* 1–33.

VAROTSOS, P.A., SARLIS, N.V., and SKORDAS, E.S. (2012), *Order parameter fluctuations in natural time and b-value variation before large earthquakes,* Natural Hazards and Earth System Sciences, doi:10.5194/nhess-12-3473-2012, 3473–3481.

WATTS, D.J., and STROGATZ, S. (1998), *Collective dynamics of 'small-world' networks,* Nature, *393,* 440–442.

ZALIAPIN, I., GABRIELOV, A., KEILIS-BOROK, V, and WONG, H.: *Clustering Analysis of Seismicity and Aftershock Identification,* Phys. Rev. Lett., *101,* 018501, 2008.

ZALIAPIN, I., and BEN-ZION, Y.: *Earthquake clusters in southern California I: Identification and stability,* Journal of Geophysical Research, *118,* 2847–2864, 2013a.

ZALIAPIN, I., and BEN-ZION, Y.: *Earthquake clusters in southern California I: Identification and stability,* Journal of Geophysical Research, *118,* 2865–2877, 2013b.

(Received March 30, 2014, revised September 11, 2014, accepted September 22, 2014, Published online November 29, 2014)

Pure Appl. Geophys. 172 (2015), 1879–1892
© 2014 Springer Basel
DOI 10.1007/s00024-014-0986-5

| Pure and Applied Geophysics

On the Use of Fractal Surfaces to Understand Seismic Wave Propagation in Layered Basalt Sequences

Catherine E. Nelson,[1] Richard W. Hobbs,[1] and Roxanne Rusch[1]

Abstract—The aim of this study is to better understand how a layered basalt sequence affects the propagation of a seismic wave, which has implications for sub-basalt seismic imaging. This is achieved by the construction of detailed, realistic models of basalt sequences, using data derived directly from outcrop analogues. Field data on the surface roughness of basaltic lava flows were captured using terrestrial laser scanning and satellite remote sensing. The fractal properties of the surface roughness were derived, and it can be shown that the lava flow surface is fractal over length scales up to approximately 2 km. The fractal properties were then used to construct synthetic lava flow surfaces using a von Karman power spectrum, and the resulting surfaces were then stacked to create a synthetic lava flow sequence. P-wave velocity data were then added, and the resulting model was used to generate synthetic seismic data. The resulting stacked section shows that the ability to resolve the internal structure of the lava flows is quickly lost due to scattering and attenuation by the basalt pile. A further result from generating wide-angle data is that the appearance of a lower-velocity layer below the basalt sequence may be caused by destructive interference within the basalt itself.

Key words: Sub-basalt seismic imaging, physical properties of lava flows, flood basalt provinces, terrestrial laser scanning.

1. Introduction

Reliable sub-basalt seismic imaging remains an outstanding problem in exploration geophysics. Several strategies have been developed to improve seismic imaging beneath basalt sequences, such as deep towed streamers to boost the low-frequency content (e.g. DAVISON *et al.* 2010). However, scattering and attenuation of the seismic energy by the basalt sequence mean that very little energy reaches a sub-basalt sedimentary succession. The greatest challenges to sub-basalt seismic imaging are caused by continental flood basalt provinces, where basalt thicknesses can range from hundreds of metres to several kilometres.

One motivating factor for research into seismic imaging below flood basalt provinces is the possible presence of hydrocarbon-bearing basins beneath. In the North Atlantic, sedimentary basins with known hydrocarbon discoveries extend below the edge of the basalt sequences of the North Atlantic Igneous Province. Additionally, one significant discovery has been made within the area covered by the basalt (HELLAND-HANSEN 2009), and oil has been observed within the basalt itself (LAIER *et al.* 1997). The Kutch Basin, onshore and offshore northwest India, is another area thought to have substantial sub-basalt prospectivity (KUMAR *et al.* 2004; ROHRMAN 2007).

A continental flood basalt province (CFBP) consists of an abnormally voluminous sequence of volcanic material, often thought to be formed from mantle plume material (e.g. SAUNDERS 2005). Examples include the Palaeogene North Atlantic Igneous Province, the Cretaceous–Tertiary Deccan Traps and the Permian–Triassic Siberian Traps. Generally, lava flows are the most common eruptive product, though there can also be substantial volumes of hyaloclastites and pyroclastic material (WHITE *et al.* 2009).

The lava flows in a CFBP are most commonly of pahoehoe type (SELF *et al.* 1998), and can be tens of metres thick, extending laterally for tens of kilometres continuously, or can be just a few metres thick or less, with an anastomosing, braided architecture (JERRAM 2002). Volumes of individual lava flows can be over 1,000 km^3 dense rock equivalent (BRYAN *et al.* 2010). Each lava flow has a three-part vertical division with a rubbly, vesicular top; a massive, relatively unfractured

[1] Department of Earth Sciences, Durham University, Durham DH1 3LE, UK. E-mail: catherine.nelson1@gmail.com

Reprinted from the journal

core; and a vesicular base. This three-part division means that the rock properties of each lava flow are highly heterogeneous: the crust and base have a low P-wave velocity and density, whereas the core has a much higher velocity and density (PLANKE 1994). A stack of lava flows therefore exhibits strong layering of its acoustic impedance. It has been shown that the attenuation of a seismic wave caused by a thick basalt sequence is due to a combination of both this layering, and scattering caused by the rough surface of the lava flows (MARESH *et al.* 2006; discussed further in NELSON *et al.* 2009a). High frequencies are preferentially attenuated, so the returned energy is predominantly at low frequency.

In this study, we seek to better understand the effect of a layered basalt sequence on seismic wave propagation, so that the results can be used to improve sub-basalt seismic imaging. This is achieved by the use of models that accurately capture the heterogeneity within a layered basalt sequence. These must have a vertical resolution of tens of centimetres to accurately model this heterogeneity (MARTINI *et al.* 2005). Ideally, these would be deterministic models constructed from data collected from flood basalt outcrops. However, there are several features of flood basalt provinces in the field that make this difficult:

- Incomplete exposure: To map the extent of a lava flow, it is necessary to know the position of all the edges of the flow. This is easy for recent, uneroded lava flows, but extremely difficult when a lava flow is buried by many subsequent flows.
- Poor exposure: Basalt weathers easily and is often covered by vegetation or scree.
- Difficulty in correlation: Lava flows within a flood basalt province can be of a similar composition, and have a similar physical appearance. If no marker horizons (such as sedimentary layers) are present, it can be extremely difficult to correlate flows between outcrops—especially as they may change in thickness between two outcrops.

Accordingly, in this work we derive the fractal properties of field exposures of basaltic lava flows, and use these to construct synthetic lava flow surfaces. The incorporation of the field data makes the models more realistic, and the surfaces can be generated at any scale required.

2. Background and Previous Work

Many geological and topographic features are scale invariant, meaning they can be described by a fractal dimension, D (TURCOTTE 1989). It is possible to test whether a medium is fractal by taking the Fourier transform of a profile across it to generate a power spectrum. A fractal medium displays a relationship between the power spectrum and wavenumber given by the following equation (HUANG and TURCOTTE 1989):

$$P(k) = Ak^{-\beta},$$

where $P(k)$ is the power spectrum, k is the wavenumber, and A is a constant. If the power spectral density is plotted against the wavenumber on a log–log scale (Fig. 1), β is given by the slope of the straight line part. The fractal dimension D can be calculated from the slope as follows (HIGUCHI 1990):

$$D = \frac{2E + 3 - \beta}{2},$$

where E is the Euclidean dimension. For a one-dimensional (1D) profile, $E = 1$; for a two-dimensional (2D) surface, $E = 2$; and so on.

For the one-dimensional case (i.e. a cross-section of a surface), the equation is therefore as follows (HUANG and TURCOTTE 1989; DOLAN and BEAN 1997):

$$D = \frac{5 - \beta}{2}.$$

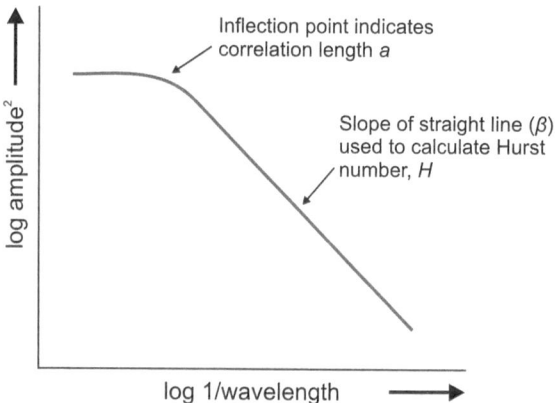

Figure 1
Determining Hurst number and correlation length from a power spectrum

The upper limit of the straight line in Fig. 1 is the correlation length *a*, and beyond this the medium is no longer fractal (FRENJE 2000). If a 1D profile across a basalt lava flow surface shows this type of plot, we know it is fractal at scales up to the correlation length.

The Hurst number is also used to quantify the fractal properties of a medium, and is useful for comparison between 1D and 2D fractal parameters. The Hurst number, *H* (or *v*), is related to *D* by the following equation (MANDELBROT 1985; FEDER 1988; DOLAN and BEAN 1997):

$$D = E + 1 - H.$$

The effect on surface roughness of altering the Hurst number is shown in Fig. 2.

Previous studies have considered the tops of basalt lava flow successions as fractal surfaces. WALIA and BULL (1997) analysed seismic data from the Rockall Trough, using spectral analysis to obtain a fractal dimension of 1.36, corresponding to a Hurst number of 0.7 for the top surface of the lava pile. MARTINI *et al.* (2005) used this value to construct three-dimensional (3D) velocity models, along with a fractal dimension for the vertical velocity distribution obtained from a large dataset of Ocean Drilling Program wells. MARESH (2004) analysed 3D seismic data from the Rockall Trough, obtaining a 2D fractal dimension of 2.49, equating to a Hurst number of 0.51. However, these results cannot be taken as representative of individual flows: because of the resolution limit (quarter wavelength) of band-limited seismic data, it is impossible to know whether the top surface of the basalt sequence is one flow or many.

BEAN and MARTINI (2010) used digital photographs of a single lava flow from the Tertiary Basalt Province of Northern Ireland to obtain a Hurst number of 0.2. This is substantially different from the results of WALIA and BULL (1997) and of MARESH (2004). This difference is likely to be because the surface chosen by BEAN and MARTINI (2010) is the interface between the base of the basalt succession and the top of a limestone sequence. The roughness of this surface is controlled by the erosion of the limestone, which is then passively infilled by the lava flow, making this result useful for modelling the base of a basalt succession. It does not represent the roughness of an individual lava flow.

Much of this work stems from analysis of borehole data to produce stochastic models (e.g. BEAN 1996; HOLLIGER 1996; DOLAN and BEAN 1997; DOLAN *et al.* 1998; FRENJE and JUHLIN 1998) and analysis of topography both on land (HUANG and TURCOTTE 1989) and on the seafloor (GOFF and JORDAN 1988). Similar approaches have been used in a number of different fields including the analysis of sedimentary cycles (BROWAEYS and FOMEL 2009) and in seismic oceanography (BUFFETT *et al.* 2010). Fractal surfaces are also observed in other geological settings, for example stylolites (e.g. EBNER *et al.* 2009, 2010). BEN-ITZHAK *et al.* (2012) used terrestrial laser scanning of stylolite surfaces up to 10 m long to derive their Hurst numbers and correlation length. The roughness of fault surfaces has also been shown to be fractal, at length scales from millimetres to thousands of kilometres (RENARD *et al.* 2013).

3. Field Data for Fractal Analysis

For this study, three datasets were available for fractal analysis. The datasets are known to be from single lava flows, at scales covering 10 cm to 10 km. Two datasets are from modern lava flows, where the top lava flow surface has not been subject to significant weathering, and the third is from part of the Palaeogene North Atlantic Igneous Province (NAIP).

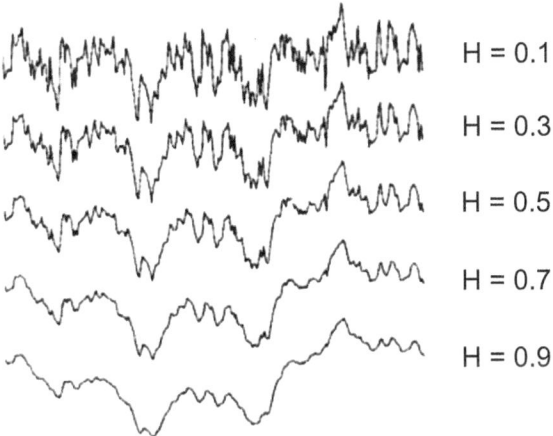

H = 0.1

H = 0.3

H = 0.5

H = 0.7

H = 0.9

Figure 2
Effect of Hurst number on surface roughness, while the standard deviation remains constant; from SAUPE (1988)

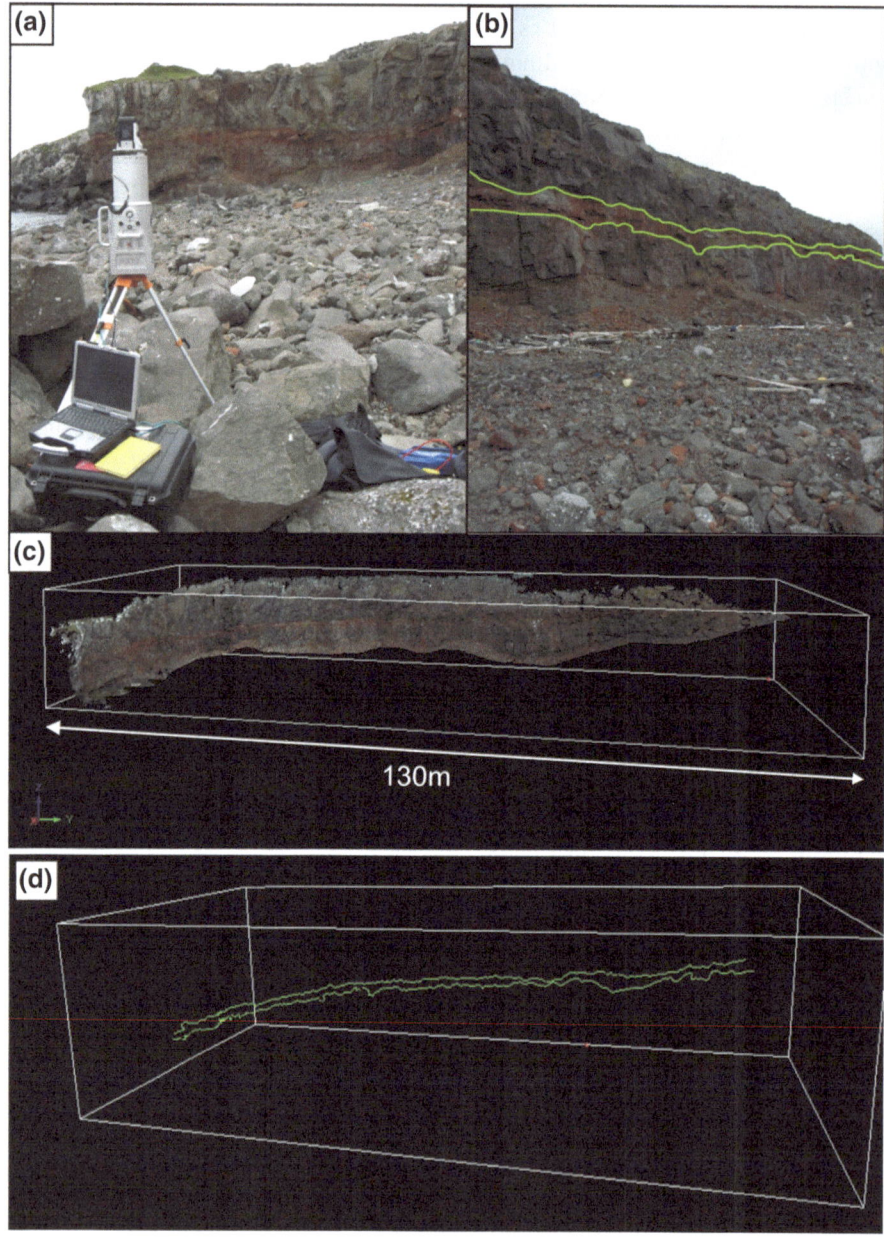

Figure 3
Laser scan data from coastal quarry at Glyvursnes. Outcrop height is approximately 10 m. **a** Acquiring the laser scan data. **b** The quarry wall with photo-interpretation of the lava flow top and base. **c** The completed scan data. **d** *Lines* interpreted from the scan data

3.1. Glyvursnes

The first dataset comes from a quarried exposure near Glyvursnes, on the Island of Streymoy, Faroe Islands. This outcrop, shown in Fig. 3, was chosen because it offers a very clear top surface of one lava flow and a base surface of another, with a sedimentary bole horizon in the middle. The top and base are very well preserved and offer an excellent 2D section for analysing the surface roughness. It is within the Enni Formation of the Faroe Islands Basalt Group, part of the NAIP.

Surface roughness data at this location were acquired using terrestrial laser scanning (TLS), allowing the outcrop geometry to be accurately captured in digital format. This equipment has become increasingly popular amongst geologists, as it allows 3D data to be captured for analysis away from the field situation. Three-dimensional point clouds thus obtained can be analysed to provide quantitative structural or geological data (e.g. McCaffrey *et al.* 2005, 2008; Nelson *et al.* 2011). The laser scanner measures the *XYZ* coordinates of points on the outcrop at specified intervals by calculating the distance from the scanner based on the return time of the laser beam. These points can then be coloured from digital photos to give an accurate representation of the outcrop, which can then be viewed from any angle and features on it measured. This is particularly useful for inaccessible parts of outcrops. Multiple scans from different angles are obtained to minimise shadow areas where parts of the outcrop hide other areas from the scanner viewpoint. Reflectors are used to provide common points of reference between scan and photo, and between scans. Previous studies have documented in detail the standard workflow for capturing and processing TLS data (e.g. Buckley *et al.* 2008; Enge *et al.* 2007).

The equipment used in this work was a Riegl LMS-Z420i terrestrial laser scanner (Fig. 3a), frequently used in geology and optimized for rapid data acquisition, a long range and usage in demanding environmental conditions. Two scans were required

Figure 4
Laser scan data from Erta Ale, Ethiopia. Photographs courtesy of Dougal Jerram. **a** Collecting the scan data. **b** Overview of the crater and surroundings. **c** Complete scan data and scan positions

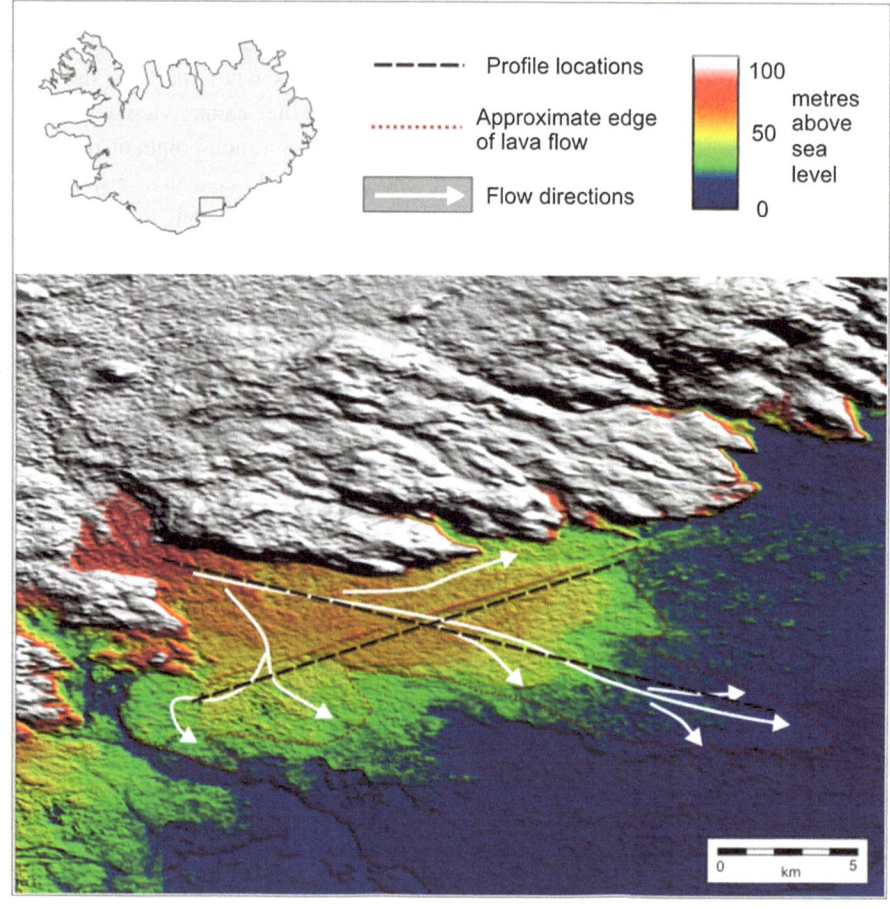

Figure 5
Eldhraun branch of the Laki lava flow: location map and satellite DEM from ASTER Global Digital Elevation Model (http://www.gdem.aster.ersdac.or.jp/). Data at 30 m resolution. Original data of ASTER GDEM are the property of the Ministry of Economy, Trade and Industry (Japan) and NASA. Flow directions and Laki extent from GUILBAUD et al. (2005)

for complete coverage of this outcrop because its geometry was relatively simple. A total of approximately 8,200,000 points were collected, and 14 digital photographs in two panoramic sequences. The flow top and base were identified on the digital photographs, and 3D lines were drawn on the scan surface following the methods described in NELSON et al. (2011), shown in Fig. 3d. The total length of the outcrop is approximately 130 m. Approximately 700 points were digitized for the base of the upper flow, giving an average point spacing of around 17 cm. Approximately 900 points were digitized for the top of the lower flow, giving an average point spacing of around 15 cm. These were converted to 2D profiles along the top and base of the flows, suitable for the generation of power spectra.

3.2. Erta Ale

The second dataset available for this work comprised terrestrial laser scanning data from Erta Ale Volcano, Ethiopia. This is a remote and rarely visited active basaltic volcano, with a lava lake and surrounding crater. The lava lake is one of the oldest known, having persisted for over 90 years (OPPENHEIMER and FRANCIS 1998). Data used here are taken from the crater floor surrounding the lava lake. The crater floor is covered by pahoehoe lava flows formed when the lava lake overflows (FIELD et al. 2012). This provides a large fresh surface with the opportunity of taking many profiles in any direction. The area of interest is approximately 80 m by 125 m.

Figure 6
Laki surface roughness. **a** Eldhraun branch near Kirkjubæjarklaustur. **b** Looking south across the Eldhraun from Fjaðrárgljúfur

Data were collected by Dougal Jerram and Steve Smith as part of filming for the BBC1 television series "The Hottest Place on Earth" (JERRAM and SMITH 2010). A total of six scans were required to completely capture the crater, and approximately 24,000,000 points and 42 digital photographs were collected. The coloured 3D point cloud is shown in Fig. 4. Two-dimensional profiles were captured directly from the scan data, and the resulting sections have length of 70–125 m and point spacing of approximately 10 cm. Two sections were chosen for further analysis: the longest possible sections in orthogonal directions, along and across the crater floor.

3.3. Laki

The third dataset is a satellite digital elevation model (DEM) covering the Laki lava flow, Iceland. This was emplaced in 1783–1784 and is regarded as the closest modern analogue to a flood basalt lava flow (SELF et al. 1998). It has been used to model the environmental impact of flood basalt eruptions (SELF et al. 2006) and has also been used as an analogue to Martian lavas (KESZTHELYI et al. 2004). THORDARSON and SELF (1993) and GUILBAUD et al. (2005) provide a full description of the flow morphology and its eruption. Cross-sections through the Laki lava flow and the Roza Member of the Columbia River flood basalt province reveal the same three-part internal structure (SELF et al. 1998).

The Laki lava field totals approximately 14.7 km^3 of basaltic lava, covering an area of approximately 599 km^2 (THORDARSON and SELF 1993). The lava erupted from fissures in the Sída Highlands of southern Iceland, part of the Grimsvötn volcanic system, and flowed south down the gorges of two rivers: the Skaftá and the Hverfisfljót. It then spread out onto the flat coastal plain formed by the earlier Eldgja lava flow (934 AD). The area of interest for this study, shown in Fig. 5, is the branch that flowed out of the Skaftá gorge onto the coastal plain—the Eldhraun branch. This comprises ~5 km^3 of lava that was emplaced directly onto the Eldgja lava, into an unconfined area. This area is therefore a useful analogue for a flood basalt lava flow.

The Eldhraun branch is also useful because it offers a very large distance to analyse—approximately 15-km profiles can be made as shown in Fig. 5. As it is so young, and remains uncovered by any later lava flows, it preserves its surface morphology very well, as shown in Fig. 6. Two cross-sections were taken from this dataset, one across the flow direction and one along the flow direction, as shown in Fig. 5. The along-flow section has a length of 23.7 km, with a point spacing of approximately 23 m, and the across-flow section has a length of 15.1 km, with a point spacing of approximately 15 m.

4. Spectral Analysis

One-dimensional profiles were taken through each of the datasets described above, and power spectra were calculated for each profile. The Generic Mapping Tools (GMT) software was used to analyse the data (WESSEL and SMITH 2009). This is an open-source set of tools for manipulating x, y and x, y, z data, available to download from http://gmt.soest.hawaii. edu/. Spectral density estimates follow the method of WELCH (1967), and error bars are produced following the method of BENDAT and PIERSOL (1986).

The resulting power spectra are shown in Fig. 7. There is good agreement between datasets at each location. The Laki power spectra (Fig. 7a) show very close agreement between the section along and across the flow, suggesting that the roughness is approximately isotropic and the same fractal properties can be used in both directions. This spectrum also displays a clear roll-over point, indicating that the surface can be treated as fractal up to around 2,000 m (the correlation length).

The data from Glyvursnes show that there is no significant difference between the roughness of a flow top and flow base. For values below approximately 20 cm (2 on the x axis in Fig. 7b) the wavelength is equal to the spacing of the data points, so the results are not useful. The same applies to the results from Erte Ale. Again, there is no significant anisotropy apparent in the surface roughness.

The data from Laki, Glyvursnes and Erta Ale are plotted together in Fig. 8. A best fit line can be drawn

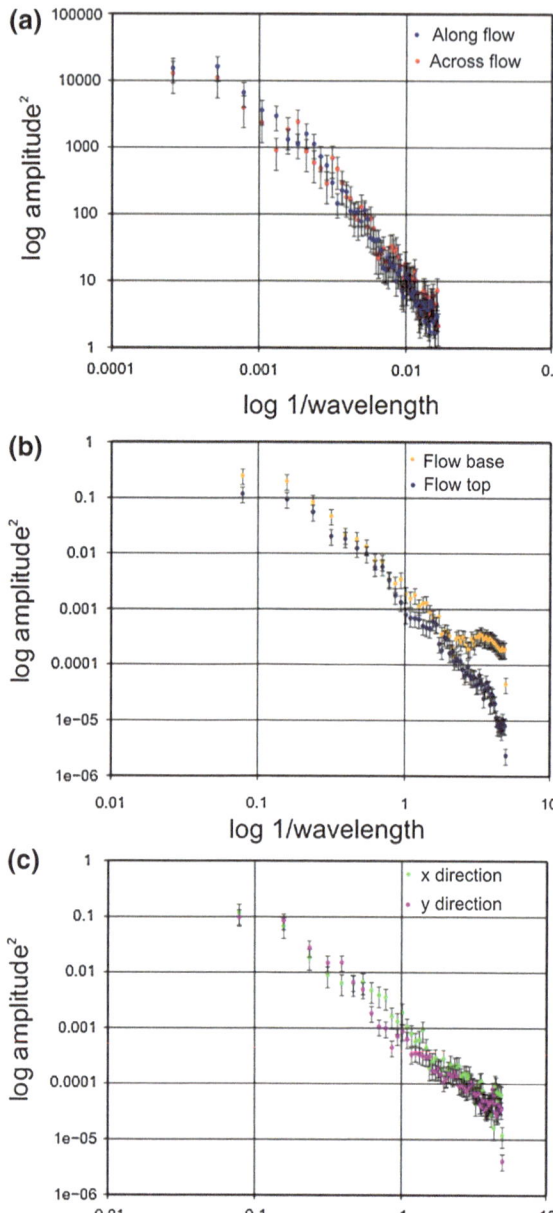

Figure 7
Power spectra of data from a Laki, b Glyvursnes and c Erte Ale

within the uncertainties of the data. From this, a β value (slope) of 2.1 is measured (derived from a 1D Fourier transform). Using the equations described earlier, a Hurst number of 0.55 is obtained. As discussed above, the correlation length is 2,000 m.

The dashed lines in Fig. 8 give an estimate of the error on the Hurst number: the likely range is around

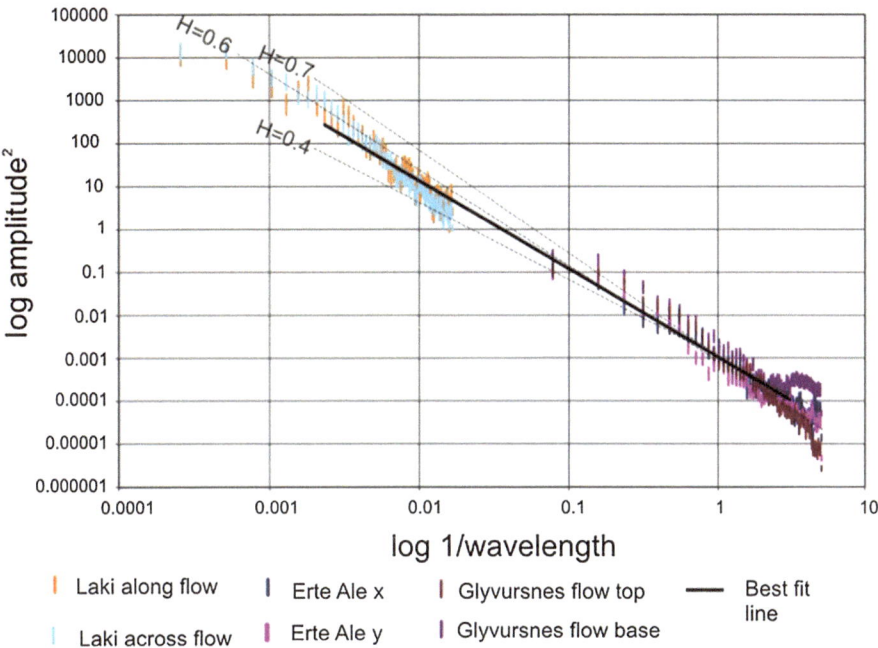

Figure 8

All power spectra plotted together, only 1 standard deviation (sd) *error bars* shown. *Dashed lines* show other Hurst numbers for comparison

0.5–0.65. It can be seen that there is a slight step between the data from Laki and the laser scan data, and this is likely to be due to the change in instrumentation. However, this is within the errors of the data, so a simple single slope was chosen to describe the surface, consistent with the individual slopes from each dataset and previous work. Data at scales in between the satellite data and the laser scan data would help to verify this.

5. Model Construction

The Hurst number of 0.55 and correlation length of 2,000 m derived in the previous section were used to build synthetic surfaces with the same statistical properties as the real surfaces. The random volume building code developed by WHITE (2009) for building velocity models of basalt sequences was adapted to produce surface topography. The code generates either 2D surfaces or 3D fractal volumes using a von Karman power spectrum with specified correlation lengths, standard deviations in the two or three Cartesian coordinates, as required, and a Hurst number. Here we used the code to generate surfaces that represent lava flow top surfaces, using the fractal properties derived in the previous section. Figure 9 shows a final surface composed of 512 nodes in both the *x* and *y* directions, with a horizontal grid node spacing of 20 m, giving an overall size of 10.22 × 10.22 km.

The surfaces can then be stacked to simulate flows of various thicknesses and types. First, the types of volcanic facies required in the model are given along with the mean flow thickness and standard deviation, as determined from mapping in the field, and the number of flows in the basalt pile. The facies types can be any of the following:

- Tabular lava flows, which tend to be the thicker layers, include a massive core with a crust and base (PLANKE 1994).
- Compound lava flows without a massive core (JERRAM 2002).
- Volcaniclastics, with a uniform structure (representing hyaloclastites or pyroclastic deposits). Hyaloclastites have been shown to have a narrow, relatively uniform velocity distribution (NELSON *et al.* 2009b).

107

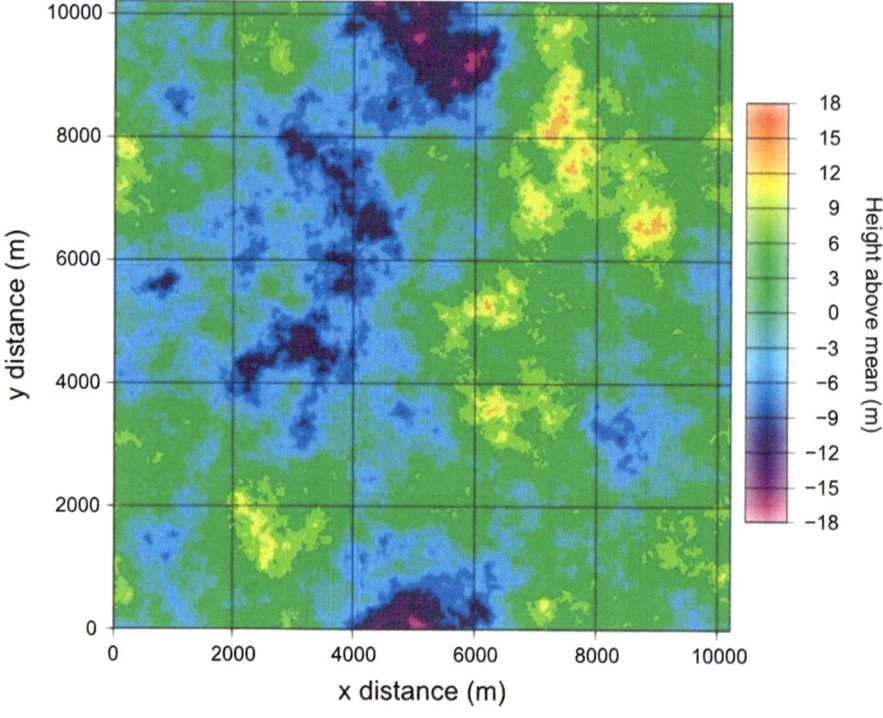

Figure 9
Fractal model produced with the random volume building code of WHITE (2009)

• Sedimentary sequences with a uniform structure.

The process is iterative, starting with the low-ermost layer. The base of the basalt pile may be horizontal representing an erosional surface, or with topography if that can be predicted from field mapping. A fractal surface is generated that represents the top of the first flow, a thickness is drawn from the specified population using a random number generator, and a facies type is selected from a weighted probabilistic list such that over a significant number of flows (>10) the number of each facies type approaches the expected probability. The interior structure of the flow is constructed depending on the facies type. Again, this internal structure is preconditioned on a set of probabilistic functions determined from the field which, as in the case of geophysical models demonstrated here, will include data from well logs (e.g. NELSON 2010). The process of generating layer top, thickness and facies type is repeated until the required number of layers is reached. At this point

the whole cycle can be repeated but with a new set of probabilistic functions for thickness, flow type and internal structure to represent the different phases in the evolution of a flood basalt province. The final model can be arbitrarily complex, as the probabilistic descriptors and fractal surfaces can be applied to any length scale. A cross-section through a possible model is shown in Fig. 10.

The geophysical model shown in Fig. 10 is then used to help understand seismic wave propagation. Two possible outputs are shown in Fig. 11. By using viscoelastic complex-screen modelling (WHITE and HOBBS 2007), it is possible to generate an ideal stacked section. The high-impedance contrasts at the top of the basalt pile form the brightest reflections. The ability to resolve the internal structure is quickly lost as scattering by the rough surfaces and internal reverberation absorbs the higher frequencies and destroys the coherence of the wavefront, which is consistent with the result seen by MARESH et al. (2006).

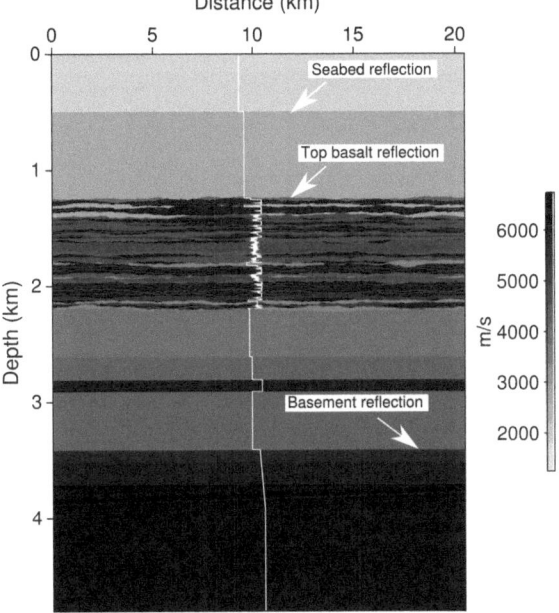

Figure 10
Two-dimensional section through a 3D probabilistic model for a sequence of flows that have built up a 2-km-thick basalt-dominated pile, which has then subsided and been buried by later sediments (similar to the Faroe–Shetland Trough). The model is *coloured* by seismic P-wave velocity, but it could equally be *coloured* by density, S-wave velocity or geological descriptors

The second and possibly more interesting result is the wide-angle response computed using finite difference code (COHEN and STOCKWELL 2012). The first refracted arrival from the basalt shows a characteristic response of an event that appears to be truncated with a step-back in time to the refraction event associated with the deeper basement. This characteristic is widely interpreted as evidence of a velocity gradient within the basalt where the magnitude of the step-back is used to estimate the thickness of the lower-velocity sediments beneath the basalt layer (e.g. RAUM *et al.* 2005). This interpretation is based on the high-frequency ray-tracing approximation with a homogeneous basalt layer where the velocity gradient is used to predict the first arrival from the basalt layer, and its truncation corresponds to the grazing ray that just stays within the basalt layer. However, as demonstrated by this model, it is possible to generate the same effect in a model with no internal gradient, but solely by destructive interference of scattered energy.

6. Discussion

This study has shown that spectral analysis is a valid method for understanding the properties of a basalt lava flow surface, and that the surface can be considered as fractal up to the correlation length of 2,000 m. Having obtained the correlation length and Hurst number (0.55), it is possible to construct random surfaces with the same statistical properties as the real surfaces.

The result of 0.55 for the Hurst number achieved here is similar to that of WALIA and BULL (1997), and also in good agreement with MARESH (2004), as discussed earlier. It is substantially different from that of Bean and Martini, but again as discussed earlier this is likely because their result is taken from the base of a flow overlying eroded limestone. The datasets presented here are known to be from fresh, unaltered top surfaces of lava flows. The result of Bean and Martini would be of use in modelling the base surface of a flood basalt province; however the data presented here are useful for modelling internal reflectors and the top surface.

Another issue for discussion is the correlation length. Here, we use the same value of 2,000 m for both along-flow and across-flow directions. WHITE (2009) used a correlation length ten times larger in one direction than the other, citing THOMSON (2005). MARTINI *et al.* (2005), working along similar lines, used a correlation length five times greater in one horizontal direction than the other. This is appropriate for lava flows emplaced onto dipping surfaces, however THOMSON (2005) also described flat-lying flows of roughly equal *x*–*y* dimensions, similar to our results from Laki. The dip of the required flows should be taken into account when developing a model of a flood basalt province, as lava flows emplaced onto a dipping surface will display longer correlation lengths along the direction of flow. The correlation length can easily be altered in the random volume code of WHITE (2009) to reflect the desired geometry.

The methods presented here for constructing random surfaces are useful in situations where the entire scale length of a feature cannot easily be captured. It is relatively easy to characterise the surface of a lava flow at the outcrop scale (tens of metres), and satellite data

109

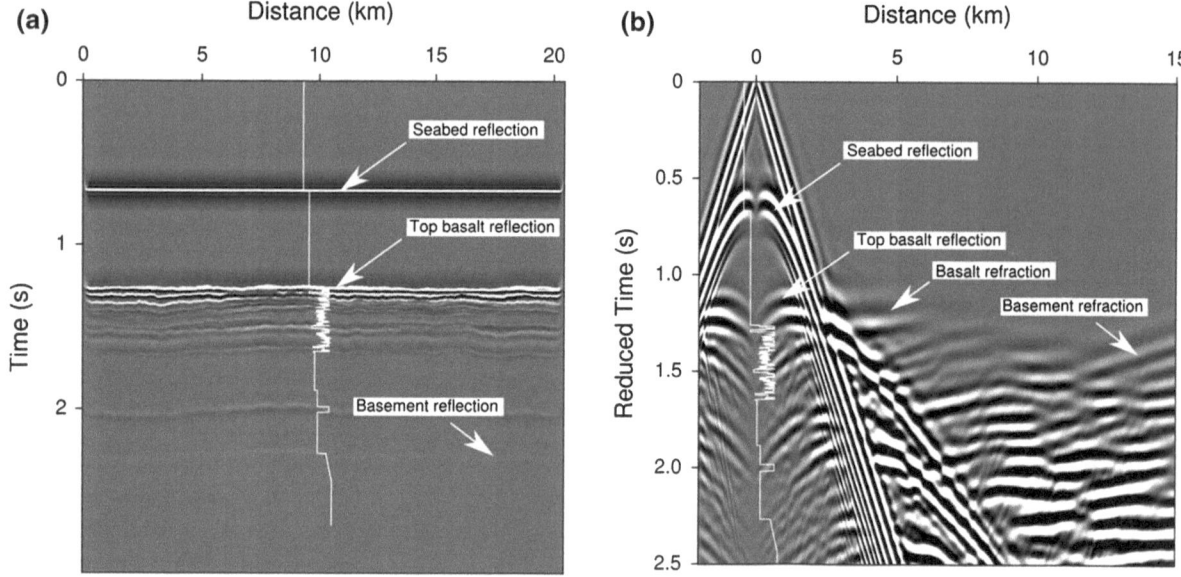

Figure 11
a Ideal primary only seismic reflection section from the model in Fig. 10 showing the rapid loss of back-scattered energy from the basalt sequence. In reality, the received image will be significantly degraded by the inclusion of the multiply scattered energy that will appear as noise on the section. **b** The corresponding long-offset refraction model. Note the termination of the basalt refraction event caused by multiple destructive interference of scattered energy which is dominated by scattering in the upper layers of the model

allow features to be measured at hundreds of metres to kilometre scales; however in between this it becomes more difficult. Satellite data are not easily obtained at resolution below 30–50 m, and outcrops larger than 100–200 m are rare and difficult to measure. The use of fractal modelling covers this scale gap. One problem with studying flood basalt provinces is that exposure is often incomplete, and it is impossible to tell the extent of a single lava flow as the edges are not preserved or are difficult to correlate. By bringing together these diverse datasets, it is possible to build realistic models of lava flow surfaces without the need for correlation between outcrops or time-consuming mapping.

Having determined the statistics of the basalt surface and probabilistic descriptors of the internal structure of the various basalt facies, it is then possible to construct constrained models that can be used to investigate the combined effects of rough surfaces and complex high-impedance internal structure on seismic wave propagation. Below the top of the basalt surface, the reflections quickly die off in amplitude and do not correspond to the layering in the model. An interesting result is that the truncation of the first arrival from the basalt layer can be created in wide-angle data by scattering

within the basalt. This could lead to error in the interpretation of the thickness of the basalt and sedimentary succession below. Further work should explore different geological scenarios using these models, varying the thickness of both the basalt layer and the sedimentary basin, to understand what interpretations are possible.

Acknowledgments

The authors would like to thank Dougal Jerram for his part in this project, and for providing the Erte Ale data. Python modelling code was developed by Sarah McMullan, undergraduate student at Durham University. The authors also thank the anonymous reviewer for a thorough and constructive review.

REFERENCES

BEAN, C.J. (1996), *On the cause of 1/f-power spectral scaling in borehole sonic logs*, Geophysical Research Letters *23*, 3119–3122.

BEAN, C.J., and MARTINI, F. (2010), *Sub-basalt seismic imaging using optical-to-acoustic model building and wave equation datuming processing*, Marine and Petroleum Geology *27*, 555–562.

BENDAT, J.S., and PIERSOL, A.G., Random data: analysis and measurement procedures (Wiley, New York, 1986).

BROWAEYS, T.J., and FOMEL, S. (2009), *Fractal heterogeneities in sonic logs and low-frequency scattering attenuation*, Geophysics *74*, WA77–WA92.

BRYAN, S. E., PEATE, I.U., PEATE, D.W., SELF, S., JERRAM, D.A., MAWBY, M.R., MARSH, J.S., and MILLER, J.A. (2010), *The largest volcanic eruptions on Earth*, Earth-Science Reviews *102*, 207–229.

BUCKLEY, S.J., HOWELL, J.A., ENGE, H.D., and KURZ, T.H. (2008), *Terrestrial laser scanning in geology: data acquisition, processing and accuracy considerations*, Journal of the Geological Society *165*, 625–638.

BUFFETT, G.G., HURICH, C.A., VSEMIRNOVA, E.A., HOBBS, R.W., SALLARES, V., CARBONELL, R., KLAESCHEN, D., and BIESCAS, B. (2010), *Stochastic heterogeneity mapping around a Mediterranean salt lens*, Ocean Science *6*, 423–429.

COHEN, J. K. and STOCKWELL, JR. J. W., (2012), CWP/SU: Seismic Un*x Release No.43r3: an open source software package for seismic research and processing, Center for Wave Phenomena, Colorado School of Mines.

DAVISON, I., STASIUK, S., NUTTALL, P., KEANE, P. (2010), Sub-basalt hydrocarbon prospectivity in the Rockall, Faroe–Shetland and Møre basins, NE Atlantic, in B. A. Vining, S. C. Pickering, ed., Petroleum Geology: From Mature Basins to New Frontiers—Proceedings of the 7th Conference, The Geological Society of London, 1025–1032.

DOLAN, S.S., and BEAN, C.J. (1997), *Some remarks on the estimation of fractal scaling parameters from borehole wire-line logs*, Geophysical Research Letters *24*, 1271–1274.

DOLAN, S.S., BEAN, C.J., and RIOLLET, B. (1998), *The broad-band fractal nature of heterogeneity in the upper crust from petrophysical logs*, Geophysical Journal International *132*, 489–507.

EBNER, M., D. KOEHN, R. TOUSSAINT and F. RENARD, (2009) *The influence of rock heterogeneity on the scaling properties of simulated and natural stylolites*, Journal of Structural Geology *31*, 72.

EBNER, M., R. TOUSSAINT, J. SCHMITTBUHL, D. KOEHN and P. BONS, (2010) *Anisotropic scaling of tectonic stylolites: a fossilized signature of the stress field?*, Journal of Geophysical Research *115*, B06403.

ENGE, H.D., BUCKLEY, S.J., ROTEVATN, A., and HOWELL, J.A. (2007), *From outcrop to reservoir simulation model: workflow and procedures*, Geosphere *3*, 469–490.

FEDER, J. Fractals (Plenum, New York, 1988).

FIELD, L., BARNIE, T., BLUDY, J., BROOKER, R.A., KEIR, D., LEWI, E. and SAUNDERS, K. (2012), *Integrated field, satellite and petrological observations of the November 2010 eruption of Erta Ale*, Bulletin of Volcanology *74*, 2251–2271.

FRENJE, L., Scattering of Seismic Waves in Random Velocity Models (PhD thesis, Uppsala University, 2000).

FRENJE, L. and JUHLIN, C. (1998), *Scattering of seismic waves simulated by finite difference modelling in random media: application to the Gravberg-1 well, Sweden*, Tectonophysics *293*, 61–68.

GOFF, J.A. and JORDAN, T.H. (1988). *Stochastic modeling of seafloor morphology: Inversion of sea beam data for second order statistics*, Journal of Geophysical Research *93*, 13589–13608. doi:10.1029/88JB03160.

GUILBAUD, M., SELF, S., THORDARSON, T., and BLAKE, S. (2005), *Morphology, surface structures, and emplacement of lavas produced by Laki, AD 1783–1784*, Special paper—Geological Society of America *396*, 81–102.

HELLAND-HANSEN, D. (2009), Rosebank—Challenges to development from a subsurface perspective, in Varming, T., and Ziska, H., (eds), Faroe Islands Exploration Conference: Proceedings of the 2nd Conference, Annales Societatis Scientiarum Faroensis, supplementum *50*, 241–245.

HIGUCHI, T. (1990). *Relationship between the fractal dimension and the power law index for a time series: a numerical investigation*, Physica D *46*, 254–264.

HOLLIGER, K. (1996), *Upper-crustal seismic velocity heterogeneity as derived from a variety of P-wave sonic logs*, Geophysical Journal International *125*, 813–829.

HUANG, J., and TURCOTTE, D.L. (1989), *Fractal mapping of digitized images: application to the topography of Arizona and comparisons with synthetic images*, Journal of Geophysical Research *94*, 7491–7495.

JERRAM, D.A. (2002), Volcanology and facies architecture of flood basalts, in Menzies, M.A., Klemperer, S.L., Ebinger, C.J., and Baker, J., eds., Volcanic rifted margins. Geological Society of America Special Paper *362*, 119–132.

JERRAM, D.A., and SMITH, S.A.F. (2010), *Earth's hottest place*, Geoscientist *20*, 12–13.

KESZTHELYI, L., THORDARSON, T., McEWEN, A., HAACK, H., GUILBAUD, M.N., SELF, S., and ROSSI, M.J. (2004), *Icelandic analogs to Martian flood lavas*, Geochemistry Geophysics Geosystems *5*, 1–32.

KUMAR, D., BASTIA, R., and GUHA, D. (2004), *Prospect hunting below Deccan basalt: imaging challenges and solutions*: First Break *22*, 35–39.

LAIER, T., NYCROFT, H.P., JØRGENSEN, O., and ISAKSEN, G.H. (1997), *Hydrocarbon traces in the Tertiary basalts of the Faeroe Islands*, Marine and Petroleum Geology *14*, 257–266.

LARONNE BEN-ITZHAK, L., E. AHARONOV, R. TOUSSAINT and A. SAGY, (2012) *Upper bound on stylolite roughness as indicator for the duration and amount of dissolution*, Earth and Planetary Science Letters, 337–338, 186–196.

McCAFFREY, K.J.W., FEELY, M., HENNESSY, R., and THOMPSON, J. (2008), *Visualization of folding in marble outcrops, Connemara, western Ireland: an application of virtual outcrop technology*, Geosphere *4*, 588–599.

McCAFFREY, K.J.W., JONES, R.R., HOLDSWORTH, R.E., WILSON, R.W., CLEGG, P., IMBER, J., HOLLIMAN, N, and TRINKS, I. (2005), *Unlocking the spatial dimension: digital technologies and the future of geoscience fieldwork*, Journal of the Geological Society *162*, 927–938.

MANDELBROT, B. (1985), *Self-affine fractals and fractal dimension*. Physica Scripta *32*, 257–260.

MARESH, J. The Seismic Expression of Paleogene Basalts on the Atlantic Margin (PhD thesis, Cambridge University, 2004).

MARESH, J., WHITE, R.S., HOBBS, R.W., and SMALLWOOD, J.R. (2006), *Seismic attenuation of Atlantic margin basalts: observations and modeling*, Geophysics *71*, B211–B221.

MARTINI, F., and BEAN, C.J. (2002), *Application of pre-stack wave equation datuming to remove interface scattering in sub-basalt imaging*, First Break *20*, 395–403.

MARTINI, F., HOBBS, R.W., BEAN, C.J., and SINGLE, R. (2005), *A complex 3-D volume for subbasalt imaging*, First Break *23*, 41–51.

NELSON, C.E., JERRAM, D.A., HOBBS, R.W., TERRINGTON, R., and KESSLER, H. (2011). *Reconstructing flood basalt lava flows in 3D using terrestrial laser scanning*, Geosphere *7*, 87–96.

Reprinted from the journal

NELSON, C.E., Methods for constructing 3D geological and geophysical models of flood basalt provinces (PhD thesis, Durham University, 2010).

NELSON, C.E., JERRAM, D.A., SINGLE, R.T., and HOBBS, R.W. (2009a), Understanding the facies architecture of flood basalts and volcanic rifted margins and its effect on geophysical properties., in Varming, T., and Ziska, H., eds., Faroe Islands Exploration Conference: Proceedings of the 2nd Conference, 84–103.

NELSON, C.E., JERRAM, D.A., and HOBBS, R.W. (2009b), *Flood basalt facies from borehole data: implications for prospectivity and volcanology in volcanic rifted margins*, Petroleum Geoscience *15*, 313–324.

OPPENHEIMER, C., and FRANCIS, P. (1998), *Implications of longeval lava lakes for geomorphological and plutonic processes at Erta Ale volcano, Afar*, Journal of Volcanology and Geothermal Research *80*, 101–111.

PLANKE, S. (1994), *Geophysical response of flood basalts from analysis of wire line logs: Ocean Drilling Program Site 642, Vøring Volcanic Margin*, Journal of Geophysical Research-Solid Earth *99*, 9279–9296.

RAUM, T., MJELDE, R., BERGE, A.M., PAULSEN, J.T., DIGRANES, P., SHIMAMURA, H., SHIOBARA, H., KODAIRA, S., LARSEN, V.B., FREDSTED, R., HARRISON, D.J. and JOHNSON, M. (2005). *Sub-basalt structures east of the Faroe Islands revealed from wide-angle seismic and gravity data*, Petroleum Geoscience *11*, 291–308.

RENARD, F., CANDELA, T., BOUCHAUD, E. (2013) *Constant dimensionality of fault roughness from the scale of micro-fractures to the scale of continents*, Geophysical Research Letters *40*, p. 83–87.

ROBERTS, A. W., WHITE, R. S., LUNNON, Z. C., CHRISTIE, P. A. F., SPITZER, R. and ISIMM TEAM (2005), Imaging magmatic rocks on the Faroes margin. In: Petroleum geology: North-west Europe and global perspectives—Proceedings of the 6th Petroleum Geology Conference. Geological Society, London, Petroleum Geology Conference series *5*, 755–766.

ROHRMAN, M. (2007). *Prospectivity of volcanic basins: Trap delineation and acreage de-risking*, AAPG Bulletin *91*, 915–939.

SAUNDERS, A.D. (2005). *Large igneous provinces: origin and environmental consequences*, Elements *1*, 259–263.

SAUPE, D., Algorithms for random fractals, in Peitgen, H., and Saupe, D., eds., The science of fractal images (Springer-Verlag, New York, 1988).

SELF, S., KESZTHELYI, L., and THORDARSON, T. (1998), *The importance of pahoehoe*, Annual Review of Earth and Planetary Sciences *26*, 81–110.

SELF, S., WIDDOWSON, M., THORDARSON, T., and JAY, A.E. (2006), *Volatile fluxes during flood basalt eruptions and potential effects on the global environment: a Deccan perspective*, Earth and Planetary Science Letters *248*, 518–532.

THOMSON, K. (2005), *Volcanic features of the North Rockall Trough: application of visualisation techniques on 3D seismic reflection data*, Bulletin of Volcanology *67*, 116–128.

THORDARSON, T., and SELF, S. (1993), *The Laki (Skaftár-Fires) and Grimsvötn eruptions in 1783–1785*, Bulletin of Volcanology *55*, 233–263.

TURCOTTE, D.L. (1989), *Fractals in Geology and Geophysics*, Pure and Applied Geophysics *131*, 171–196.

WALIA, R.K., and BULL, J.M. (1997), *Modelling rough interfaces on seismic reflection profiles—The application of fractal concepts*, Geophysical Research Letters *24*, 2067–2070.

WELCH, P. (1967), *The use of fast Fourier transform for the estimation of power spectra: a method based on time averaging over short, modified periodograms*, IEEE Transactions on Audio and Electroacoustics *15*, 70–73.

WESSEL, P., and SMITH, W.H.F. (2009), The Generic Mapping Tools (GMT) version 4.5.0 Technical Reference & Cookbook, SOEST/ NOAA, http://gmt.soest.hawaii.edu/.

WHITE, J.C., Development and application of the phase-screen seismic modelling code (PhD thesis, Durham University, 2009).

WHITE, J. C. & HOBBS, R. W. (2007), *Extension of forward modelling phase-screen code in isotropic and anisotropic media up to critical angle*, Geophysics *72*, SM107–SM114.

WHITE, J.D.L., BRYAN, S.E., ROSS, P.-S., SELF S., and THORDARSON, T. (2009). Physical volcanology of continental large igneous provinces: update and review, in, Thordarson, T., Self, S., Larsen, G., Rowland, S.K., Hoskuldsson, A. (eds), Studies in Volcanology: The Legacy of George Walker. Special Publications of IAVCEI *2*, 291–321.

(Received March 3, 2014, revised October 29, 2014, accepted November 8, 2014, Published online November 27, 2014)

Pure Appl. Geophys. 172 (2015), 1893–1908
© 2014 Springer Basel
DOI 10.1007/s00024-014-0860-5

Pure and Applied Geophysics

Influence of High Energy Electromagnetic Pulses on the Dynamics of the Seismic Process Around the Bishkek Test Area (Central Asia)

Teimuraz N. Matcharashvili,[1] Tamaz L. Chelidze,[1] and Natalia N. Zhukova[1]

Abstract—Investigation of dynamical features of the seismic process as well as the possible influence of different natural and man-made impacts on it remains one of the main interdisciplinary research challenges. The question of external influences (forcings) acquires new importance in the light of known facts on possible essential changes, which occur in the behavior of complex systems due to different relatively weak external impacts. Seismic processes in the complicated tectonic system are not an exclusion from this general rule. In the present research we continued the investigation of dynamical features of seismic activity in Central Asia around the Bishkek (Kyrgyzstan) test area, where strong electromagnetic (EM) soundings were performed in the 1980s. The unexpected result of these experiments was that they revealed the impact of strong electromagnetic discharges on the microseismic activity of investigated area. We used an earthquake catalogue of this area to investigate dynamical features of seismic activity in periods before, during, and after the mentioned man-made EM forcings. Different methods of modern time series analysis have been used, such as wavelet transformation, Hilbert Huang transformation, detrended fluctuation analysis, and recurrence quantification analysis. Namely, inter-event (waiting) time intervals, inter-earthquake distances and magnitude sequences, as well as time series of the number of daily occurring earthquakes have been analyzed. We concluded that man-made high-energy EM irradiation essentially affects dynamics of the seismic process in the investigated area in its temporal and spatial domains; namely, the extent of order in earthquake time and space distribution increase. At the same time, EM influence on the energetic distribution is not clear from the present analysis. It was also shown that the influence of EM impulses on dynamical features of seismicity differs in different areas of the examined territory around the test site. Clear changes have been indicated only in areas which, according to previous researches, have been characterized by anomalous increase of average rates of strain release and thus can be regarded as close to the critical state.

1. Introduction

The problem of induced seismicity that may be caused by different natural as well as man-made

impacts remains an important domain of earthquake physics. Moreover, because of complicated character of related questions, this type of seismic activity became the subject of intense interdisciplinary researches (see e.g., Kisslinger 1976; Grasso and Sornette 1998; Guha 2004; Prejean et al. 2004; Peinke et al. 2006; Telesca et al. 2013).

The main characteristic of induced seismicity is that it is triggered by external forcing(s), which are weak compared to large tectonic forces acting in the lithosphere. Earlier observations of induced effects in natural seismicity were not quite reliable due to both technical and data analysis problems. The breakthrough in such studies is due to the introduction of new equipment, new seismic networks, and contemporary methods of signal processing, as well as recent field and laboratory experiments, carried out in fully or partly controllable conditions, enabling us to model statistical and dynamical features of the seismic process under different external influences.

The question of changes induced by external influences (forcings) in the seismic process acquires new importance in the light of facts established in the last several years. It was shown that the behavior of complex systems really may undergo essential changes due to different relatively weak external impacts (see e.g., Strogatz 2000; Rundle et al. 2000; Pikovsky et al. 2001; Meyers 2009; Chelidze and Matcharashvili 2007; etc.). Seismic processes taking place in the complicated tectonic system are not an exclusion from this general rule. Indeed, earthquakes occurring in the complex system of the earth crust are considered as dynamic instabilities generated in the process of unstable friction (stick–slip) between the faces of geological faults. The main driving force of the seismic process is tectonic stress. In addition to tectonic stress, relatively weak forcing may be

[1] M. Nodia Institute of Geophysics, Tbilisi State University, 1, Alexidze str., 0171 Tbilisi, Georgia. E-mail: matcharashvili@gtu.ge

Reprinted from the journal

exerted by various external impacts: Coulomb stress change by previous strong earthquakes (KISSLINGER 1976; GUHA 2004), mining activity and water reservoir loading (KISSLINGER 1976; TALWANI 1997; PEINKE et al. 2006; GUHA 2004), water injection in boreholes for the production of geothermal energy (KISSLINGER 1976; GUHA 2004), precipitation and snow melt (ROTH et al. 1992; SAAR and MANGA 2003; HAINZL et al. 2006; METIVIER et al. 2009), magnetic storms (SOBOLEV and ZAKRZHEVSKAYA 2002), explosions, including nuclear ones (KISSLINGER 1976; GUHA 2004), strong electrical pulses (JONES 2001; TARASOV 2010; CHELIDZE et al. 2006a, b, 2010; TARASOV and TARASOVA 2011; SMIRNOV and ZAVYALOV 2012; ADUSHKIN 2013), tides (GUHA 2004; AGNEW 2007; METIVIER et al. 2009), wave trains of remote strong earthquakes (PREJEAN et al. 2004), oceanic microseisms, solar activity (SIMPSON 1968) atmospheric pressure (HAINZL et al. 2006; BOLLINGER et al. 2007; BETTINELLI et al. 2008), etc.

As it was mentioned above, there are plenty of observations supporting the idea that these relatively weak perturbations can qualitatively and quantitatively influence features of seismic process driven by enormous tectonic forces to a certain critical state. Assessments show that the intensity of such external forcing may indeed be quite low, of the order of tens of kPa (GRASSO and SORNETTE 1998) or, according to BOGOMOLOV et al. (2011), activating forcing stress can be as small as of the order of 10^{-6}–10^{-7} of the main acting (tectonic) stress.

In the present research we focus on the interesting example of the dynamical effects of specific man-made impact on the seismic process. Specifically, we mean field experiments on the influence of strong electromagnetic (EM) discharges on temporal, spatial, and energetic characteristics of the seismic process. Such experiments started in the 1980s in Central Asia. Here, we focus on the experimental work carried out by researchers from the Joint Institute of High Temperatures of the Russian Academy of Sciences (IVTAN RAS), since the end of 1983 to beginning of 1990 at the test area close to Bishkek, Kyrgyzstan (TARASOV 1997; TARASOV et al. 1999; JONES 2001). In these experiments a considerable amount of electrical energy has been released during deep electrical soundings of the Earth's crust. The

source of electrical energy was a magnetohydrodynamic (MHD) generator or another high voltage source and a load was an electrical dipole of 0.4-Ohm resistance with electrodes located at a distance of 4.5 km from each other. When the generator was fired, the load current was 0.28–2.8 kA, the sounding pulses had durations of 1.7–12.1 s, and the generated energy was mostly in the range of 1.2–23.1 MJ (experimental setup was described in series of reports e.g., VOLYKHIN et al. 1993; TARASOV 1997; TARASOV and TARASOVA 2004). Results of these experiments have been carefully analyzed and several important facts have been established. For example, it was shown statistically that changes occurring in the local seismic process are indeed caused by strong EM discharges. Exact variations of seismotectonic strains and the release of seismic energy in the form of comparatively small earthquakes caused by electromagnetic pulses were documented (TARASOV and TARASOVA 2004, 2011; TARASOV 2010; BOGOMOLOV et al. 2011). CHELIDZE et al. (2006a, b) found evident changes in the correlation dimension of waiting time sequences during EM experiments. Further, SMIRNOV and ZAVYALOV (2012) investigated the statistical characteristics of the seismic process in the above-mentioned test area and reported increases of b values during EM soundings. On the other hand, the authors point out that "clear anomalies" in other statistical parameters of seismicity have not been found.

Independently, the results of these and similar analyses are very important, while at the same time they point to the necessity of further detailed studies of the process to understand the characteristics of EM-induced seismicity. Modern methods of complex data analysis can be very helpful for this purpose. Indeed, features of the seismic process and earthquake triggering phenomena cannot be fully understood in the framework of a traditional linear approach; high sensitivity to weak impact implies essentially nonlinear interactions occurring in the system close to the critical state. Presently, it is clear that qualitative descriptions and quantitative understandings of the processes related to induced seismicity and earthquake triggering necessitate the use of modern data analysis methods. Such methods developed in the last decade have already successfully been used for qualitative and quantitative analysis of dynamical structures of different complex processes including the seismic process.

In this research we continue our analysis, already started in previous publications (CHELIDZE *et al.* 2006a, 2010), of changes in the dynamical features of the seismic process under strong EM soundings.

As a source of data we used a catalogue of earthquakes that occurred around the Bishkek test area in the Northern Tien Shan and adjacent territories of Central Asia from the beginning of 1975 to the end of 1996 (MIKHAILOVA 1990). This catalogue covers time periods before, during, and after man-made impact (EM forcing). The main subject of interest was to compare features of (presumed) induced seismicity triggered by EM soundings with the intact seismic process, as well as with seismic processes after the cessation of MHD soundings.

To achieve targeted goals we used modern methods of data analysis (time–frequency, scaling and recurrence quantification analysis), resulting in non-overlapping, consecutive windows of analyzed data sets. Namely, data sets of sequences of inter-event (waiting) time intervals, inter-earthquake distances, magnitudes, as well as time series of the

number of earthquakes that occurred daily were obtained from the above-mentioned seismic catalogue. We investigated dynamical features of seismicity of the whole territory, covered by the catalogue. Similar analysis for certain areas of interest, throughout the entire catalogue territory, has also has been performed.

2. Data and Analysis

We conducted analyses on time series retrieved from a catalogue of Central Asian earthquakes, specifically focusing on the North Tien Shan and adjacent areas (Fig. 1) of Central Asia (1975–1996) (MIKHAILOVA 1990). In this area during a period of MHD soundings, a relatively strong earthquake with Mw = 6.3 occurred on 24 January in 1987.

As shown in Fig. 2 (and as it was already described in our previous articles), according to analysis of the Guttenberg–Richter relationship, the catalogue was considered complete for $M \geq 1.7$ (CHELIDZE *et al.* 2006a, 2010). At the same time, taking into account the results of time completeness

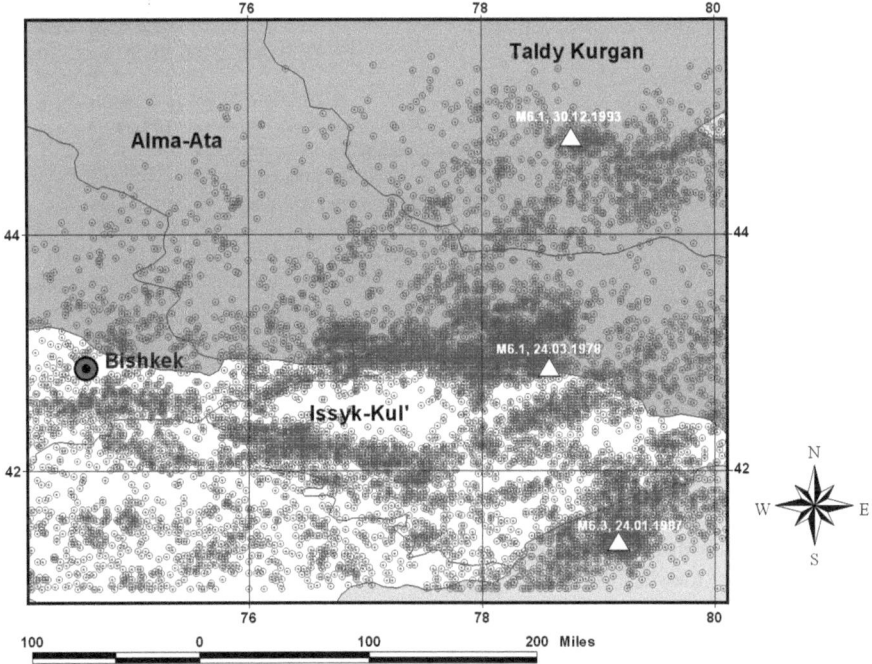

Figure 1
Map of earthquake distribution in the investigated area of North Tien Shan and adjacent areas in Central Asia (1975–1996)

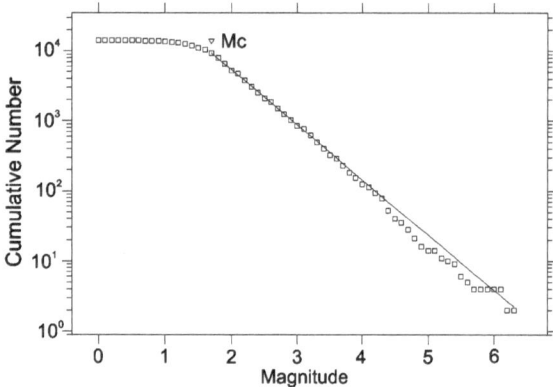

Figure 2
Results of the completeness analysis, where Log cumulative number of earthquake vs. magnitude is given for the entire catalogue

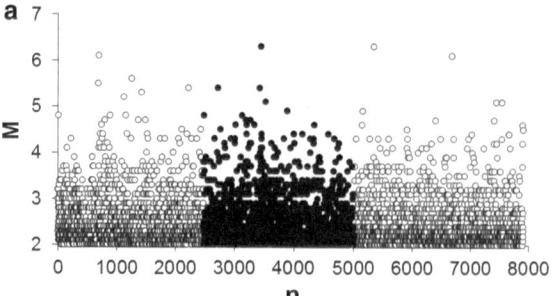

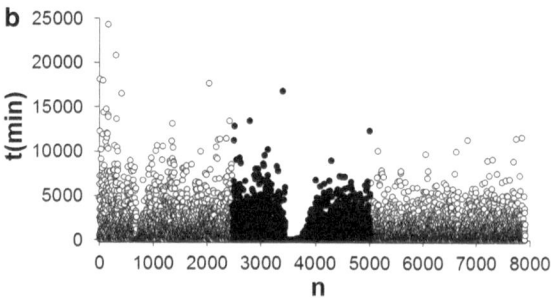

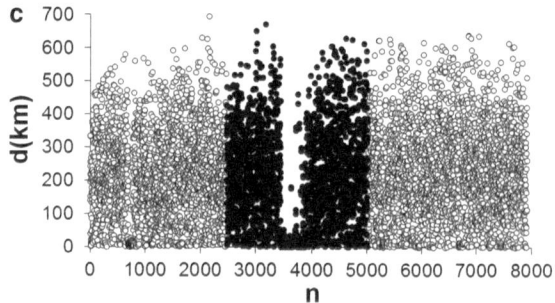

Figure 3
Sequence of **a** earthquakes magnitudes, **b** inter-event time intervals, **c** distances between consecutive events obtained from the earthquake catalogue of North Tien Shan and adjacent areas in Central Asia 1975–1996, $M \geq 2.0$. *Dark circles* correspond to time period of MHD experiments (1983–1990). Noticeable decrease in temporal and spatial domains occurred approximately in the first half year of 1987 (3,400–4,000 data in **b** and **c**)

analysis (not shown here), the $M \geq 2.0$ completeness threshold was used when we considered the whole catalogue for the entire time period under investigation (see also CHELIDZE *et al.* 2006a, 2010).

The time span of the available catalogue enabled us to consider the seismic process during time periods before, during, and after active EM experiments (correspondingly 1975–1983, 1983–1990, and 1990–1996). As it was already mentioned in the previous section, data sets of inter-event (waiting) time intervals, distances between earthquakes' epicenters and magnitude sequences, as well as time series of the number of daily occurred earthquakes, both from complete as well as declustered [according to REASENBERG (1985)] catalogues were analyzed (typical plots of some used data sets are shown in Fig. 3).

During MHD firing experiments in the test area located in the Northern Tien Shan near Bishkek, Kyrgyzstan, more than 100 start-ups were carried out (VOLYKHIN *et al.* 1993).

According to TARASOV (1997) and TARASOV and TARASOVA (2004), the EM energy from the MHD generator was transmitted into the crust through an electrical dipole, which was installed within Paleozoic crystalline structures in the North Tien Shan (42.69°N, 74.68°E); detailed information about the test area can be seen in series of works by Russian colleagues, e.g., VOLYKHIN *et al.* (1993), TARASOV (1997), TARASOV and TARASOVA (2004, 2011),

SMIRNOV and ZAVYALOV (2012), and BOGOMOLOV *et al.* (2011), etc.

The time distribution of MHD soundings was not uniform for the whole period of experiments. As we see in Fig. 4a, time intervals between consecutive MHD firings varied in a wide interval from about 20 to 2,000 h, though about one-third of all discharges started in an approximately consecutive two-week period—Fig. 4b.

Time distribution of the amount of EM energy released during MHD experiments also was not

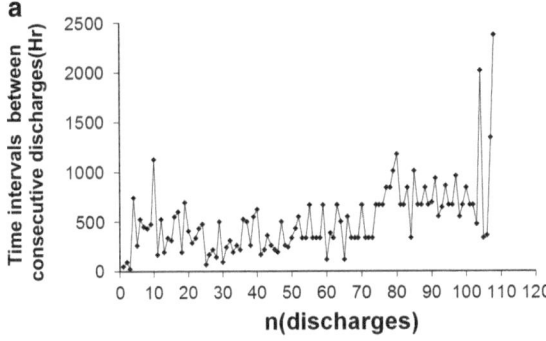

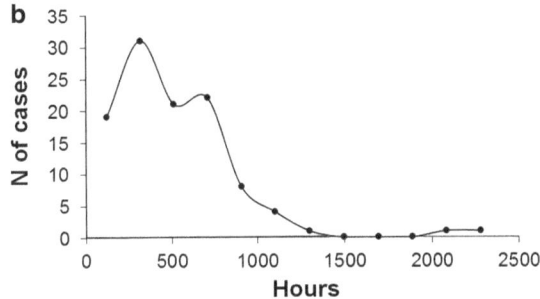

Figure 4

a Sequence of time intervals between consecutive MHD discharges, **b** distribution of time intervals between MHD soundings

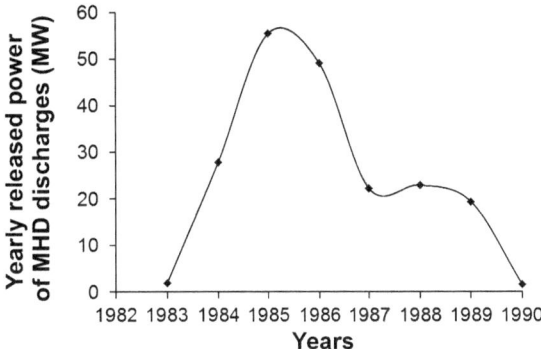

Figure 5

Integral power of EM energy, released yearly during MHD discharges

uniform, reaching a maximum in 1985–1986 (Fig. 5). It is important to note that a drastic fall in the length of the inter-event time intervals as well as inter-earthquake distances (data with sequential numbers of about 3,400–4,000 in Fig. 3) occurred during the first several months in 1987, following those two-years of experiments, when the largest amount of EM energy has been released (Fig. 5).

As it was said in the previous section, induced seismicity is a nonlinear phenomenon occurring in a complex earth crust system. Conclusions about the dynamic features of the processes taking place in such systems, being based only on linear methods of the analysis of data should be treated with caution. At the same time, methods based on linear concepts in combination with nonlinear data analysis may provide valuable new information.

In the present study we indeed combined linear and nonlinear data analysis methods to test whether the dynamical features (extent of regularity) of the seismic process triggered by strong (on human scale) EM impacts undergo changes qualitatively and/or quantitatively.

In selecting appropriate linear tools for analysis, we chose wavelet and Hilbert Huang transformations

(correspondingly WT and HHT). These tools, which are based on different underlying concepts, enabled us to assess the frequency content of the investigated process. As a first approximation such analyses may provide important information about the temporal structure of a targeted complex process.

In order to argue why these methods were selected and used, we provide a very short overview. WT is a well-known, efficient tool for the analysis of data containing non-stationary power distribution at different frequencies (DAUBECHIES 1990). It enables us to decompose a time series into time–frequency domain and to determine both the dominant modes of variability as well as to see how those modes vary in time. This method is based on data decomposition by analyzing functions localized in both frequency and time domains. These basic or analyzing functions—mother wavelets, slide along the time direction and contracts or stretches in the high or low frequency region, respectively. In the recent years the wavelet transformation (WT) has been applied to analysis of many different data sets from geophysics to biology, economics, etc. (see e.g., ASHKENAZY *et al.* 1998; TELESCA *et al.* 2004; ISSAC *et al.* 2004; MATCHARASHVILI *et al.* 2013). WT analysis was useful for our purposes, enabling us to find out whether the distribution of cyclic constituents of the investigated seismic process was uniform throughout the used data sets or if it underwent changes in different periods of observation (what may be regarded as a signature of increased extent of order in the analyzed process).

We used also HHT transformation (HUANG *et al.* 1998). We added HHT to the toolbox analysis because of the well-known time and frequency limitations of WT, as well as the difficulties related to choosing a reliable mother wavelet function (HUANG et al. 1998; HUANG and SHEN 2005), which forces us to be especially cautious when we deal with complicated seismic data sets. HHT presents a relatively new approach to the analysis of non-stationary series of data. This approach is not limited by the above-mentioned time–frequency uncertainty because it uses an adaptive time–frequency decomposition that does not impose a fixed-basis set on the data.

HHT, as a time–frequency analysis technique, consists of two stages. At the first one, the so-called empirical mode decomposition (EMD) stage, the original time series is decomposed into implicit mode functions (IMFs) putting forward the scale characteristics imbedded in the signal (HUANG and SHEN 2005; HUANG and WU 2008). These IMFs represent simple oscillatory modes playing the role similar to a simple harmonic function for spectral analysis, though they are much more general because they have variable amplitudes and frequencies as functions of time, instead of the constant amplitude and frequency of simple harmonic components (HUANG and WU 2008). At the second stage of HHT, Hilbert transformation is applied to the IMFs, yielding a time–frequency representation (Hilbert spectrum). As a result of HHT, a measure of total amplitude (or energy) of contribution from each frequency component to the Marginal spectrum can be computed.

At present there are many examples of successful usage of HHT in different fields of science (DING et al. 2007; RAO and HSU 2008; HUANG and SHEN 2005; TANG et al. 2007; MATCHARASHVILI et al. 2013).

Thus, using the WT and HHT analysis we began to assess of the character of time variation of the strength of cyclic components in analyzed seismic data for the entire period of observation, including periods of the MHD soundings. As it already was mentioned, by these methods we aimed to investigate changes in the extent of regularity of the complex seismic process, through the analysis of the behavior of cyclic constituents in periods prior, during, and after strong EM impacts.

Proposed variations in the extent of regularity of the complex seismic process, besides changes in the strength of cyclic components assessed by WT and HHT, should be revealed by changes in its long-range features. To investigate such possible changes of the power law,

scaling features in the correlation characteristics of seismic process, we used detrended fluctuation analysis (PENG et al. 1993, 1994, 1995). In practice, DFA provides insight into long-term variation features of process through analysis of local variability and was introduced specifically to address nonstationary data sets.

According to the standard DFA procedure, by integration, the "profile" of the original time series is determined. Then the "profile" is divided into boxes of the size n and in each box of the length n, the polynomial local trend is calculated and removed from the profile. After this, a root mean square fluctuation of the integrated and detrended series is calculated. Finally, from the relation between the square root fluctuation and box size, the DFA scaling exponent α can be found (PENG et al. 1993). The DFA scaling exponent α gives information about the correlation properties of the nonstationary time series: $\alpha = 0.5$ corresponds to a white noise (noncorrelated signal); $\alpha < 0.5$ means that the correlation in the signal is anti-persistent. When $\alpha > 0.5$, the correlation in the signal is persistent; $\alpha = 1$ corresponds to a uniform power law behavior of $1/f$ noise and $\alpha = 1.5$ represents a Brownian motion (PENG et al. 1993, 1995). The value $\alpha > 1.5$ corresponds to long-range correlations that may be related to both stochastic and deterministic correlations (PENG et al. 1994, 1995). DFA is a popular tool of time series scaling properties analysis in different fields (e.g., PENG et al. 1993, 1995; RUNDLE et al. 2000; RODRIGUEZ et al. 2007; TELESCA et al. 2008; MATCHARASHVILI et al. 2012).

Next, in order to quantify possible changes that occurred in the dynamical structure of the induced seismic process under strong EM discharges, we used a recurrence quantification analysis (RQA) approach (ZBILUT and WEBBER 1992; WEBBER and ZBILUT 1994; MARWAN et al. 2007), which helps in studying the temporal dynamics of a different time series obtained from nonstationary processes.

In general, RQA is a quantitative extension of the recurrent plot (RP) construction method, which is based on the observation that returns (recurrence) to a certain system condition or state space location is a fundamental property of any deterministic dynamical system (ECKMANN et al. 1987). Recurrence property holds also for systems, which are not exactly deterministic but have nonrandom dynamical structures.

RQA calculations, to be successfully fulfilled, at first necessitate reconstruction of the phase space trajectory from given scalar data sets. The proximity of the phase trajectory points should be tested and marked by the condition that the distance between them is less than a specified threshold ε (Eckman *et al.* 1987). In this way, we achieved a two-dimensional representation of the recurrence features of dynamics embedded in a high-dimensional phase space. Then, the small-scale structure of recurrence plots can be quantified by the recurrence quantification method (Zbilut and Webber 1992; Webber and Zbilut 1994, 2005; Marwan *et al.* 2007; Webber *et al.* 2009). This technique quantifies visual features in an $N \times N$ distance matrix recurrence plot and defines several measures of complexity. Specifically, RQA provides measures of complexity based on the quantification of diagonally and vertically oriented lines in the recurrence plot. In this study, we calculated several such measures, but because conclusions from these measures as a rule are not contradicting, in figures below only two of them are presented. In the figures, we show (1) determinism (DET), the ratio of recurrence points forming diagonal structures to all recurrence points, and (2) laminarity (LAM), the fraction of recurrence points forming vertical lines. This last measure is physically related to the amount of laminar (stable) states in the system because a vertical/horizontal line marks a period in which the state does not change or changes very slowly. In other words, LAM is a measure of system intermittency.

Thus, in this research we combined linear and nonlinear data analysis methods, taking into consideration problems that linear methods face in dealing with complex data. Results of linear analysis have been used mainly to distinguish transitions between more and less ordered states in the local seismic process under the impact of strong EM discharges, and then nonlinear methods were used to make the final conclusion about the character of the observed changes.

3. Results and Discussions

Results of our analysis are presented in Figs. 6–19. We start from the wavelet analysis results, which are shown in Fig. 6. We see that the power of

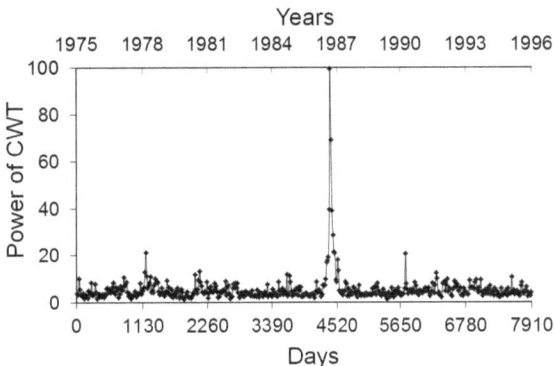

Figure 6
CWT of the time series of number of earthquakes that occurred daily in the entire period of observation (1975–1996), including periods before, during, and after MHD soundings

continuous wavelet transform (CWT) of time series of the number of daily occurred earthquakes essentially increases from about 4,100–4,800 days, starting from the beginning of the observation period (January 1975); thus, the increase took place approximately in 1985–1986, i.e., in the period when the total amount of yearly released MHD energy reached a maximum, see Fig. 5. This result obtained for entire period of observation was further tested separately for the above mentioned three periods: before, during and after active EM experiments (1975–1983, 1983–1990 and 1990–1996). It was confirmed that cyclic components in the time series of earthquakes daily occurrence essentially increased in the period of MHD soundings from the end of 1984 to the end of 1986 (not shown here).

Such increase in the strength of cyclic constituents in the time series of daily earthquake occurrence may point to the changes that occurred in the dynamics of induced seismicity, triggered by man-made influence—EM discharges.

Next, we performed an HHT analysis of time series of the number of earthquakes that occurred daily. For this we used an EMD pre-processing procedure of time series before application of the Hilbert transform. In Fig. 7, IMFs and residues of daily data are presented. We see in this figure that though the frequency range is wide, in most of the IMFs the segments with high amplitudes are localized between 4,000 and 5,000 data, i.e., almost in the same period, which was indicated by WT analysis (approximately

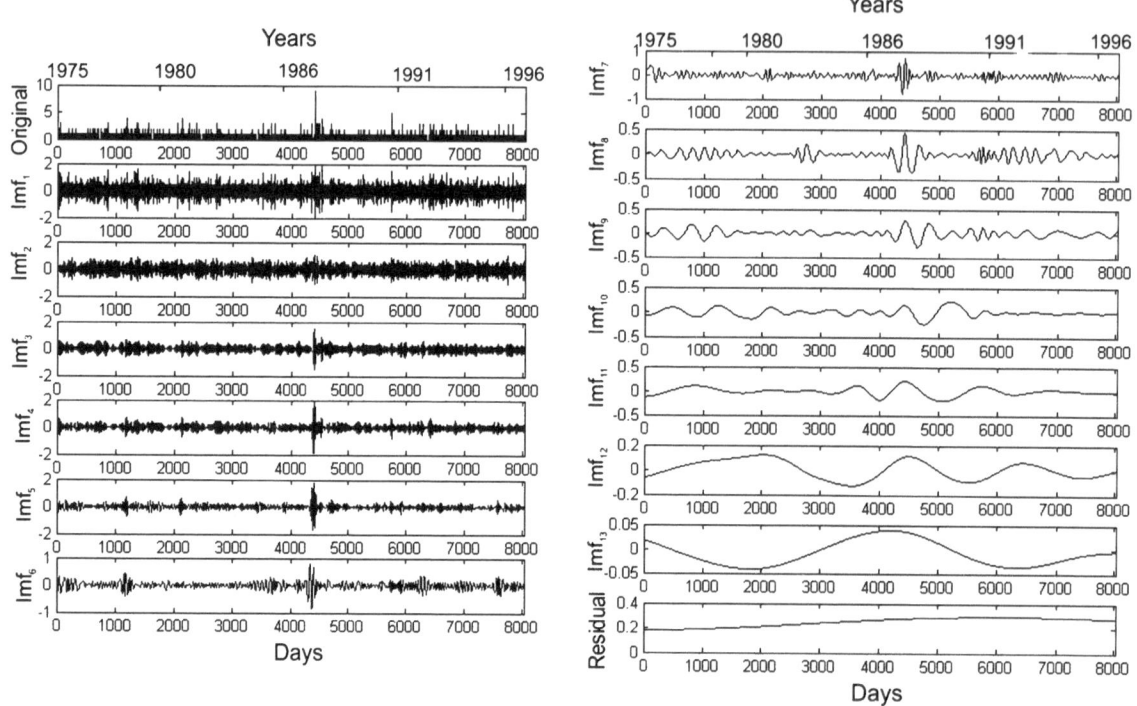

Figure 7
Original time series of the number of earthquakes that occurred daily (*first row*) and their EMD decompositions, vs. time (days) elapsed from the start of catalogue. *Last row* corresponds to a residual

1985–1987). Calculated from these IMFs, the Hilbert power graph confirms that the contribution of cyclic components to time evolution of daily earthquake occurrence indeed increases for this period (Fig. 8).

Thus, using WT and HHT we found that the impact of cyclic constituents or oscillatory modes in the variation of the earthquakes daily occurrence is not uniform for the entire period of observation and undergoes noticeable changes. These changes were revealed as a clear increase of power at time periods of EM discharges from 1985–1987. It can be suggested that these changes in the time–frequency relation are directly related to some inherent features of seismicity triggered by external EM forcing.

For further analysis of the dynamics of the seismic process and the influence of MHD soundings, we assessed long-range correlation properties of daily earthquake occurrence, using DFA at different polynomial fits (p).

As we see in Fig. 9, scaling features of the seismic process also change in the analyzed period

(results for $p = 3$ and $p = 4$ are similar to $p = 2$ and are not shown in Fig. 9). Specifically, due to their self-affine properties the earthquakes' daily distributions are close to random, but become clearly persistent in the period of MHD discharges (1985–1986). Taking into consideration the above-mentioned increase in the contribution of cyclic components, this result can be regarded as an indication of increase in long-range temporal correlations between seismic events triggered by EM influence.

Next, we performed recurrence quantitative analysis, which is a useful tool for the reliable quantification of extent of regularity in complex high-dimensional processes. First, similar to the above considerations we analyzed the longest available data sets: time series of sequences of the daily number of earthquake occurrences. Results of these calculations for the entire time period of observations are presented in Fig. 10. We observed an increase in %DET and %LAM RQA characteristics of daily earthquake occurrence data sets from approximately 1984–1987, i.e., in the period when the maximal (yearly) amount

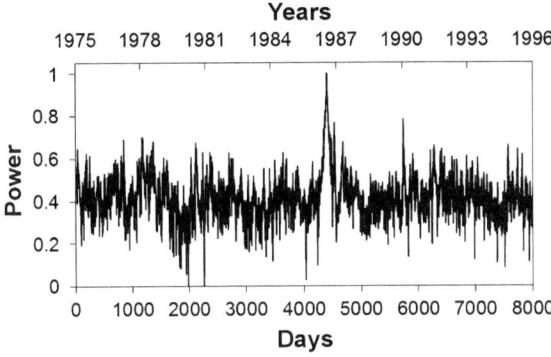

Figure 8

Log power vs. time relation of the Hilbert energy spectrum of time series of the number of earthquakes that occurred daily, vs. time (days) elapsed from the start of catalogue

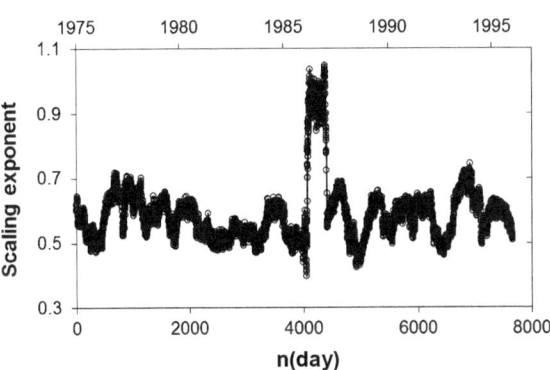

Figure 9

DFA scaling exponents (for $p = 2$) of time series of the number of earthquakes that occurred daily for the entire period of observations, vs. time (days) elapsed from the start of catalogue. Results for $p = 3$ and $p = 4$ are similar to $p = 2$

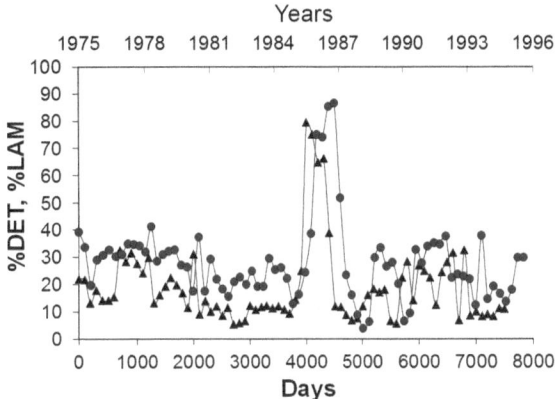

Figure 10

RQA %DET (*triangles*) and %LAM (*circles*) of time series of the number of earthquakes that occurred daily retrieved from the catalogue of Northern Tien Shan and adjacent territories of Central Asia (1975–1996), at an M2.5 representative threshold, vs. time (days) elapsed from the start of catalogue; 500 data length sliding window and 100 data step (in the abscissa, starting points of consecutive sliding windows are shown)

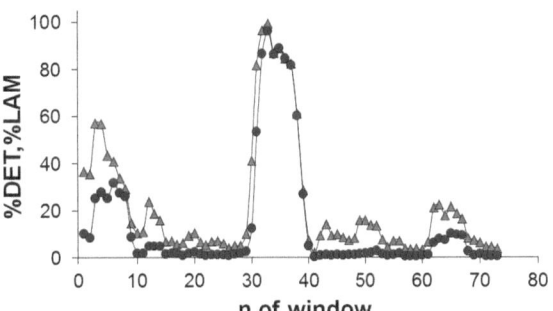

Figure 11

RQA %DET (*circles*) and %LAM (*triangles*) of sequence of inter-event times from the catalogue of Northern Tien Shan and adjacent territories of Central Asia (1975–1996), above an M2.0 representative threshold; 600 data length sliding window and 100 data step (in the abscissa, starting points of consecutive sliding windows are shown)

of EM energy was released, which corresponds to 40–50 sliding windows of analyzed time series. In this period, the daily earthquake distribution became more regular compared to the previous (background) and post-experimental time periods.

Next, we proceeded to analyze the sequences of inter-event (waiting) time intervals between consecutive earthquakes for the M2.0 representative threshold. In Fig. 11, we present the results of %DET and %LAM measure calculations for these data sets. Both RQA measures calculated for a 600 data window shifted by a 100 data step essentially increase from 30–40 windows, i.e., in the period of the ongoing EM experiments. The results of other RQA measure calculations do not contradict these results and are not shown here.

As in the previous analyses, in order to avoid mistakes, which can be caused by problems related to the registration of small earthquakes, we increased the representative threshold to M2.5. Results of this analysis are presented in Fig. 12. Here, we again observed a clear increase in ordering between 5–13 sliding windows, which for this data set and used time step corresponds to the period of EM discharges.

Similar increases of %DET and %LAM values have been observed in the declustered catalogue. Through our analysis of the declustered catalogue we

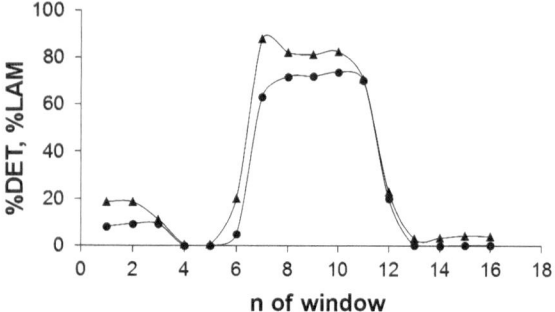

Figure 12

RQA %DET (*circles*) and %LAM (*triangles*) of time series of inter-event times from the catalogue of Northern Tien Shan and adjacent territories of Central Asia (1975–1996), at an M2.5 representative threshold; 500 data length sliding window and 100 data step

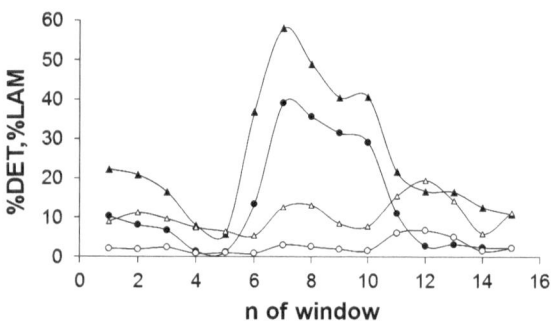

Figure 13

RQA %DET (*circles*) and %LAM (*triangles*) of sequence of inter-event times from declustered catalogue of Northern Tien Shan and adjacent territories of Central Asia (1975–1996), at an M2.5 representative threshold; 500 data length sliding window and 100 data step. *Open figures* correspond to randomized data sets

ensured that the revealed changes in RQA measures were not caused by aftershocks of strong earthquakes ($M \approx 6.1 - 6.3$) that occurred in the periods before, during, and after EM soundings (Fig. 13).

All of the above results, presented in Figs. 6, 7, 8, 9, 10, 11, 12 and 13, could be considered as a clear indication of qualitative and quantitative changes occurring in the temporal domain of the seismic process throughout the considered territory, generated by EM discharges.

At the next step of analysis, we investigated spatial patterns of seismicity, both natural and induced by EM discharges. In particular, we performed the RQA analysis of data sets containing sequences of distances between the epicenters of

consecutive earthquakes in the considered catalogue.

Figure 14 presents the %DET and %LAM characteristics of our RQA analysis data set of inter-earthquake distances. We see strong increases in determinism and laminarity between the 30 and 40 windows. This approximately corresponds to the period from the end of 1986 to the end of 1988. We also see relatively low increases from the 4 to 8 sliding windows, which can be related to an M6.1 earthquake that occurred in March of 1978 in the area of investigation.

In order to exclude the influence of aftershock activity and diminish possible errors caused by small earthquake uncertainty, we continued analysis of inter-earthquake distances for a higher M2.5 threshold as well as for M2.5 events in the declustered catalogue. In Fig. 15, the results of the RQA analysis of these data sets are presented. We clearly see that extent of regularity in the spatial distribution of earthquakes increase in the 7–13 sliding windows, i.e., in 1983–1990. Approximately the same result is obtained for the declustered catalogue (Fig. 15b); %DET and %LAM characteristics increase during the period of 1983–1988.

Thus, it can be said that the seismic process under MHD discharges also becomes noticeably regular (ordered) in the spatial domain compared to periods before and after man-made influence.

Finally, in order to analyze how man-made impacts influence the energy domain of the seismic process, we analyzed data sets of magnitude sequences.

In Fig. 16, we present the results of %DET calculation of magnitude sequences. We see that the randomized data gives results which are hardly different from the %DET results of the original magnitude sequences; randomization of original time series is one of the often-used surrogate data generation methods that are useful in trying to understand whether results of nonlinear analysis express internal features of investigated dynamics or are caused by different random effects (see e.g., KANTZ and SCHREIBER 2000; STROGATZ 2000). At the same time, we observe some increase in %DET characteristics for the time period of MHD discharges (7–9 windows, Fig. 16).

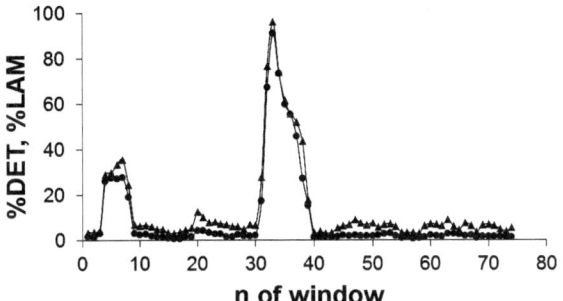

Figure 14

RQA %DET (*black circles*) and %LAM (*white circles*) of inter-earthquake distances sequence of the original catalogue of Northern Tien Shan and adjacent territories of Central Asia, at M1.8; 500 data length sliding window and 100 data step

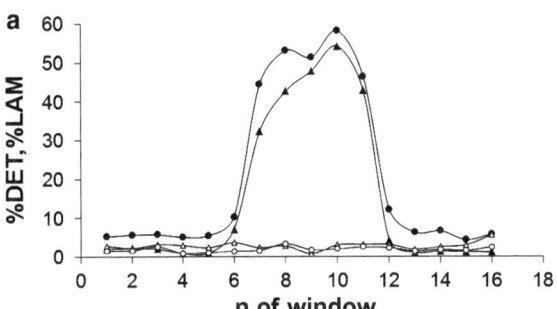

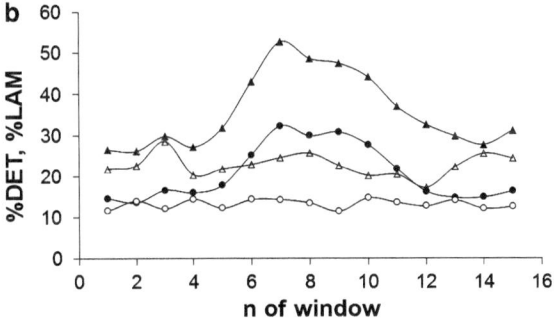

Figure 15

RQA %DET (*black triangles*) and %LAM (*black circles*) of inter-earthquake distances sequence from **a** original and **b** declustered catalogue of Northern Tien Shan and adjacent territories above an M2.5 representative threshold; 500 data length sliding window and 100 data step. *Open figures* correspond to randomized data sets

To have stronger arguments we accomplished the same analysis for the magnitude sequences from the declustered catalogue. Our results, shown in Fig. 17, provide additional arguments to say that MHD discharges in the Bishkek test area have not induced changes in the energy domain of the local seismic process. At the same time we cannot exclude that in Fig. 17 we observed the effects of distortion of the dynamical process caused by declustering. Indeed it is well-known that filtration usually damages the intact structure of a time series (KANTZ and SCHREIBER 2000; STROGATZ 2000). In our case, the effects of catalogue declustering, which in fact is a kind of filtering, cannot be excluded. Thus, further careful analysis should be carried out in the future to make final conclusions about the dynamical features of the process of earthquake energy distribution, which is recognized as high-dimensional in comparison with earthquake temporal and spatial distributions (GOLTZ 1998; MATCHARASHVILI *et al.* 2000).

Here, we have to point out that at present the mechanism of how EM pulses may induce seismicity is not fully understood (TARASOV and TARASOVA 2004, 2011; BOGOMOLOV *et al.* 2011). Generally, according to the present views of the theory of complex systems, small compared to tectonic force influences may lead to essential changes in the dynamics of the seismic process when the tectonic system is close to the critical state (GRASSO and SORNETTE 1998; SCHOLZ 1998; RUNDLE *et al.* 2000). At the same time, an increased number of cited literature convinces us that the energy transmitted by EM impulses into the Earth's crust can trigger induced seismicity, dynamical features of which may be different from those of natural seismicity. The possibility of such dynamical changes was demonstrated in laboratory experiments on the triggering of the stick–slip process by series of external EM and mechanical impacts (CHELIDZE *et al.* 2006b).

The presented results show that the local seismicity induced by the EM discharges is quantifiably different from the natural seismic process by its characteristic dynamical features.

It should be taken into consideration that the results presented in this work have been obtained for data sets derived from the entire area of the available catalogue and for the whole period of observation. At the same time, it is known that triggering factors can induce seismicity only in cases where the medium is at certain stress/strain conditions; in other words, the medium should be pre-stressed to a substantial fraction of its breaking strength (GRASSO and SORNETE 1998; KISSLINGER 1976; RUNDLE *et al.* 2000;

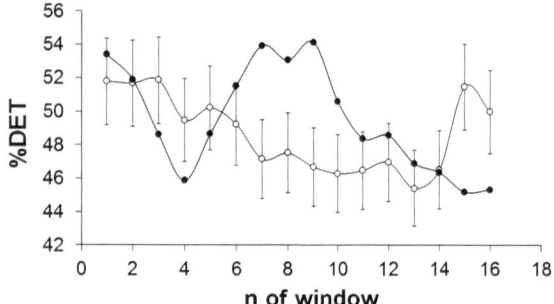

Figure 16

RQA %DET of magnitude sequence from the catalogue of Northern Tien Shan and adjacent territories of Central Asia, above an M2.5 representative threshold (*black circles*). RQA %DET of randomized magnitude sequence are shown by *open circles*; 500 data length windows and 100 data step were used

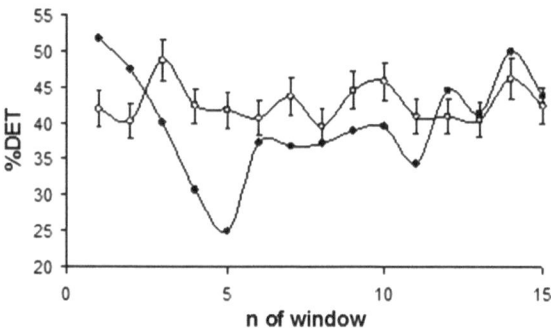

Figure 17

RQA %DET of magnitude sequence from declustered catalogue of Northern Tien Shan and adjacent territories of Central Asia, above an M2.5 representative threshold (*black circles*). RQA %DET of randomized magnitude sequence are shown by *open circles*; 500 data length windows and 100 data step were used

BOGOMOLOV *et al.* 2011). It is obvious that important factors in earthquake generation, such as stress concentrations, the presence of old faults, inhomogeneities in the material properties, etc., cannot be similar over the whole catalogue area (KISSLINGER 1976). Thus, the seismic process cannot be uniformly influenced by MHD impulses throughout the entire area of observation. It is more realistic to suppose that if such influence exists it should be revealed in the certain parts of the area that are closest to the critical state.

Such areas, with anomalous increases of average rates of strain release (by a factor of 10–20 times), were indeed revealed by TARASOV and TARASOVA (2004, 2011) in the territory covered by the catalogue. Namely, the authors investigated features of

seismotectonic strain variations in the Northern Tien Shan for the period of EM excitations and showed that the strongest observed M6.3 earthquake occurred (24.01.1987) in the area with an anomalously high rate of strain release.

Taking this result into consideration, in order to better discern features of natural seismicity and possible changes caused by man-made influences, we decided to analyze separately the areas with different levels of stress/strain release. For this purpose we compiled sub-catalogues (parts of the whole considered catalogue) around epicenters of three strong earthquakes that occurred in areas with different stress release characteristics according to TARASOV and TARASOVA (2011). These earthquakes include: (1) those that occurred before the start of MHD experiments (M6.1, 24.03.1978, lat 42.87, long 78.579), (2) the above-mentioned earthquake (M6.3, 24.01.1987, 41.40, 79.171) that occurred during EM experiments, and (3) an earthquake that occurred long after cessation of the MHD firings (M6.1, 30.12.1993; 44.82, 78.77).

From the whole available catalogue of Northern Tien Shan and adjacent territories (MIKHAILOVA 1990), we formed a catalogues of events that occurred in the 100 km-radius area around epicenters of these three earthquakes for the time period from 1975 to 1996. From these catalogues the sequences of waiting times, magnitudes of consecutive earthquakes, and inter-earthquake distances have been compiled and analyzed by the methods described above.

The most important was the situation with M6.3 earthquake (M6.3, 24.01.1987, 41.40, 79.171), which occurred in the area with increased stress release in the Southeast part of the considered area (see Fig. 6, in TARASOV and TARASOVA 2011). Results of RQA analysis for the data sets retrieved from the catalogue of this area for the entire observation period are shown in Figs. 18 and 19. We see a clear increase of ordering in the temporal and spatial domains during this period, when EM pulses were transmitted into the Earth's crust. Waiting time intervals and inter-earthquake distance sequences for the 100 km-radius areas around the M6.1 earthquakes both before and after the period of MHD runs have also been analyzed. It was established that in these cases, RQA characteristics (including %DET and %LAM) do not show any

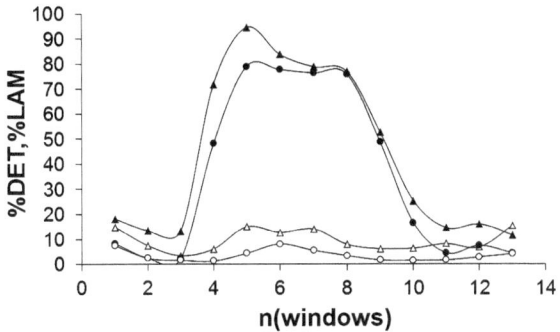

Figure 18

RQA %DET (*black circles*) and %LAM (*black triangles*) of waiting time sequence from the declustered catalogue in a 100-km area around an M6.3 (41.40, 79.171, 24.01.1987) earthquake above the M1.8 representative threshold for the entire observation period. *Open circles* and *triangles* correspond to randomized data sets

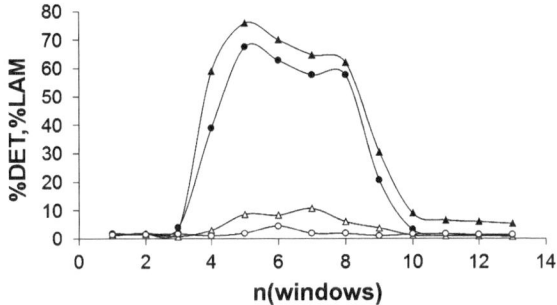

Figure 19

RQA %DET (*black circles*) and %LAM (*black triangles*) of inter-earthquake distances sequence from the declustered catalogue in a 100-km area around an M6.3 (41.40, 79.171, 24.01.1987) earthquake above the M1.8 representative threshold for the entire observation period. *Open circles* and *triangles* correspond to randomized data sets

difference from randomized data sets, so these results are not shown here.

The presented results indicate that the changes described above in the dynamical features of the seismic process (based on data sets derived from the whole catalogue) are generally caused by triggering effects of EM discharges that occurred in the areas (of the whole observation territory) previously recognized as critical in terms of stress release anomalies (TARASOV and TARASOVA 2011).

As it was said in the introduction, significant spatial and temporal variations of the seismic regime in the area of observation have been established (TARASOV 2010; TARASOV and TARASOVA 2011; AVAGIMOV *et al.* 2005; CHELIDZE *et al.* 2006a;

BOGOMOLOV *et al.* 2011; SMIRNOV and ZAVYALOV 2012). It was shown in particular that the weak seismicity after the MHD runs was significantly higher than before the runs, and that the seismicity within the upper 5 km of the crust was more sensitive to electromagnetic impacts. Strong activation of local seismic activity took place with several days' delay after the run. Later influence of EM discharges on seismotectonic strain rate variation was also established (TARASOV and TARASOVA 2004, 2011).

Based on these and similar findings on the approximate spatial and temporal coincidences of MHD discharges and changes in the features of local earthquake generation enables authors to speak about seismic triggering due to EM influence.

At the same time, the exact physical mechanism of such influence is still not known and as we pointed out above, the main attention in the present analysis has been paid to the proof of influence of strong EM discharges on the dynamical aspects of seismic processes in the test area.

At present we can only speculate that EM discharges, in addition to regional stress anomalies, may cause significant changes in local mechanical stresses according to the suggested earlier mechanical percolation model (CHELIDZE 1987; BENGUIGUI 1984). The same model for the electric breakdown (WIESMANN 1989) predicts that near the percolation threshold p_c the breakdown voltage V_a follows power law: $V_a = (p - p_c)^v$, where p is the concentration of conducting paths and p_c is its threshold value, when the infinite cluster of conducting bonds spans the whole system. It is evident that at $p \rightarrow p_c$ the breakdown voltage can be infinitely small. We presume that a similar percolation relation can be suggested for electrostriction forces (CHELIDZE *et al.* 2002); this can help to get a general understanding of the high sensitivity of the seismic process to a weak EM forcing.

At the same time, the triggering of earthquakes, whatever the triggering mechanism, does not necessarily indicate changes in dynamical features of the induced seismic process compared to natural seismicity. The possibility of such changes is demonstrated in the present work.

In our previous research we showed that the time distribution of earthquakes was more regular during the influence of EM discharges than before and after the sounding operations (CHELIDZE *et al.* 2006a, b). As it follows from our new results (Figs. 6, 7, 8, 9, 10, 11, 12, 13, 14, 15, 16, 17, 18, 19), under the influence of strong EM impulses the extent of order increases both in temporal and spatial distributions of earthquakes. At the same time, we do not see changes in the energy domain of the seismic process.

It should be underlined again that the observed changes in spatial and temporal domains of the seismic process have been found in those areas, which according to TARASOV and TARASOVA (2011) are close to the critical state with an increased stress release ratio.

Most important in our findings is that the waiting time series became much more ordered during EM sounding experiments even in the case of EM pulse firings not being quasiperiodic (Fig. 4a). Here, a question arouse: what is the source of the ordering in temporal distribution of seismicity that was strong enough to produce activity during EM experiments. We can suggest some guesses to explain this phenomenon. Though the entire sequence of EM discharges is not periodic, there are some dominant firing intervals in the corresponding EM pulse time distribution (Fig. 4b), at approximately 300 and 800 h. It is possible that this relatively small share of repeating sequences may have imparted some order in the whole seismic process through some synchronization mechanisms (PIKOVSKY *et al.* 2001). On the other hand, influencing the Earth's crust in the areas close to the critical state, we may deal with a situation close to "bang–bang chaos control" described by STARRETT (2003), who shows that the stabilization of a periodic saddle orbit in a strange attractor can be achieved even by the application of a fixed or several fixed perturbations.

Additional experimental and theoretical modeling is needed to choose between suggested explanations of the increase of regularity in temporal and spatial earthquake distributions under strong EM influences.

4. Conclusion

Sequences of waiting times, magnitudes and interearthquake distances, as well as time series of the number of daily occurred earthquakes have been investigated using an earthquake catalogue of Northern Tien Shan and adjacent territories in Central Asia. This catalogue included earthquakes registered around the Bishkek test area, where experiments on deep soundings by strong electromagnetic discharges were carried out in 1975–1996; later, it was discovered that EM impact affects local seismicity.

We investigated features of temporal, spatial, and energetic distributions of earthquakes and compared periods, when the seismic process was affected by strong EM soundings with periods without such influence.

The results of our analysis indicate that MHD discharges strongly affected the dynamics of the seismic process in the test area; namely, they imparted some order in temporal and spatial domains. This conclusion is confirmed by the close similarity of results, obtained by the application of different (linear and nonlinear) methods of time series analysis, such as Wavelet and Hilbert Huang transformations, detrended fluctuation analysis, and recurrence quantitative analysis. At the same time, in the energy domain we do not have clear evidence of the influence of strong EM discharges on dynamics of the seismic process.

It was also shown that the found changes occur only in the areas close to the critical state, for which essential increases in the average rates of strain release increase have been recently confirmed.

Acknowledgments

The research was conducted in the frame of a Rustaveli Scientific Foundation grant FR/567/9-140/12: Dynamically triggered seismicity—a new avenue for earth crust stress state assessment.

REFERENCES

ADUSHKIN, V. V. (2013), *Blasting-induced seismicity in the European part of Russia, Izvestiya*, Physics of the Solid Earth 49(2), 258–277.

AGNEW, D. C., Earth Tides, In Treatise on Geophysics: Geodesy (ed. Herring. T. A.) (Elsevier, New York 2007) pp. 163–195.

ASHKENAZY, Y., LEWKOWICZ, M., LEVITAN, J., MØLGAARD, H., THOMSEN, P. E. B., SÆRMARK, K. (1998), *Discrimination of the healthy and sick cardiac autonomic nervous system by a new wavelet analysis of heartbeat intervals*, Fractals 6, 197–203.

AVAGIMOV, A. A., ZEIGARNIK, V. A., FAINBERG, E. B. (2005), *Electromagnetically Induced Spatial–Temporal Structure of Seismicity, Izvestiya*, Physics of the Solid Earth 41(6), 475–484.

BENGUIGUI L. (1984), *Experimental study of the elastic properties of a percolating system*. Phys. Rev. Lett, *53*, 2028–2032.

BETTINELLI, P., AVOUAC, J. P., FLOUZAT, M., BOLLINGER, L., RAMILLIEN, G., RAJAURE, S., SAPKOTA, S., (2008), *Seasonal variations of seismicity and geodetic strain in the Himalaya induced by surface hydrology*, Earth Planet. Sc. Lett., *266*, 332–344.

BOLLINGER, L., PERRIER, F., AVOUAC, J. P., SAPKOTA, S., GAUTAM, U., TIWARI, D. R., (2007), *Seasonal modulation of seismicity in the Himalaya of Nepal*. Geophys. Res. Lett. *34*. doi:10.1029/2006GL029192.

BOGOMOLOV, L., ZAKUPIN, A., SICHEV, V., Electric impact on the Earth crust and variations of weak seismicity (Lambert Academic Publishing, Saarbrucken 2011). ISBN 978-3-8465-1436-8 (In Russian).

CHELIDZE, T. (1987), Methods of percolation theory in mechanics of geomaterials. Nauka, Moscow, (In Russian).

CHELIDZE T., VARAMASHVILI, N., DEVISDZE, M., CHELIDZE, Z.,CHIKHLADZE, V., MATCHARASHVILI, T. (2002), *Laboratory study of electromagnetic initiationof slip*. Annals of Geophysics, *45*, 1–8.

Chelidze, T. De Rubeis, V., Matcharashvili, T., Tosi, P. (2006a), *Influence of Electro Magnetic strong discharges on the dynamics of earthquakes time distribution at the Bishkek test area*, Annals of Geophysics *49*(4/5), 989–1003.

CHELIDZE, T., LURSMANASHVILI, O., MATCHARASHVILI, T., DEVIDZE, M. (2006b), *Triggering and synchronization of stick slip: waiting times and frequency-energy distribution*, Tectonophysics. *424*, 139–155.

CHELIDZE, T., and MATCHARASHVILI, T. (2007), *Complexity of seismic process; measuring and applications—a review*, Tectonophysics 431, 49–60.

CHELIDZE, T. DE RUBEIS, V. MATCHARASHVILI, T., TOSI, P., Dynamical Changes Induced by Strong Electromagnetic Discharges in Earthquakes' Waiting Time Distribution at the Bishkek Test Area (Central Asia), In Synchronization and Triggering: from Fracture to Earthquake Processes (ed. de Rubeis. V., Czechowski. Z., Teisseyre. R.) (Geoplanet: Earth and Planetary Sciences, Springer 2010) *1*(3), pp. 339–360.

DAUBECHIES, I. (1990), *The wavelet transform, time-frequency localization and signal analysis*, IEEE Transactions on Information Theory 36(5), 961–1005.

DING, H., HUANG, Z., SONG, Z., YAN, Y. (2007), *Hilbert–Huang transform based signal analysis for the characterization of gas–liquid two phase flow*, Flow measurement and instrumentation *18*, 37–46.

ECKMANN, J. P., KAMPHORST, S., RUELLE, D. (1987), *Recurrence plots of dynamical systems*, Europhysics Letters. *4*, 973–977.

GOLTZ, C., Fractal and chaotic properties of earthquakes (Springer, Berlin 1998).

GRASSO, J. R., and SORNETTE, D. (1998), *Testing Self-Organized Criticality by Induced Seismicity*, Journal of Geophysical Research 103, 29965–29987.

GUHA, S.K. (2004), *Induced Earthquakes*, Natural Hazards *31*, 289–290.

HAINZL, S., KRAFT, T., WASSERMANN, J., SCHMEDES, E., (2006), *Evidence for Rainfall-Triggered Earthquake Activity*, Geophys. Res. Lett., *33*, L19303.

HUANG, N.E, SHEN, Z., LONG, S.R., WU, M.C., SHIH, H.H., ZHENG, Q., YEN, N. C., TUNG, C. C., LIU, H. H. (1998), *The Empirical mode decomposition and the Hilbert spectrum for nonlinear and nonstationary time-series analysis*, Proc. R Soc. Lond. *A 454*, 903–995.

HUANG, N., SHEN, S., The Hilbert-Huang Transform and Its Applications (B & JO Enterprise., Singapore 2005).

HUANG, N., WU, Z. (2008), *A review on Hilbert-Huang transform: method and its applications to geophysical studies*, Reviews of Geophysics 46, RG2006, 1–23.

ISSAC, M., RENUKA, G. VENUGOPAL, C. (2004), *Wavelet analysis of long period oscillations in geomagnetic field over the magnetic equator*, Journal of Atmospheric and Solar-Terrestrial Physics *66*, 919–925.

JONES, N. (2001), *The quake machine*, New Scientist. *30*, 34-37.

KANTZ, H., SCHREIBER, T., Nonlinear time series analysis (Cambridge 2000).

KISSLINGER, C. (1976), *A review of theories of mechanisms of induced seismicity*, Engineering Geology 10(2–4), 85–98.

MARWAN, N., ROMANO, M. C., THIEL, M., KURTHS, J. (2007), *Recurrence plots for the analysis of complex system*, Phys. Rep. *438*, 237–329.

MATCHARASHVILI, T., CHELIDZE, T., JAVAKHISHVILI, Z. (2000), *Nonlinear analysis of magnitude and interevent time interval sequences for earthquakes of Caucasian region*, Nonlinear Processes in Geophysics 7, 9–19.

MATCHARASHVILI, T., TELESCA, L., CHELIDZE, T., JAVAKHISHVILI, Z., ZHUKOVA, N. (2013), *Analysis of temporal variation of earthquake occurrences in Caucasus from 1960 to 2011*, Tectonophysics, doi:10.1016/j.tecto.2013.07.033.

MATCHARASHVILI, T., CHELIDZE, T., JAVAKHISHVILI, Z., JORJIASHVILI, N, and ZHUKOVA, N. (2012), *Scaling features of ambient noise at different levels of local seismic activity: A case study for the Oni seismic station*, Acta Geophysica 60(3), 809–832.

METIVIER, L., VIRON, O., CONRAD, C. P., RENAULT, S., DIAMENT, M., PATAU, G. (2009), *Evidence of earthquake triggering by the solid earth tides*, Earth and Planetary Science Letters *278*, 370–375.

MIKHAILOVA, N.N., *Catalogue of Earthquake in the Northern Tien Shan and Adjacent Areas (in Russian)*, Alma Ata, Nauka, 1990.

MEYERS, R., Encyclopedia of Complexity and Systems Science (Springer 2009).

PEINKE, J., MATCHARASHVILI, T., CHELIDZE, T., GOGIASHVILI, J., NAWROTH, A., LURSMANASHVILI, O., JAVAKHISHVILI, Z. (2006), *Influence of Periodic Variations in Water Level on Regional Seismic Activity Around a Large Reservoir: Field and Laboratory Model*, Physics of the Earth and Planetary Interiors *156/1-2*, 130–142.

PENG, C. K., MIETUS, J., HAUSDORFF, J., HAVLIN, S., STANLEY, H. E., and GOLDBERGER, A. L. (1993), *Long-Range Anticorrelations and Non-Gaussian Behavior of the Heartbeat*, Phys. Rev. Lett. *70*, 1343–1346.

PENG, C. K., BULDYREV, S. V., HAVLIN, S., SIMONS, M., STANLEY, H. E. and GOLDBERGER, A. L. (1994), *Mosaic organization of DNA nucleotides*, Phys Rev E *49*, 1685–1689.

PENG, C. K., HAVLIN, S., STANLEY, H. E., and GOLDBERGER, A. L. (1995), *Quantification of scaling exponents and crossover phenomena in nonstationary heartbeat time series*, Chaos 5, 82–87.

PIKOVSKY, A., ROSENBLUM, M., and KURTHS, J., SYNCHRONIZATION, A Universal Concept in Nonlinear sciences (Cambridge University Press 2001).

PREJEAN, S. G., HILL, D. P., BRODSKY, E. E., HOUGH, S. E., JOHNSTON, M. J. S., MALONE, S. D., OPPENHEIMER, D. H., PITT, A. M., RICHARDS-DINGER, K. B. (2004), *Remotely triggered seismicity on the United States west coast following the Mw 7.9 Denali Fault earthquake*, BSSA 6B, *12*, S348–S359.

RAO, R., HSU E. C., Hilbert-Huang transform analysis of hydrological and environmental time series (Water Science and Technology Library 60, Springer 2008).

REASENBERG, P. (1985), *Second-order moment of central California seismicity, 1969-1982*, J. Geophys. Res. *90*, 5479–5495.

RODRIGUEZ, E., ECHEVERRIA, J. C., and ALVAREZ-RAMIREZ, J. (2007), *Detrended fluctuation analysis of heart intrabeat dynamics*, Physica A: Statistical Mechanics and its Applications *384*(2), 429–438.

ROTH, P. PAVONI, N., DEICHMANN, N., (1992), *Seismotectonics of the eastern Swiss Alps and evidence for precipitation-induced variations of seismic activity*, Tectonophysics, *207*, 1–2, 183–197.

RUNDLE, J. B., TURCOTTE, D. L., and KLEIN, W., GeoComplexity and the Physics of Earthquakes, AGU Monograph 120 (American Geophysical Union, Washington, DC 2000).

SCHOLZ, C. H. (1998), Earthquakes and friction laws. Nature 391, 37–42.

SAAR, M. O., MANGA, M., (2003), *Seismicity induced by seasonal groundwaterr echarge at Mt. Hood, Oregon*, Earth and Planetary Science Letters *214*, 605–618.

SIMPSON, J. F., (1968), *Solar activity as a triggering mechanism for earthquakes*, Earth and Planetary Science Letters, *3*, 417–425.

SMIRNOV, V. B., and ZAVYALOV, A. D. (2012), *Seismic Response to Electromagnetic Sounding of the Earth's Lithosphere, Izvestiya*, Physics of the Solid Earth *48*(7–8), 615–639.

SOBOLEV, G. A. and ZAKRZHEVSKAYA, N. A., *On the Seismicity Effect of Magnetic Storms*, Izv. Earth Phys., 2002, vol. *38*, no. 4, pp. 249–261.

STARRETT, J. (2003), *Control of chaos by occasional bang–bang*, Physical Review E *67*, 036203.

STROGATZ, S. H., Nonlinear Dynamics and Chaos with Applications to Physics, Biology, Chemistry and Engineering (Perseus Books Publishing 2000).

TANG, J., ZOU, Q., TANG, Y., LIU, B., ZHANG, X. (2007), *Hilbert-Huang Transform for ECG De-Noising*, Bioinformatics and Biomedical Engineering, 664–667.

TARASOV, N. T. (1997), *Crustal seismicity variation under electric action*, Transactions (Doklady) of the Russian Academy of Sciences *353A*(3), 445–448.

TARASOV, N. G., TARASOVA, N. V., AVAGIMOV, A.A., ZEIGARNIK, V. A. (1999), *The effect of high-power electromagnetic pulses on the seismicity of the central Asia and Kazakhstan*, Vulkanologia i seismologia. *4–5*, 152–160 (in Russian).

TARASOV, N. T., and TARASOVA, N. V. (2004), *Spatial-Temporal Structure of Seismicity of the North Tien Shan and Their Change Under Effect of High Energy Electromagnetic Pulses*, Ann. Geophys. *47*(1), 199–212.

TARASOV, N. T. (2010), *Influence of Strong Electromagnetic Fields on the Seismotectonic Strain Rate*, ISSN 1028_334X, Doklady RAS, Earth Sciences *433*(2), 1088–1091.

TARASOV, N. T., and TARASOVA, N. V. (2011), *Influence of Electromagnetic Fields on the Seismotectonic Strain Rate; Relaxation and Active Monitoring of Elastic Stresses, Izvestiya*, Physics of the Solid Earth *47*(10), 937–950.

TALWANI, P. (1997), *On nature of reservoir-induced seismicity*, Pure Appl. Geophys. *150*, 473–492.

TELESCA, L., LAPENNA, V., ALEXIS, N. (2004), *Multiresolution wavelet analysis of earthquakes*, Chaos, Solitons and Fractals *22*, 741–748.

TELESCA, L., LOVALLO, M., LAPENNA, V., MACCHIATO, M. (2008), *Space-magnitude dependent scaling behaviour in seismic inter-event series revealed by detrended fluctuation analysis*, Physica A. *387*, 3655–3659.

TELESCA, L., MATCHARASHVILI, T., CHELIDZE, T., ZHUKOVA, N., JAVAKHISHVILI, Z. (2013), *Investigating the dynamical features of the time distribution of the reservoir-induced seismicity in Enguri area (Georgia)*, Nat Hazards, doi:10.1007/s11069-013-0855-z.

VOLYKHIN, A. M., BRAGIN, V. D., ZUBOVICH, A. P. (1993), Geodynamic Processes in Geophysical Fields, Moscow, Nauka.

WEBBER, C. L., and ZBILUT, J. P. (1994), *Dynamical assessment of physiological systems and states using recurrence plot strategies*, J. Appl. Physiol. *76*, 965–973.

WEBBER, C. L., and ZBILUT, J. P., Recurrence quantification analysis of nonlinear dynamical systems, In Tutorials in contemporary nonlinear methods for the behavioral sciences (ed. Riley M. A. and Van Orden G. C.) (2005) pp. 26–94.

WEBBER, C.L., MARWAN, N., FACCHINI, A., GIULIANI, A. (2009), *Simpler methods do it better: Success of Recurrence Quantification Analysis as a general purpose data analysis tool*, Physics Letters A *373*, 3753–3756.

WIESMANN, H. Realistic Models of Dielectric Breakdown, In: Fractals' Physical Origin and Properties, Ettore Majorana International Science Series pp 243–257, 1989.

ZBILUT, J. P., and WEBBER, C. L. (1992), *Embeddings and delays as derived from quantification of recurrence plots*, Phys. Lett. A *171*, 199–203.

(Received November 21, 2013, revised May 8, 2014, accepted May 19, 2014, Published online June 8, 2014)

Pure Appl. Geophys. 172 (2015), 1909–1921
© 2014 The Author(s)
This article is published with open access at Springerlink.com
DOI 10.1007/s00024-014-0875-y

❙ Pure and Applied Geophysics

Dynamic Multifractality in Earthquake Time Series: Insights from the Corinth Rift, Greece

GEORGIOS MICHAS,[1] PETER SAMMONDS,[1] and FILIPPOS VALLIANATOS[1,2]

Abstract—Earthquake time series are widely used to characterize the main features of seismicity and to provide useful insights into the dynamics of the seismogenic system. Properties such as intermittency and non-stationary clustering are common in earthquake time series such that multifractal concepts seem essential to describe the temporal clustering variability. Here we use a multifractal approach to study the time dynamics of the recent earthquake activity in the Corinth rift. The results indicate the degree of heterogeneous clustering and correlations acting at all time scales that suggest non-Poissonian behavior. Additionally, the multifractal analysis in different time periods showed that the degree of multifractality exhibits strong variations with time, which are associated with the dynamic evolution of the earthquake activity in the rift and the transition between periods of high and low seismicity.

1. Introduction

Earthquakes are classic examples of complex phenomena that exhibit scale-invariance and fractality in their collective properties. These properties are revealed both in nature and laboratory experiments where the spatial, temporal and size distributions of earthquakes or laboratory acoustic emissions display structures that are invarient in scale (e.g., SAMMONDS et al., 1992; MAIN, 1996; TURCOTTE, 1997). The emergence of these properties is indicative of complexity and nonlinear dynamics in the earthquake generation process (KAGAN, 1994), such that concepts like fractals and multifractals are becoming increasingly fundamental for understanding geophysical processes and estimating seismic hazard more efficiently. The general objective of this kind of analysis

is to characterize the structure of the earthquake activity in a seismic region and evaluate qualitatively and, if possible, quantitatively its dynamical evolution. Following this approach, the fractal dimension has been used to characterize the degree of clustering in seismic sequences and laboratory acoustic emissions (KAGAN and KNOPOFF,1980; SMALLEY et al., 1987; HIRATA et al., 1987; HIRATA, 1989) and to map the dynamical evolution of the states in the seismogenic system (SINGH et al., 2008).

In nature, though, most fractals in complex dynamical systems are known to be heterogeneous (MANDELBROT, 1989). Since earthquakes are the result of complex interactions operating in different scales, it is reasonable to expect the need for multifractal concepts to describe seismicity. In particular, properties like non-stationarity and intermittency are common in earthquake sequences such that the clustering degree varies with time. These variations can be identified by multifractal approaches that can enlighten the local fluctuations in the scaling properties of seismicity and then provide an appropriate tool for mapping the dynamical changes that take place in the physical process of seismogenesis (GEILIKMAN et al., 1990; GODANO et al., 1997). This ability has been in fact used in various studies in the detection of possible temporal changes in the multifractal dimension prior to large earthquakes (HIRABAYASHI et al., 1992; NAKAYA and HASHIMOTO, 2002; KIYASHCHENKO et al., 2003).

Earthquake temporal clustering is revealed in aftershock sequences, where the Omori's scaling relation states that the aftershock production rate decays as a power law with time (OMORI, 1894; UTSU et al., 1995). Short-term and long-term clustering has also been exhibited in various earthquake sequences (KAGAN and JACKSON, 1991) and recently in the West Corinth rift, where the presence of scaling and two

---
[1] Institute for Risk and Disaster Reduction, University College London, Gower Street, London WC1E 6BT, UK. E-mail: georgios.michas.10@ucl.ac.uk
[2] Laboratory of Geophysics and Seismology, Technological Educational Institute of Crete, Chania, Greece.

power law regions in short and long time intervals has been shown (MICHAS et al., 2013). These properties are indicative of non-stationary temporal clustering, multifractality and long-term memory in the earthquake generation process (e.g., LIVINA et al., 2005). The multifractal structure of interevent time series has been exhibited in various earthquake sequences (GODANO and CARUSO, 1995; GODANO et al., 1997; ENESCU et al., 2006; TELESCA and LAPENNA, 2006) and in the West Corinth rift (TELESCA et al., 2002). These properties have led recently to the consideration of non-extensive statistical mechanics as an appropriate framework for studying the collective properties of earthquake time series based on the principle of entropy (e.g., TSALLIS, 2009; VALLI-ANATOS et al., 2012; PAPADAKIS et al., 2013; MICHAS et al., 2013).

In this work we study the earthquake time series of the recent activity in one of the most seismically active areas in Europe, the Corinth rift. The earthquake activity in the Corinth rift is typically characterized by fluctuating behavior, where periods of low to moderate activity are interspersed by sudden seismic bursts, which are related to frequent earthquake swarms and the occurrence of stronger events, followed by aftershock sequences. This intermittent behavior constitutes the multifractal approach as an appropriate tool to study the local fluctuations, the degree of clustering, and the dynamic variability of the seismotectonic activity in this region.

2. 2008–2013 Earthquake Activity in the Corinth Rift

The area of Greece represents the most seismically active area in Europe due to its location on a tectonically active plate boundary in the convergence of the Eurasian and African lithospheric plates (Fig. 1) (e.g., LE PICHON and ANGELIER, 1979). In the back-arc region significant extension is taking place in the Aegean (e.g., MERCIER et al., 1989; LE PICHON et al., 1995) that is accommodated across a series of extending grabens, such as the North Aegean trough, the Euboea graben, and the Corinth rift. In the latter area, a rapid continental extension of the order of 1.5 to 1 cm/year is taking place in the west and east part,

respectively, (BRIOLE et al., 2000) that classifies Corinth rift among the fastest extending continental rifts of the world. Deformation is accommodated by a S-dipping and N-dipping active normal fault system at the north and south margin of the rift respectively of an en echelon E-W system, creating an asymmetric tectonic graben (e.g., ARMIJO et al., 1996).

The high earthquake activity of the area is revealed from both historic and instrumental records, where several earthquakes of magnitude greater than 6 have occurred in the past (AMBRASEYS and JACKSON, 1990, 1997; LATOUSSAKIS et al., 1991; DRAKATOS and LATOUSSAKIS, 1996; PAPAZACHOS and PAPAZACHOU, 1997; PAPADOPOULOS et al., 2000). The last major earthquake was the 1995 Aigion earthquake ($M_s = 6.2$) that occurred in the west part of the rift (BERNARD et al., 1997) and since then the activity is dominated by low to intermediate size earthquakes and frequent earthquake swarms. In January 2010 an $M_w = 5.1$ earthquake, followed four days later by another strong event ($M_w = 5.1$) and numerous aftershocks, occurred in the west part of the rift near the city of Efpalion (Fig. 1), ending a 15-year period of relative seismic quiescence in the area (KARAKOSTAS et al., 2012; SOKOS et al., 2012; GANAS et al., 2013).

In this work we study the 2008–2013 earthquake activity in the Corinth rift, as has been recorded by the Hellenic Unified Seismological Network (HUSN) (http://www.gein.noa.gr/en/networks/husn). HUSN started operating in 2007 as a nationwide unified network, linking the main seismological networks that operate in Greece and providing a better monitoring coverage for the wider area. Evaluation of the seismic network through simulations indicated that the magnitude of completeness (M_c) of HUSN catalog is approximately equal to 2 in most parts of continental Greece, with minimum 1.6 in the West Corinth rift (D'ALESSANDRO et al., 2011). The spatiotemporal evolution of the completeness magnitude (M_c) of HUSN catalog has also been studied in a recently published work (MIGNAN and CHOULIARAS, 2014). In this work, and by using the Bayesian magnitude of completeness (BMC) method (MIGNAN et al., 2011), it was shown that M_c varies between 2 and 2.5 for 2008–2011 in the area of Corinth rift, while an upgrade of the magnitude determination software in 2011 resulted in the improvement of the catalog's

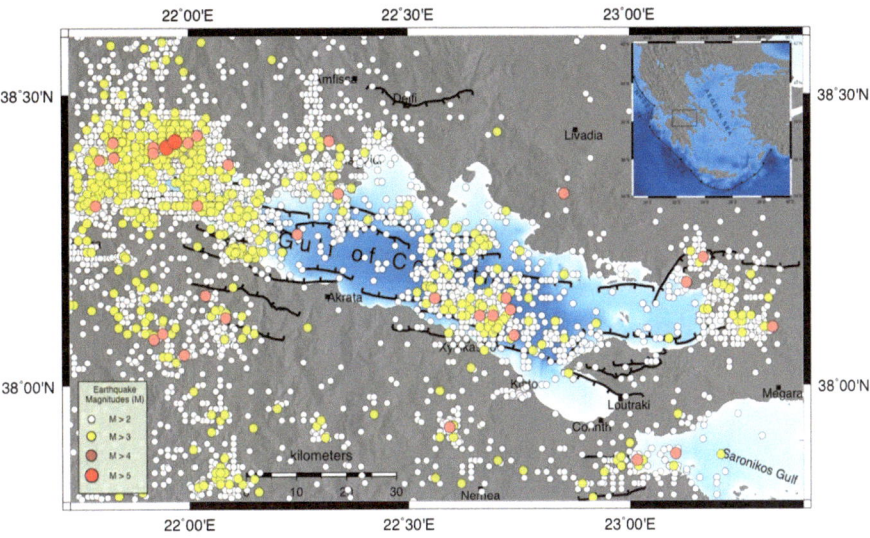

Figure 1

The 2008–2013 earthquake activity in the Corinth rift for $M_L \geq 2$. *Solid black lines* represent faults in the broader area. *Inset* map of the broader area of Greece and the main tectonic features (*SHSZ* South Hellenic Subduction Zone, *NHSZ* North Hellenic Subduction Zone, *KT* Kephalonia Transform Fault). The *rectangle marks* the area of study

quality by shifting M_c one magnitude unit lower, being around 1.5 for the area of central Greece.

The catalogue used in this study was extracted from the database of HUSN, available at (http://bbnet.gein.noa.gr/HL/). In Fig. 1 the crustal earthquake activity (depth ≤ 30 km) of magnitude $M_L \geq 2$ that corresponds to 8,320 events is shown. As can be seen from the seismicity map (Fig. 1), the majority of earthquakes are occurring offshore and in the west part of the rift, while two other clusters can be recognized in the central offshore part and in the northeastern end of the rift. This image of the earthquake activity is consistent with geodetic surveys that indicate that high strain rates are currently accommodated offshore (BRIOLE *et al.*, 2000). In the northwestern part of the rift, the 2010 Efpalion earthquakes are highlighted as the largest earthquakes during this period (Fig. 1). The Efpalion earthquake sequence is also apparent in Fig. 2 where the seismicity rate and the magnitude rate per day are plotted. After the occurrence of the first event on 18 January 2010 there is a sudden increase in the seismicity rate due to the production of numerous aftershocks. Then the seismicity rate continues to fluctuate till the beginning of 2011 where a period of low earthquake activity and a rather constant seismicity rate per day initiates in the rift (Fig. 2a). This

period lasts for more than 2 years where an increase in the seismicity rate is observed due to the occurrence of two earthquake swarms during 22/5/13–26/6/13 that both lasted approximately two weeks. The first one occurred near the city of Aigion and the second one at the northeastern end of the rift near Kapareli that was the epicentral area of the third strong earthquake ($M_s = 6.4$) of the 1981 earthquake sequence (JACKSON *et al.*, 1982).

3. Multifractal Analysis

While several methods have been used in the literature to perform multifractal analysis in non-stationary fluctuating signals, the two most widely used are the wavelet transform modulus maxima method (WTMM) (MUZY *et al.*, 1991) and the multifractal detrended fluctuating analysis (MF-DFA) (KANTELHARDT *et al.*, 2002). The main advantage of these two methods is that they can eliminate polynomial trends from the signal and avoid artifacts due to non-stationarity. Although both methods can reliably estimate the multifractal structure of the analyzed signal, MF-DFA is preferred due to its simpler implementation and computational effort. MF-DFA also seems to perform slightly better than

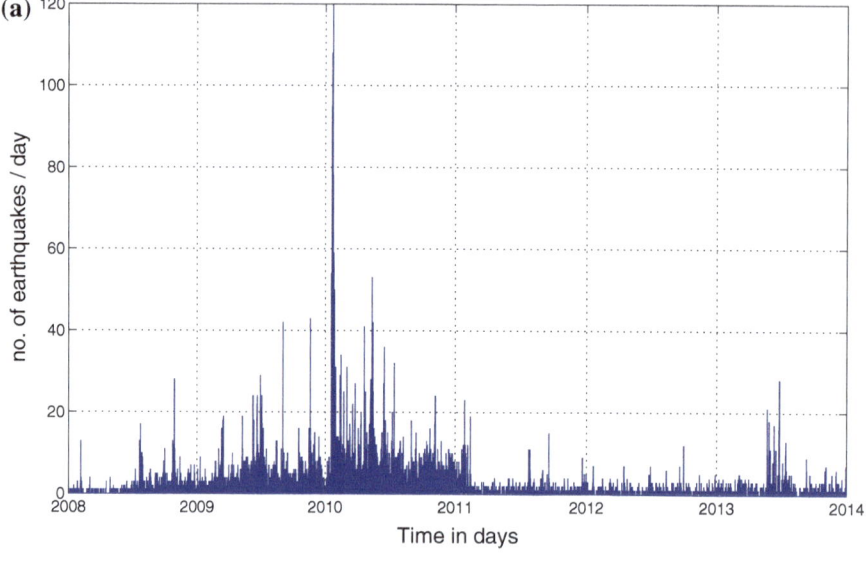

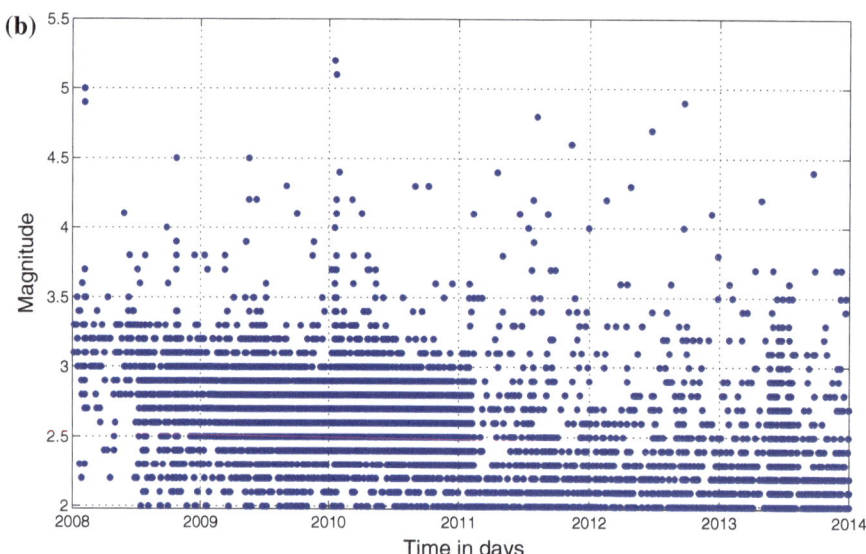

Figure 2
a Seismicity rate per day and **b** magnitude (M_L) rate per day

WTMM for short series (KANTELHARDT *et al.*, 2002), and it is recommended in the majority of situations in which the fractal character of the signal is unknown a priori (OŚWIECIMKA *et al.*, 2006).

Multifractal detrended fluctuation analysis (MF-DFA) consists of the following steps. Let's consider a fluctuating signal $u(i)$ of total length N ($i = 1,...,N$) and of compact support, where $u(i) = 0$ has an insignificant fraction in the series. The signal is first shifted by the average $\langle u \rangle$ and integrated,

$$y(k) = \sum_{i=1}^{k} [u(i) - \langle u \rangle], \qquad (1)$$

with $k = 1,2,...,N$. Then the integrated signal $y(k)$ is divided into non-overlapping segments of equal size n resulting in $N_n = (N/n)$ segments. Since the length N of the signal is not always a multiple of the segments size n, a short part at the end of the integrated signal $y(k)$ is not included in the analysis. In order not to disregard this part, the procedure is repeated starting from the end of $y(k)$, resulting in this way into $2N_n$ total segments (KANTELHARDT et al., 2001). In each of the $2N_n$ segments, the integrated signal $y(v)$ ($v = 1, 2, \ldots 2N_n$) is fitted to a polynomial function $y_n(v)$ that represents the local trend. For different orders l of the polynomial function, different trends can be removed from the signal. Thus, the polynomial trend is linear when $l = 1$, quadratic when $l = 2$, and cubic when $l = 3$. Then the integrated signal $y(v)$ is "detrended" by subtracting the local trend $y_n(v)$ in each window, and the root mean-square fluctuation $F(n, v)$ is calculated,

$$F(n, v) = \sqrt{\frac{1}{2N_n} \sum_{v=1}^{2N_n} [y(v) - y_n(v)]^2}. \qquad (2)$$

Then, the qth order fluctuation function is obtained after averaging over all segments,

$$F_q(n) = \left\{ \frac{1}{2N_n} \sum_{v=1}^{2N_n} [F^2(n, v)]^{q/2} \right\}^{1/q}, \qquad (3)$$

where the index variable q can take any real value. The standard detrended fluctuation analysis (DFA) corresponds to $q = 2$. The previous steps have to be repeated for various time scales n to obtain the relationship between the generalized q dependent fluctuation functions $F_q(n)$ and the time scales n. Then, the scaling behavior of the fluctuation functions can be determined by analyzing the plots of $F_q(n)$ and n for each value of q. If the series $u(i)$ are long-range power law correlated, $F_q(n)$ will increase as a power law for the various values of n,

$$F_q(n) \sim n^{h(q)}. \qquad (4)$$

For a stationary series, $h(2)$ is identical to the Hurst exponent H (HURST, 1951; FEDER, 1988). Thus, $h(q)$ may be called the generalized Hurst exponent or alternatively the q-order Hurst exponent. In the limit $q \to 0$, the value of $h(0)$ cannot be determined directly from Eq. (3) due to the diverging exponent and a logarithmic averaging procedure is employed,

$$F_0(n) \equiv \exp\left\{ \frac{1}{4N_n} \sum_{v=1}^{2N_n} \ln[F^2(n, v)] \right\}. \qquad (5)$$

For multifractal series, the exponent $h(q)$ will depend on the various values of q. In this case small and large fluctuations scale differently. For positive values of q, the segments v with large variance $F^2(n, v)$ will dominate the average $F_q(n)$ and $h(q)$ will describe the scaling behavior of the segments v with large fluctuations. For negative values of q, $h(q)$ will describe the scaling behavior of the segments v with small fluctuations, as the segments v with small variance $F^2(n, v)$ will dominate the average $F_q(n)$ (KANTELHARDT et al., 2002). On the other hand, if the series is monofractal, the scaling behavior of the variances $F^2(n, v)$ is similar for all segments v and $h(q)$ will be independent of q.

The q-order Hurst exponents $h(q)$ is only one of various types of scaling exponents that are used to characterize the multifractal structure of a signal. In fact, it has been shown (KANTELHARDT et al., 2002) that the multifractal scaling exponents $h(q)$ defined in Eq. (4) can be directly related to the q-order mass exponents $\tau(q)$ that correspond to the multifractal formalism based on the standard partition function as,

$$\tau(q) = qh(q) - 1. \qquad (6)$$

Typically, for random and monofractal series the mass exponent $\tau(q)$ has a linear dependency on q, while for multifractal series this linearity breaks.

Another way to characterize multifractal series is the singularity spectrum $f(a)$ that can be directly obtained from the mass exponents $\tau(q)$ by using the Legendre transform (FEDER, 1988):

$$\alpha = \frac{d\tau}{dq}. \qquad (7)$$

Then the singularity spectrum $f(a)$ can be obtained as

$$f(\alpha) = q\alpha - \tau(q), \qquad (8)$$

where a is the singularity strength or Hölder exponent and $f(a)$ indicates the dimension of the subset of the series that has the same singularity strength a. In a monofractal series the singularity strength is the same in the entire range of the set so that the singularity spectrum collapses into a single point. The singularity spectrum $f(a)$ and the generalized Hurst exponents

$h(q)$ can be considered as fundamental characteristics of a multifractal set.

4. Results

We applied MF-DFA in the interevent time series, i.e., the series of the time intervals between the successive earthquake events, defined as $\tau = \tau (i + 1) - \tau (i)$ ($i = 1, 2,..., N - 1$, where N is the total length). We performed the analysis for earthquakes with magnitude $M \geq M_{th} = 2.6$, considering that the selection of this threshold magnitude (M_{th}) insures magnitude completeness for the studied area and the considered period (see discussion in Sect. 2). The interevent time series of total length $N = 5,233$ was initially divided into non-overlapping segments of minimum size $n = 10$ events up to the maximum size of $N/4$. Figure 3 shows the logarithm of the fluctuation functions $F_q(n)$ versus the logarithm of the segment size n that resulted from the analysis for $q \in [-5, 5]$ and step size 0.2. Generally $F_q(n)$ grows as a power-law with n, indicating the presence of scaling. After performing the analysis for various orders l of the "detrending" polynomial function $y_n(v)$, a second order ($l = 2$) polynomial function was considered sufficient to remove any possible trends from the series. For the segments of size greater than 285 events, considerable fluctuations in $F_q(n)$ for $q < -2$ are appearing due to the small number of segments that are formed in greater sizes (Fig. 3). These fluctuations can influence the least square fitting procedure for the low values of q. Thus, we restricted the fitting procedure up to the segment size n of 285 events (Fig. 3). For the various values of q, the fluctuation curves have different slopes in the range 0.623–1.073 for $q = 5$ and $q = -5$, respectively, indicating that small and large fluctuations scale differently. The entire range of the generalized Hurst exponents $h(q)$ is shown in Fig. 4. The exponents $h(q)$ are decreasing monotonically with increasing q that is typical for multifractal sets.

Another way to identify multifractality in a series is the mass exponent $\tau(q)$ that can be estimated directly from Eq. (6). Figure 5 shows that the mass exponent $\tau(q)$ grows nonlinearly with q, exhibiting different behavior for $q < 0$ and $q > 0$ as can be

expected for a multifractal set. Then, by using the Legendre transform (Eq. 7), we estimated the singularity spectrum $f(a)$ from Eq. (8) for the various Hölder exponents a. The singularity spectrum $f(a)$ indicates the fractal dimensions of the subsets that have the same singularity strength a and gives information about the relative importance of each fractal dimension. The singularity strength a can be considered as a local measure of self-similarity in a time series or equivalently as a global indicator of the local differentiability in the time series. Figure 6 shows the singularity spectrum $f(a)$ for the various Hölder exponents a. The spectrum is wide, indicating once again multifractality in the interevent time series. Additionally, the width W of the spectrum that can be estimated as $W = a_{max} - a_{min}$ and in our case is $W = 0.6$, is a measure of how wide the range of fractal dimensions in the series is. It can be then considered as a measure of the degree of multifractality.

In order to evaluate the effect of the threshold magnitude M_{th} on the observed multifractality, we performed the analysis for various M_{th} in the range 2–2.8 and present the results in Fig. 7, where the spectrum's width W versus M_{th} is plotted. In Fig. 7, we can observe that the selection of M_{th} in the range 2–2.6 and the possible incompleteness of the catalog in the lower magnitudes does not affect the results of the analysis, while for greater M_{th} a wider spectrum is observed, indicating a wider range of fractal dimensions.

Next we look into the origin of multifractality in the interevent time series. While different long-term correlations for small and large fluctuations in the earthquake time series are most likely the cause of the observed multifractality, a broad probability distribution can also engender multifractality (KANTELHARDT et al., 2002). The simplest way to distinguish the type of multifractality in the series is by analyzing the corresponding randomly shuffled series. By randomly shuffling the series, any correlations due to the order of the successive earthquakes are destroyed, while the probability distribution remains unchanged. Hence, the shuffled series will exhibit non-multifractal scaling and a random behavior ($h_{shuf}(q) = 0.5$) in the case of long-term correlations. In the other hand, the generalized Hurst

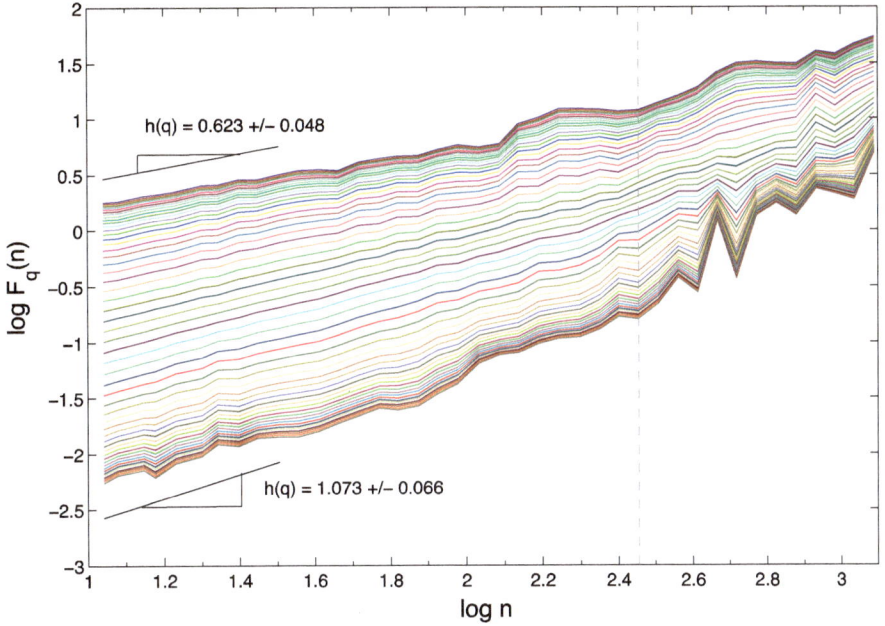

Figure 3

The logarithm of the fluctuation function $F_q(n)$ versus the logarithm of the segment size n for various values of q in the interval $[-5, 5]$ and step size 0.2. A second order polynomial ($l = 2$) was used for detrending the interevent time series

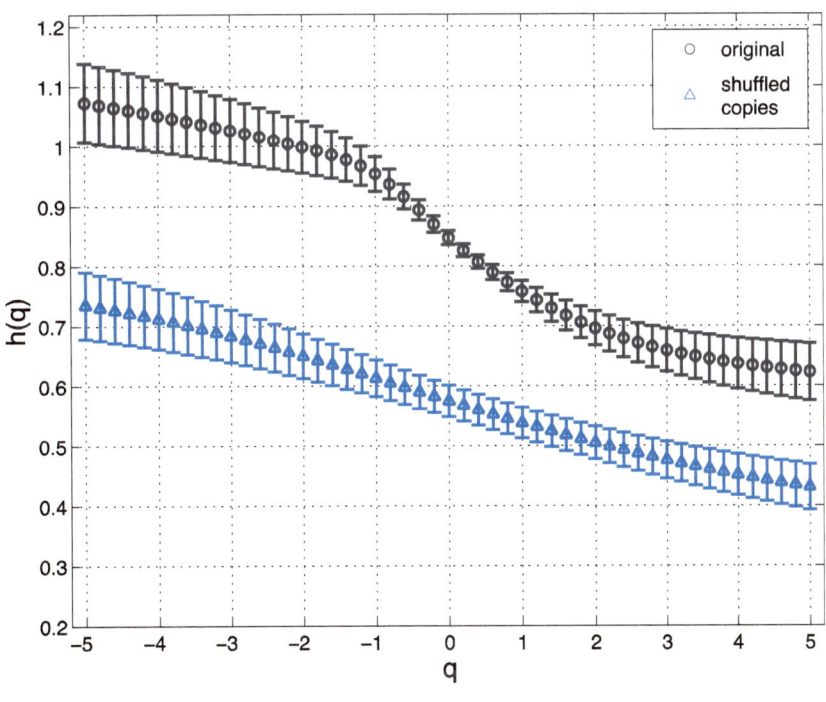

Figure 4

The spectrum of the generalized Hurst exponents $h(q)$ for various values of $q \in [-5, 5]$ and step size 0.2. The corresponding confidence intervals are plotted as *error bars*. The q-dependence of $h(q)$ is typical for multifractal sets. The mean $h(q)$ that resulted from ten randomly shuffled copies of the original interevent time series is also plotted

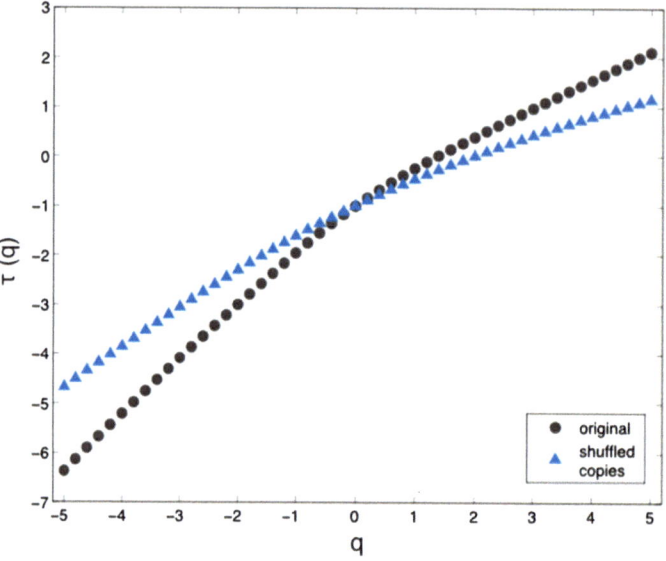

Figure 5

The mass exponent $\tau(q)$ versus q. For the original interevent time series $\tau(q)$ is nonlinear indicating multifractality. The mean $\tau(q)$ that resulted from ten randomly shuffled copies of the original interevent time series is also presented

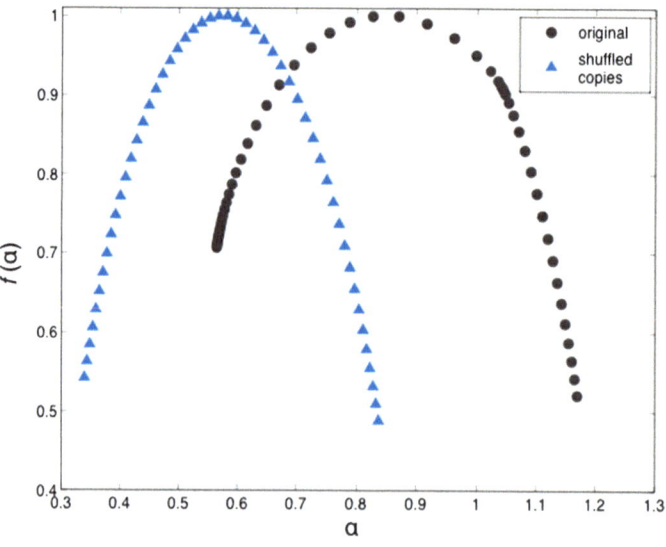

Figure 6

The singularity spectrum $f(a)$ versus the Hölder exponent a. The original interevent time series exhibit a wide spectrum indicating multifractality. The mean spectrum of the randomly shuffled series exhibits a less wide spectrum indicating weaker multifractality

exponents will not be affected by the shuffling procedure ($h(q) = h_{\text{shuf}}(q)$) if multifractality appears due to a broad probability distribution. In the case where both types of multifractality are present in the time series, then the randomly shuffled series will exhibit a weaker multifractality than the original.

We performed this analysis by applying MF-DFA in ten randomly shuffled copies of the original interevent time series. Then we averaged the resulted values to obtain the mean $h(q)$, $\tau(q)$, α and $f(\alpha)$ for the ten randomly shuffled copies, which are shown in Figs. 4, 5, and 6, respectively, along with the results

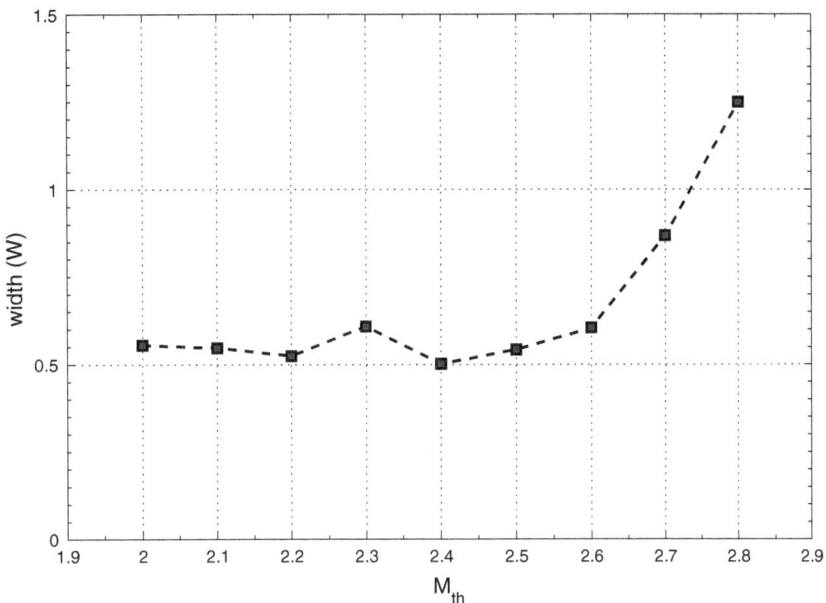

Figure 7
Singularity spectrum's width W for various threshold magnitudes M_{th}

of the analysis on the original series. In Fig. 4 we can see that the shuffling procedure affects the values of the generalized Hurst exponents $h(q)$, which are reduced to the range of values between 0.73 and 0.43 for $q = -5$ and $q = 5$, respectively, exhibiting weaker multifractality than the original series. The corresponding DFA scaling exponent $h(2)$ is approximately equal to 0.5, indicating random behavior. Although the fluctuations of $h(q)$ around the value of 0.5 indicate the loss of long-term correlations in the shuffled series, a q-dependence of $h(q)$ still appears. This can also be observed in the plot of the mass exponents $\tau(q)$ (Fig. 5), where $\tau(q)$ is not linear with q and is made more apparent in the singularity spectrum $f(\alpha)$ (Fig. 6). Although is less wide than the spectrum of the original series, its width is still significant. Then the contribution of a broad probability distribution in the observed multifractality in the time series cannot be excluded. It has been in fact shown that the probability distribution of the interevent times in the West Corinth rift for the period 2001–2008 exhibits both short-term and long-term clustering effects and the presence of scaling at short time and long time intervals (MICHAS et al., 2013). This is a property shared by the dataset that we

consider in the present study that covers the broader area of the Corinth rift (work under preparation).

The multifractal analysis described previously gives us important insights about the time dynamics of the earthquake activity during the considered period, but it does not give us any information about the dynamical changes that may occur in the evolution of the earthquake activity. In order to perform such an analysis, we apply MF-DFA at different time intervals by using a sliding temporal window F that depends on the width w and the sliding factor Δ (GAMERO et al., 1997). In each temporal window the set of interevent times τ is defined according to:

$$F_m = \{\tau_j, j = 1 + m\Delta, \ldots, w + m\Delta\}, m = 0, 1, 2, \ldots, M,$$

where m controls the time displacement of the sliding window. The width of $w = 10^3$ and the sliding factor of $\Delta = 10$ events was used, resulting in 99 % overlapping between the successive windows. This choice of w and Δ assures that in each time window there is a sufficient number of data to perform the analysis and a good smoothing and resolution among the estimated values and the time. Then, in each time window, we apply MF-DFA on the interevent time series as

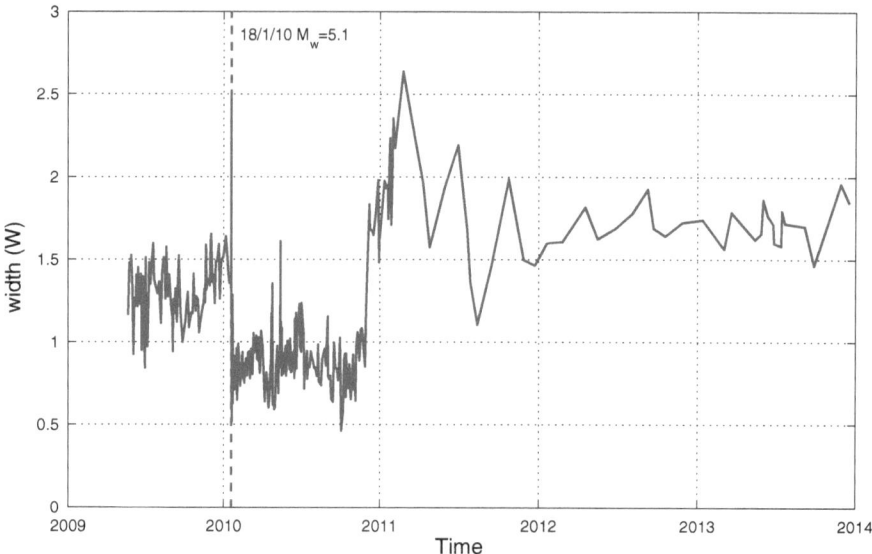

Figure 8
The width (*W*) of the singularity spectrum *f(α)* as a measure of multifractality in time for a window of 1,000 events sliding in time every ten events

previously and estimate the width *W* of the singularity spectrum *f(α)* as a measure of the degree of multifractality in each time period. The estimated value was then associated with the time of the last earthquake in each window. The results of this analysis are presented in Fig. 8. The degree of multifractality exhibits strong variations in time that are associated with the range of fractal dimensions present in each time period. In particular, these variations are more intense during 2009–2011, where the seismicity rate exhibits sharp increases (Fig. 2a). The most abrupt change in the multifractal behavior of the series occurs during the aftershock sequence that followed the 18/1/2010 Efpalion earthquake. During this period, when the aftershock sequence dominates the earthquake activity in the rift, the degree of clustering increases and the time dynamics are more homogeneous, resulting in a loss of multifractality and into a tendency towards monofractality. TELESCA and LAPENNA (2006) have observed a similar behavior in central Italy, where a loss of multifractality in the interevent time series occurred during the aftershock sequence that followed a strong event ($M_D = 5.8$) on September 26, 1997. During the period of 2011–2013, where the earthquake activity is relatively low (Fig. 2a), the degree of multifractality

is nearly constant and the width of the singularity spectrum takes values around 1.5–2. In periods where both low and increased earthquake activity are incorporated in the dataset, the spectrum is "richer" in structure, with a wider range of fractal dimensions. We can observe this effect in the beginning of 2011 and immediately after the occurrence of the 18/1/10 Efpalion earthquake, where a sharp increase of *W* is observed, before the interevent times of the aftershock activity starting to dominate the series, causing the sharp decrease of *W*.

5. Discussion and Conclusions

In the present work we studied the structure and the clustering properties of the 2008–2013 earthquake time series in the Corinth rift by using multifractal detrended fluctuation analysis (MF-DFA), which is a suitable method to study the clustering variability of non-stationary fluctuating signals. The multifractal analysis showed that small time and large time intervals scale differently, indicating a multifractal structure in the interevent time series and a heterogeneous degree of clustering. These results imply non-Poissonian temporal evolution of the earthquake activity in the Corinth rift. The presence of a broad probability distribution on the other hand, can enhance multifractality in the analyzed series. To distinguish this effect, we analyzed the randomly shuffled series, which showed the loss of correlations and a weaker multifractality than the original series, indicating that the observed multifractality is mainly due to different scaling of the subsets and to a lower extent due to a broad probability distribution.

The analysis of the singularity spectrum's width as a degree of multifractality in time exhibited high variability over a large range of scales for the different time periods. The large oscillations are related to variations between high and low earthquake activity, indicating an internal instability in the geodynamic behavior of the system. The most prominent effect is the loss of multifractality after the Efpalion earthquake due to greater homogeneity in the time series of the aftershock sequence that followed the main event. Coulomb stress analysis of the Efpalion earthquake showed static stress triggering of the second strong event four days later (KARAKOSTAS *et al.*, 2012; GANAS *et al.*, 2013), while the spatiotemporal analysis of the entire aftershock sequence indicated a slow stress diffusion process as the most possible mechanism for the slow migration of the aftershock zone after the occurrence of the second strong event (MICHAS *et al.*, 2014). This particular process corresponds to a slow subdiffusive process, in accordance with anomalous stress diffusion in earthquake triggering on a global scale (HUC and MAIN, 2003) and aftershocks diffusion in California (HELMSTETTER *et al.*, 2003). Diffusion is known to generate hierarchical clustering in very short times and small spatial scales (e.g., GODANO *et al.*, 1997). In this process, the influence of small earthquakes on local stress changes and stress diffusion cannot be considered negligible (HELMSTETTER *et al.*, 2005), especially when the spatial clustering is strong like in the Corinth rift. In addition, high pore fluid pressures can reduce the effective normal stress in fault zones and trigger earthquakes. This is a mechanism that might be quite important in the seismogenic process in the Corinth rift, especially during sudden earthquake swarms (BOUROUIS and CORNET, 2009; PACCHIANI and LYON-CAEN, 2010). In a critically stressed crust, these processes can influence

substantially the geodynamic behavior of the system and mark the critical transitions that occur between periods of high and low earthquake activity.

Concluding, the dynamic multifractality in earthquake time series in the Corinth rift indicated strong clustering variability that is inherently related to the seismotectonic complexity in this region, where pore pressure induced seismicity and anomalous stress diffusion can be important factors in the seismogenic process.

Acknowledgments

G. Michas wishes to acknowledge the financial support from the Greek State Scholarships Foundation (IKY). This work has been accomplished in the framework of the postgraduate program and cofunded through the action "Program for scholarships provision I.K.Y. through the procedure of personal evaluation for the 2011–2012 academic year" from resources of the educational program "Education and Life Learning" of the European Social Register and NSRF 2007–2013.

Open Access This article is distributed under the terms of the Creative Commons Attribution License which permits any use, distribution, and reproduction in any medium, provided the original author(s) and the source are credited.

REFERENCES

AMBRASEYS, N.N., and JACKSON, J.A. (1990), *Seismicity and associated strain of central Greece between 1890 and 1988*, Geophys. J. Int. *101*, 663–708.

AMBRASEYS, N.N., and JACKSON, J.A. (1997), *Seismicity and strain in the Gulf of Corinth (Greece)*, J. Earthquake Eng. *1*, 433–474.

ARMIJO, R., MEYER, B., KING, G.C.P., RIGO, A., and PAPANASTASSIOU, D. (1996), *Quaternary evolution of the Corinth Rift and its implications for the Late Cenozoic evolution of the Aegean*, Geophys. J. Int. *126*, 11–53.

BERNARD, P., BRIOLE, P., MEYER, B., LYON-CAEN, H., GOMEZ, J.-M., TIBERI, C., BERGE, C., CATTIN, R., HATZFELD, D., LACHET, C., LEBRUN, B., DESCHAMPS, A., COURBOULEX, F., LARROQUE, C., RIGO, A., MASSONNET, D., PAPADIMITRIOU, P., KASSARAS, J., DIAGOURTAS, D., MAKROPOULOS, K., VEIS, G., PAPAZISI, E., MITSAKAKI, C., KARAKOSTAS, V., PAPADIMITRIOU, E., PAPANASTASSIOU, D., CHOULIARAS, M., and STAVRAKAKIS, G. (1997), *A low angle normal fault earthquake: the Ms = 6.2, June 1995 Aigion earthquake (Greece)*, J. Seismol. *1*, 131–150.

BOUROUIS, S., and CORNET, F.H. (2009), *Microseismic activity and fluid fault interactions: Some results from the Corinth Rift Laboratory (CRL), Greece*, Geophys. J. Int. *178*, 561–580.

BRIOLE, P., RIGO, A., LYON-CAEN, H., RUEGG, J., PAPAZISSI, K., MISTAKAKI, C., BALODIMOU, A., VEIS, G., HATZFELD, D., and DESCHAMPS, A. (2000), *Active deformation of the gulf of Korinthos, Greece: results from repeated GPS surveys between 1990 and 1995*, J. Geophys. Res. *105*, 25605–25625.

D'ALESSANDRO, A., PAPANASTASSIOU, D., and BASKOUTAS, I. (2011), *Hellenic Unified Seismological Network: an evaluation of its performance through SNES method*, Geophys. J. Int. *185*, 1417–1430.

DRAKATOS, G., and LATOUSSAKIS, J. (1996), *Some features of aftershock patterns in Greece*, Geophys. J. Int. *126*, 123–134.

ENESCU, B., ITO, K., and STRUZIK, Z.R. (2006), *Wavelet-based multiscale resolution analysis of real and simulated time-series of earthquakes*, Geophys. J. Int. *164*, 63–74.

FEDER, J., Fractals (Plenum Press, New York 1988).

GAMERO, L., PLASTINO, A., and TORRES, M.E. (1997), *Wavelet analysis and nonlinear dynamics in a nonextensive setting*, Physica, A *246*, 487–509.

GANAS, A., CHOUSIANITIS, K., BATSI, E., KOLLIGRI, M., AGALOS, A., CHOULIARAS, G., and MAKROPOULOS, K. (2013), *The January 2010 Efpalion earthquakes (Gulf of Corinth, central Greece): Earthquake interactions and blind normal faulting*, J. Seismol. *17*, 465–484.

GEILIKMAN, M.B., GOLUBEVA, T.V., and PISARENKO, V.F. (1990), *Multifractal patterns of seismicity*, Earth Planet. Sci. Lett. *99*, 127–132.

GODANO, C., and CARUSO, V. (1995), *Multifractal analysis of earthquake catalogues*, Geophys. J. Int. *121*, 385–392.

GODANO, C., ALONZO, M.L., and VILARDO, G. (1997), *Multifractal Approach to Time Clustering of Earthquakes. Application to Mt. Vesuvio Seismicity*, Pure Appl. Geophys. *149*, 375–390.

HELMSTETTER, A., OUILLON, G., and SORNETTE, D. (2003), *Are aftershocks of large California earthquakes diffusing?* J. Geophys. Res. B *108*, 2483.

HELMSTETTER, A., KAGAN, Y.Y., and JACKSON, D.D. (2005), *Importance of small earthquakes for stress transfers and earthquake triggering*, J. Geophys. Res. B *110*, 1–13.

HIRABAYASHI, T., ITO, K., and YOSHII, T. (1992), *Multifractal Analysis of Earthquakes*, Pure Appl. Geophys. *138*, 591–610.

HIRATA, T. (1989), *A correlation between the b-value and the fractal dimension of earthquakes*, J. Geophys. Res. *94*, 7507–7514.

HIRATA, T., SATOH, T., and ITO, K. (1987), *Fractal structure of spatial distribution of microfracturing in rock*, Geophys. J. R. Astr. Soc. *90*, 369–374.

HUC, M., and MAIN, I.G. (2003), *Anomalous stress diffusion in earthquake triggering: Correlation length, time dependence, and directionality*, J. Geophys. Res. B, *108*, 2324.

HURST, H.E. (1951), *Long-term storage capacity of reservoirs*, T. Am. Soc. Civ. Eng. *116*, 770–808.

JACKSON, J.A., GAGNEPAIN, J., HOUSEMAN, G., KING, G.C.P., PAPADIMITRIOU, P., SOUFLERIS, C., and VIRIEUX, J. (1982), *Seismicity, normal faulting and the geomorphological development of the Gulf of Corinth (Greece): the Corinth earthquakes of February and March 1981*, Earth Planet. Sci. Lett. *57*, 377–397.

KAGAN, Y.Y. (1994), *Observational evidence for earthquakes as a nonlinear dynamic process*, Physica D *77*, 160–192.

KAGAN, Y.Y., and JACKSON, D.D. (1991), *Long-term earthquake clustering*, Geophys. J. Int. *104*(1), 117–133.

KAGAN, Y.Y., and KNOPOFF, L. (1980), *Spatial distribution of earthquakes: the two-point correlation function*, Geophys. J. R. Astr. Soc. *62*, 303–320.

KANTELHARDT, J.W., KOSCIELNY-BUNDE, E., REGO, H.A.A., HAVLIN, S., and BUNDE, A. (2001), *Detecting long-range correlation with detrended fluctuation analysis*, Physica A *295*, 441–454.

KANTELHARDT, J.W., ZSCHIEGNER, S.A., KOSCIELNY-BUNDE, E., HAVLIN, S., BUNDE, A., and STANLEY, H.E. (2002), *Multifractal detrended fluctuation analysis of nonstationary time series*, Physica A *316*, 87–114.

KARAKOSTAS, V., KARAGIANNI, E., and PARADISOPOULOU, P. (2012), *Space-time analysis, faulting and triggering of the 2010 earthquake doublet in western Corinth gulf*, Nat.Haz. *63*(2), 1181–1202.

LATOUSSAKIS, J., STAVRAKAKIS, G., DRAKOPOULOS, J., PAPANASTASSIOU, D., and DRAKATOS, G. (1991), *Temporal characteristics of some earthquake sequences in Greece*, Tectonophysics *193*, 299–310.

LE PICHON, X., and ANGELIER, J. (1979), *The Hellenic arc and trench system: a key to the neotectonic evolution of the Eastern Mediterranean region*, Tectonophysics *60*, 1–42.

LE PICHON, X., CHAMOT-ROOKE, N., LALLEMANT, S., NOOMEN, R., and VEIS, G. (1995), *Geodetic determination of the kinematics of central Greece with respect to Europe: Implications for eastern Mediterranean tectonics*, J. Geophys. Res. *100*, 12675–12690.

KIYASHCHENKO, D., SMIRNOVA, N., TROYAN, V., and VALLIANATOS, F. (2003), *Dynamics of multifractal and correlation characteristics of the spatio-temporal distribution of regional seismicity before the strong earthquakes*, Nat. Haz. Earth Syst. Sci. *3*, 285–298.

LIVINA, V.N., HAVLIN, S., and BUNDE, A. (2005), *Memory in the occurrence of earthquakes*, Phys. Rev. Lett., *95*, 208501.

MAIN, I. (1996), *Statistical physics, seismogenesis, and seismic hazard*, Rev. Geophys., *34*, 433–462.

MANDELBROT, B.B. (1989), *Multifractal Measures, Especially for the Geophysicist*, Pure Appl. Geophys. *131*, 5–42.

MERCIER, J.L., SOREL, D., VERGELY, P., and SIMEAKIS, K. (1989), *Extensional tectonic regimes in the Aegean basins during the Cenozoic*, Basin Res. *2*, 49–71.

MICHAS, G., VALLIANATOS, F., and SAMMONDS, P. (2013), *Non-extensivity and long-range correlations in the earthquake activity at the West Corinth rift (Greece)*, Nonlin. Processes Geophys. *20*, 713–724.

MICHAS, G., VALLIANATOS, F., KARAKOSTAS, V., PAPADIMITRIOU, E., and SAMMONDS, P. (2014), *Anomalous stress diffusion, Omori's law and Continuous Time Random Walk in the 2010 Efpalion aftershock sequence (Corinth rift, Greece)*, Geophys. Res. Abstracts *16*, EGU2014-6552.

MIGNAN, A., and CHOULIARAS, G. (2014), *Fifty years of seismic network performance in Greece (1964–2013): Spatiotemporal evolution of the completeness magnitude*, Seismol. Res. Lett. *85*, 657–667.

MIGNAN, A., WERNER, M. J., WIEMER, S., CHEN, C.-C., and WU, Y.-M. (2011), *Bayesian estimation of the spatially varying completeness magnitude of earthquake catalogs*, Bull. Seismol. Soc. Am. *101*, 1371–1385.

MUZY, J.F., BACRY, E., and ARNEODO, A. (1991), *Wavelets and multifractal formalism for singular signals: Application to turbulence data*, Phys. Rev. Lett. *67*, 3515–3518.

NAKAYA, S., and HASHIMOTO, T. (2002), *Temporal variation of multifractal properties of seismicity in the region affected by the mainshock of the October 6, 2000 Western Tottori Prefecture, Japan, earthquake (M = 7.3)*, Geophys. Res. Lett. *29*, 133–1.

OMORI, F. (1894), *On the aftershocks of earthquakes*, J. Coll. Sci. Imp. Univ. Tokyo *7*, 111–216.

OŚWIECIMKA, P., KWAPIEŃ, J., and DROZDZ, S. (2006), *Wavelet versus detrended fluctuation analysis of multifractal structures*, Phys. Rev. E *74*, 1.

PACCHIANI, F., and LYON-CAEN, H. (2010), *Geometry and spatio-temporal evolution of the 2001 Agios Ioanis earthquake swarm (Corinth rift, Greece)*, Geophys. J. Int. *180*, 59–72.

PAPADAKIS, G., VALLIANATOS, F., and SAMMONDS, P. (2013), *Evidence of Nonextensive Statistical Physics behavior of the Hellenic Subduction Zone seismicity*, Tectonophysics *608*, 1037–1048.

PAPADOPOULOS, G.A., DRAKATOS, G., and PLESSA, A. (2000), *Foreshock activity as a precursor of strong earthquakes in Corinthos Gulf, Central Greece*, Phys. Chem. Earth *25*, 239–245.

PAPAZACHOS, B., and PAPAZACHOU, K., Earthquakes in Greece (Ziti, Thessaloniki 1997).

SAMMONDS, P.R., MEREDITH, P.G., and MAIN, I.G. (1992), *Role of pore fluids in the generation of seismic precursors to shear fracture*, Nature *359*, 228–230.

SINGH, C., BHATTACHARYA, P.M., and CHADHA, R.K. (2008), *Seismicity in the Koyna-Warna reservoir site in western India: Fractal and b-value mapping*, Bull. Seismol. Soc. Am. *98*, 476–482.

SMALLEY, R.F., JR. CHATELAIN, J.-L., TURCOTTE, D.L., and PREVOT, R. (1987), *A Fractal Approach to the Clustering of Earthquakes: Applications to Seismicity of the New Hebrides*, Bull. Seismol. Soc. Am. *77*, 1368–1381.

SOKOS, E., ZAHRADNÍK, J., KIRATZI, A., JANSKÝ, J., GALLOVIČ, F., NOVOTNY, O., KOSTELECKÝ, J., SERPETSIDAKI, A., and TSELENTIS, G.-A. (2012), *The January 2010 Efpalio earthquake sequence in the western Corinth Gulf (Greece)*, Tectonophysics *530–531*, 299–309.

TELESCA, L., and LAPENNA, V. (2006), *Measuring multifractality in seismic sequences*, Tectonophysics *423*, 115–123.

TELESCA, L., LAPENNA, V., and VALLIANATOS, F. (2002), *Monofractal and multifractal approaches in investigating scaling properties in temporal patterns of the 1983–2000 seismicity in the western Corinth graben, Greece*, Phys. Earth Planet. Int. *131*, 63–79.

TSALLIS, C., Introduction to nonextensive statistical mechanics: Approaching a complex world (Springer, Berlin 2009).

TURCOTTE, D.L., Fractals and Chaos in Geology and Geophysics (Cambridge University Press, Cambridge, UK, 2nd ed. 1997).

UTSU, T., OGATA, Y., MATSUURA R.S. (1995), *The centenary of the Omori formula for a decay law of aftershock activity*, J. Phys. Earth *43*, 1–33.

VALLIANATOS, F., MICHAS, G., PAPADAKIS, G., and SAMMONDS, P. (2012), *A non-extensive statistical physics view to the spatio-temporal properties of the June 1995, Aigion earthquake (M = 6.2) aftershock sequence (West Corinth rift, Greece)*, Acta Geophys. *60*(3), 758–768.

(Received April 1, 2014, revised May 26, 2014, accepted June 4, 2014, Published online July 9, 2014)

Reprinted from the journal

Pure Appl. Geophys. 172 (2015), 1923–1931
© 2014 The Author(s)
This article is published with open access at Springerlink.com
DOI 10.1007/s00024-014-0876-x

| Pure and Applied Geophysics

A Nonextensive Statistical Physics Analysis of the 1995 Kobe, Japan Earthquake

GIORGOS PAPADAKIS,[1] FILIPPOS VALLIANATOS,[1,2] and PETER SAMMONDS[1]

Abstract—This paper presents an analysis of the distribution of earthquake magnitudes for the period 1990–1998 in a broad area surrounding the epicenter of the 1995 Kobe earthquake. The frequency–magnitude distribution analysis is performed in a nonextensive statistical physics context. The nonextensive parameter q_M, which is related to the frequency-magnitude distribution, reflects the existence of long-range correlations and is used as an index of the physical state of the studied area. Examination of the possible variations of q_M values is performed during the period 1990–1998. A significant increase of q_M occurs some months before the strong earthquake on April 9, 1994 indicating the start of a preparation phase prior to the Kobe earthquake. It should be noted that this increase coincides with the occurrence of six seismic events. Each of these events had a magnitude $M = 4.1$. The evolution of seismicity along with the increase of q_M indicate the system's transition away from equilibrium and its preparation for energy release. It seems that the variations of q_M values reflect rather well the physical evolution towards the 1995 Kobe earthquake.

1. Introduction

The Kobe (Hyogo-ken Nanbu) earthquake ($M = 7.2$) occurred on January 17, 1995 (5:46 a.m. Japan local time) in the southwestern part of Japan (Fig. 1a). This earthquake substantially damaged the city of Kobe and its surrounding areas claiming more than 6,000 lives (KIKUCHI and KANAMORI 1996).

In the present study we examine possible variations of the thermostatistical parameter q_M related to the 1995 Kobe earthquake. This parameter, which is derived from the fragment-asperity model (SOTOLON-GO-COSTA and POSADAS 2004), is related to the

frequency-magnitude distribution and can be used as an index of the stability of a seismic area. The aforementioned model comes from first principles and describes the earthquake generation mechanism in a nonextensive statistical physics (NESP) framework (TSALLIS 1988, 2009).

NESP was proposed by TSALLIS (1988) and refers to the nonadditive entropy which is a generalization of Boltzmann–Gibbs (BG) statistical physics.

According to nonextensive formalism (TSALLIS 2009 and references therein), entropy is given by $S_q = k_B(1 - \sum_{i=1}^{W} p_i^q)/(q - 1)$, where k_B is Boltzmann's constant, p_i is a set of probabilities and W is the total number of microscopic configurations. The nonadditive entropy Sq is proposed (TSALLIS 2009) based on simple physical principles and multifractal concepts. The entropic index q introduces a bias in probabilities. Given that $0 < p_i < 1$, we have $p_i^q > p_i$ if $q < 1$, and $p_i^q < p_i$ if $q > 1$. Therefore, $q < 1$ enhances the rare events that have probabilities close to zero, whereas $q > 1$ enhances the frequent events having probabilities close to unity. Following (TSALLIS 2009 and references therein) the proposed entropic form is based on p_i^q. In addition, the entropic form must be invariant under permutation. The simplest expression consistent with the latter statement is $S_q = F(\sum_{i=1}^{W} p_i^q)$, where $F(x)$ is a continuous function. Moreover, the simplest form of $F(x)$ is the linear one. That leads to $S_q = C_1 + C_2 \sum_{i=1}^{w} p_i^q$. As any entropic expression Sq must be a measure of disorder. Thus, $C_1 + C_2 = 0$ (TSALLIS 2009) and $S_q = C_1(1 - \sum_{i=1}^{W} p_i^q)$. In the limit $q \to 1$ Sq approaches the Boltzmann–Gibbs entropy. The simplest way for this approach is when $C_1 = k_B/(q - 1)$. The index q has been interpreted as the degree of nonadditivity and is inherent in systems where many non-independent, long-range interacting subsystems, memory effects and (multi) fractality are present

[1] Institute for Risk and Disaster Reduction, University College London, Gower Street, London WC1E 6BT, UK. E-mail: georgios.papadakis.10@ucl.ac.uk; f.vallianatos@ucl.ac.uk; fvallian@chania.teicrete.gr; p.sammonds@ucl.ac.uk
[2] Laboratory of Geophysics and Seismology, Technological Educational Institute of Crete, Crete, Greece.

Reprinted from the journal

(LYRA and TSALLIS 1998; TSALLIS 2001; VALLIANATOS et al. 2011; VALLIANATOS 2012).

For any two probabilistically independent systems A and B Tsallis entropy satisfies $\frac{S_q(A+B)}{k} = \frac{S_q(A)}{k} + \frac{S_q(B)}{k} + (1-q)\frac{S_q(A)}{k}\frac{S_q(B)}{k}$. The last term on the right hand side of this equation brings the origin of nonadditivity.

NESP formalism has been applied in many nonlinear dynamical systems (TSALLIS 2009) and seems an appropriate framework for the study of complex phenomena including various types of natural hazards such as earthquakes, forest fires and landslides (VALLIANATOS 2009). Recently, its applicability in seismicity from local (MICHAS et al. 2013; VALLIANATOS et al. 2012, 2013) to regional (ABE and SUZUKI 2003, 2005; TELESCA 2010a; PAPADAKIS et al. 2013) and global scale (VALLIANATOS and SAMMONDS 2013) reveals its usefulness in the investigation of phenomena exhibiting nonlinearity, fractality and long-range interactions (VALLIANATOS and TELESCA 2012; VALLIANATOS et al. 2012).

In general long-range correlations originate from two processes: the process' memory (temporal correlations) and the process increments' "infinite" variance (heavy tails in the distribution). In this study we solely focus on the latter origin by employing NESP formalism.

Recent studies (SARLIS et al. 2010) based on natural time analysis (VAROTSOS et al. 2001, 2002), from which we deduce the maximum information from a given time series (ABE et al. 2005) and which identifies the critical time before the mainshock occurrence (SARLIS et al. 2008), reveal that a combination of nonextensivity with natural time analysis leads to results that satisfactorily describe the seismicity of Japan (VAROTSOS et al. 2011). In this study we focus on the variations of the nonextensive parameter prior to the Kobe earthquake.

Analysis of the magnitude distribution is performed for the period 1990–1998 in a broad area surrounding the epicenter of the 1995 Kobe earthquake. The temporal variations of q_M values reveal a significant increase of the nonextensive parameter months before the Kobe earthquake. Moreover, the evolution of seismicity is consistent with the observed variations of the nonextensive parameter leading us to recover the main characteristics of the Kobe earthquake dynamics.

2. Data

The dataset used in this study concerns shallow earthquakes (focal depth, $h \leq 60$ km) and is based on the earthquake catalog provided by the Japan University Network Earthquake Catalog (https://wwweic.eri.u-tokyo.ac.jp/CATALOG/junec/monthly.html—last accessed October 2013). It covers the period between 03:32:56.09 January 3, 1990 and 20:51:56.24 December 31, 1998 in the region spanning 34.35°N–35.60°N latitude and 134.00°E–136.00°E longitude (Fig. 1b). This corresponds to a total of 5,811 seismic events with threshold magnitude equal to $m_0 = 2$. The study area is chosen to be a large area surrounding the epicenter and the northeast trending earthquake clusters that define the physical evolution of seismicity related to the 1995 Kobe earthquake. Furthermore, the cluster method introduced by REASENBERG (1985) is used in this study for declustering of the earthquake catalog. This method identifies the aftershocks by linking earthquakes to clusters according to spatial and temporal interaction zones (VANSTIPHOUT et al. 2012). The spatial extent of the interaction zone is chosen according to stress distribution near the mainshock area. The temporal extent of the interaction zone is based on Omori's law. Decluster analysis is performed using the ZMAP program (WIEMER 2001).

Several parameters are used for the declustering procedure. These parameters are chosen based on the results of the applied method and the effect of varying their values. The errors of the epicenter location and depth were taken to be equal to 1.5 and 2, respectively. The minimum value of the look-ahead time for building clusters, when the first event is not clustered (τ_{min}), and the maximum value of the look-ahead time for building clusters (τ_{max}) have been set equal to 0.5 day and 15 days, respectively. The probability of detecting the next clustered event (P_1) and the factor for the interaction radius of dependent events (R_{fact}) have been set equal to 0.95 % and 10, respectively.

Figure 2 shows the time distribution of seismicity of the declustered catalog for the period 1990–1998. We observe the absence of earthquakes with magnitude $M \geq 4$ between 1991 and 1993 and the occurrence of many significant events as we move towards the 1995 Kobe earthquake.

(a)

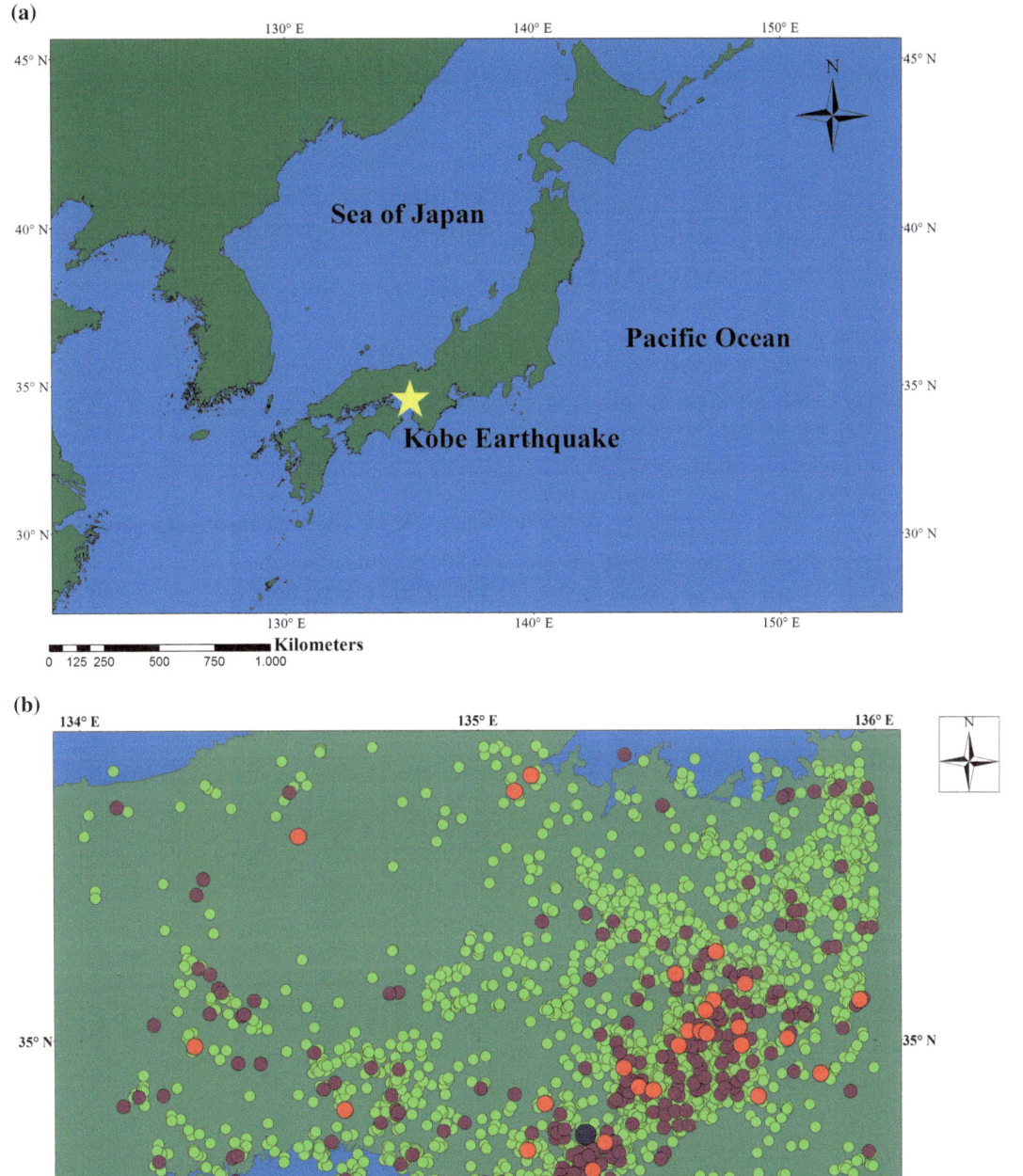

Figure 1
a The epicenter of the 1995 Kobe earthquake ($M = 7.2$) indicated by the *yellow star;* **b** The seismic events (*colored circles*) of shallow earthquakes (focal depth, $h \leq 60$ km) of the declustered earthquake catalog. The *yellow star* indicates the epicenter of the 1995 Kobe earthquake ($M = 7.2$)

Reprinted from the journal

3. The Fragment-Asperity Model for Earthquakes: A Nonextensive Approach of the Frequency–Magnitude Distribution of Earthquakes

The fragment-asperity model for earthquakes was developed by SOTOLONGO-COSTA and POSADAS (2004) and describes the earthquake generation mechanism in an NESP context.

This model leads to the earthquake triggering mechanism considering the interaction between two rough surfaces (fault planes) and the fragments filling the space between them. Stress accumulates until a fragment is displaced or an asperity is broken resulting in fault plane slip and the release of energy. The aforementioned authors introduced an energy distribution function (EDF) that shows the influence of the size distribution of fragments on the energy distribution of earthquakes, including the Gutenberg-Richter law as a particular case. Furthermore, SILVA et al. (2006) revised the fragment-asperity model and calculated an EDF that allows us to determine the relative number of earthquakes as a function of magnitude.

TELESCA (2011) considered the relationship between magnitude (M) and released relative energy (ε) as:

$$M \sim \frac{2}{3}\log(\varepsilon), \tag{1}$$

and, taking into account the threshold magnitude m_0 (TELESCA 2012), proposed a modified function that relates the cumulative number of earthquakes with magnitude, given as:

$$\log\left(\frac{N(>M)}{N}\right) = \frac{2-q_M}{1-q_M}\log\left(\frac{1-\left(\frac{1-q_M}{2-q_M}\right)\left(\frac{10^M}{A^{2/3}}\right)}{1-\left(\frac{1-q_M}{2-q_M}\right)\left(\frac{10^{m_0}}{A^{2/3}}\right)}\right), \tag{2}$$

where M is the earthquake magnitude, m_0 is the threshold magnitude and A is proportional to the volumetric energy density.

The fragment-asperity model has been applied to various earthquake catalogs (SILVA et al. 2006; VILAR et al. 2007; TELESCA 2010a, b, c, 2011, 2012; MICHAS et al. 2013;

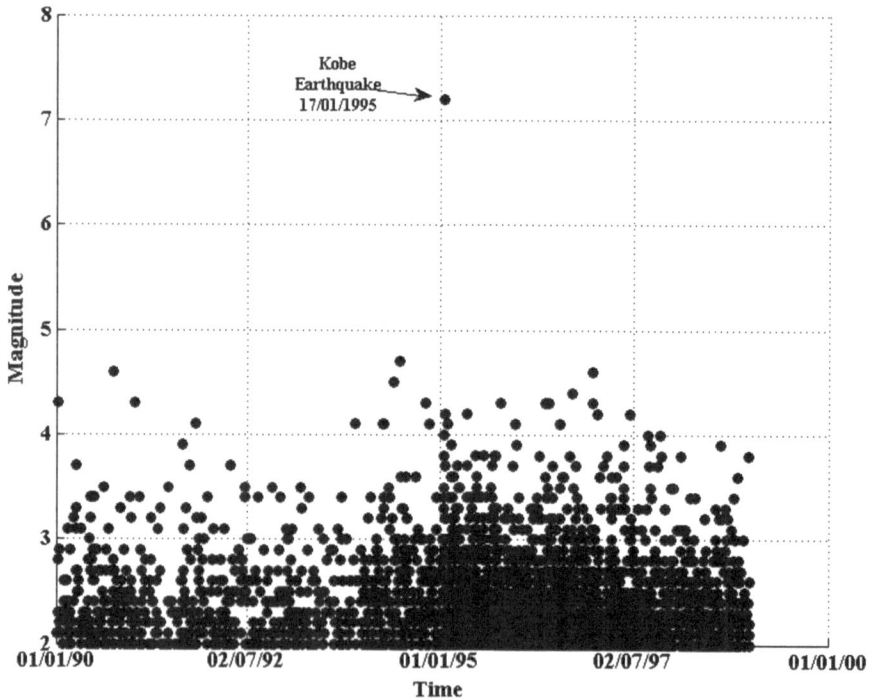

Figure 2
Time distribution of seismicity of the declustered catalog

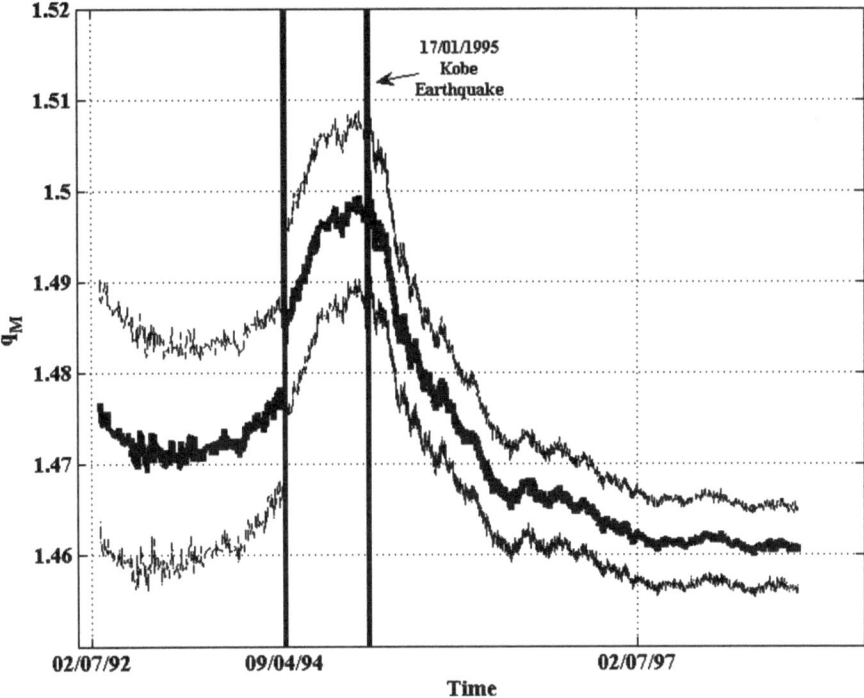

Figure 3
Time variations of q_M values (*black continuous line*) over increasing (cumulative) time windows and the associated standard deviation (*black dashed lines*). On April 9, 1994 the nonextensive parameter increases significantly indicating the start of a transition phase towards the 1995 Kobe earthquake

PAPADAKIS *et al.* 2013) including volcano related seismicity (TELESCA 2010b; VALLIANATOS *et al.* 2013).

In the fragment-asperity model framework the nonextensive parameter q_M informs us about the scale of interactions between the fault planes and the fragments (MATCHARASHVILI *et al.* 2011; PAPADAKIS *et al.* 2013; TELESCA 2010b, c; VALVERDE-ESPARZA *et al.* 2012). Thus, the increase of q_M indicates that the physical state of a seismic area moves away from equilibrium.

4. Results

In this study our interest concerns the temporal variations of the thermostatistical parameter q_M during the period 1990–1998 in regards to the 1995 Kobe earthquake. TELESCA (2010c) investigated the variations of this parameter regarding the seismicity of the L'Aquila area (central Italy) and estimated that the nonextensive parameter q_M increases in a time interval starting days before the occurrence of the strong earthquake on April 6, 2009 ($M_L = 5.8$).

Calculation of the nonextensive parameters q_M and A is performed using the maximum likelihood estimation (MLE) method as this is proposed by SHALIZI (2007) for the q-exponential (TSALLIS) distributions and by TELESCA (2012) for the earthquake cumulative magnitude distribution. Standard deviation and confidence intervals of the estimated parameters are calculated using the bootstrap method (ZOUBIR and BOASHASH 1998) by taking 500 bootstrap samples.

For the detection of possible variations of the nonextensive parameter q_M we calculate Eq. (2) in different time windows. Figure 3 shows the q_M variations over increasing (cumulative) time windows. The initial time window has a 400-event width increasing per 1 event. The cumulative estimate of the q_M parameter reveals that on April 9, 1994 the parameter increases significantly indicating the start of a transition phase towards the 1995 Kobe earthquake. Moreover, the q_M parameter peaks ($q_M = 1.5$) during the Kobe earthquake and starts decreasing afterwards having a value equal to $q_M = 1.46$ in 1997.

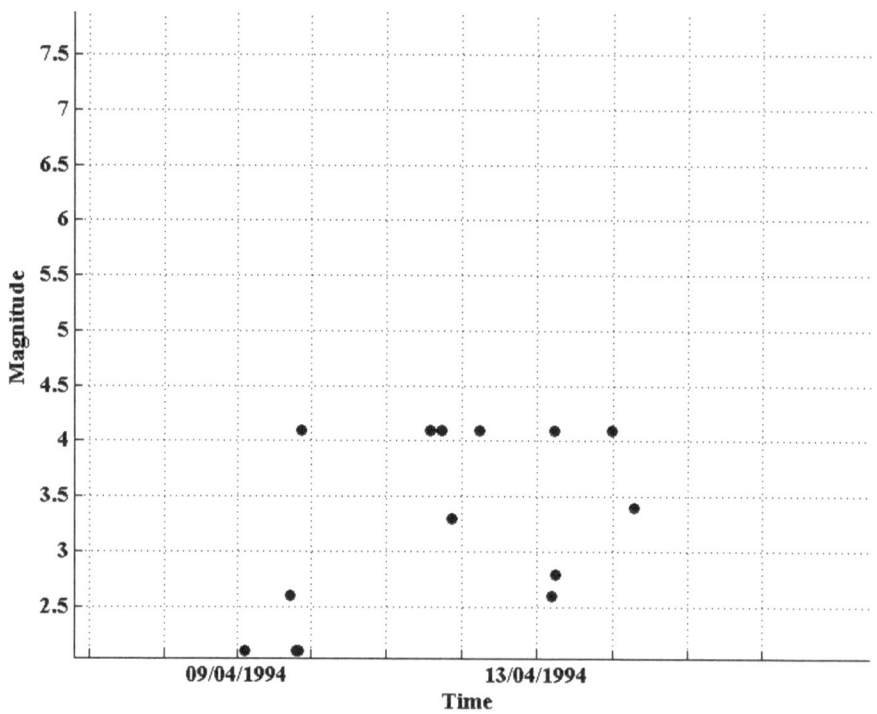

Figure 4
Magnification of a portion of Fig. 2, showing six seismic events equal to $M = 4.1$ between April 9, 1994 and April 13, 1994

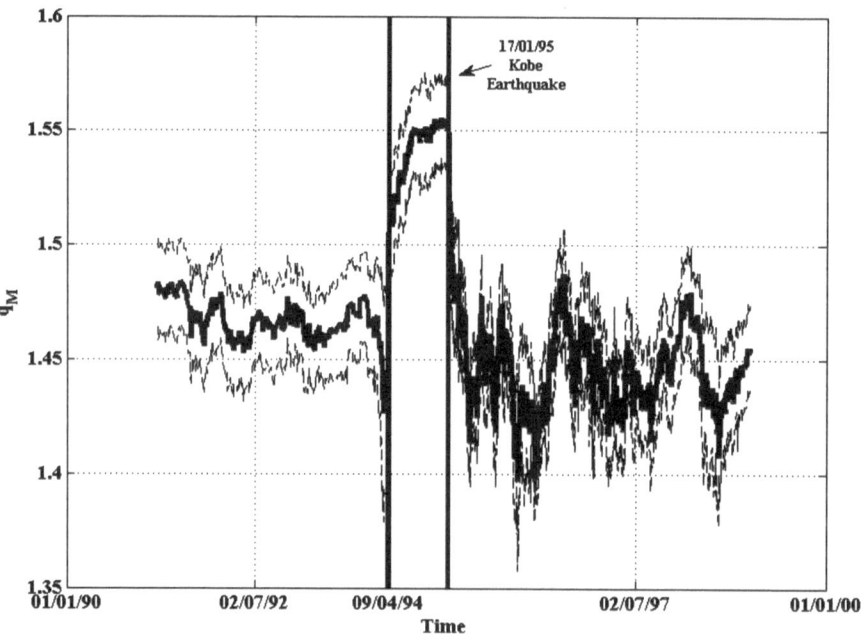

Figure 5
Time variations of q_M values (*black continuous line*) over 200-event moving windows (overlapping) having a sliding factor equal to 1 and the associated standard deviation (*black dashed lines*). On April 9, 1994 the nonextensive parameter increases significantly indicating the start of a transition phase towards the 1995 Kobe earthquake

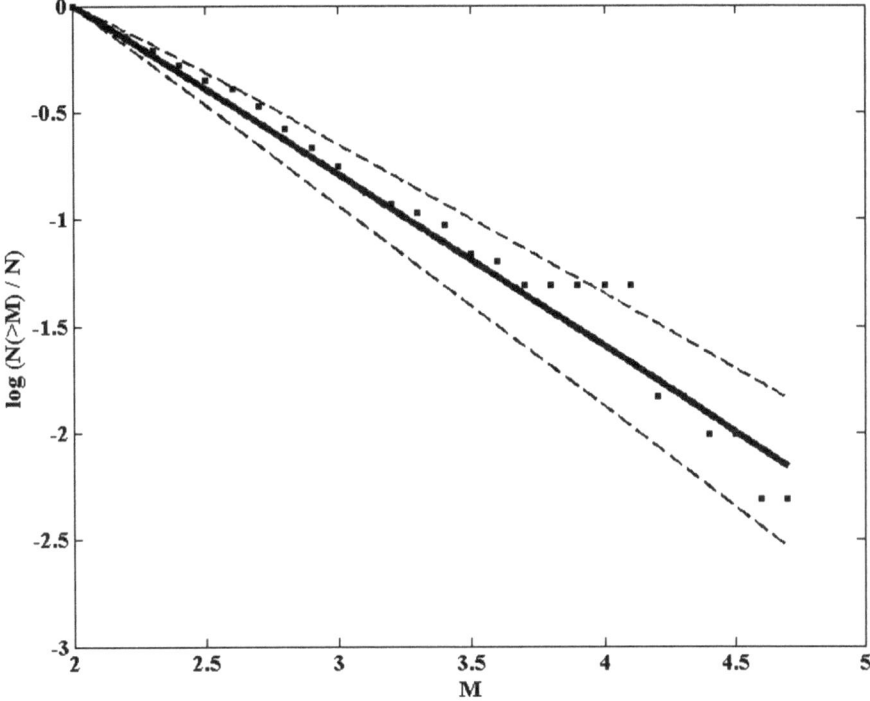

Figure 6

Normalized cumulative magnitude distribution and the fitting curve (*black continuous line*) according to Eq. (2) for the period between April 9, 1994 and January 16, 1995 (1 day before the 1995 Kobe earthquake). The nonextensive parameters are estimated equal to $q_M = 1.55$ and $A = 11.42$, respectively. The *dashed black lines* indicate the 95 % confidence intervals

Furthermore, a more detailed inspection of Fig. 2 reveals the occurrence of an earthquake equal to $M = 4.1$ on April 9, 1994 followed by five more earthquakes of the same magnitude. Figure 4, which is a magnification of a portion of Fig. 2, shows occurrence of these seismic events between April 9, 1994 and April 13, 1994. We can clearly see (Figs. 3, 4) that there is agreement between the variations of q_M values and the evolution of seismicity. Occurrence of many significant events with magnitude $M > 4$ breaks the seismicity pattern and this causes the increase of q_M. Change in the values of the nonextensive parameter and in the seismicity pattern indicates a tendency for the physical state to move away from equilibrium.

Figure 5 shows the variations of q_M values over 200-event moving windows (overlapping) having a sliding factor equal to 1. As in the increasing (cumulative) time windows estimation (Fig. 3), the q_M parameter increases significantly on April 9, 1994

and peaks ($q_M = 1.55$) as we move towards the 1995 Kobe earthquake. After the strong event the nonextensive parameter starts decreasing rapidly.

Figure 6 shows the distribution of the relative cumulative number of earthquakes as a function of magnitude and the associated fitting curve according to Eq. (2) for the period between April 9, 1994 and January 16, 1995 (1 day before the 1995 Kobe earthquake). The nonextensive parameters are estimated equal to $q_M = 1.55$ and $A = 11.42$, respectively. This high q_M value also supports the fact that during this period the studied area is in a preparatory stage progressing to the strong event.

It should be noticed that ENESCU and ITO (2001) studied the evolution of seismic activity and possible precursory changes associated with the 1995 Kobe earthquake and found a significant decrease of the fractal dimension (correlation dimension, D2) in 1994.

5. Conclusions

Using NESP the analysis of seismicity is performed in a broad region surrounding the 1995 Kobe earthquake for the period 1990–1998. In the framework of the fragment-asperity model the nonextensive parameter q_M informs us about the physical state of the studied area.

This parameter reflects the scale of the interaction between fault planes and the fragments filling the space between them. Thus, as q_M increases the physical state moves from equilibrium in a statistical physics sense.

For the examination of possible distinct variations of the nonextensive parameter q_M we use different time windows (cumulative and moving windows). We detect a significant increase of the nonextensive parameter on April 9, 1994 which coincides with the occurrence of six seismic events equal to $M = 4.1$. The occurrence of these events breaks the magnitude pattern and along with the observed q_M variations indicates a transition phase towards the 1995 Kobe earthquake.

It seems that the examination of q_M variations in time is a useful index of the physical state of a seismic area and it becomes clear that using NESP, we can recover the main characteristics of the dynamic evolution of seismicity. The observed thermostatistical variations allows us to distinguish different dynamical regimes and to decode the physical processes towards a strong event. We conclude that for various study cases and for different tectonic regimes the analysis of the nonextensive parameter q_M behavior in time is of crucial importance.

Acknowledgments

G. Papadakis acknowledges the financial support from the Greek State Scholarships Foundation (IKY). This work has been accomplished in the framework of the postgraduate program and co-funded through the action "Program for scholarships provision I.K.Y. through the procedure of personal evaluation for the 2011–2012 academic year" from resources of the educational program "Education and Life Learning" of the European Social Register and NSRF 2007–2013.

Open Access   This article is distributed under the terms of the Creative Commons Attribution License which permits any use, distribution, and reproduction in any medium, provided the original author(s) and the source are credited.

REFERENCES

ABE, S., and SUZUKI, N. (2003), *Law for the distance between successive earthquakes*, J. Geophys. Res. *108*, 2113.

ABE, S., and SUZUKI, N. (2005), *Scale-free statistics of time interval between successive earthquakes*, Physica A *350*, 588–596.

ABE, S., SARLIS, N.V., SKORDAS, E.S., TANAKA, H.K., and VAROTSOS, P.A. (2005), *Origin of the usefulness of the natural-time representation of complex time series*, Phys. Rev. Lett. *94*, 170601.

ENESCU, B., and ITO, K. (2001), *Some premonitory phenomena of the 1995 Hyogo-Ken Nanbu (Kobe) earthquake: seismicity, b-value and fractal dimension*, Tectonophysics *338*, 297–314.

KIKUCHI, M., and KANAMORI, H. (1996), *Rupture process of the Kobe, Japan, earthquake of Jan. 17, 1995, determined from teleseismic body waves*, J. Phys. Earth *44*, 429–436.

LYRA, M.L., and TSALLIS, C. (1998), *Nonextensivity and multifractality in low-dimensional dissipative systems*, Phys. Rev. Lett. *80*, 53–56.

MATCHARASHVILI, T., CHELIDZE, T., JAVAKHISHVILI, Z., JORJIASHVILI, N., and PALEO, U.F. (2011), *Non-extensive statistical analysis of seismicity in the area of Javakheti, Georgia*, Comput. Geosci. *37*, 1627–1632.

MICHAS, G., VALLIANATOS, F., and SAMMONDS, P. (2013), *Non-extensivity and long-range correlations in the earthquake activity at the West Corinth rift (Greece)*, Nonlinear Proc. Geophys. *20*, 713–724.

PAPADAKIS, G., VALLIANATOS, F., and SAMMONDS, P. (2013), *Evidence of nonextensive statistical physics behavior of the Hellenic subduction zone seismicity*, Tectonophysics *608*, 1037–1048.

REASENBERG, P. (1985), *Second-order moment of central California seismicity, 1969–82*, J. Geophys. Res. *90*, 5479–5495.

SARLIS, N.V., SKORDAS, E.S., and VAROTSOS, P.A. (2010), *Nonextensivity and natural time: The case of seismicity*, Phys. Rev. E *82*, 021110.

SARLIS, N.V., SKORDAS, E.S., LAZARIDOU, M.S., and VAROTSOS, P.A. (2008), *Investigation of seismicity after the initiation of a seismic electric signal activity until the main shock*, Proc. Jpn. Acad., Ser. B *84*, 331–343.

SHALIZI, C. R. (2007), *Maximum likelihood estimation for q-exponential (Tsallis) distribution, arXiv:math/0701854v2 [math.ST] 1 February 2007*, http://arxiv.org/pdf/math/0701854v2.pdf (last accessed January 2014).

SILVA, R., FRANCA, G.S., VILAR, C.S., and ALCANIZ, J.S. (2006), *Nonextensive models for earthquakes*, Phys. Rev. E *73*, 026102.

SOTOLONGO-COSTA, O., and POSADAS A. (2004), *Fragment-asperity interaction model for earthquakes*, Phys. Rev. Lett. *92*, 048501.

TELESCA, L. (2010a), *Analysis of Italian seismicity by using a nonextensive approach*, Tectonophysics *494*, 155–162.

TELESCA, L. (2010b), *Nonextensive analysis of seismic sequences*, Physica A *389*, 1911–1914.

TELESCA, L. (2010c), *A non-extensive approach in investigating the seismicity of L' Aquila area (central Italy), struck by the 6 April 2009 earthquake (M_L = 5.8)*, Terra Nova *22*, 87–93.

TELESCA, L. (2011), *Tsallis-based nonextensive analysis of the southern California seismicity*, Entropy *13*, 1267–1280.

TELESCA, L. (2012), *Maximum likelihood estimation of the nonextensive parameters of the earthquake cumulative magnitude distribution*, Bull. Seismol. Soc. Am. *102*, 886–891.

TSALLIS, C. (1988), *Possible generalization of Boltzmann–Gibbs Statistics*, J. Stat. Phys. *52*, 479–487.

TSALLIS, C., *Nonextensive statistical mechanics and thermodynamics: Historical background and present status, In Nonextensive statistical mechanics and its applications*, (eds. ABE. S. and OKAMOTO. Y.) (Springer, Berlin 2001) pp. 3–98.

TSALLIS, C., *Introduction to nonextensive statistical mechanics: approaching a complex world* (Springer, Berlin 2009).

VALLIANATOS, F. (2009), *A non-extensive approach to risk assessment*, Nat. Hazards Earth Syst. Sci. *9*, 211–216.

VALLIANATOS, F. (2012), *On the non-extensive nature of the isothermal depolarization relaxation currents in cement mortars*, J. Phys. Chem. Solids *73*, 550–553.

VALLIANATOS, F., and TELESCA, L. (2012), *Statistical mechanics in earth physics and natural hazards*, Acta Geophys. *60*, 499–501.

VALLIANATOS, F., and SAMMONDS, P. (2013), *Evidence of nonextensive statistical physics of the lithospheric instability approaching the 2004 Sumatran- Andaman and 2011 Honsu mega-earthquakes*, Tectonophysics *590*, 52–58.

VALLIANATOS, F., TRIANTIS D., and SAMMONDS, P. (2011), *Non-extensivity of the isothermal depolarization relaxation currents in uniaxial compressed rocks*, Europhys. Lett. *94*, 68008.

VALLIANATOS, F., MICHAS, G., PAPADAKIS, G., and SAMMONDS, P. (2012), *A non-extensive statistical physics view to the spatiotemporal properties of the June 1995, Aigion earthquake (M6.2) aftershock sequence (West Corinth rift, Greece)*, Acta Geophys. *60*, 758–768.

VALLIANATOS, F., MICHAS, G., PAPADAKIS, G., and TZANIS, A. (2013), *Evidence of non-extensivity in the seismicity observed during the 2011–2012 unrest at the Santorini volcanic complex, Greece*, Nat. Hazards Earth Syst. Sci. *13*, 177–185.

VALVERDE-ESPARZA, S.M., RAMIREZ-ROJAS, A., FLORES-MARQUEZ, E.L., and TELESCA, L. (2012), *Non-extensivity analysis of seismicity within four subduction regions in Mexico*, Acta Geophys. *60*, 833–845.

vanSTIPHOUT, T., ZHUANG, J., and MARSAN, D. (2012), *Seismicity declustering, Community Online Resource for Statistical Seismicity Analysis*, http://dx.doi.org/10.5078/corssa-52382934, Available at http://www.corssa.org (last accessed January 2014).

VAROTSOS, P.A., SARLIS, N.V., and SKORDAS, E.S. (2001), *Spatio-temporal complexity aspects on the interrelation between Seismic Electric Signals and seismicity*, Practica of Athens Academy *76*, 294–321.

VAROTSOS, P.A., SARLIS, N.V., and SKORDAS, E.S. (2002), *Long-range correlations in the electric signals that precede rupture*, Phys. Rev. E *66*, 011902.

VAROTSOS, P.A., SARLIS, N.V., and SKORDAS, E.S., *Natural time analysis: The new view of time, Precursory seismic electric signals, earthquakes and other complex time series* (Springer–Verlag, Berlin Heidelberg 2011) p. 285.

VILAR, C.S., FRANCA, G.S., SILVA, R., and ALCANIZ J.S. (2007), *Nonextensivity in geological faults*, Physica A *377*, 285–290.

WIEMER, S. (2001), *A software package to analyse seismicity: ZMAP*, Seismol. Res. Lett. *72*, 373–382.

ZOUBIR, A.M., and BOASHASH, B. (1998), *The bootstrap and its applications in signal processing*, IEEE Signal Process. Mag. *15*, 55–76.

(Received May 5, 2014, revised May 27, 2014, accepted June 4, 2014, Published online July 4, 2014)

Reprinted from the journal

Pure Appl. Geophys. 172 (2015), 1933–1943
© 2014 Springer Basel
DOI 10.1007/s00024-014-0862-3

Pure and Applied Geophysics

Investigating the Tsunamigenic Potential of Earthquakes from Analysis of the Informational and Multifractal Properties of Seismograms

LUCIANO TELESCA,[1] ASHUTOSH CHAMOLI,[2,3] MICHELE LOVALLO,[4] and TONY ALFREDO STABILE[1]

Abstract—Revealing the tsunamigenic potential of an earthquake is very challenging in regards to minimizing the casualties a tsunami can provoke. Thus, development of methodologies that can reliably furnish a early warnings of a tsunami is crucial. In order to accomplish this aim it is important to preliminarily identify the characteristics of seismograms that can be used to distinguish tsunamigenic (TS) earthquakes from non-tsunamigenic (NTS) earthquakes. In this paper P-wave time dynamic of 17 seismograms of TS earthquakes and 26 NTS seismograms are analysed by means of two advanced statistical tools: the Fisher–Shannon method and the multifractal detrended fluctuation analysis (MFDFA). Both methods are well suited to disclosing the inner time properties of complex signals, as seismograms appear to be. Using these two methods jointly, we defined a classifier, the performance of which was tested by means of the receiver-operating characteristic curve that plots true positive rate versus false positive rate. This classifier shows a discrimination power that can be considered acceptable in comparison with the devastating effects caused by a non-alarmed tsunami. Our findings indicate that proper choice of the classifier's threshold allows correctly identification of approximately 69 % of the NTS seismograms and approximately 76 % of the TS seismograms. The presented results presented may be helpful in addressing the complex problem of early tsunami warning.

Key words: Tsunami, major earthquakes, Fisher–Shannon method, multifractal properties, seismograms.

1. Introduction

Seismograms are seismic signals that can be utilized to investigate revealing features related to the basic characteristics of earthquakes. YULMETYEV *et al.* (2001) studied the discreteness, long-range memory and local time behavior of seismograms by using methodologies generally applied in statistical mechanics, like the discrete non-Markov random processes and the generalized Hurst exponent. Via these methods they were able to discriminate between weak earthquake seismograms and those generated by explosions. TELESCA *et al.* (2011) applied the Fisher information measure (FIM) to seismograms registered in the Vertex Hills (Hungary) to distinguish between small earthquakes and quarry blasts. LYUBUSHIN *et al.* (2013) applied the multifractal singularity spectrum to provide a classification procedure for the discrimination between earthquakes and quarry blasts in Egypt; the application of a linear Bayesian discriminator revealed correct classification for 93 % of earthquake records and 99 % of quarry blasts. TELESCA *et al.* (2013) applied informational methods to discriminate between TS and NTS earthquakes by analysing the seismic signals recorded at several stations worldwide.

Therefore, the application of advanced non-standard statistical techniques to seismogram investigations can reveal features and patterns not evidenced by standard methodologies. In fact, in the present paper we aim to investigate the informational and multifractal properties of earthquake seismograms that have or have not generated a tsunami worldwide by using the FIM method and the multifractal detrended fluctuation analysis (MFDFA). Using these two methods we intend to investigate how the tsunamigenic potential of an earthquake can be revealed by looking only at the P-wave of the whole seismogram. Such a goal is directly connected with the potential for enhancing early tsunami warning systems.

[1] Consiglio Nazionale delle Ricerche, Istituto di Metodologie per l'Analisi Ambientale, C.da S.Loja, 85050 Tito, PZ, Italy. E-mail: luciano.telesca@imaa.cnr.it
[2] Department of Earth Sciences, Indian Institute of Technology Roorkee, Roorkee 247667, Uttarakhand, India.
[3] CSIR-National Geophysical Research Institute, Hyderabad 500 007, India.
[4] ARPAB, 85100 Potenza, Italy.

Reprinted from the journal

Table 1

Earthquake events used in the present study

Category	Date	Place	Hypocenter location			Iris station used	MAG (M_w)	Type of faulting
			Lat (°N)	Lon (°E)	Depth (km)			
NTS	2011-04-11	Fukushima	37.00	140.40	13.1	JHJ2	7.3	Normal
	2008-11-16	Sulawesi	1.35	122.16	30.0	KAPI; MBWA-00; MBWA-10	7.3	Reverse
	2002-11-02	Sumatra	2.98	96.11	33.0	COCO-00; DAV	7.3	Reverse
	2011-04-07	Miyagi	38.28	141.59	49.0	ERM; JNU; MAJO-10; MAJO-00	7.4	Reverse
	2009-08-10	Andaman	14.05	92.87	33.1	CHTO; PALK	7.5	Normal
	2009-03-19	Tonga	−23.16	−174.59	34.0	MSVF	7.6	Reverse
	2009-09-30	Sumatra	−0.71	99.97	81.0	CHTO-00; CHTO-10	7.9	Oblique Reverse
	2007-01-13	Kuril	46.23	154.50	10.0	ERM-00; MA2-00; MAJO-00; MAJO-10; PET-00; YSS-00	7.9	Normal
TS	2012-04-11	Sumatra	2.33	93.06	22.9	CHTO-00; PALK-00; PALK-10; PSI	8.6	Oblique Reverse
	2005-03-28	Sumatra	2.10	97.11	30.0	PALK-00	8.7	Reverse
	2010-10-25	Sumatra	−3.52	100.10	20.6	BTDF; PSI	7.7	Reverse
	2006-11-15	Kuril	46.68	153.21	30.3	ERM-00; PET-00; YSS-00	7.8	Reverse
	2003-09-25	Hokkaido	41.75	143.87	27.0	INCN-00	8.3	Reverse
	2007-09-12	Sumatra	−4.46	101.40	34.0	KAPI-00; PSI	8.5	Reverse
	2004-12-26	Sumatra	3.3	95.98	30.0	CHTO-00; COCO-00; DGAR-00; DGAR-10; INCN-00; INCN-10; MBWA-10; TATO-00; TATO-10	9.3	Reverse

NTS non-tsunamigenic, *TS* tsunamigenic

The subscripts '00' and '10' are for different sensors at same site

2. Data

The importance of warning systems in minimizing casualties gained recognition after the Sumatra earthquake of December 26, 2004 that resulted in severe hazards in numerous countries. The high quality recording of the seismograms of this earthquake at stations in different countries is valuable in understanding the physical characteristics and developing new methodologies (CHAMOLI *et al.* 2010). Other global TS and NTS events are also analysed. Details of the seismic events used in this study are presented in Table 1, including hypocentral coordinates and faulting style. The events studied are major earthquakes capable of producing a tsunami ($M_w > 6.5$) (Fig. 1). TS and NTS categories are listed in Table 1. Though there are some events with a moderate tsunami in the near field only, these events are considered TS. We analysed earthquake events from Andaman Nicobar, Indonesia, Japan, the Kuril Islands and the Tonga islands. Vertical component data is used and instrument correction is applied. The length of waveforms used consists of a truncated P-wave train from the first P-wave to the theoretical

$T_s - T_p$ travel time difference, evaluated by using the IASP91 model (KENNETT and ENGDAHL 1991) for given station coordinates and hypocentral earthquake coordinates. This selection overcomes the difficulties related to identifying reliable first S-wave arrivals on the vertical component of seismograms. Depending on the hypocentral distance of the station, the length of truncated waveforms ranges from 37.7 s ($M_w = 7.4$, 2011 Miyagi earthquake recorded at the MAJO station) to 392.6 s ($M_w = 9.3$, 2004 Sumatra earthquake recorded at the INCN station), as reported in Table 2. Figure 2 shows, as an example, the $T_s - T_p$ selected windows (blue boxes) on vertical component recordings of the $M_w = 7.4$, 2011 Miyagi earthquake at MAJO, ERM, and JNU station, respectively.

3. Methods

3.1. The Fisher–Shannon Method

The Fisher–Shannon method is comprised of two statistical methods related to each other: the FIM and the Shannon entropy method, both efficient tools

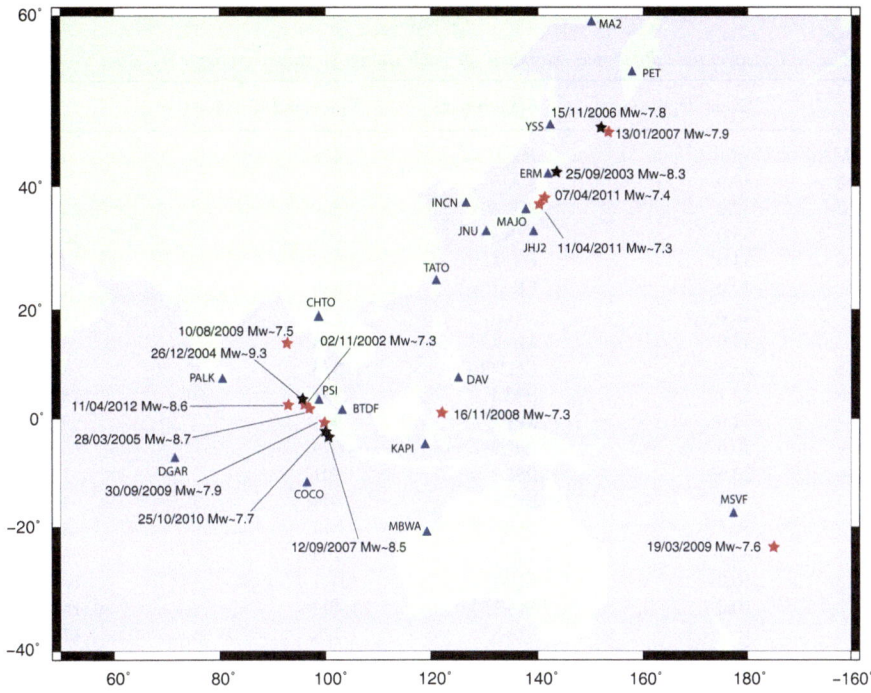

Figure 1

The map shows location of earthquakes (*stars*) and stations (*triangle*) used in the study. The tsunamigenic and non-tsunamigenic events are marked with *black and red stars*, respectively

in investigating complex temporal fluctuations of nonstationary signals. The FIM quantifies the amount of organization or order in a system. It was developed by FISHER (1925) in the context of statistical estimation. Then, it was utilized for different aims. FRIEDEN (1990) used the FIM to describe the evolution laws of physical systems. MARTIN *et al.* (1999, 2001) applied it to characterize the temporal fluctuations of electroencephalograms (EEG) and to detect significant dynamical changes. Analysis of complex geophysical and environmental phenomena, such as volcano-related signals, earthquake-related electromagnetic signals and atmospheric particulate matter, benefited by the application of the FIM methodology by gaining insight into their inner time dynamics and the mechanisms underlying their temporal fluctuations and by revealing the precursory signatures of critical phenomena (LOVALLO and TELESCA 2011; TELESCA and LOVALLO 2011; TELESCA *et al.* 2009, 2010, 2011). Recently TELESCA *et al.* (2013) employed the FIM to discriminate between TS and NTS earthquake seismograms.

The Shannon entropy method is generally employed in defining the degree of uncertainty involved in predicting the output of a probabilistic event. For discrete distributions this means that if one predicts the outcome exactly before it happens the probability will be a maximum value and, as a result, the Shannon entropy will be minimal. If one is absolutely able to predict the outcome of an event the Shannon entropy will be zero. Such is not the case for distributions (probability density functions) on a continuous variable ranging over the real line. In this case, the Shannon entropy can reach any arbitrary value, positive or negative. Therefore, the use of the power entropy (defined below) avoids the difficulty of dealing with negative values. The Shannon entropy method provides a scientific method for understanding the essential state of things.

Let $f(x)$ be the probability density function of a signal x, then its FIM I is given by

$$I = \int_{-\infty}^{+\infty} \left(\frac{\partial}{\partial x} f(x) \right)^2 \frac{dx}{f(x)}, \qquad (1)$$

Table 2

Theoretical Ts − Tp travel time difference evaluated for each station-earthquake couple by using the IASP91 model

M_w	Place	Event depth (km)	Station name	Epicentral distance (km)	$T_s - T_p$ (s)	Category
7.3	FUKUSHIMA	13.1	JHJ2	436	46.8	NTS
7.3	SULAWESI	30.0	KAPI	748	76.4	NTS
7.3	SULAWESI	30.0	MBWA	2,509	247.0	NTS
7.3	SUMATRA	33.0	COCO	1,693	168.4	NTS
7.3	SUMATRA	33.0	DAV	3,286	293.6	NTS
7.4	MIYAGI	49.0	ERM	438	45.5	NTS
7.4	MIYAGI	49.0	JNU	1,126	112.9	NTS
7.4	MIYAGI	49.0	MAJO	358	37.7	NTS
7.5	ANDAMAN	33.1	CHTO	835	84.8	NTS
7.5	ANDAMAN	33.1	PALK	1,531	152.5	NTS
7.6	TONGA	34.0	MSVF	961	97.1	NTS
7.7	SUMATRA	20.6	BTDF	675	69.7	TS
7.7	SUMATRA	20.6	PSI	701	72.3	TS
7.8	KURIL	30.3	ERM	949	96.1	TS
7.8	KURIL	30.3	PET	812	82.7	TS
7.8	KURIL	30.3	YSS	798	81.3	TS
7.9	KURIL	10.0	ERM	1,017	104.1	NTS
7.9	KURIL	10.0	MA2	1,499	151.1	NTS
7.9	KURIL	10.0	MAJO	1,728	173.3	NTS
7.9	KURIL	10.0	PET	809	83.7	NTS
7.9	KURIL	10.0	YSS	896	92.3	NTS
7.9	SUMATRA	81.0	CHTO	2,175	217.2	NTS
8.3	HOKKAIDO	27.0	INCN	1,552	154.9	TS
8.5	SUMATRA	34.0	KAPI	2,037	204.4	TS
8.5	SUMATRA	34.0	PSI	849	86.1	TS
8.6	SUMATRA	22.9	CHTO	1,944	194.8	NTS
8.6	SUMATRA	22.9	PALK	1,476	147.9	NTS
8.6	SUMATRA	22.9	PSI	650	67.2	NTS
8.7	SUMATRA	30.0	PALK	1,897	189.5	NTS
9.3	SUMATRA	30.0	CHTO	1,756	174.9	TS
9.3	SUMATRA	30.0	COCO	1,728	172.0	TS
9.3	SUMATRA	30.0	DGAR	2,857	267.7	TS
9.3	SUMATRA	30.0	INCN	4,924	392.6	TS
9.3	SUMATRA	30.0	MBWA	3,764	323.1	TS
9.3	SUMATRA	30.0	TATO	3,651	316.1	TS

Event depths and epicentral distances used to compute travel-times are also reported

and the Shannon entropy H_X is given by

$$H_X = - \int_{-\infty}^{+\infty} f_X(x) \log f_X(x) dx \ (4). \qquad (2)$$

For convenience the alternative notion of entropy power (MARTIN *et al.* 2001)

$$N_X = \frac{1}{2\pi e} e^{2H_X} \qquad (3)$$

will be used rather than the entropy H_X to deal with negative Shannon entropy that can be obtained with continuous distributions.

The product between the FIM and the Shannon entropy power is called complexity (ESQUIVEL *et al.* 2010) (C_X).

Calculation of the FIM and N_X depends on calculation of the probability density function $f(x)$ (pdf). The pdf can be estimated by means of the kernel density estimator technique (DEVROYE *et al.* 1987; JANICKI and WERON 1994) that approximates the density function as

$$\hat{f}_M(x) = \frac{1}{Mb} \sum_{i=1}^{M} K\left(\frac{x - x_i}{b}\right), \qquad (4)$$

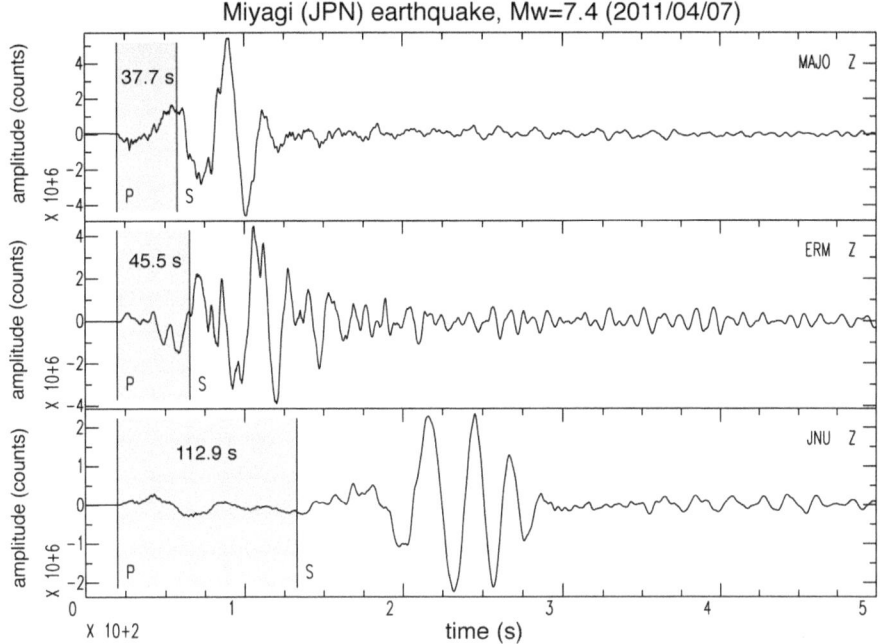

Figure 2

Vertical component recordings of the $M_w = 7.4$, 2011 Miyagi earthquake at MAJO, ERM, and JNU station, respectively. Traces are aligned respect to the first P-wave arrival read on seismograms, and the begin time has been fixed to 20 s before the first P-wave arrival. *Blue boxes* indicate the selected $T_s - T_p$ window used for our statistical analyses, and the time length of each truncated waveform is reported in the figure

with b the bandwidth, M the number of data and $K(u)$ the kernel function, a continuous non-negative and symmetric function satisfying the two following conditions

$$K(u) \geq 0 \quad \text{and} \quad \int_{-\infty}^{+\infty} K(u)du = 1. \quad (5)$$

In our study, we estimated the pdf $f(x)$ by means of the algorithm developed in TROUDI *et al.* (2008) combined with that developed in RAYKAR and DURAISWAMI (2006), that uses a Gaussian kernel with zero mean and unit variance:

$$\hat{f}_M(x) = \frac{1}{M\sqrt{2\pi b^2}} \sum_{i=1}^{M} e^{-\frac{(x-x_i)^2}{2b^2}} \quad (6)$$

3.2. Multifractal Detrended Fluctuation Analysis (MFDFA)

The main feature of multifractals is high variability on a wide range of temporal scales, associated with intermittent fluctuations and long-range power-law correlations (CHAMOLI *et al* 2007). MFDFA has given useful information in several seismological studies (TELESCA and LAPENNA 2006; CHAMOLI and YADAV 2013).

The MFDFA (KANTELHARDT *et al.* 2002) operates on the time series $x(i)$, where $i = 1,2,...,N$ and N is the length of the series. With x_{ave} we indicate the mean value

$$x_{ave} = \frac{1}{N} \sum_{k=1}^{N} x(k) \quad (7)$$

We assume that $x(i)$ are increments of a random walk process around the average x_{ave}, thus the "trajectory" or "profile" is given by the integration of the signal

$$y(i) = \sum_{k=1}^{i} [x(k) - x_{ave}]. \quad (8)$$

Furthermore, integration will reduce the level of measurement noise present in observational and finite records. Next, the integrated time series is divided into $N_S = \text{int}(N/s)$ nonoverlapping segments of equal

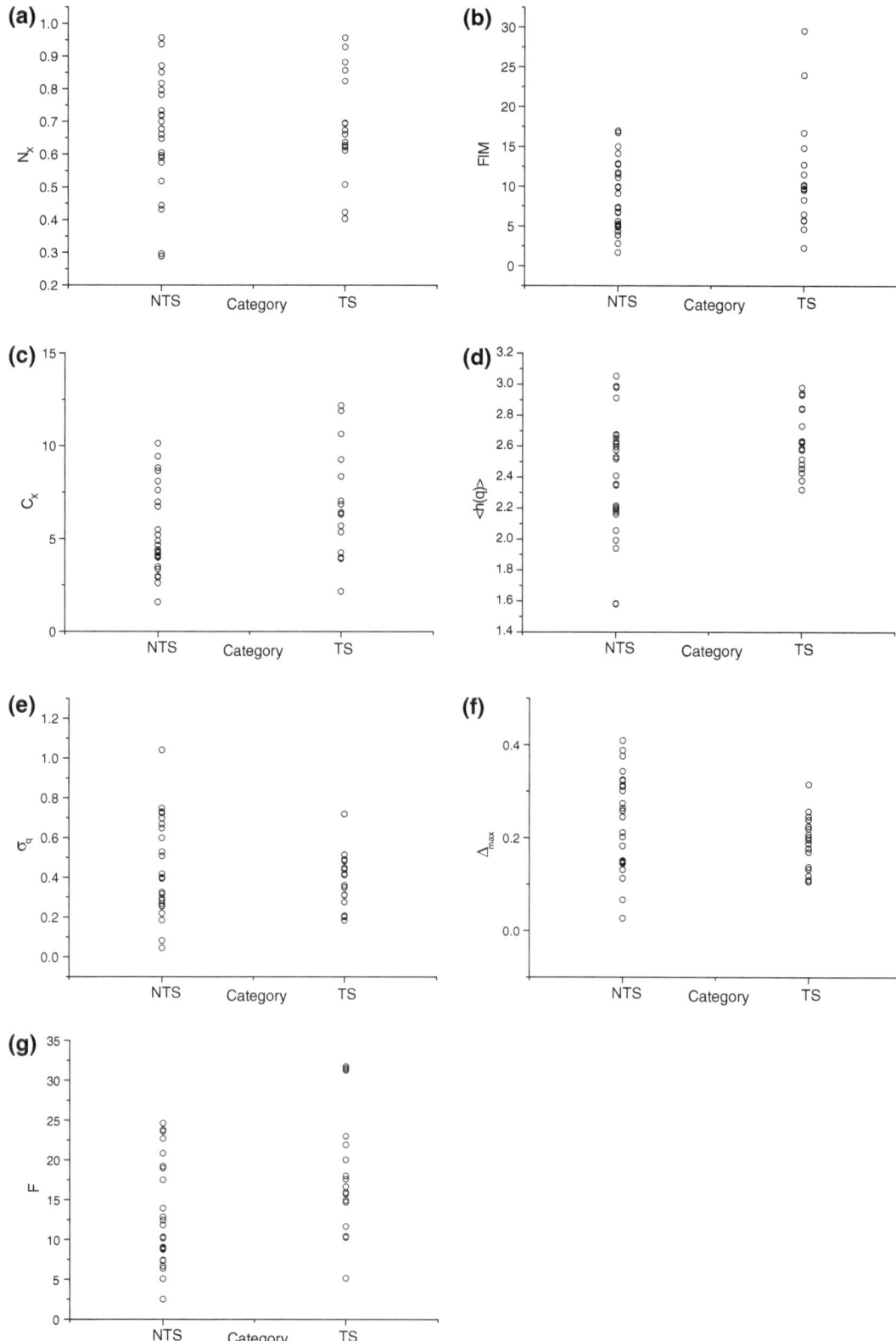

Figure 3
Values of the informational and multifractal parameters calculated for the seismograms indicated in Table 1

length s. Since the length N of the series is often not a multiple of the considered time scale s, a short part at the end of the profile $y(i)$ may remain. In order to not disregard this part of the series, the same procedure is repeated starting from the opposite end. Thereby, $2N_S$ segments are obtained altogether. Then we calculate the local trend for each of the $2N_S$ segments by a least square fit of the series. Then we determine the variance

$$F^2(s, v) = \frac{1}{s} \sum_{i=1}^{s} \{y[(v-1)s+i] - y_v(i)\}^2 \qquad (9)$$

for each segment v, $v = 1,...,N_S$ and

$$F^2(s, v) = \frac{1}{s} \sum_{i=1}^{s} \{y[N - (v - N_S)s + i] - y_v(i)\}^2$$

$$\qquad (10)$$

for $v = N_{S+1},...,2N_S$. Here, $y_v(i)$ is the fitting line in segment v. Then, after detrending the series, we average over all segments to obtain the q-th order fluctuation function

$$F_q(s) = \left\{ \frac{1}{2N_S} \sum_{v=1}^{2N_S} [F^2(s, v)]^{\frac{q}{2}} \right\}^{\frac{1}{q}} \qquad (11)$$

where, in general, the index variable q can take any real value except zero.

Repeating the procedure described above, for several time scales s, $F_q(s)$ will increase with increasing s. Then analyzing log–log plots $F_q(s)$ versus s for each value of q, we determine the scaling behaviour of the fluctuation functions. If the series x_i is long-range power-law correlated, $F_q(s)$ increases for large values of s as a power-law

$$F_q(s) \propto s^{h(q)}. \qquad (12)$$

In general the exponent $h(q)$ will depend on q. For stationary time series, $h(2)$ is the well defined Hurts exponent H. Thus, we call $h(q)$ the generalized Hurst exponent. Monofractal time series with compact support are characterized by $h(q)$ independent of q. The different scaling of small and large fluctuations will yield a significant dependence of $h(q)$ on q. For positive q, the segments v with large variance (i.e. large deviation from the corresponding fit) will dominate the average $F_q(s)$. Therefore, if q is positive, $h(q)$ describes the scaling behaviour of the

segments with large fluctuations; and generally, large fluctuations are characterized by a smaller scaling exponent $h(q)$ for multifractal time series. For negative q, the segments v with small variance will dominate the average $F_q(s)$. Thus, for negative q values, the scaling exponent $h(q)$ describes the scaling behaviour of segments with small fluctuations, usually characterized by a larger scaling exponents.

The value $h(0)$ corresponds to the limit $h(q)$ for $q \to 0$, and cannot be determined directly using the averaging procedure of Eq. 12 because of the diverging exponent. Instead, a logarithmic averaging procedure has to be employed,

$$F_0(s) \equiv \exp\left\{ \frac{1}{4N_S} \sum_{v=1}^{2N_S} \ln[F^2(s, v)] \right\} \approx s^{h(0)}. \qquad (13)$$

4. Results and Discussion

Each P-wave window was first normalized to its own average and standard deviation (each normalized seismogram has zero mean and unitary standard deviation), in order to ensure that the results obtained for the different signals are not biased by different scale of variation. We calculated several parameters for each normalized P-wave, some based on the Fisher–Shannon method, and others based on the multifractal method.

In order to test the performance of the used parameters we employed receiver operating characteristic (ROC) analysis. ROC curves have long been used to evaluate the performance of binary classifiers (KHARIN and ZWIERS 2003). ROC curves display the true positive rate (ratio between the number of correctly classified positives and the total number of positives) versus the false positive rate (ratio between the number of negatives incorrectly classified and the total number of negatives). The true positive rate is also called sensitivity (SE); the false positive rate is equal to 1-Specificity (1-SP), where the SP is the probability of classifying a negative occurrence as negative. Depending on the position in the ROC space, the classifier is good or not. The point (0, 1) represents perfect classification. One point in ROC space is better than another if it is to the northwest of

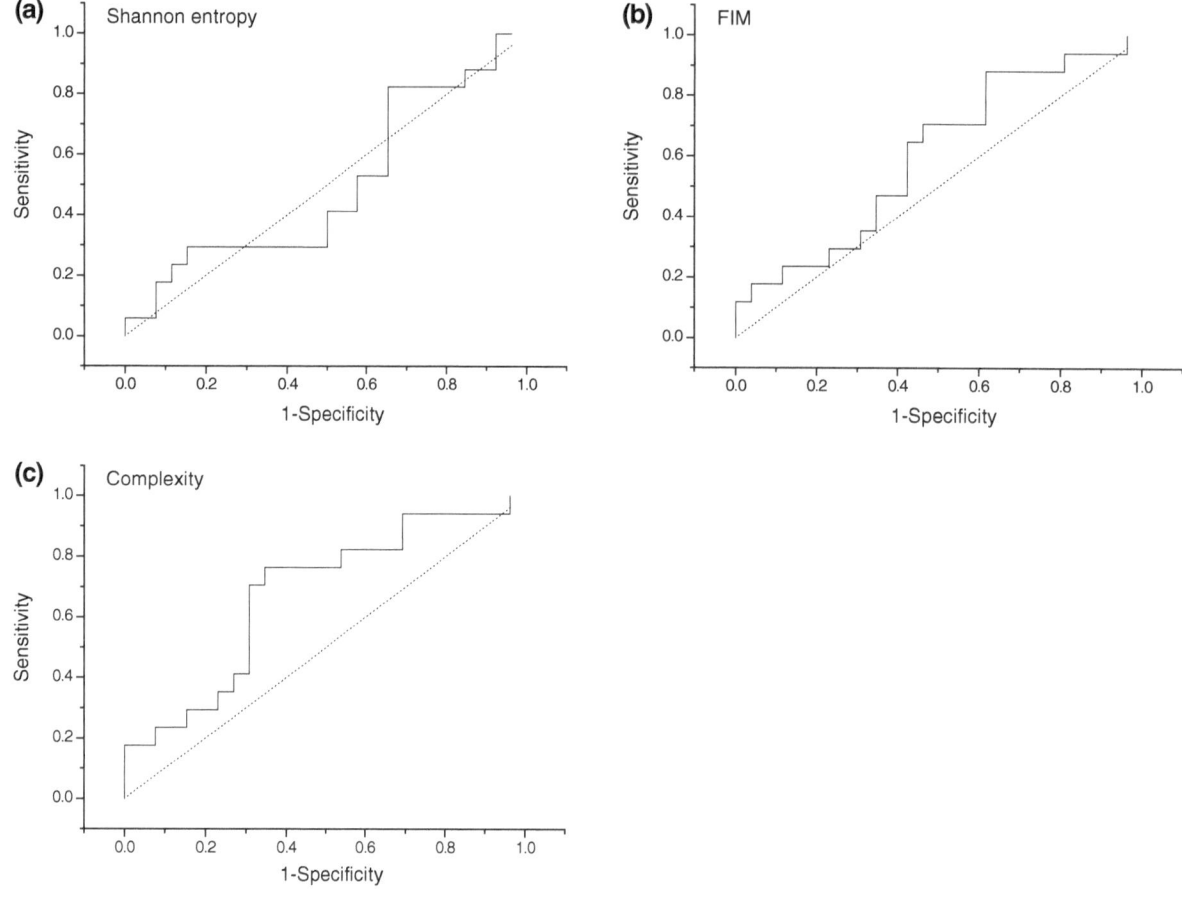

Figure 4
ROC curves for the Fisher–Shannon parameters

the first. The diagonal line $y = x$ represents the strategy of randomly guessing a class. A classifier is better than another if its ROC graph is located above that of the other, or if the area under the ROC curve (AUC) is larger. Therefore, the performance of the classifier can be judged from the value of the AUC.

Based on the Fisher–Shannon method, we investigated the performance of the FIM, the Shannon entropy (N_X) and the complexity (C_X) in discriminating TS NTS P-waves. Figure 3a–c show the values of the three Fisher–Shannon based parameters. Figure 4 shows the ROC curve of the Shannon entropy (Fig. 4a), FIM (Fig. 4b) and complexity (Fig. 4c). The ROC curve of the Shannon entropy is almost overlapping that corresponding to the random guess $(y = x)$ and its AUC is about 0.45; this indicate that the Shannon entropy should not be used as a classifier

to discriminate between NTS and TS seismograms. The ROC curve of the FIM is a little above the diagonal $y = x$ and its AUC is about 0.58. The ROC curve of the complexity is well above the diagonal $y = x$ and its AUC is about 0.66.

Based on the multifractal method, we investigated the performance of three parameters defined below. We first applied the MFDFA using a detrending polynomial of the 2nd degree and a q between -10 and 10 with a step of 1; for each P-wave we calculated the generalized Hurst exponents and calculated the mean generalized Hurst exponent $(<h(q)>)$, the standard deviation of the generalized H exponents (σ_q) and the deviation from the maximum Hurst exponent $(\Delta_{max} = |max(h(q))-<h(q)>|/<h(q)>)$. These three parameters were recently defined for testing the performance of the multifractal detrended

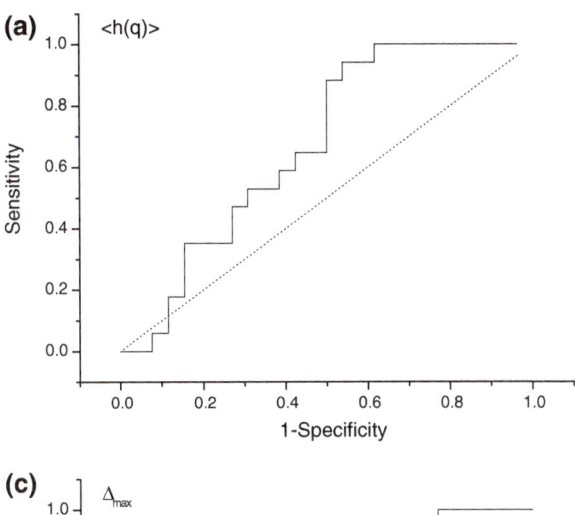

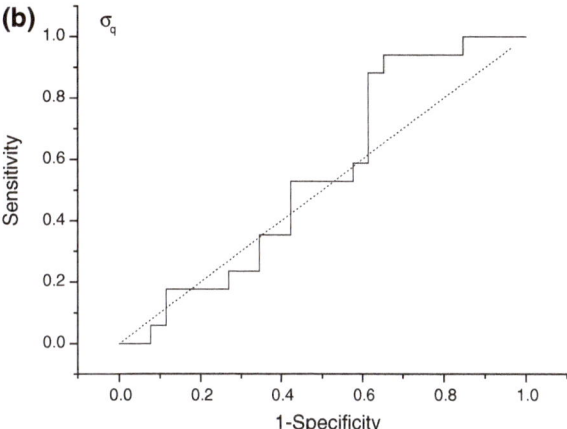

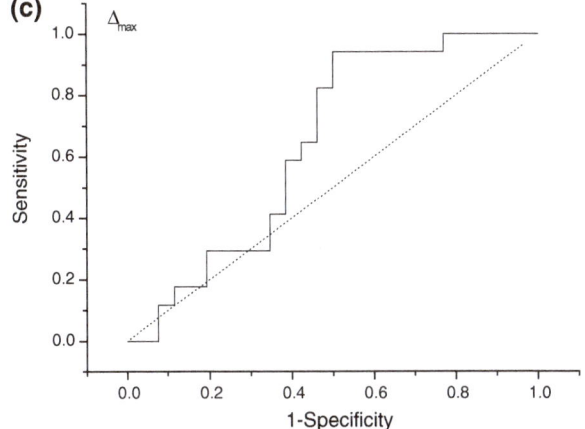

Figure 5
ROC curves for the MFDFA parameters

fluctuation analysis in short time series. The values of the multifractal parameters are shown in Fig. 3d–f.

Figure 5 shows the ROC curve of the mean Hurst exponent (Fig. 5a), σ_q (Fig. 5b) and Δ_{max} (Fig. 5c). The $<h(q)>$ is characterized by the best performance, having the largest AUC, which is about 0.66; while the AUC corresponding to σ_q and Δ_{max} is about 0.52 and 0.62, respectively.

Figure 6 shows comparison of the two best classifiers derived from the Fisher-Shannon and MF-DFA methods, namely the complexity and $<h(q)>$, respectively. Even if the two ROC curves have almost the same value of AUC they are quite different.

From the ROC results for the $<h(q)>$ and complexity, we defined a new discriminator (F) given simply by the product of C_X and $<h(q)>$ (Fig. 3g). The ROC curve of F is shown in Fig. 7. The AUC of the new parameter is 0.70, which is larger than that of

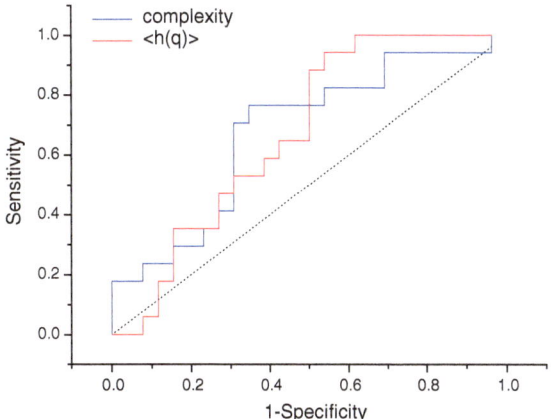

Figure 6
Comparison between the ROC curves of complexity and $<h(q)>$

$<h(q)>$ and complexity. Therefore, its performance in discriminating the TS and NTS seismograms is better.

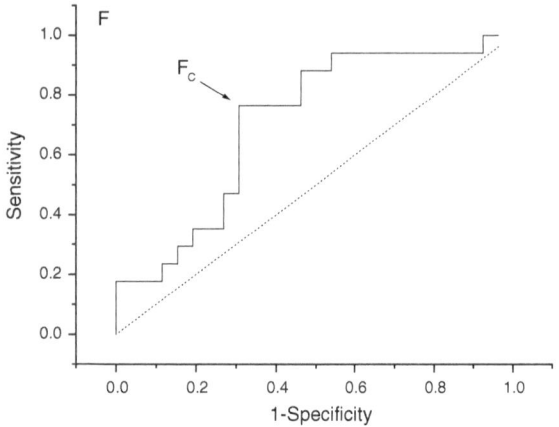

Figure 7
ROC curve of the parameter F (see text for details)

classifier can be considered acceptable when considering the devastating effects caused by a tsunami for which no warning is given. Our findings indicate that the threshold value of $F_C = 13.93$ allows for correctly identifying about 69 % of NTS seismograms and about 76 % of TS ones. Of course, based on the ROC analysis, we have to say that the procedure described in the present paper does not recognize 100 % of the TS events by just using the first arrival of the seismic waves; however, the results obtained envisage a new promising approach in investigating the reliability of tsunami warnings via application of advanced statistical methodologies.

Considering the last parameter F, the best operating point, that is the threshold value optimizing the discrimination power between NTS and TS seismograms, is given by the point of the ROC curve closest to (0,1). In our case, the threshold value $F_C = 13.93$ can be considered as the best operating point; in particular, for $F < F_C$, about 69 % of NTS seismograms are correctly identified while about 31 % of NTS seismograms are misclassified as TS. For $F > F_C$, about 76 % of TS are correctly identified while about 24 % of TS are misclassified as NTS.

5. Conclusions

In the present study we analysed seismogram P-wave time dynamics of several strong earthquakes, some of which generated a tsunami and some of which dod not. The adopted methodologies, namely the Fisher–Shannon method and MFDFA, represent advanced tools in obtaining information regarding the inner dynamics of complex signals. Seismograms are very complex signals because they contain information about characteristics of the source that has generated the seismic wavefield and the properties of the complex medium in which the wavefield has propagated. Using the two methods jointly we defined a new classifier (F) that showed good discrimination between TS and NTS earthquakes based just on analysis of the first arrival of the seismic waves (the P-wave). The performance of such

Acknowledgments

The present study was supported by the project "Development of time series analysis tools for the earthquakes and their tsunamigenic behaviour" in the framework of the Italy-India Bilateral Agreement 2012–2014 between the Consiglio Nazionale delle Ricerche (CNR) and the Council of Scientific and Industrial Research (CSIR).

REFERENCES

CHAMOLI A. and YADAV R.B.S., 2013 Multifractality in Seismic Sequences of NW Himalaya, Natural Hazards. doi:10.1007/s11069-013-0848-y.

CHAMOLI A., BANSAL A.R. and DIMRI V.P., 2007. Wavelet and rescaled range approach for the Hurst coefficient for short and long time series, Computer & Geosciences, 33, 83–93.

CHAMOLI A., SWAROOPA RANI V., SRIVASTAVA K., SRINAGESH D. and DIMRI V.P., 2010. Wavelet analysis of the Seismograms for Tsunami warning, Nonlinear Processes in Geophysics, 17(5), 569–574.

DEVROYE, L. A Course on Density Estimation, Birkhauser, Boston, 1987.

ESQUIVEL, R.O., ANGULO, J.C., ANTOLIN, J., DEHESA, J.S., LOPEZ-ROSA, S., FLORES-GALLEGOS, N., 2010. Analysis of complexity measures and information planes of selected molecules in position and momentum spaces. Phys. Chem. Chem. Phys. 12, 7108–7116.

FISHER, R. A., 1925. Theory of statistical estimation. Proc. Cambridge Philos. Soc. 22, 700–725.

FRIEDEN, B. R., 1990. Fisher information, disorder, and the equilibrium distributions of physics. Phys. Rev. A 41, 4265–4276.

JANICKI A, WERON A. Simulation and Chaotic Behavior of Stable Stochastic Processes. Marcel Dekker, New York, 1994.

KANTELHARDT, J.W., ZSCHIEGNER, S.A., KOSCIELNY-BUNDE, E., HAVLIN, S., BUNDE, A., STANLEY, H.E., 2002. Multifractal detrended fluctuation analysis of nonstationary time series, Physica A, *316*, 87–114.

KENNETT, B.L.N., ENGDAHL, E.R., 1991. Traveltimes for global earthquake location and phase identification. Geophys. J. Int., *105*(2), 429–465. doi:10.1111/j.1365-246X.1991.tb06724.x.

KHARIN, V. V., ZWIERS, F.V., 2003. On the ROC Score of Probability Forecasts, J. Climate, *16*, 4145–4150.

LOVALLO, M., TELESCA, L., 2011. Complexity measures and information planes of X-ray astrophysical sources, J. Stat. Mech., P03029.

LYUBUSHIN, A. A., KALÁB, Z., LEDNICKÁ, M., HAGGAG, H.M., 2013. Discrimination of earthquakes and explosions using multi-fractal singularity spectrums properties, J. Seismol., *17*, 975–983.

MARTIN, M. T., PENNINI, F., PLASTINO, A., 1999. Fisher's information and the analysis of complex signals. Phys. Lett. A *256*, 173–180.

MARTIN, M. T., PEREZ, J., PLASTINO, A., 2001. Fisher information and nonlinear dynamics. Physica A *291*, 523–532.

RAYKAR VC, DURAISWAMI R. Fast optimal bandwidth selection for kernel density estimation. In Proceedings of the sixth SIAM International Conference on Data Mining, Bethesda, April 2006, 524–528.

TELESCA, L., LAPENNA, V., 2006. Measuring multifractality in seismic sequences, Tectonophysics, *423*, 115–123.

TELESCA, L., LOVALLO, M., 2011. Analysis of the time dynamics in wind records by means of multifractal detrended fluctuation analysis and the Fisher–Shannon information plane, J. Stat. Mech., P07001.

TELESCA, L., LOVALLO, M., RAMIREZ-ROJAS, A., ANGULO-BROWN, F., 2009. A nonlinear strategy to reveal seismic precursory signatures in earthquake-related self-potential signals. Physica A *388*, 2036–2040.

TELESCA, L., LOVALLO, M., CARNIEL, R., 2010. Time-dependent Fisher Information Measure of volcanic tremor before 5 April 2003 paroxysm at Stromboli volcano, Italy. J. Volcanol. Geoterm. Res. *195*, 78–82.

TELESCA, L., LOVALLO, M., HSU, H.-L., CHEN, C.-C., 2011. Analysis of dynamics in magnetotelluric data by using the Fisher-Shannon method. Physica A *390*, 1350–1355.

TELESCA, L., LOVALLO, M., CHAMOLI, A., DIMRI, V. P., 2013. Fisher-Shannon analysis of seismograms of tsnunamigenic and non-tsnunamigenic earthquakes, Physica A., *392*, 3424–3429.

TROUDI, M., ALIMI, A. M., SAOUDI, S., 2008. Analytical Plug-In Method for kernel density estimator applied to genetic neutrality study, EURASIP Journal on Advances in Signal Processing, Vol. *2008*, Article ID 739082, 8 pages.

YULMETYEV, R. M., GAFAROV, F., HANGGI, P., NIGMATULLIN, R., KAYUMOV, SH., 2001. Possibility between earthquake and explosion seismogram differentiation by discrete stochastic non-Markov processes and local Hurst exponent analysis, Phys. Rev. E. *64*, 066132 (2001).

(Received April 4, 2014, revised May 9, 2014, accepted May 20, 2014, Published online June 18, 2014)

Pure Appl. Geophys. 172 (2015), 1945–1957
© 2014 Springer Basel
DOI 10.1007/s00024-014-0895-7

Is It Possible to Predict Strong Earthquakes?

Y. S. POLYAKOV,[1,2] G. V. RYABININ,[3] A. B. SOLOVYEVA,[4] and S. F. TIMASHEV[5,6]

Abstract—The possibility of earthquake prediction is one of the key open questions in modern geophysics. We propose an approach based on the analysis of common short-term candidate precursors (2 weeks to 3 months prior to strong earthquake) with the subsequent processing of brain activity signals generated in specific types of rats (kept in laboratory settings) who reportedly sense an impending earthquake a few days prior to the event. We illustrate the identification of short-term precursors using the groundwater sodium-ion concentration data in the time frame from 2010 to 2014 (a major earthquake occurred on 28 February 2013) recorded at two different sites in the southeastern part of the Kamchatka Peninsula, Russia. The candidate precursors are observed as synchronized peaks in the nonstationarity factors, introduced within the flicker-noise spectroscopy framework for signal processing, for the high-frequency component of both time series. These peaks correspond to the local reorganizations of the underlying geophysical system that are believed to precede strong earthquakes. The rodent brain activity signals are selected as potential "immediate" (up to 2 weeks) deterministic precursors because of the recent scientific reports confirming that rodents sense imminent earthquakes and the population-genetic model of KIRSHVINK (Soc Am 90, 312–323, 2000) showing how a reliable genetic seismic escape response system may have developed over the period of several hundred million years in certain animals. The use of brain activity signals, such as electroencephalograms, in contrast to conventional abnormal animal behavior observations, enables one to apply the standard "input-sensor-response" approach to determine what input signals trigger specific seismic escape brain activity responses.

Key words: Earthquake precursors, Abnormal animal behavior, Flicker-noise spectroscopy, Nonstationarity factor, Earthquake prediction, Hydrogeochemical precursors.

[1] New Jersey Institute of Technology, Newark, NJ, USA. E-mail: polyakov@njit.edu

[2] USPolyResearch, Ashland, PA, USA.

[3] Kamchatka Branch, Geophysical Survey of Russian Academy of Sciences, Petropavlovsk-Kamchatsky, Russia.

[4] Semenov Institute of Chemical Physics, Russian Academy of Sciences, Moscow, Russia.

[5] Institute of Laser and Information Technologies, Russian Academy of Sciences, Troitsk, Moscow Region, Russia.

[6] National Research Nuclear University MEPhI, Moscow, Russia.

1. Introduction

Earthquake prediction within a time frame of several months to less than an hour before the catastrophic event, which is often referred to in the literature as "short-term" prediction, has been a subject of extensive research studies and controversial debates in both academia and the mass media in the past 2 decades (CICERONE *et al.* 2009; GELLER 1997; GELLER *et al.* 1997; UYEDA *et al.* 2009; WYSS *et al.* 1997). A recent striking example was when Italian seismologists were prosecuted in court and given a 6-year jail sentence for providing "inaccurate, incomplete and contradictory" statements regarding the risk of a major earthquake near L'Aquila, with a series of foreshocks recorded a few days prior to the deadly seismic event (HALL 2011; SHOCK and LAW 2012). One of the key areas in earthquake prediction is the study of precursors, physical phenomena that reportedly precede at least some earthquakes. The precursory signals are usually grouped as electromagnetic, hydrological/hydrochemical, gas geochemical, geodetic, and seismic (BORMANN 2011; CICERONE *et al.* 2009; GELLER 1997; HARTMANN and LEVY 2005; UYEDA *et al.* 2009; RYABININ *et al.* 2011). Abnormal animal behavior prior to strong earthquakes, which was typically discussed in the past based on anecdotal and retrospective evidence, has recently been reported and studied in scholarly journals and may now be considered as a potential precursor as well (GRANT and HALLIDAY 2010; GRANT *et al.* 2011; LI *et al.* 2009; LU-LU *et al.* 2010; YOKOI *et al.* 2003).

Despite the large number of earthquake precursors reported in the literature, most of which are summarized by HARTMANN and LEVY (2005) and CICERONE *et al.* (2009), an International Commission on Earthquake Forecasting for Civil Protection concluded on 2 October 2009, "the search for precursors that are diagnostic of an impending earthquake has not yet produced a successful short-term prediction scheme" (ICEFCP 2009). The reports of the International Association of Seismology and Physics of the Earth's Interior contain similar findings (WYSS and BOOTH 1997). The lack of confidence can be attributed to several reasons. First, some fundamental aspects of many non-seismic signals, for example, lithosphere-atmosphere–ionosphere coupling and propagation of ultra-low-frequency electromagnetic signals in the conductive earth, are unresolved, and many of the proposed physical models are questionable (UYEDA et al. 2009). Second, the experimental data on precursory signals are often limited to a few earthquakes and few measurement sites, and they frequently contain gaps and different types of noise (CICERONE et al. 2009; HARTMANN and LEVY 2005; UYEDA et al. 2009). Third, different techniques of identifying the anomalies are used for different signals or even in different studies for the same signal. In some cases, the anomalous changes are determined by analyzing the signals themselves (CICERONE et al. 2009; HARTMANN and LEVY 2005; UYEDA et al. 2009), while in other cases they are identified by studying the derived statistics or functions, such as Fisher information or scaling parameters (TELESCA et al. 2009a, b). Moreover, seasonal changes and instrumentation or other background noise often need to be filtered out prior to the identification of precursors.

To overcome some of these challenges, we proposed a phenomenological approach to searching for earthquake precursors based on the analysis of signals of different types in the same local geographic region (RYABININ et al. 2011, 2012). We assume that a large earthquake may be preceded by a reconfiguration of a geophysical system on different time and space scales, which manifests itself in qualitative changes of various signals within relatively short time intervals. Our previous studies performed using flicker-noise spectroscopy (FNS), a phenomenological framework for extracting information from time series with stochastically varying components (TIMASHEV and POLYAKOV 2007, 2008; TIMASHEV 2007; TIMASHEV et al. 2010a), show that the peak values in FNS nonstationarity factors, which correspond to the time

moments of major rearrangements (within relatively short time intervals) of a complex geophysical system (preceding a future strong earthquake), may be considered as "precursors" of the upcoming earthquakes (DESCHEREVSKY et al. 2003; TELESCA et al. 2004; VSTOVSKY et al. 2005; HAYAKAWA and TIMASHEV 2006; IDA et al. 2007; RYABININ et al. 2011). These studies demonstrate that the timing of the peaks in FNS nonstationarity factors with respect to strong earthquakes may dramatically vary depending on the nature of the signals under study. In this regard, it was suggested to introduce "short-term" (weeks to months prior to the earthquake) and "immediate" (minutes to weeks) precursors (RYABININ et al. 2011). The short-term precursors may be identified by examining the nonstationarity factor for daily time series of a certain parameter recorded in a seismically active zone; for instance, it can be water salinity in a borehole within that geographical area (RYABININ et al. 2011, 2012). The immediate precursors can be found by analyzing the nonstationarity factor for hourly and then minute-level sequences of another appropriately selected dynamic variable, such as electrotelluric or electrochemical potential, and intensity of geoacoustic or electromagnetic signals in the ultra-low-frequency band. A combined analysis of these two kinds of precursors could be used to forecast a strong earthquake (RYABININ et al. 2011, 2012).

One can argue that such a forecast would only be probabilistic rather then deterministic. The occurrence of the peak value in the nonstationarity factor reflects some major reconfiguration in the geophysical system, which may be treated only as a necessary but not a sufficient condition of a strong earthquake. For instance, the geophysical medium may simultaneously experience relaxation rearrangements, which would also be seen as peak values in the nonstationarity factor. This reasoning can be applied to interpret Figures 2 and 3 in (RYABININ et al. 2011). The question is how to differentiate between the abrupt changes preceding a strong earthquake from these relaxation phenomena and deterministically predict the earthquake. In other words, what could be used as a "sufficient" condition of the upcoming strong earthquake? Can such a sufficient condition be formulated to enable earthquake prediction?

In this study, we attempt to answer these questions. First, we illustrate the procedure for identifying a necessary condition for a strong earthquake based on the analysis of hydrogeochemical time series for the Kamchatka Peninsula in the time frame from 2010 to 2014 (a major earthquake was recorded on 28 February 2013), and then we discuss our hypothesis regarding the selection of an appropriate sufficient condition.

The article is structured as follows. In Sect. 2, we provide the fundamentals of FNS and present the nonstationarity factor. Section 3 describes the experimental setup used to illustrate the selection of a necessary condition. Section 4 deals with the nonstationarity analysis of the hydrogeochemical data. Section 5 discusses the results and elaborates on the mechanism for selecting the sufficient condition. Section 6 presents the concluding remarks.

2. FNS Nonstationarity Factor

Here, we will only deal with the basic FNS relations needed to understand the nonstationarity factor. FNS is described in more detail elsewhere (POLYAKOV et al. 2012; TIMASHEV and POLYAKOV 2007, 2008; TIMASHEV 2007; TIMASHEV et al. 2010a, 2012).

In FNS, all introduced parameters for signal $V(t)$, where t is time, are related to the autocorrelation function

$$\psi(\tau) = \langle V(t)V(t+\tau)\rangle_T \qquad (1)$$

where τ is the time lag parameter ($0 \leq \tau \leq T_M$) and T_M is the upper bound for τ ($T_M \leq T/2$). This function characterizes the correlation in the values of dynamic variable V at higher $t + \tau$ and lower t values of the argument. The angular brackets in relation (1) stand for the averaging over time interval $[0, T]$.

$$\langle(\ldots)\rangle_T = \frac{1}{T}\int_0^T (\ldots)dt. \qquad (2)$$

The averaging over interval $[0, T]$ implies that all the characteristics that can be extracted by analyzing functions $\psi(\tau)$ should be regarded as average values on this interval.

To extract the information contained in $\psi(\tau)$ ($\langle V(t)\rangle = 0$ is assumed), the following transforms or "projections" of this function are analyzed: cosine transforms ("power spectrum" estimates) $S(f)$, where f is the frequency,

$$S(f) = 2\int_0^{T_M} \langle V(t)V(t+t_1)\rangle_{T-\tau}\cos(2\pi f t_1)\,dt_1 \qquad (3)$$

and its difference moments (Kolmogorov transient structure functions) of the second order $\Phi^{(2)}(\tau)$

$$\Phi^{(2)}(\tau) = \left\langle [V(t) - V(t+\tau)]^2 \right\rangle_T \qquad (4)$$

Here, we use the quotes for the power spectrum because according to the Wiener-Khinchin theorem the cosine (Fourier) transform of the autocorrelation function is equal to the power spectral density only for wide-sense stationary signals at infinite integration limits.

The information contents of $S(f)$ and $\Phi^{(2)}(\tau)$ are generally different, and the parameters for both functions are needed to solve the parameterization problems. By considering the intermittent character of the signals under study, interpolation expressions for the stochastic components $S_s(f)$ and $\Phi_s^{(2)}(\tau)$ of $S(f)$ and $\Phi^{(2)}(\tau)$, respectively, were derived using the theory of generalized functions by TIMASHEV (2006). It was shown that structural functions $\Phi_s^{(2)}(\tau)$ are formed only by jump-like (random-walk) irregularities corresponding to a dissipative process of anomalous diffusion, and functions $S_s(f)$, which characterize the "energy side" of the process, are formed by spike-like (inertial) and jump-like irregularities. It should be noted that τ in Eqs. (1)–(4) is considered as a macroscopic parameter exceeding the sampling period by at least one order of magnitude. This constraint is required to derive the expressions and separate out the contributions of dissipative jump-like and inertial (non-dissipative) spike-like components.

The analysis of experimental stochastic series often requires the original data to be smoothed. In this study, we will apply the "relaxation" procedure proposed for nonstationary signals by TIMASHEV and VSTOVSKII (2003) based on the analogy with a finite-difference solution of the diffusion equation, which allows one to split the original signal into low-frequency $V_R(t)$ and high-frequency $V_F(t)$ components. The iterative procedure finding the new values of the signal at every "relaxation" step using its values for the previous step allows one to determine the low-frequency component V_R [see expressions (7)–(9) in Appendix A]. The high-frequency component V_F is obtained by subtracting V_R from the original signal. This algorithm progressively reduces the local gradients of the "concentration" variable, causing the points in every triplet to come closer to each other. The smoothing procedure is described in more detail in Appendix A.

To analyze the effects of nonstationarity in real processes, we study the dynamics of changes in $\Phi^{(2)}(\tau)$ for consecutive "window" intervals $[t_k, t_k + T]$, where $k = 0, 1, 2, 3 \ldots$ and $t_k = k\Delta T$, which are shifted within the total time interval T_{tot} of the experimental time series $(t_k + T < T_{\text{tot}})$. The averaging interval T and difference ΔT are chosen based on the physical understanding of the problem in view of the suggested characteristic time of the process, which is the most important parameter of the system evolution. The FNS nonstationarity factor is defined as:

$$C_J(t_k) = 2 \cdot \frac{Q_k^J - P_k^J}{Q_k^J + P_k^J} \bigg/ \frac{\Delta T}{T}, \qquad (5)$$

$$Q_k^J = \frac{1}{\alpha T^2} \int\limits_0^{\alpha T} \int\limits_{t_k}^{t_k+T} [V_J(t) - V_J(t+\tau)]^2 \, dt \, d\tau, \quad (6)$$

$$P_k^J = \frac{1}{\alpha T^2} \int\limits_0^{\alpha T} \int\limits_{t_k}^{t_k+T-\Delta T} [V_J(t) - V_J(t+\tau)]^2 dt \, d\tau. \quad (7)$$

Here, J indicates which function $V_J(t)$ ($J = R, F$ or G) is used to evaluate $\Phi_J^{(2)}(\tau)$, and the subscripts R, F, and G refer to the low-frequency component, high-frequency component, and unfiltered signal, respectively. Expressions (5)–(7) are given in discrete form in Appendix B. Note that functions $\Phi_J^{(2)}(\tau)$ can be reliably evaluated only on the τ interval of $[0, \alpha T]$, which is less than half the averaging interval T, i.e., $\alpha < 0.5$.

The phenomenon of "precursor" occurrence is assumed to be related to abrupt changes in functions $\Phi^{(2)}(\tau)$ when the upper bound of the interval

$[t_k, t_k + T]$ approaches the time moment t_c of a catastrophic event accompanied by total system reconfiguration on all space scales. Graphically, this corresponds to peaks in the plots of the nonstationarity factor.

3. Data

The data were recorded in the southeastern part of the Kamchatka Peninsula located at the Russian Far East. The eastern part of the peninsula is one of the most seismically active regions in the world. The area of highest seismicity localized in the depth range between 0 and 40 km represents a narrow stripe with a length of approximately 200 km along the east coast of Kamchatka, which is bounded by a deep-sea trench on the east (FEDOTOV et al. 1985).

Specialized measurements of groundwater characteristics were started in 1977 to find and study possible hydrogeochemical precursors of Kamchatka earthquakes. Currently, the observation network includes four stations in the vicinity of Petropavlovsk-Kamchatsky: Pinachevo, Moroznaya, Khlebozavod, and Verkhnyaya Paratunka. The Pinachevo station includes five water reservoirs: four warm springs and one borehole, GK-1, with a depth of 1,261 m. The Moroznaya station has a single borehole, No. 1, with a depth of 600 m. The Khlebozavod station also includes a single borehole, G-1, with a depth of 2,540 m, which is located in Petropavlovsk-Kamchatsky. The Verkhnyaya Paratunka station comprises four boreholes (GK-5, GK-44, GK-15, and GK-17) with depths in the range from 650 to 1,208 m.

The system of hydrogeochemical observations includes the measurement of atmospheric pressure and air temperature, measurement of water discharge and temperature of boreholes and springs, and collection of water and gas samples for further laboratory analyses. For water samples, the following parameters are determined: pH; ion concentrations of chlorine (Cl^-), bicarbonate (HCO_3^-), sulfate (SO_4^{2-}), sodium (Na^+), potassium (K^+), calcium (Ca^{2+}), and magnesium (Mg^{2+}); concentrations of boric (H_3BO_3) and silicone (H_4SiO_4) acids. For the samples of gases dissolved in water, the following

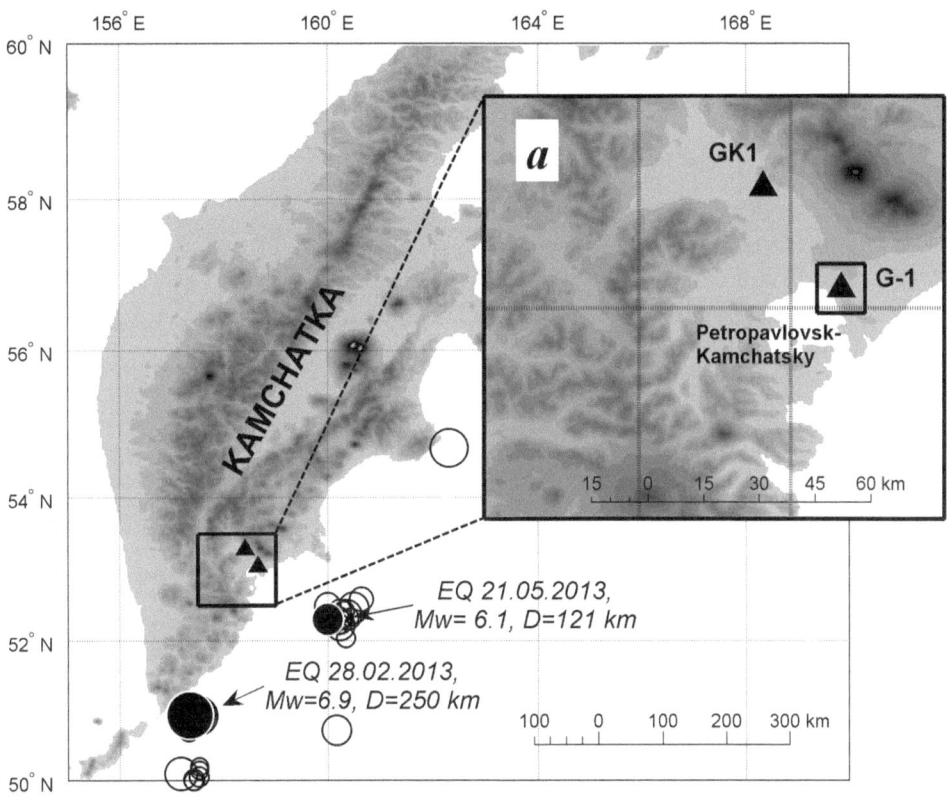

Figure 1

Schematic of the measurement area, observation points, and epicenters of the strongest earthquakes in 2013 ($M \geq 5.0$, $H \leq 50$ km, $D \leq 350$ km), where M earthquake magnitude, H depth, D distance from the epicenter to Petropavlovsk-Kamchatsky. *Solid black circles* denote the major earthquake of 28 February 2013 ($M_w = 6.9$, $D = 250$ km) and strongest earthquake ($M_w = 6.1$, $D = 121$ km) in a series of seismic events in May 2013. Frame *a* shows an enlarged view of the positions of boreholes G-1 and GK-1. The earthquakes were selected using the catalog of the US Geological Survey: http://earthquake.usgs.gov/earthquakes/search/

concentrations are determined: methane (CH_4), nitrogen (N_2), oxygen (O_2), carbon dioxide (CO_2), helium (He), hydrogen (H_2), hydrocarbon gases: ethane (C_2H_6), ethylene (C_2H_4), propane (C_3H_8), propylene (C_3H_6), butane ($C_4H_{10}n$), and isobutane ($C_4H_{10}i$). The data are recorded at nonuniform sampling intervals with one dominant sampling frequency. For the Pinachevo, Moroznaya, and Khlebozavod stations, this average sampling frequency is one measurement per 3 days; for the Verkhnyaya Paratunka station, it is one measurement per 6 days. Multiple studies of the hydrogeochemical data and corresponding seismic activity for the Kamchatka peninsula reported anomalous changes in the chemical and/or gas composition of groundwater prior to several large earthquakes (BELLA *et al.* 1998; BIAGI *et al.* 2000, 2004, 2006; KHATKEVICH and RYABININ 2006; KOPYLOVA *et al.* 1994).

In this study, we analyze the time series at GK-1 (Pinachevo station) and G-1 (Khlebozavod station). Our goal is to examine the buildup for the major earthquake of 28 February 2013 and subsequent earthquake swarm in May 2013, which are illustrated in Fig. 1.

4. Results

To illustrate the procedure for identifying a necessary condition of the impending strong earthquake, we examined the signals in the time interval from 1 January 2010 to January 2014. This period is characterized by one major earthquake with a magnitude above 6.5, which took place on 28 February 2013. According to our hypothesis that any large earthquake is preceded by a major reconfiguration of the

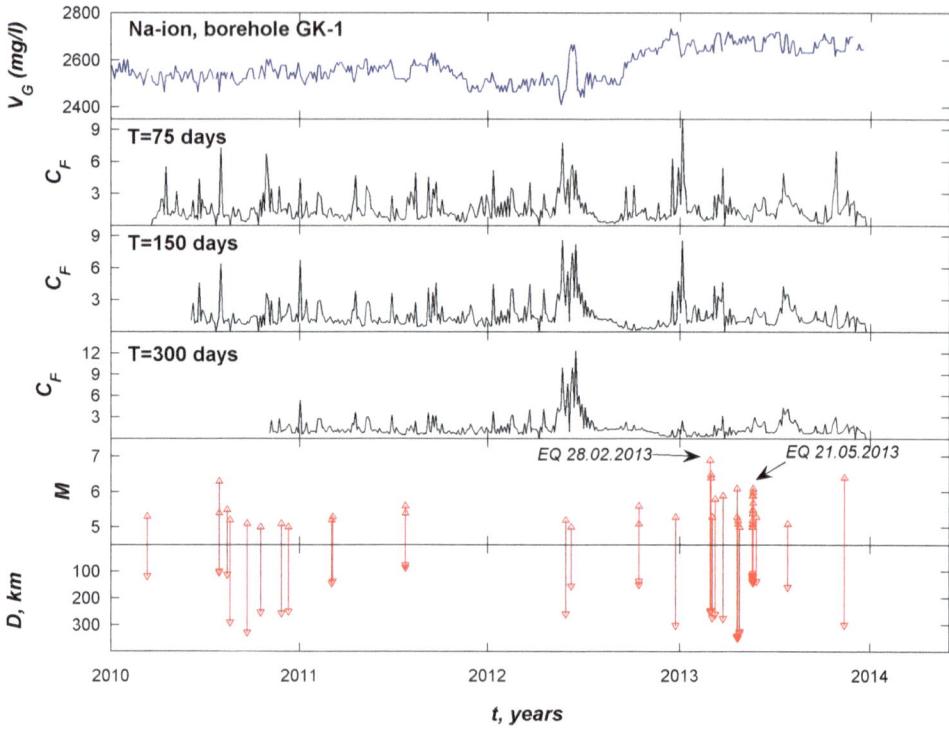

Figure 2
Comparison of nonstationarity factor C_F for the high-frequency component of the sodium-ion concentration in the water of borehole GK-1 (Pinachevo station) with seismic activity ($M \geq 5.0$): V_G unfiltered time series of the sodium-ion concentration; M earthquake magnitude; D distance of the epicenter from Petropavlovsk-Kamchatsky

geophysical system in the affected region (RYABININ et al. 2011), we expected to observe strong qualitative changes in at least some of the hydrogeochemical parameters occurring simultaneously at multiple measurement stations in the Kamchatka region. Our preliminary analysis showed that the nonstationarity factor for the high-frequency component of the sodium-ion concentration shows a striking peak at multiple stations 52–54 days prior to the strongest earthquake (Figs. 2, 3). As a result, we selected this hydrogeochemical parameter for further analysis.

Figures 2 and 3 display the variation of sodium-ion concentration in boreholes GK-1 and G-1, respectively, along with the nonstationarity factor C_F evaluated at three different values of the averaging interval T and seismic activity data during this period. The average sampling frequency of the sodium-ion concentration time series is (3 day^{-1}). The high-frequency component was obtained by applying ten iterations of the smoothing procedure (7)–(9) at $\omega = 0.25$, which corresponds to the effective high-pass cutoff frequency of approximately (18 day^{-1}) (where the value of power spectrum density for the component reaches its maximum).

It can be seen in Fig. 2 that the highest peak in the nonstationarity factor C_F at $T = 75$ days for the sodium-ion concentration in borehole GK-1 occurs 52 days prior to the earthquake of 28 February 2013. As we increase the averaging interval T, the relative magnitude of the peak becomes less pronounced because of the contribution of other local reorganizations in the high-frequency range during the current averaging interval $[t_k, t_k + T]$. This implies that the averaging interval of 75 days is most adequate for identifying single local reorganizations in the high-frequency component of the sodium-ion concentration data for GK-1, at least for the earthquake under study. It should also be noted that there is a relatively high peak in the nonstationarity factor ~ 9 months prior to the major earthquake. This peak is not followed by any major seismic event, which suggests that not every local reorganization leads to a strong earthquake.

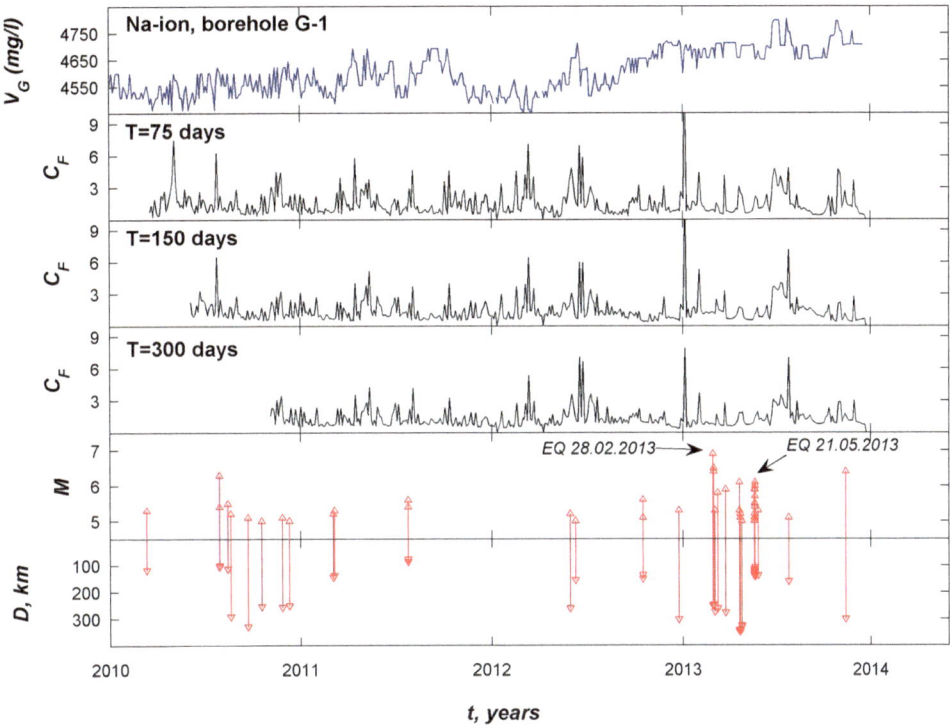

Figure 3

Comparison of nonstationarity factor C_F for the high-frequency component of the sodium-ion concentration in the water of borehole G-1 (Khlebozavod station) with seismic activity ($M \geq 5.0$): V_G unfiltered time series of sodium-ion concentration; M earthquake magnitude; D distance of the epicenter from Petropavlovsk-Kamchatsky

The nonstationarity factor C_F in Fig. 3 has the highest peak 54 days prior to the strongest earthquake at all three values of T. Considering that the average sampling interval for both time series is 3 days, we can conclude that the nonstationarity factors for both boreholes point to the same local reorganization, which happens to be the most significant one in the time period under study. It is reasonable to relate this synchronized local reorganization with a buildup for some major event, such as the earthquake of 28 February. It is noteworthy that there is another noticeable peak in the nonstationarity factor C_F, which occurs approximately at the same time (9 months prior to the major earthquake) as in the data for GK-1.

5. Discussion

The largest peaks in FNS nonstationarity factors for the high-frequency component of groundwater sodium-ion concentration simultaneously occurring at two stations 32 km apart from each other suggest that a relatively large-scale reorganization may have preceded the earthquake of 28 February 2013. The peaks were observed 52–54 days prior to the major seismic event, which is in agreement with our previous results for the variation of groundwater chlorine-ion concentration where the peaks were observed in the range from 50 to 70 days at GK-1 and GK-44 boreholes prior to the major earthquake of 8 October 2001 (RYABININ et al. 2011). Large peaks in the nonstationarity factor before strong earthquakes were also observed for other types of data, including geoelectrical signals (TELESCA et al. 2004), electrochemical potentials (DESCHEREVSKY et al. 2003), and ultra-low-frequency electromagnetic transmissions (HAYAKAWA and TIMASHEV 2006). This implies that local reorganizations in the geophysical system of a particular seismic region may be related to a buildup for impending earthquakes.

At the same time, such local reorganizations (peaks in the nonstationarity factor) do not always lead to a strong earthquake, as illustrated by Figs. 2 and 3 in (RYABININ et al. 2011) and the other non-stationarity factor peaks observed in Figs. 2 and 3 of this study. These nonstationarity factor peaks may also be related to certain rearrangements when the geophysical system responds to abrupt changes in external conditions. It is also possible that a series of local reorganizations has to take place prior to triggering a strong earthquake, and only the last local reorganization could be considered in this case as a real precursor. The question is how to differentiate between the abrupt changes preceding a strong earthquake from these non-precursory local rearrangements. Taking into account the low number of major earthquakes for which the accompanying non-seismic time series are available and the absence of reliable physical models explaining the correlation between the variability in the data sets and seismic events, it can be concluded that such a distinction cannot be made at this time and the peaks in non-stationarity factors of certain signals cannot be regarded as deterministic precursors. The occurrence of the peak can only imply that a major earthquake is possible within a certain interval of time after the peak [in the case of hydrogeochemical signals with a sampling frequency of ($3 \, \text{day}^{-1}$), the strong earthquake may be expected to happen between 50 and 70 days after the peak]. In other words, a high peak in the nonstationarity factor can only be regarded as a necessary condition for a strong seismic event. Is it possible to complement this necessary condition by a sufficient condition to enable true earthquake prediction?

We believe that a truly reliable system for predicting strong earthquakes may have developed over the period of several hundred million years in certain animals, which is supported by several recent scientific studies reporting abnormal behavior in mice, rats, and toads in the range from 1 to 6 days before strong earthquakes (GRANT et al. 2011; GRANT and HALLIDAY 2010; LI et al. 2009; Lu-Lu et al. 2010; YOKOI et al. 2003). Kirschvink (2000) argues that the evolutionary mechanism of exaptation (the ability to adapt or link a genetic pattern that evolved for one function for another role) may have produced a seismic-response system through the process of random mutation and natural selection. The facts that the first burrowing animals appeared at least 540 million years ago and the plate tectonics on the earth have operated for at least 2 billion years suggest that from the distant ancestors of all mammals (synapsids) up to modern age mammals, the evolution dealt with a very large sample of strong earthquakes associated with increased mortality of the animals without appropriate seismic response gene(s). Using the concept of episodic selection and Monte-Carlo simulations, KIRSCHVINK (2000) demonstrates that such a hypothetical signal response gene could easily reach fixation within a population only after 1,000 generations, with every seismic event occurring once in 50 generations and resulting in 10 % mortality of the individuals without the gene. In contrast, the sample of strong earthquakes and accompanying seismic and non-seismic data that are available to the scientific community is by many orders of magnitude smaller, which makes many seismologists believe that earthquake prediction is impossible (GELLER 1997; GELLER et al. 1997). Another possible explanation discussed in the literature is that the anomalous animal behavior observed prior to strong earthquakes is a direct response of animals to certain adverse stimuli caused by the changes in the environment shortly before the seismic event (GRANT et al. 2011; FREUND and STOLC 2013). These stimuli may be brought about by changes in the concentrations of H_2O_2 in water, CO in air, or positive ions in air; shifts in electromagnetic fields; or certain variations in other physical/chemical characteristics (FREUND and STOLC 2013).

Up to the present, abnormal animal behavior has not been considered as a precursor candidate by many seismologists and was not submitted for evaluation to the IASPEI Sub-Commission on Earthquake Prediction because it cannot be permanently monitored in a controlled way and unambiguously assessed (BORMANN 2011). It is generally believed that similar anomalous animal behavior may have different causes and cannot be reproduced using the standard "input-sensor-response" approach. This reasoning is based on the fact that many past studies relied on subjective behavioral observations, such as anomalous locomotive activities in the circadian rhythms of mice (YOKOI et al. 2003; LI et al. 2009) or the quiescence or lack of spawning in toads (GRANT and HALLIDAY 2010). However, if we rely on more objective biomedical signals and study the responses to certain stimuli believed to be related to earthquake buildup, we can apply the standard input-sensor-response method and identify precursors by analyzing the signals rather than the general behavior of the animals under study. One potential approach to obtaining more objective signals may be based on the computer-aided measurements of animal behavior with specialized video cameras, automated logging machines, motion-triggered cameras, or the use of other automated technologies for recording anomalous animal responses. However, such automatically recorded behavioral responses would only capture the increase or decrease in certain parameters of locomotive activity, thus ignoring the complexity of the biological system of the animal and greatly reducing the number of degrees of freedom. It is rather likely that two different stimuli (precursory and non-precursory) may trigger the same change in the selected parameter of locomotive activity. Therefore, it is necessary to look beyond behavioral "signals" and select candidate signals that better represent the complexity of the biological system.

In our view, the brain activity signals in animals can be considered as the main biomedical data because the brain, operating as an integral system, processes any external signals, including those related to imminent danger, and generates specific responses that can be identified from the analysis of cortical brain activity. In this regard, electroencephalogram (EEG) measurements of relatively large rodents, such as California kangaroo rats, obtained with two-channel implantable EEG sensor transmitters (manufactured by BIOPAC Systems, Inc.) or four-channel tethered systems with rat headmounts (manufactured by Pinnacle Technology, Inc.), appear to be the most practical option for seismological studies. Small laboratories with large enough rat samples in each could be set up at multiple sites of a specific seismically active region, and the online analysis of the EEG data could be performed in a central signal processing facility. The EEG measurements in rats have to be performed in accordance with established ethical guidelines for animals. A potential challenge of this method is the placement of electrodes (NOBLER *et al.* 1993), which may require initial calibration to capture earthquake-relevant responses in the rodents.

It should be noted that the FNS framework looks promising for analyzing these EEG data because it recently demonstrated its effectiveness in the analysis of human EEG and magnetoencephalogram (MEG) signals. The FNS parameterization procedure and cross-correlation function were used to classify the susceptibility of 84 children/adolescents (11–14 years of age) to schizophrenia based on the analysis of EEG signals (TIMASHEV *et al.* 2012). As part of this study, three new quantitative diagnostic markers were suggested. In another study, the FNS parameterization procedure and cross-correlation function were effective in studying neuromagnetic responses (MEG signals) from a group of healthy human subjects (nine volunteers) and a photosensitive epilepsy (PSE) patient while they were viewing equiluminant flickering stimuli of different color combinations (TIMASHEV *et al.* 2009, 2010b). The purpose of that study was to develop diagnostic biomarkers for PSE. These results along with the findings reported by RYABININ *et al.* (2011, 2012) imply that all three major FNS tools (parameterization procedure, cross-correlation function, nonstationarity factor) may be applied in the analysis of rat EEG signals.

For an earthquake prediction method to be successful, it should specify: (1) the time window, (2) the spatial window, and (3) the magnitude window (GELLER 1997). According to Kirschvink (2000), Tributsch (1984), and GRANT and HALLIDAY (2010), animals can generally sense an impending strong earthquake at most 1–2 weeks prior to the event, with the highest abnormal activity occurring 1–2 days before the earthquake (here, we ignore the ability of animals to sense the P waves traveling immediately before the event because the time interval of a few seconds is of no practical interest). This provides a relatively narrow time window for the imminent seismic event. The spatial window can be determined by processing the EEG data from multiple sites; we expect there should be some synchronization between the EEG patterns of rats in the geographical zone where the effect of the strong earthquake would be detrimental. The magnitude window in this case could not be given directly in terms of magnitude and depth, but the presence of specific EEG patterns related to the seismic escape behavior implies that the mortality of the rodents under study could be affected. In other words, only relatively strong earthquakes (with magnitude $M > 5$) would be predicted by this method. The above suggests that if such a system can be built and appropriate EEG patterns identified, it may be adequate for the deterministic prediction of strong earthquakes.

It is evident that the nature of external signals warning the animals about the upcoming seismic event is yet to be determined. KIRSCHVINK (2000) suggested that there are four kinds of "precursory" signals that may be sensed by animals: (1) ground tilting, (2) humidity changes, (3) electrical currents, and (4) magnetic fields variations. There are no experimental data supporting that ground tilting could be the main input signal. Humidity changes can only be sensed in arid environments, which does not explain why mice, rats, and toads in non-desert areas demonstrated abnormal behavior prior to a number of strong earthquakes. Electrical currents could probably be sensed only by aquatic animals, whose electrical sensitivity is much higher than in terrestrial animals. The only relatively universal mechanism observed in many vertebrate animals is a highly sensitive magnetoreception system that is capable of sensing ultra-low-frequency magnetic activity known to precede certain earthquakes because this system probably evolved prior to the divergence of major subgroups of vertebrates (KIRSCHVINK 2000). This system was most likely developed for navigational and/or circadian purposes, and later may have been adapted through exaptation for the seismic escape response role (KIRSCHVINK 2000). If the precursory signals sensed by the animals are of biomagnetic nature, then the proposed EEG experiments should consider the negative effect of radiofrequency noise on the magnetoreception system of the rodents (KIRSCHVINK 2014). GRANT et al. (2011) and FREUND and STOLC (2013) suggested that the activation of mobile electronic charge carriers caused by the buildup of tectonic stress may generate electromagnetic signals in the ultra-low-frequency (ULF) and extremely low-frequency (ELF) ranges. These signals, along with resulting positively charged ions in air and elevated levels of hydrogen peroxide in water, may cause unusual reactions in certain terrestrial and aquatic animals. There may be another kind of precursory signals that could be sensed by certain animals: low-frequency geoacoustic noise. KIRSCHVINK (2000) discussed in detail the ability of many animals to sense acoustic P (primary) waves that are detected a few seconds before the destructive S (secondary) waves arrive. At the same time, some studies show that geoacoustic precursors, such as peaks in the nonstationary factors for geoacoustic emissions in the frequency range from 25 to 1,400 Hz recorded in a borehole of the Khlebozavod station (RYABININ et al. 2011), are observed several days prior to certain earthquakes. If animals can sense the P waves, maybe they could also detect the anomalous changes in such low-frequency geoacoustic noise.

In contrast to behavioral observations, the analysis of rat EEG signals also enables one to perform the laboratory experiments where brain activity responses to specific types of input are studied and categorized in an objective way. First, controlled shake-table experiments could be conducted using laboratory populations of rats from seismically active zones, for example, California kangaroo rats. These experiments would establish a baseline of EEG response patterns for comparison with reactions to other stimuli. Then a series of field-based experiments could be performed on the same species in which candidate precursory geophysical and geochemical signals are applied artificially to determine whether any of the input signals would produce similar EEG patterns. The exaptation model developed by KIRSCHVINK (2000) would suggest some similarity in the evoked EEG response between shaking and other stimuli reported to precede certain strong earthquakes. This input-sensor-output analysis would help understand what earthquake precursory signals are sensed by the rats.

6. Concluding Remarks

The above discussion suggests that an earthquake prediction system might be built using a combined approach comprising (1) the identification of short-term candidate precursors (2 weeks to several months prior to major seismic event) by locating the peaks in FNS nonstationarity factors of relatively low-frequency (hourly, daily) signals manifesting local reorganizations in the underlying geophysical system prior to strong earthquakes and (2) the subsequent processing of EEG signals of the rats living in seismic areas, for instance, California kangaroo rats, collected from their laboratory populations at multiple sites within the same seismically active region. We show that the groundwater sodium-ion concentration with a sampling frequency of (3 day^{-1}) can be chosen as a short-term candidate precursor for the earthquakes similar to the one of 28 February 2013 in the Kamchatka peninsula region. It should be noted that not every candidate precursor is followed by a strong earthquake and that the monitoring of rat EEG signals does not need to be performed continuously. The occurrence of the short-term candidate precursor should trigger a detailed EEG analysis within the 3 months following the peak in the nonstationarity factor (for the case studied in this article). If no EEG patterns corresponding to a specific animal seismic response are detected within that time frame, the EEG signal monitoring can be stopped until the next peak in the nonstationarity factor for the short-term candidate precursory signal has been identified.

Laboratory experiments where EEG responses to specific types of input are studied and compared to the responses in controlled shake-table experiments should be conducted to identify the signals detected by the sensory system of the rats prior to strong seismic events. The brain activity responses and input sensory signals may vary depending on the characteristics of impending strong earthquakes. For instance, shallow-focus, mid-focus, and deep-focus earthquakes may each be characterized by different precursory signals (for example, electromagnetic emissions in different frequency ranges). The ultimate goal of laboratory experiments would be to learn from the seismic escape response system of rats and build artificial signal-receiving systems with sensors and centralized signal processing.

Appendix A

Consider the one-dimensional diffusion equation for V_R:

$$\frac{\partial V_R}{\partial \tau} = \chi \frac{\partial^2 V_R}{\partial t^2} \quad (8)$$

with symmetry boundary conditions

$$\frac{\partial V_R}{\partial t} = 0 \quad \text{at } t = 0, \quad (9)$$

$$\frac{\partial V_R}{\partial t} = 0 \quad \text{at } t = T \quad (10)$$

and initial condition

$$V_R(0) = V(t), \quad (11)$$

where χ is a constant diffusion coefficient.

Writing a forward difference for the local term and second-order central difference for the diffusion term, Eq. (8) gets transformed to

$$\frac{V_k^{i+1} - V_k^i}{\Delta \tau} = \chi \frac{V_{k+1}^i - 2 V_k^i + V_{k-1}^i}{(\Delta t)^2}, \quad (12)$$

where i is the "time" index and k is the "spatial" index. Here, the subscript "R" is dropped for simplicity.

After introducing $\omega = \frac{\chi \Delta \tau}{(\Delta t)^2}$, Eq. (12) can be further transformed to the following explicit finite difference expression:

$$V_k^{i+1} = \omega V_{k+1}^i + (1 - 2\omega)V_k^i + \omega V_{k-1}^i. \quad (13)$$

Analogously, the finite difference formulation for the complete problem (8)–(11) can be written as

$$V_1^{i+1} = (1 - 2\omega)V_1^i + 2\omega V_2^i, \quad (14)$$

$$V_k^{i+1} = \omega V_{k+1}^i + (1 - 2\omega)V_k^i + \omega V_{k-1}^i \text{ for } 1 < k < N, \quad (15)$$

$$V_N^{i+1} = (1 - 2\omega)V_N^i + 2\omega V_{N-1}^i. \quad (16)$$

Here, i is the current iteration number and N is the length of the time series. The smoothing procedure (14)–(16) is unconditionally stable for $\omega < 1/2$, which is the maximum allowed value for ω in the smoothing algorithm. There are two input parameters: the number of iterations $i_{\max}$ (largest value for i) and the value for ω. The smoothing procedure evaluates expressions (14)–(16) at each iteration.

Appendix B

The FNS nonstationarity factor in discrete form is written as ($b = \lfloor \Delta T/\Delta t \rfloor$, $N_1 = \lfloor \alpha N \rfloor$):

$$C_J(t_k) = 2 \cdot \frac{Q_k^J - P_k^J}{Q_k^J + P_k^J} \bigg/ \frac{\Delta T}{T}, \qquad (18)$$

$$Q_k^J = \frac{1}{N_1} \sum_{n_\tau=1}^{N_1} \frac{1}{N - n_\tau} \sum_{m=1+kb}^{N-n_\tau+kb} [V_J(m) - V_J(m + n_\tau)]^p, \qquad (19)$$

$$P_k^J \frac{1}{N_1} \sum_{n_\tau=1}^{N_1} \frac{1}{N - n_\tau} \sum_{m=1+kb}^{N-n_\tau+(k-1)b} [V_J(m) - V_J(m + n_\tau)]^p. \qquad (20)$$

REFERENCES

BELLA, F., BIAGI, P.F., CAPUTO, M., COZZI, E., MONICA, G.D., ERMINI, A., GORDEEV, E.I., KHATKEVICH, Y.M., MARTINELLI, G., PLASTINO, W., SCANDONE, R., SGRIGNA, V., and ZILPIMIANI, D. (1998), *Hydrogeochemical anomalies in Kamchatka (Russia)*, Phys. Chem. Earth 23, 921–925, doi:10.1016/S0079-1946(98)00120-7.

BIAGI, P.F., CASTELLANA, L., PICCOLO, R., MINAFRA, A., MAGGIPINTO, G., ERMINI, A., CAPOZZI, V., PERNA, G., KHATKEVICH, Y.M., AND GORDEEV, E.I. (2004), *Disturbances in groundwater chemical parameters related to seismic and volcanic activity in Kamchatka (Russia)*, Nat. Hazards Earth Syst. Sci. 4, 535–539, doi:10.5194/nhess-4-535-2004.

BIAGI, P.F., CASTELLANA, L., MINAFRA, A., MAGGIPINTO, G., MAGGIPINTO, T., ERMINI, A., MOLCHANOV, O., KHATKEVICH, Y.M., and GORDEEV, E.I. (2006), *Groundwater chemical anomalies connected with the Kamchatka earthquake (M = 7.1) on March 1992*, Nat. Hazards Earth Syst. Sci. 6, 853–859, doi: 10.5194/nhess-6-853-2006.

BIAGI, P.F., ERMINI, A., COZZI, E., KHATKEVICH, Y.M., and GORDEEV, E.I. (2000), *Hydrogeochemical precursors in Kamchatka (Russia) related to the strongest earthquakes in 1988–1997*, Nat. Hazards 21, 263–276, doi: 10.1023/A:1008178104003.

BORMANN, P. (2011), *From earthquake prediction research to time-variable seismic hazard assessment applications*, Pure Appl. Geophys. 168 (2011), 329–366, doi:10.1007/s00024-010-0114-0.

CICERONE, R.D., EBEL, J.E., and BRITTON, J. (2009), *A systematic compilation of earthquake precursors*, Tectonophysics 476, 371–396, doi:10.1016/j.tecto.2009.06.008.

DESCHEREVSKY, A.V., LUKK, A.A., SIDORIN, A.YA., VSTOVSKY, G.V., TIMASHEV, S.F. (2003), *Flicker-noise spectroscopy in earthquake prediction research*, Nat. Hazards Earth Syst. Sci.. 3 (3/4), 159–164, doi:10.1016/j.pce.2003.09.017.

FEDOTOV, S.A., GUSEV, A.A., SHUMILINA, L.S., and CHERNYSHOVA, V.G. (1985), *The seismofocal zone of Kamchatka* (in Russian), Vulkanologiya i Seismologiya (4), 91–107.

FREUND, F., and STOLC, V. (2013), *Nature of pre-earthquake phenomena and their effects on living organisms*, Animals 3(2), 513–531, doi:10.3390/ani3020513.

GELLER, R.J. (1997), *Earthquake prediction: A critical review*, Geophys. J. Int. 131, 425–450, doi:10.1111/j.1365-246X.1997.tb06588.x.

GELLER, R.J., JACKSON, D.D., KAGAN, Y.Y., and MULARGIA, F. (1997), *Earthquakes cannot be predicted*, Science 275, 1616–1617, doi:10.1126/science.275.5306.1616.

GRANT, R. A., and HALLIDAY, T. (2010), *Predicting the unpredictable; evidence of pre-seismic anticipatory behaviour in the common toad*, J. Zool. 281, 263–271, doi:10.1111/j.1469-7998.2010.00700.x.

GRANT, R.A., HALLIDAY, T., BALDERER, W.P., LEUENBERGER, F., NEWCOMER, M., CYR, G., FREUND, F.T. (2011), *Ground water chemistry changes before major earthquakes and possible effects on animals*, Int. J. Environ. Res. Public Health 8, 1936–1956.

HALL, S.S. (2011), *Scientists on trial: At fault?*, Nature 477, 264–269, doi:10.1038/477264a.

HARTMANN, J. and LEVY, J.K. (2005), *Hydrogeological and gas-geochemical earthquake precursors - A review for application*, Nat. Hazards 34, 279–304, doi:10.1007/s11069-004-2072-2.

HAYAKAWA, M. and TIMASHEV, S.F. (2006), *An attempt to find precursors in the ULF geomagnetic data by means of Flicker Noise Spectroscopy*, Nonlin. Processes Geophys 13, 255–263, doi:10.5194/npg-13-255-2006.

ICEFCP (2009), *Operational Earthquake Forecasting: State of Knowledge and Guidelines for Utilization*, http://www.iaspei.org/downloads/Ex_Sum_v5_THJ9_A4format.pdf, access: 4 February 2014.

IDA, Y., HAYAKAWA, M., and TIMASHEV S. (2007), *Application of different signal analysis methods to the ULF data for the 1993 Guam earthquake*, Nat. Hazards Earth Syst. Sci. 7, 479–487, doi:10.5194/nhess-7-479-2007.

KHATKEVICH, Y. and RYABININ, G. (2006), *Geochemical an groundwater studies in Kamchatka in the search for earthquakes precursors* (in Russian), Vulkanologiya i Seysmologiya (4), 34–42.

KIRSCHVINK, J.L. (2000), *Earthquake prediction by animals: evolution and sensory perception*, Bull. Seismol. Soc. America 90, 312–323, doi: 10.1785/0119980114.

KIRSCHVINK, J.L. (2014), *Sensory biology: Radio waves zap the biomagnetic compass*, Nature 509, 296–297, doi: 10.1038/nature13334.

KOPYLOVA, G., SUGROBOV, V., and KHATKEVICH, Y. (1994), *Variations in the regime of springs and hydrogeological boreholes in the Petropavlovsk polygon (Kamchatka) related to earthquakes* (in Russian), Vulkanologiya i Seysmologiya (2), 53–70.

LI, Y., LIU, Y., JIANG, Z., GUAN, J., YI, G., CHENG, S., YANG, B., FU, T., and WANG, Z. (2009), *Behavioral change related to Wenchuan devastating earthquake in mice*, Bioelectromagnetics 30, 613–620, doi:10.1002/bem.20520.

LU-LU, CH., XIANG, H., JUAN, ZH., HAO-HAO, ZH., WEN, K., WEI-HONG, Y., TIAN-SHU, Z., JIAO-YUE, ZH., and LING, Y. (2010), *Increases in energy intake, insulin resistance and stress in rats before Wenchuan earthquake far from the epicentre*, Exp. Biol. Med. 235, 1216–1223, doi:10.1258/ebm.2010.010042.

NOBLER, M. S., SACKEIM, H.A., SOLOMOU, M., LUBER, B., DEVANAND, D.P, AND PRUDIC, J. (1993), *EEG manifestations during ECT: effects of electrode placement and stimulus intensity*, Biol. Psych. 34, 321–330, doi:10.1016/0006-3223(93)90089-V.

POLYAKOV, YU. S., NEILSEN, J., and TIMASHEV, S. F. (2012), *Stochastic variability in x-ray emission from the black hole binary GRS 1915 + 105*, Astron. J. 143, 148, doi:10.1088/0004-6256/143/6/148.

RYABININ, G., GAVRILOV, V.A., POLYAKOV, YU.S., and TIMASHEV, S.F. (2012), *Cross-correlation earthquake precursors in the hydrogeochemical and geoacoustic signals for the Kamchatka peninsula*, Acta Geophys. 60, 874–893.

RYABININ, G., POLYAKOV, YU.S., GAVRILOV, V.A., and TIMASHEV, S.F. (2011), *Identification of earthquake precursors in the hydrogeochemical and geoacoustic data for the Kamchatka peninsula by flicker-noise spectroscopy*, Nat. Hazards Earth Syst. Sci. 11, 541–548, doi:10.5194/nhess-11-541-2011.

SHOCK AND LAW (2012), Editorial, Nature 490, 446, doi:10.1038/490446b.

TELESCA, L., LAPENNA, V., TIMASHEV, S., VSTOVSKY, G., MARTINELLI, G. (2004), *Flicker-Noise spectroscopy as a new approach to investigate the time dynamics of geoelectric signals measures in seismic areas*, Phys. Chem. Earth 29, 389–395.

TELESCA, L., LOVALLO, M., RAMIREZ-ROJAS, A., and ANGULO-BROWN, F. (2009a), *A nonlinear strategy to reveal seismic precursory signatures in earthquake-related self-potential signals*, Physica A 388, 2036–2040, doi:10.1016/j.physa.2009.01.035.

TELESCA, L., LOVALLO, M., RAMIREZ-ROJAS, A., and ANGULO-BROWN, F. (2009b), *Scaling instability in self-potential earthquake-related signals*, Physica A 388, 1181–1186, doi: 10.1016/j.physa.2008.12.029.

TIMASHEV, S. F. (2006), *Flicker noise spectroscopy and its application: Information hidden in chaotic signals*, Russ. J. Electrochem. 42, 424–466, doi:10.1134/S102319350605003X.

TIMASHEV, S. F., PANISCHEV, O. YU., POLYAKOV, YU. S., DEMIN, S. A., and KAPLAN, A. YA. (2012), *Analysis of cross-correlations in electroencephalogram signals as an approach to proactive diagnosis of schizophrenia*, Physica A 391, 1179–1194, doi:10.1016/j.physa.2011.09.032.

TIMASHEV, S. F., POLYAKOV, YU. S., YULMETYEV, R. M., DEMIN, S. A., PANISCHEV, O. YU., SHIMOJO, S., and BHATTACHARYA, J. (2009), *Analysis of biomedical signals by flicker-noise spectroscopy: Identification of photosensitive epilepsy using magnetoencephalograms*, Laser Phys. 19, 836–854, doi:10.1134/S1054660X09040434.

TIMASHEV, S. F., POLYAKOV, YU. S., YULMETYEV, R. M., DEMIN, S. A., PANISCHEV, O. YU., SHIMOJO, S., and BHATTACHARYA, J. (2010b), *Frequency and phase synchronization in neuromagnetic cortical responses to flickering-color stimuli*, Laser Physics 20, 604–617, doi:10.1134/S1054660X10050208.

TIMASHEV, S.F. (2007), *Fliker-Shumovaya Spektroskopiya: Informatsiya v khaoticheskikh signalakh* (Flicker-Noise Spectroscopy: Information in Chaotic Signals), Fizmatlit, Moscow.

TIMASHEV, S.F. and POLYAKOV, Y.S. (2007), *Review of flicker noise spectroscopy in electrochemistry*, Fluct. Noise Lett. 7, R15–R47, doi:10.1142/S0219477507003829.

TIMASHEV, S.F. and POLYAKOV, Y.S. (2008), *Analysis of discrete signals with stochastic components using flicker noise spectroscopy*, Int. J. Bifurcation Chaos 18, 2793–2797, doi:10.1142/S0218127408022020.

TIMASHEV, S.F. and VSTOVSKII, G.V. (2003), *Flicker-noise spectroscopy in analysis of chaotic time series of dynamic variables and the problem of signal-to-noise ratio*, Elektrokhimiya 39, 149.

TIMASHEV, S.F., POLYAKOV, Y.S., MISURKIN, P.I., and LAKEEV, S.G. (2010a), *Anomalous diffusion as a stochastic component in the dynamics of complex processes*, Phys. Rev. E 81, 041128, doi:10.1103/PhysRevE.81.041128.

TRIBUTSCH, H., *When the Snakes Awake: Animals and Earthquake Prediction* (MIT Press, Cambridge, MA, 1984).

UYEDA, S., NAGAO, T., and KAMOGAWA, M. (2009), *Short-term earthquake prediction: Current status of seismo-electromagnetics*, Tectonophysics 470, 205–213, doi:10.1016/j.tecto.2008.07.019.

VSTOVSKY, G.V., DESCHEREVSKY, A.V., LUKK, A.A., SIDORIN, A.YA., TIMASHEV, S.F. (2005), *Search for electric earthquake precursors by the method of Flicker-noise spectroscopy*, Izvestiya: Phys. Solid Earth 41, 513–524.

WYSS, M., ACEVES, R.L., PARK, S.K., GELLER, R.J., JACKSON, D.D., KAGAN, Y.Y., and MULARGIA, F. (1997), *Cannot earthquakes be predicted?*, Science 278, 487–490, doi:10.1126/science.278.5337.487.

WYSS, M., and BOOTH, D.C. (1997), The IASPEI procedure for the evaluation of earthquake precursors, Geophys. J. Int. 131, 423–424, doi:10.1111/j.1365-246X.1997.tb06587.x.

YOKOI, S., IKEYA, M., YAGI, T., and NAGAI, K. (2003), *Mouse circadian rhythm before the Kobe earthquake in 1995*, Bioelectromagnetics 24, 289–291, doi:10.1002/bem.10108.

(Received March 1, 2014, revised June 27, 2014, accepted June 30, 2014, Published online July 18, 2014)

Pure Appl. Geophys. 172 (2015), 1959–1973
© 2014 Springer Basel
DOI 10.1007/s00024-014-0877-9

❙ Pure and Applied Geophysics

Scale-Invariance in the Spatial Development of Landslides in the Umbria Region (Italy)

LUISA LIUCCI,[1] LAURA MELELLI,[1] and CRISTIAN SUTEANU[2]

Abstract—Understanding the spatial distribution of mass movements is a major issue in the management and forecasting of landslide risk. In this context, the present study examines the most widespread types of landslide in the Umbria region (central Italy), that is, slides and flows, in order to establish if it is possible to identify a well-defined structure in their spatial pattern. By using the landslide inventory map available for the area and by resorting to the principles of fractal theory, the scaling properties of the landslide sample were investigated. The application of the box-counting algorithm to the maps of landslide triggering points and landslide areas allowed for the identification of a clear scale-invariant structure. Two distinct types of fractal behaviour were recognized, separated by a scale value of 1 km and characterized by capacity dimensions of 1.35 and 1.76, in the ranges of 25 m–1 km and 1–16 km, respectively. The comparison between the scaling exponents obtained from a map of points and one of areas, and the elaboration of the cumulative frequency distributions of landslide areas supported the interpretation of this result: the higher capacity dimension describes the spatial distribution of landslides in the Umbria region, while the lower contains additional information about their geometries, suggesting that the latter also possess scaling properties. Based on the finding of two different types of behaviour of landslides in space, the hypothesis is discussed that the contribution of each causal factor (i.e., predisposing and triggering factors) to the occurrence of landslide events and to their spatial development could be different in the two scale ranges identified, depending on its spatial variability at local and regional scale. According to this hypothesis, factors with high local variability (i.e., topographic attributes) would mainly affect the assortment of landslide geometries, while those with high regional variability (e.g., rainfalls and lithology) would mainly affect the pattern of the landslides.

Key words: Landslides, fractal analysis, box counting, spatial analysis, spatial scaling.

1. Introduction

Mass movement is caused by complex processes controlled by the interaction of numerous factors, and is difficult to manage and forecast. Increasing efforts are being made to identify causes responsible for the spatial development of landslides in an area. The use of geographical information systems (GIS) and remote sensing data is making significant contributions in pursuing this aim (CHACHÓN *et al.* 2006; VAN WESTEN *et al.* 2008). Several methods have been proposed to assess landslide susceptibility. Heuristic (RUFF and CZURDA 2008), statistic (GUZZETTI *et al.* 2006; VERGARI *et al.* 2011), and deterministic analyses (GOKCEOGLU and AKSOY 1996) are different ways to answer the same question: where could a landslide occur? Except for deterministic methods that are based on the analysis of the physical properties of slope materials (i.e., geomechanical and hydraulic properties), the basic idea of all these procedures is to identify the environmental parameters discriminating areas prone to landslides and controlling their spatial distribution; i.e., the causal factors. The ability of these models to predict the spatial occurrence of landslide events may also be very high, for instance around 85 % (MEZUGHI *et al.* 2011; PRADHAN 2011), but this is not enough to ensure an accurate forecast. Moreover, results from the models are not fully congruent (SÜZEN and DOYURAN 2004; YESILNACAR and TOPAL 2005; KAYASTHA *et al.* 2013).

With the aim of contributing to the understanding of the spatial development of this natural phenomenon, this study investigates landslide distribution. Is there a structure in the spatial distribution of landslides? Are they spatially organized? We try to answer these questions by making use of fractal theory. In particular, we explore the scaling properties of landslides in the Umbria region (central Italy),

[1] Department of Physics and Geology, University of Perugia, Via Alessandro Pascoli snc, Perugia 06123, Italy. E-mail: luisa.liucci.ll@gmail.com; laura.melelli@unipg.it

[2] Department of Geography and Department of Environmental Science, Saint Mary's University, 923 Robie St., Halifax, NS B3H 3C3, Canada. E-mail: Cristian.Suteanu@smu.ca

Reprinted from the journal

where the mass movements are a critical issue partly responsible for the high geological risk of the area.

Fractal theory has become of great importance in the understanding of natural processes and the evolution of the systems in which processes occur (PERUGINI et al. 2003) and it is based on the concept of statistical self-similarity (MANDELBROT 1967). If a natural object or a natural phenomenon is statistically self-similar, its statistical properties will not change with the scale of investigation; i.e., it is scale-invariant and may be defined as "fractal". Scale-invariance can occur both in space and in time (LIEBOVITCH 1998), and it implies the nonlinearity of the process. Nonlinearity is common in nature. Fractal scaling has been identified; for instance, in the surface phenomena governing landscape evolution, e.g., in the spatial development of drainage networks (NYKANEN et al. 1998; DEL MONTE et al. 1999), and various types of analyses and applications related to the fractal geometry of landscape morphology have been carried out (MARK and ARONSON 1984; CHASE 1992; DELLA SETA et al. 2003; PHILLIPS 2006; PERUGINI et al. 2007).

Fractal theory is also used in the field of natural hazards. Various approaches and techniques have been applied to earthquakes (SUTEANU et al. 2005; SUTEANU and IOANA 2007), floods (e.g., TURCOTTE 1994), wildfires (e.g., RICOTTA et al. 2001) and landslides.

The fractal behaviour of landslides has been investigated from different points of view. Scale-invariance has been identified in the landslide size-frequency distributions (PELLETIER et al. 1997; GUZ-ZETTI et al. 2002; BRUNETTI et al. 2009), and a possible link between frequency and geological parameters of the material has been suggested. In this regard, IWAHASHI et al. (2003) postulate that the finer the geological features, the higher the fractal dimension is. In POURGHASEMI et al. (2013), the fractal approach is used to characterize the shape of landslides. They demonstrate that analytical relations exist between the fractal dimension of a landslide and the length–width ratio of its area. Fractals have also been used to search for a link between the occurrence of landslides in a mountain ridge and the morphological evolution of the ridge itself. Starting from the evidence of the fractal structure of the mountain surfaces, CZIRÒK

et al. (1997) use a micro-model of mountain ranges to reproduce the water erosion on the surface. The saturated portions of the terrain slide down forming an avalanche, similarly to landslides on a real slope. They demonstrate that the roughness of the experimental mountain profile and its temporal evolution manifest self-affine fractal behaviour, and that the calculated fractal dimension in space is in agreement with that calculated for a real transect profile. They therefore conclude that the roughening of ridges seems to be controlled by power-law distributed landslides. The theory of the avalanche process is at the basis of the sandpile model, a particular self-organized criticality (SOC) model (BAK et al. 1988) widely used to describe landslide evolution. If a system is SOC, it will approach a critical state and the avalanche process will become scale-invariant (a detailed discussion of SOC is provided by TURCOTTE 1999). Several SOC models have been applied to landslides. In HERGARTEN (2003), the ability of some of these models to qualitatively and quantitatively describe propagation of landslides is evaluated. Further evidence to support the self-organized critical behaviour of landslides has also been provided by LI et al. (2011), who found that the cumulative frequency of landslide occurrence is power-law-related to landslide-triggering rainfall level.

The outcomes of all these studies seem to converge toward the same conclusion: landslides are the result of scale-invariant processes. As a direct implication, their spatial distribution should also exhibit scale-invariance. Evidence of fractal scaling of the pattern of landslides has also been observed in various areas of the world; for instance, in Japan (GOLTZ 1996) and Taiwan (YANG and LEE 2006). In this paper, results obtained from the investigation of the scaling properties of landslide triggering points and landslide areas in the Umbria region are shown, and hypotheses on the physical meaning of these properties are made, by analyzing and comparing results.

2. Study Area: Umbria Region (ITALY)

The Umbria region (Fig. 1) is located in central Italy and covers an area of 8,456 km^2. Despite its

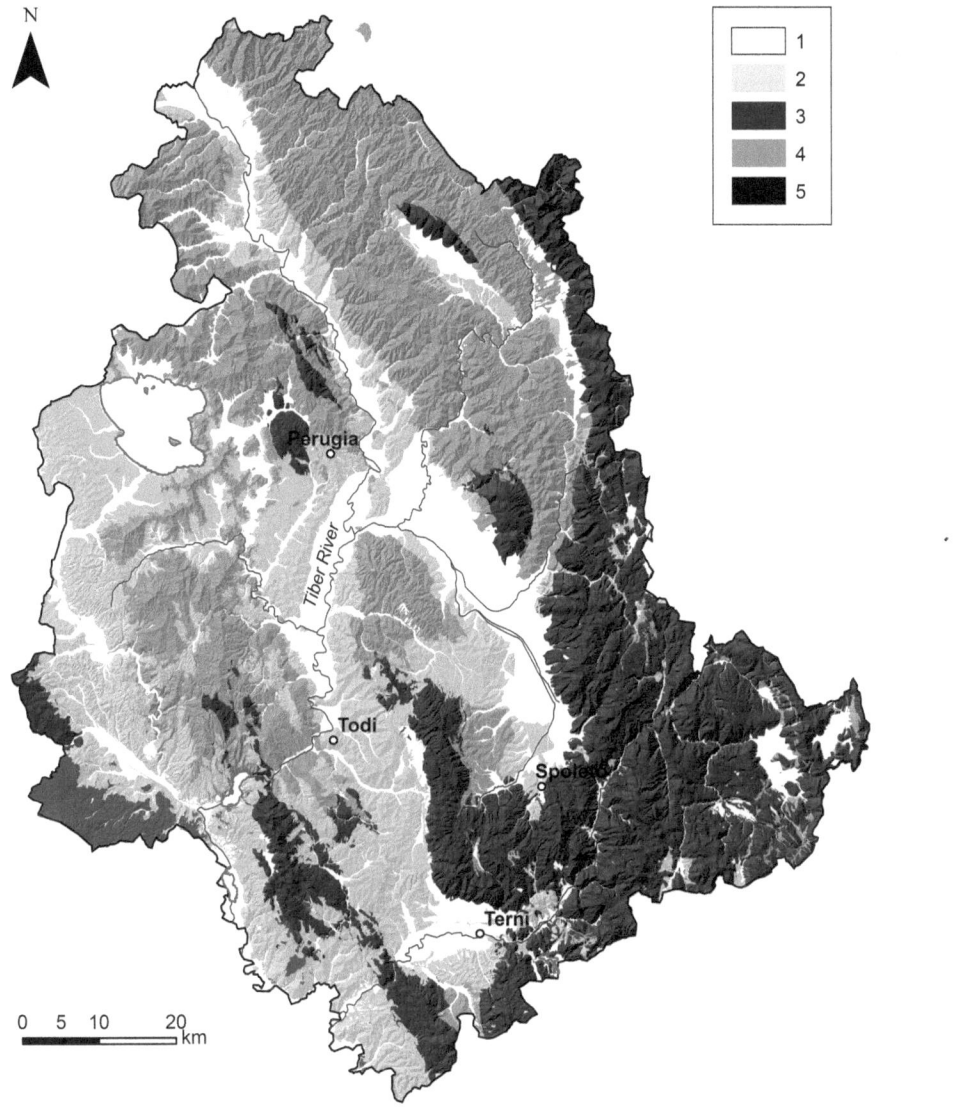

Figure 1
Umbria region. Map of the lithological groups. *1* Recent deposits, *2* Fluvial–lacustrine deposits, *3* Volcanites, *4* Terrigenous rocks, *5* Carbonate rocks

limited extent, the region is characterized by a remarkable variety of morphological and geological contexts. The landscape is dominated by hilly topography, with low (200–500 m.a.s.l.) and high (500–800 m.a.s.l.) hills occupying 50 and 24 % of the area, respectively. The rest of the region is covered by mountains (accounting for about 14 %) that extend along the eastern regional boundary with altitudes >800 m.a.s.l., and flat lands (accounting for about 10 %) with altitudes lower than 200 m.a.s.l. The distribution of the morphometric parameters reflect a well-defined morphological structure (CATTUTO and MELELLI 2006), with an increase in altitude and slope moving eastward up to the Apennine chain, which delimits the eastern regional boundary. In this zone, there are the maximum elevation values (up to 2,448 m.a.s.l.), together with flat summit areas, steep slopes, narrow and deep river valleys. The hilly topography is located in the western and northern part of the area and it is longitudinally interrupted, in the central part of the region, by the Tiber River valley, which develops like a narrow band from the northern

181

regional boundary up to the town of Perugia, where the valley divides in two branches. The western one follows the Tiber River up to Todi and then continues up to the Terni basin, and the eastern one represents the Umbrian Valley, which extends up to Spoleto. Between these two valleys, the ridge of the Martani Mountains develops.

The topographic structure of the region reflects its geology. The different lithotypes outcropping in the area can be grouped in four main geological complexes, according to their origin, composition and geomechanical behaviour: Carbonate, Terrigenous, Volcanic and post-orogenic. In this study, the latter was further subdivided into fluvial–lacustrine deposits and Recent deposits, because of their different susceptibility to landslide phenomena. In Fig. 1, the spatial distribution of the five lithological groups is shown.

The processes occurring in each complex and the corresponding landforms are conditioned by specific factors, both passive (lithotypes, geological structures) and active (neotectonic activity), which also influence the amount and type of landslides. The main geological and geomorphological characteristics of the various complexes are described below.

The Carbonate complex (upper Trias–lower Miocene) covers the eastern part of the region, with isolated relief mostly present in the central sector of the study area. Limestone prevails in the lower part of the sedimentary succession, while the marly limestone lithotypes increase moving upward. The depositional environment evolved from an evaporitic basin of shallow water to a carbonate platform and finally to a pelagic basin, as a consequence of a tectonic fragmentation generating high and low structural domains. The carbonate sequence is organized in wide anticlines, which alternate with narrow synclines with NW–SE or N–S direction. Folds are frequently cut by fault systems having two main directions, according to the Apennine and anti-Apennine trends. Mountain chains have wide and flat tops due both to the geological structures (the top of the slopes corresponds to the top of the anticlines) and to the presence of palaeosurfaces inherited from surface processes started around Late Miocene–Early Pliocene. Eluvial deposits and karst landforms (dolines) are quite frequent. The upper part of the slopes

is characterized by a convex-creep zone followed by a convex-straight profile. Gentle slope and thick colluvial deposits are instead present in bedrock with a high marly fraction. Fluvial erosion prevails in zones where the evolution of the drainage network is controlled by the regional fault system or by lithological discontinuities, thus generating narrow and deep river valleys and rectangular drainage patterns. Alluvial deposits are instead common in the riverbed of streams flowing along the syncline axes. Overall, the Carbonate complex exhibits low drainage density.

The Terrigenous complex (Oligocene–medium Miocene) outcrops in the northern and central part of the region, and it is characterized by bedrock with heterogeneous behaviour. The depositional environment varied from pelagic basin, to continental slope, to foredeep basin, thus generating a syn-orogenic turbidite sequence consisting of arenaceous and limestone layers interbedded with clays or marls. The entire sedimentary multilayer sequence was involved in a compressive tectonic phase (upper Miocene–lower Pliocene) resulting in folds and thrusts dipping eastward. Then, an extensional period generated sets of normal faults producing valleys and intermountain basins. The response to weathering and erosion, and the permeability of this sequence vary according to its lithological composition, thus causing different dynamics in terms of geomorphological evolution. High values of amplitude of relief are present where sandstone and limestone rocks prevail, gentle slopes are instead frequent in bedrock with a high marly fraction. Slope geometry is not homogeneous and it is strongly affected by the drainage density, which increases with the clay percentage, thus generating dendritic drainage patterns that are characteristic of this complex. Fluvial processes and surface runoff are the main shaping processes, together with slides and flows.

The Volcanic complex (age of 600–130 ky) occupies the southwestern part of the region and represents the most northeastern edge of the Alfina plateau related to the Vulsini District. The main lithotypes consist of ignimbrite deposits and stratified tuffs (PECCERILLO 2005; GREGORI and MELELLI 2012). Low reliefs and gently dipping summit surfaces characterize this area, and the main morphological features are linked to the past volcanic activity (such as, for example, the Bolsena Lake, which occupies an ancient caldera) and to the post-Miocene extensional tectonics, which generated N–S and NW–SE fault systems. Evidences of the ancient tectonics can be found in the numerous scarps affecting the slopes and following the direction of the fault systems, and in the geometric configuration of the drainage network, which shows rectangular pattern controlled by the tectonic lineaments (CIOTOLI *et al.* 2003).

The post-orogenic complex (Pliocene–Holocene) is composed by sediments of marine and continental origin with a great compositional heterogeneity because of the different paleoenvironments in which sediments were deposed. The fluvial–lacustrine environment is the most represented. Conglomerates, sands and clays alternate in sedimentary sequences of widely variable thicknesses. Generally, at the top of the hills the coarser fraction outcrops, while sands with a lower conglomerate fraction are abundant on the side-slope. Clays and fine sands prevail at the foot-slope. The larger part of these sediments fills the wide valleys and the intermountain basins, while the most shallow and recent sediments (Holocene) are constituted by alluvial material characterizing the flat areas of the region, and by eluvial, colluvial and debris deposits covering the transition zones between the mountainous areas and the adjacent plains, and filling the hollows along the slopes and on the top of the reliefs.

2.1. Landsliding in Umbria: Type of Movements and Inventory Map

Landscape evolution is strongly affected by the geological setting of the Umbria region. Each lithological complex is characterized by a well-defined proneness to landslide events, depending on its composition, degree and type of fracturing and morphological characteristics (CARDINALI *et al.* 2002), which also influence the type of mass movement within the complex.

The Carbonate complex shows frequent fractures and fault systems. As a consequence of the sudden Pleistocene uplift, the area contains a large number of sub-vertical slopes, so a high amplitude of relief (i.e., maximum difference in elevation, per area unit; DELLA SETA *et al.* 2004) is preserved. Due to this, the drainage network is characterized by strong linear erosion. As a result, the main types of landslide movements, for this complex, are falls and topples, mostly triggered by earthquakes events. Where the slopes are cut by steep streams and the bedrock is intensely fractured, debris flows are present too.

In the Terrigenous complex, the contrast of permeability between the different lithotypes and the packed layering represent the optimal condition for the occurrence of slides and slumps. Where the clay component prevails, slow earth flows take place. A fundamental factor controlling for the occurrence of landslide events within this complex is the strata dip direction. The dip slope zones are prone to slide, mainly with translational movement, characterized by high values of area, but not necessarily by huge volumes of bedrock involved. The larger events are affected, in most cases, by smaller landslides seasonally reactivated by intense rainfall events. On the contrary, the anti-dip slope zones are often affected by slump or falls, statistically smaller and confined to the intersection between strata and faults or joint systems.

In the fluvial–lacustrine complex, there is a great variety of landslide types, because of the wide range in size and composition of the deposits. At the top of the hills where the coarse fraction outcrops and if the infiltration is enough to increase the cohesion, there are steep hillsides with sub-vertical slopes, which allow for the triggering of falls and topples. At the midslope where the sand is the main component, slides and slumps are the most frequent events. Where the clayey and fine-sandy fractions prevail, slow flows occur.

In the Volcanic complex, the mass movements are bonded to the overlapping of a thick and hard volcanic caprock on a landscape constituted by clayey slopes. Falls and topples occur at the top of the sequence while flows, slides and slumps are frequent lower down. In these areas, the slope evolution is the final result of the interaction between the retreat of the caprock and the undermining of the clays at the foot of the slope, and the greater the erosion at the foot, the faster and more efficient the landslide mechanisms are.

Finally, the occurrence of landslides is very low or null in the areas occupied by the Recent deposits complex, because of the very low slope of the topographic surface.

The link between type of mass movement and geology is well represented by the landslide inventory map used in this study. This map was created by the Geological Service of the Umbria Region for the IFFI project (Inventory of Landslide Phenomena in Italy) (ISPRA 2006) by collecting information from multiple sources; that is, from the inventory map of 1:10,000 scale built by the CNR-IRPI (IRPI-National Research Council), from geological and geothematic maps of 1:50,000 scale, and from several studies and surveys relating to projects for the planning and management of the territory, or subsequent to extreme natural events. The inventory map is the result of photo interpretations and field surveys, and it was produced both as a point map (Fig. 2a), in which landslides are represented through their triggering points, and as a polygon map, in which the extent of each landslide area is displayed as a single polygon including both crown and depositional zone. On the whole, the inventory map highlights a high concentration of events in the northeastern and southwestern part of the region, while the flat areas corresponding to the main valley floors and to the intermountain basins are completely free of mass movements (Fig. 2a). Figure 2b shows the percentage of each type of mass movement in the data set. The most widespread are slides (68 %), followed by flows (slow and rapid) and complex movement, which constitute 12 and 11 % of the sample, respectively. Falls, topples and lateral spreads are rare events in the area (in Fig. 2b, lateral spreads take a value of 0 % because of the use of integer values; the actual value is 0.02 %). More than 70 % of the mass movements inventoried are reactivations of older and wider dormant landslide bodies, caused by heavy rains. More than 66 % are shallow landslides, mainly characterized by translational–rotational slow-movement (only 4.5 % are fast moving slides), and about 10 % are deep-seated landslides. Figure 3 shows the density of each mass movement (number of landslides N_L per 100 km^2) within each lithological group, calculated by overlapping the triggering point map (Fig. 2a) to the map of lithological groups (Fig. 1). A logarithmic scale is used for the y-axis, in order to improve visualization of the data. The density values obtained reflect the above-described link between mass movements and geological setting.

Terrigenous and fluvial–lacustrine complexes have the greater concentration of mass movements (about 590 and 670 N_L per 100 km^2, respectively), and they also exhibit higher density of slides, slow flows and complex movements. Rock falls are instead mainly present in the Volcanic and Carbonate complexes, the latter also showing the higher concentration of rapid flow movements. As expected, only a few events are recorded in the complex of the Recent deposits.

The present study focuses on flows (slow and rapid) and slides, which are the main gravitational processes in the study area, thus representing the most critical phenomena in the context of landslide risk. These two types of event were treated and processed as a single sample, since their occurrence and evolution are strongly linked to the occurrence and intensity of rainfall, which can be considered one of the principal triggering factors for both of them. The sample comprises 24,122 landslides.

3. Fractal Analysis Method

Several methods are available in the literature for the investigation of the scale-invariant properties of a sample in space. If the spatial pattern of the sample exhibits power-law scaling (or "fractal scaling"), all of these methods allow for the estimate of the fractal dimension describing its structure; that is, its geometrical complexity, the way in which it fills the embedded space. Capacity dimension (KOLMOGOROV 1958), information dimension (RÉNYI 1959) and correlation dimension (GRASSBERGER and PROCACCIA 1983) are the fractal dimensions most commonly estimated in studying spatial patterns. Capacity dimension provides basic information on the spatial distribution of the elements of the sample, based on a criterion of presence/absence of elements in space, taking into account neither their concentration nor their inter-correlations; its value has a purely geometric fractal meaning. Information dimension describes the scaling behaviour in space of the probability distribution entropy of the data sample. Correlation dimension analyses the spatial correlation between elements. Given their analytical formulation, information and correlation dimensions can only be calculated for point data sets.

(A)

N

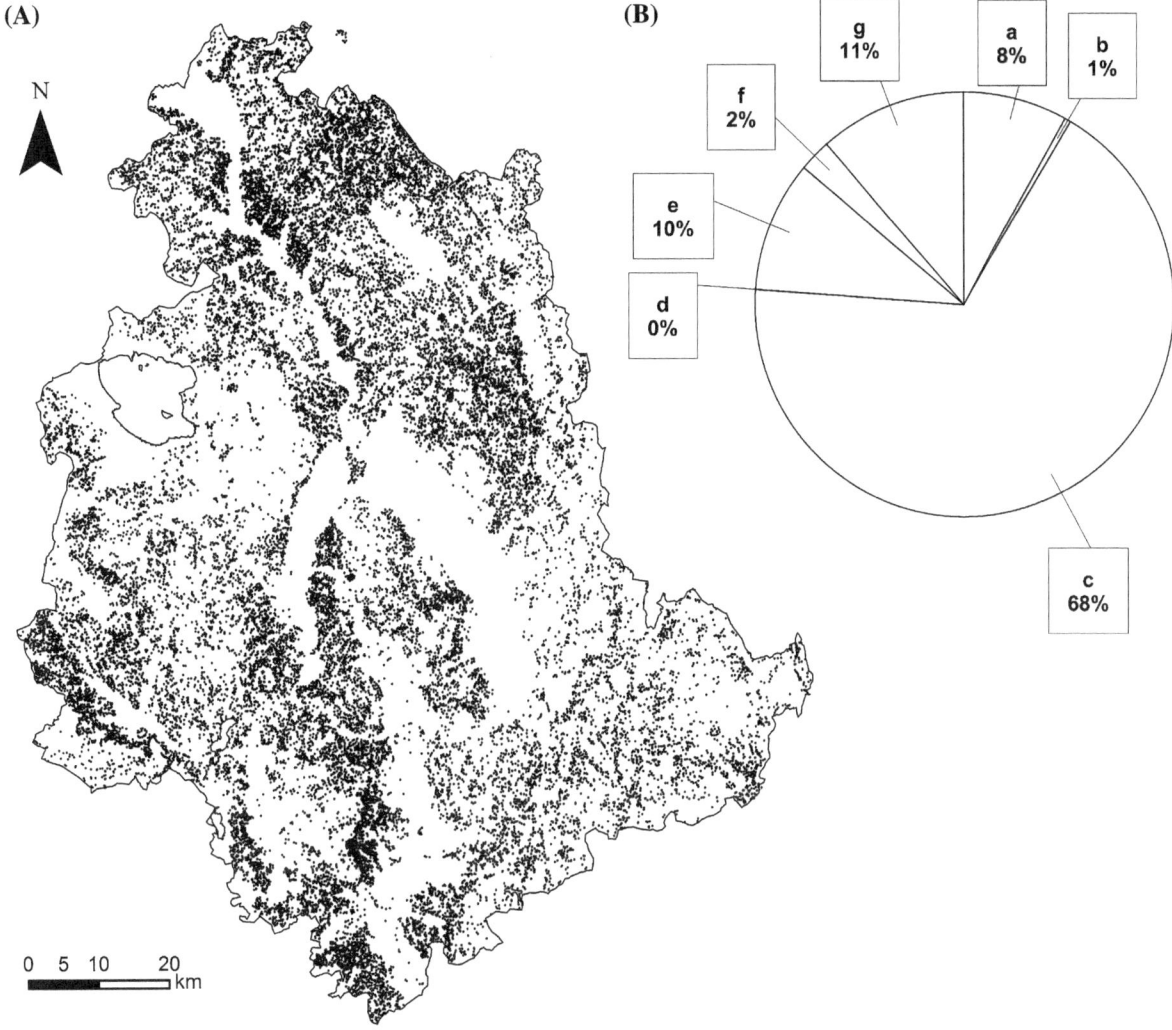

0 5 10   20
km

(B)

g
11%

a
8%

b
1%

f
2%

e
10%

d
0%

c
68%

Figure 2

a Umbria region. Inventory map of landslide triggering points; **b** Pie chart of the percentage presence of the various types of mass movements in the data set; *a* unclassifiable, *b* fall and topple, *c* slide, *d* lateral spread, *e* slow flow, *f* rapid flow, *g* complex

In this research, the scaling properties of landslide triggering points and landslide areas are analysed by evaluating the capacity dimension D_0 by means of the box-counting algorithm. The reason for the choice of this method is that it can be applied both to a point set and to a polygon set, thus allowing for the comparison of the fractal dimensions identified. One of the advantages of box-counting, when compared to other methods, consists of the strong correlation in log–log graphs obtained for self-similar features, which supports a reliable detection of distinct scaling regimes (SUTEANU 2000). Moreover, the implementation of this algorithm in two-dimensional space is rather

simple, and the fractal dimension thus estimated is not affected by the problem of the edge effect, that is, the problem caused by the boundaries of the embedded space of the sample, which instead produces inaccuracy in the estimate of the correlation dimension (THEILER 1990). The capacity dimension is defined as follows:

$$D_0 = \lim_{r \to 0} \frac{\log N(r)}{\log \frac{1}{r}} \qquad (1)$$

where $N(r)$ is the number of square boxes of side r necessary to cover the data set. If D_0 is calculated for a real data set, Eq. 1 cannot be applied as such.

185

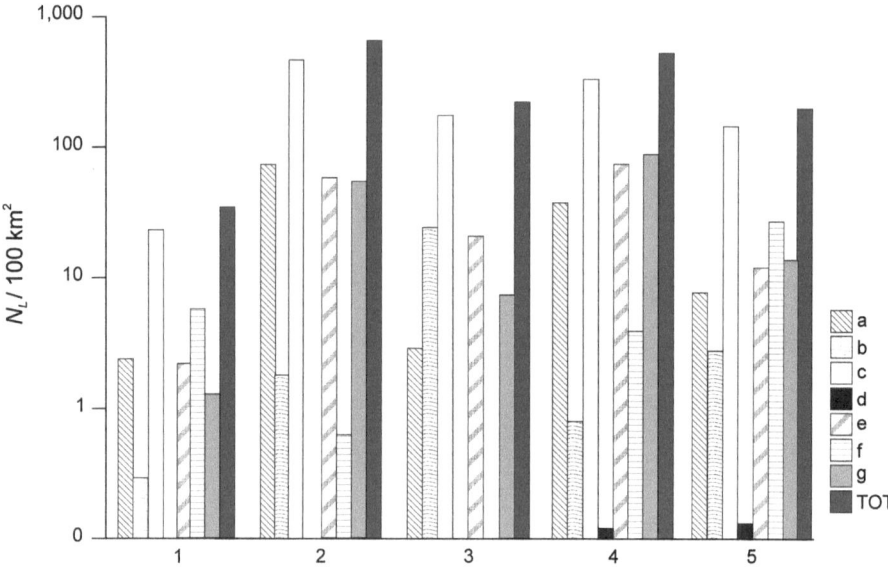

Figure 3

Number of landslides (N_L) per 100 km^2 (y-axis) versus the lithological groups (x-axis). *1* Recent deposits, *2* Fluvial–lacustrine deposits, *3* Volcanites, *4* Terrigenous rocks, *5* Carbonate rocks; *a* unclassifiable, *b* fall and topple, *c* slide, *d* lateral spread, *e* slow flow, *f* rapid flow, *g* complex, *TOT* total

A lower and an upper limit for *r* have to be identified within which D_0 exists, and which define the scaling range for D_0. Equation 1 can thus be written as:

$$N(r) = k \cdot r^{-D_0}; \quad r_1 \leq r \leq r_2 \qquad (2)$$

The box-counting algorithm was implemented to the maps of landslides triggering points and landslide areas, respectively, by using the ArcGIS software 10.0 ©Esri. By following the procedure, the two maps were covered with a grid of square boxes of side *r* and the number of boxes containing elements was counted. The procedure was then repeated by progressively reducing *r*, and the results of the analysis were plotted in log–log graphs. The ability of the fractal power law (Eq. 2) to fit the data was assessed for each of the two samples (i.e., triggering points and areas), and once verified, the values of D_0 were established from the functions identified. For consistency, the same orientation was used, in both cases, for the grid, and the southern and western coordinates of each data extent was used to fix its spatial position.

In Fig. 4, a detail of the map of landslide areas with boxes used for the application of the box-counting algorithm is shown.

4. Results and Discussion

The results of the analysis are shown in Fig. 5. For the sample of landslide triggering points (hereafter, A) (Fig. 5a), the existence of fractal power-law scaling is verified within the range of 1–16 km, with a coefficient of determination (R^2) of 0.999. The scaling exponent is 1.74, with a standard error of 0.03 (in statistical terms, the standard errors of the various D_0 refer to standard errors of the slopes of the regressions of the log–log plots). Conversely, for values of *r* lower than 1 km, the sample does not preserve its fractal properties. At these scales the points follow a curved trajectory, and the slope progressively decreases with *r*. For the sample of landslide areas (hereafter, B) (Fig. 5b) the result is partly different. Two scaling regimes are identified, separated by a scale threshold of 1 km. In the range from 1 to 16 km, D_0 is 1.76 (R^2 of 0.999; standard error of D_0 equal to 0.01), while in the range from 25 m to 1 km, D_0 is 1.35 (R^2 of 0.997; standard error of D_0 equal to 0.02). Note that the scaling exponents, respectively, obtained for A and B within the "upper scale range" (i.e., the range corresponding to the larger sizes of the boxes, that is: 1–16 km) are very similar.

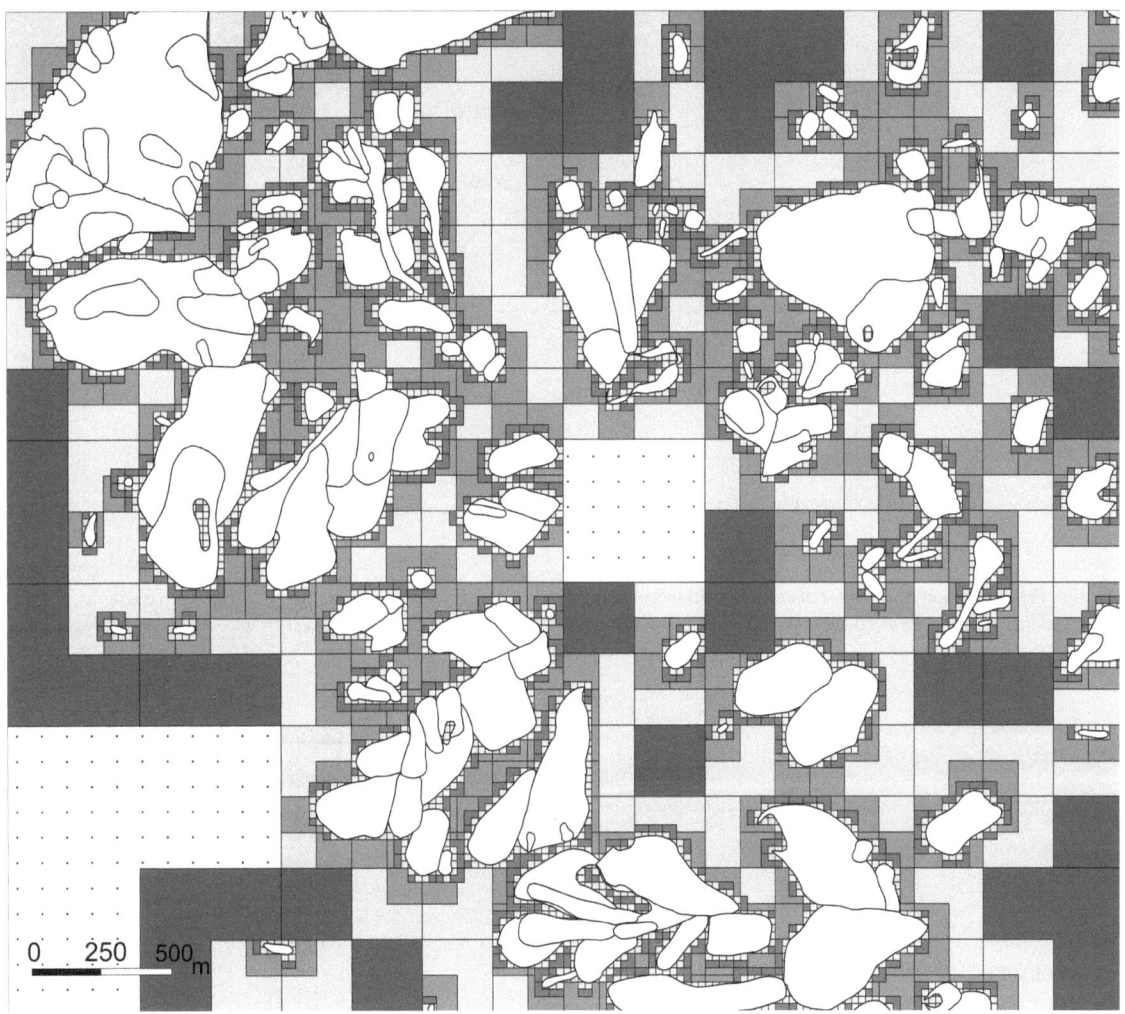

Figure 4
Detail of the map of landslide areas, with *boxes* used for the application of the box-counting algorithm

By analyzing the results, we can infer that the rate of decrease observed in the slope of the curve fitting point for sample A within the "lower scale range" (i.e., the range corresponding to the smaller sizes of the boxes, that is: 25 m–1 km) depends on the distances between landslide triggering points in the inventory map. In detail, below the scale value of 1 km, there are few pairs of landslide triggering points with inter-distances of a lower value. This implies that when also further reducing the size of the boxes, the number of new cells occupied by points does not significantly increase. Based on this information, we can consider the value of 1 km as the characteristic minimum distance between landslide triggering points in the study area. However, it should

be noted that this value depends on the scale of the original data used to construct the inventory map (e.g., aerial photos and geological maps). Sample B does not present this situation. As the size of boxes is reduced, an ever-increasing number of cells is needed to cover landslide areas.

This suggests that the reason for the different D_0 obtained in the lower scale range for sample B must be sought in the features of landslide areas. With the aim to investigate this possibility, the cumulative frequency distribution of landslide areas was calculated (Fig. 6).

In Fig. 6, the y-axis is the cumulative number of landslides (N_{CL}) [i.e., those with area lower than the area value in the x-axis (A_L)]. The two dashed lines

Reprinted from the journal

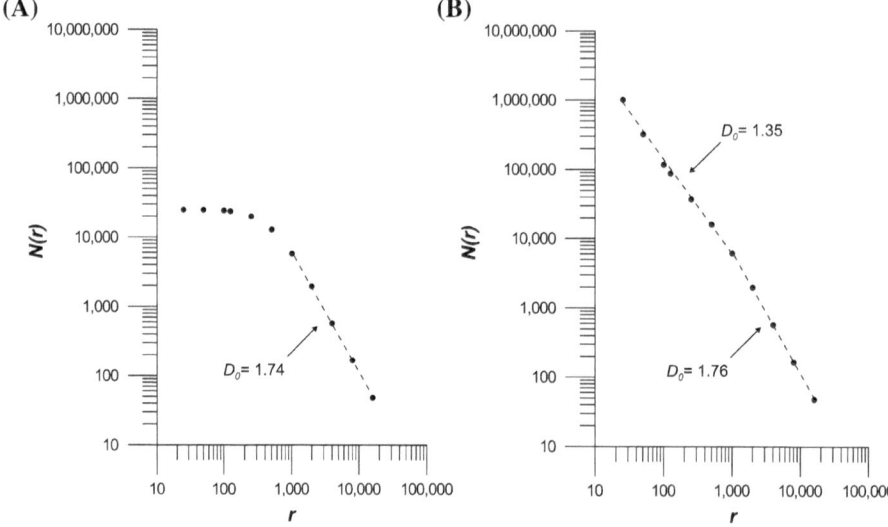

Figure 5
Results from the application of the box-counting algorithm to the inventory maps **a** and **b**. $N(r)$ number of boxes with elements, r side of the boxes, *dashed lines* best-fit lines obtained. D_0 capacity dimension estimated from the corresponding best-fit line. **a** Landslide triggering points; **b** landslide areas

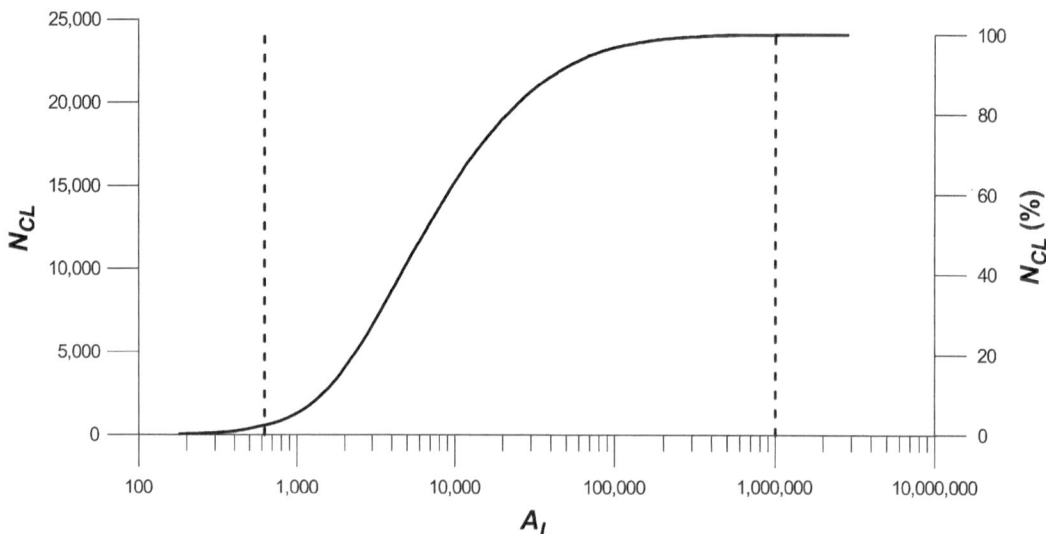

Figure 6
Solid line cumulative frequency distribution of landslide areas. A_L landslide area (m^2), N_{CL} number of landslides with area lower than A_L. *Dashed lines* area of the boxes for the lower (*left*) and upper (*right*) scale boundaries of the scale range 25–1,000 m

correspond to the sizes of the boxes for the upper and lower scaling boundary of the lower scale range; that is 1,000,000 m^2 (r equal to 1 km) and 625 m^2 (r equal to 25 m), respectively. By analyzing the graph we observe that, regardless of the shape of landslides, their areas are numerically analogous to the areas of the boxes used in the lower scale range.

In fact, about 98 % of landslide areas fall between the dotted lines, only 0.03 % have an area greater than the maximum size of the boxes and only 2.34 % have an area lower than the minimum size of the boxes. This demonstrates that the lower scale range is within the range of landslide areas of the sample, thus suggesting that, compared to what occurs within the upper scale range, for r values between 25 m and 1 km not only is the spatial distribution of landslides analysed, but also their geometric features, since the count of boxes is also constrained by their areas and shapes. For instance, boxes with an r equal to 4 km (i.e., r within the upper scale range) have an area of 16 km^2, a value greater than the maximum landslide area in the sample. This implies that one box may contain more than one landslide, and a single landslide in the box may have any size and any shape. We cannot know anything about its geometry, and the same applies for all the box sizes in the range of 1–16 km. Thus, the D_0 value calculated for map B within the upper scale range (equal to 1.76) only gives information on how landslides are distributed in the study area, and this is the reason for which we get about the same D_0 as for map A (equal to 1.74). Conversely, when we use boxes with an r equal, for example, to 250 m (r within the lower scale range), their area is equal to 62,500 m^2. In this case, the count of the boxes is affected not only by the areal distribution of the phenomenon, but also by the geometry of landslides, which have an area $>62,500$ m^2. For these landslides, the boxes begin to "trace" their shapes, and further reducing the size of the boxes, they will depict the shapes in an increasingly accurate way. The same applies for all the box sizes in the range between 25 m and 1 km (Fig. 4 facilitates the understanding of this concept). Thus, within the lower scale range, we get a different D_0 (equal to 1.35) because it contains information not only about the spatial distribution of landslides, but also about their geometry. The fractal dimension of 1.35 can be interpreted as the result of the various spatial and geometrical features characterizing landslide polygons, that is: their area, their shape and their mutual spatial relationships. Indeed, if we consider several landslides, each with its own area and shape, the count of the boxes necessary to cover them will differ, depending on whether landslides are in contact with each other or not.

The results obtained and the above theoretical analysis suggest that not only the landslide distribution, but also the landslide geometry, possesses scaling properties. Power-law scaling in landslide frequency-area and frequency–volume distributions has already been detected in the Umbria region

(GUZZETTI *et al.* 2002; MALAMUD *et al.* 2004; BRUNETTI *et al.* 2009). These results support our hypothesis.

A possible interpretation of the findings of this research could be sought in the instability factors driving the dynamics of the gravitational processes and, consequently, the spatial evolution of the system. These factors are generally divided into triggering factors (e.g., earthquakes, rainfalls) and predisposing factors (GRIFFITHS 1999), the latter mainly morphological (amplitude of relief, slope, curvature) and geological (lithology, layering, fracture intensity and orientation) (CARRARA *et al.* 1995). The relative contribution of each of these factors to the occurrence of landslide events is commonly referred to as "weight of factors" (VAN WESTEN *et al.* 2003). Both the pattern of landslides and their geometries are areal properties describing the spatial development of the same phenomenon, hence controlled by the same geomorphological process for which the main driving force is gravity. As a consequence, we suggest that the reason for the two scaling exponents identified for map B must be searched for in the factors mentioned above, the effects of which could differ from one scale range to another. In particular, we formulate the hypothesis that the relative weights of the various factors on the spatial development of landslide events might be dissimilar below and above the scale of 1 km: below this value (i.e., from 25 m to 1 km), their relative weights would control the assortment of landslide geometries in the area, while above it (i.e., from 1 to 16 km), they would control the spatial pattern of the phenomenon. Below we discuss this hypothesis.

Geometry is a specific property of each landslide, closely linked to the type of movement characterizing the landslide itself. Indeed, it is well known that the shape of a landslide strongly depends on the generating mechanism; for instance, flows can have a little crown, closely spaced flanks and a fan-shaped body, while slumps have a more compact structure that approximates the shape of an ellipse. The geometry of a landslide is thus characterised by the morphology that is specific to the type of movement associated to it and it is also affected by the specific topographic properties of the area of emplacement. Because of the distinctiveness of landslide geometries, they can be considered as "local" properties of the system describing the spatial development of the phenomenon in the lower scale range. We therefore argue that the environmental factors mainly affecting morphological and morphometric features of landslide geometries are those related to the topographic attributes of the slopes, which have themselves strong local variability; that is, the slope angle, the length of the slope and its planar and radial curvature.

The landslide pattern describes the spatial distribution of the landslide sample in the area, when projected in a two-dimensional space. Thus, the pattern can be considered as a "global" property of the system describing the spatial development of the phenomenon in the upper scale range. As a consequence, we believe that the environmental factors affecting landslide pattern must be mainly searched among those having a spatial variability that is more significant at regional scale than at local scale; that is, rainfalls and lithotypes. In fact, regarding rainfalls, it is reasonable to expect that their frequency and intensity will have a greater spatial variability when analysed at regional scale and concerning lithotypes, we have already discussed in Sect. 2 the strong link between outcropping lithologies and the spatial density of landslides.

We believe that the hypothesis formulated above could help us distinguish between the causal factors that promote instability of an area, on the basis of the scale of investigation.

With regard to the outcomes of this research, a further discussion can address the spatial evolution of the system. Investigating the spatial pattern of landslides means to analyse the spatial fingerprint of a dynamic system manifesting itself through mass movement events. Starting from this observation, we suggest that the scale-invariance identified in the landslide distribution—the "effect"—could be a clue of scale-invariant gravity-driven processes—the "cause". This hypothesis is supported by the SOC sandpile model (BAK et al. 1988), which simulates the behaviour of a pile of sand under local perturbations. The system described by this model tends toward a self-organized critical state acting as an attractor for its dynamics. Local perturbations generate avalanches of slides as in a "domino effect", thus producing self-similar fractal structure in space, the temporal evolution of which also obeys power-law fractal scaling. If we looked at this system, we would

only see the products of the avalanche processes, we could not evaluate how the system evolved in time; however, we would discover that this products exhibit statistical self-similarity in space just as for landslides in the study area.

Although research has not yet been done to investigate the possibility that the Umbrian landscape evolves as a SOC system, analogies to this model can be drawn. The Umbria region is a tectonically active area. At present, its evolution is controlled by the Pleistocene uplift and an eastward migration has been affecting the extensional and compressional domains of the area since the Middle Miocene, due to the effect of the regional stress field. As a consequence of the neotectonic activity, seismic events are common in the region (COLLETTINI et al. 2000), and the landscape undergoes a progressive rejuvenation consequent to the deepening of the rivers caused by linear erosion. The deepening of the drainage network mainly affects the mountainous areas corresponding to the apical parts of the river basins, and generates high values of amplitude of relief and frequent convex profiles of the slopes, which push the landscape far from equilibrium. In response to this, gravity processes generate mass movements by which the system dissipates energy, thus restoring metastable equilibrium.

In other words, just as in the sandpile model, in the Umbria Region there are sites in a metastable critical state. The action of endogenous forces (tectonic activity and earthquakes) combined with the effect of exogenous forces (rainfalls) locally perturb the critical state of these zones, similar to what occurs by adding a grain of sand or by locally increasing the slope of the pile in the sandpile model, thus driving sites of the system out of the metastable condition. Gravitational processes occur in these sites and cause slides, which allow the perturbation to propagate within the system, analogous to the avalanche processes in the sandpile model. This process is repeated in time under the effect of the neotectonic activity, again bringing other sites of the area out of the metastable state, which is similar to the effect of perturbing randomly selected sites in the pile of sand.

A strong analogy is also observed between the scaling exponent obtained from the sandpile model and that obtained from the sample of Umbrian landslides. In the three-dimensional simulation of a pile of sand far from equilibrium with closed boundary conditions, BAK *et al.* (1988) obtain a scaling exponent of 1.37 for the distribution function of landslide sizes. In our study, the fractal dimension obtained in the lower scale range, that is, the range in which the avalanche sizes are measured as number of affected sites, is of 1.35. In introducing results from their model, BAK *et al.* (1988) claim that "...it is certainly possible that quantitative properties (such as scaling exponents) may apply to more realistic situation, since the system operates at a critical point where universality may apply". The similarity between the two scaling exponents supports this statement.

By following the conclusion drawn by BAK *et al.* (1988), the system described by the sandpile model "stays there", that is, it reaches a sort of stationary self-organized critical state: its configuration continues to change in space and in time but the way in which it changes obeys the same scaling law. In other words, the system continuously organizes itself toward a critical state and this criticality is independent from the local randomness. The reasons above support the idea that the Umbrian landscape evolves as described by this model. Thus, we believe that the hypothesis of self-organized critical processes generating landslides in the study area is plausible.

The analogies observed for the Umbria region suggest that this model could also be applied in areas where there are few or no recorded landslides, provided that the area in question shows the characteristics that are fundamental for the model to work: the presence of metastable sites, which implies the presence of active dynamic processes generating perturbations, and the predominant role of gravity-driven processes among those responsible for the modeling of the landscape.

5. Conclusion

In areas characterised by high slope instability, landslide events have a predominant role in shaping the landscape, but they also represent a serious risk for the population, the infrastructures and the agricultural areas, such as in the Umbria region (Italy) where landslides are widespread due to gravitational processes. In order to improve forecasting, numerous techniques are used to identify areas prone to landslides on the basis of the information acquirable from past events. Up to now, however, no single model has been able to fully simulate the complexity of the distribution of this phenomenon in space.

In spite of this and despite the extreme unpredictability of landslide occurrence, results presented in this paper support the idea that mass movements are not random events. The processing of the inventory maps of landslide triggering points and areas of the Umbria region revealed that the spatial pattern of this phenomenon possesses its own characteristic geometric structure, which can be described as "fractal structure". Scale-invariance has been identified in the spatial distribution of landslides, and in particular, a scaling regime boundary has been found at the scale of 1 km, which separates two scaling regimes, with clearly different scaling exponents (1.35 in the range of 25 m and 1 km and 1.76 in the range of 1 and 16 km). With reference to the theory of the SOC sandpile model, we hypothesize that the spatial fractal scaling identified for landslides in the study area could be the consequence of scale-invariant processes governing their spatial development.

Further analyses demonstrated that the fractal dimensions calculated contain information both on the spatial organization of the phenomenon and on their geometry; i.e., shape and area. It is noted that each type of mass movement generates specific landslide morphologies. Our results suggest that in the Umbria region, these morphologies also possess scaling properties.

Based on these findings, we suppose that the way in which environmental causal factors responsible for the mass movement combine, could be different in the two scale ranges. According to this hypothesis, the relative weights of the factors in the scale range of 25–1,000 m would control the assortment of landslide geometries, while those in the scale range of 1–16 km would control the spatial distribution of the phenomenon; i.e., the "local" and "global" spatial development of landslides in the area, respectively. However, more thorough studies are necessary to determine the exact causes of the two types of scaling behaviour identified. We believe that similar studies conducted in other areas of the world might contribute to our understanding.

Acknowledgments

The authors would like to thank the Civil Protection Department of the Umbria Region for providing the data used in this study, which is part of the PhD research of Luisa Liucci, XXVIII PhD cycle in "Earth Sciences and Geotechnologies", University of Perugia, Italy.

REFERENCES

BAK, P., TANG, C., and WIESENFELD, K., (1988), *Self-organized criticality*, Phys. Rev. A *38*(1), 364–374.

BRUNETTI, M.T., GUZZETTI, F., and ROSSI, M., (2009), *Probability distribution of landslide volumes*, Nonlin. Processes Geophys. *16*, 179–188.

CARDINALI, M., REICHENBACH, P., GUZZETTI, F., ARDIZZONE, F., ANTONINI, G., GALLI, M., CACCIANO, M., CASTELLANI, and M., SALVATI, P., (2002), *A geomorphological approach to the estimation of landslide hazards and risks in Umbria, Central Italy*, Nat. Hazard Earth Syst. Sci. *2*, 57–72.

CARRARA, A., CARDINALI, M., GUZZETTI, F., and REICHENBACH, P., Gis Technology in Mapping Landslide Hazard, In Geographical Information Systems in Assessing Natural Hazards (eds. Carrara A., and Guzzetti F.) (Kluwer Academic Publishing, Dordrecht 1995), pp. 135–176.

CATTUTO, C., and MELELLI, L., L'origine fisica dei grandi contrasti. In Campagne umbre. Contributo allo studio dei paesaggi rurali dell'Italia centrale (ed. H. Desplanques) (Quattroemme, Perugia 2006) pp. 1165–1207.

CHACÓN, J., IRIGARY, C., FERNÁNDEZ, T., and EL HAMDOUNI, R. (2006), *Engineering geology maps: landslides and geographical information systems*, Bull Eng Geol Environ *65*, 341–411.

CHASE, C.G., (1992), *Fluvial landsculpting and the fractal dimension of topography*, Geomorphology *5*(1–2), 39–57.

CIOTOLI, G., DELLA SETA, M., DEL MONTE, M., FREDI, P., LOMBARDI, S., LUPIA PALMIERI, E., and PUGLIESE, F., (2003), *Morphological and geochemical evidence of neotectonics in the volcanic area of Monti Vulsini (Latium, Italy)*, Quatern. Int. *101–102*, 103–113.

COLLETTINI, C., BARCHI, M., PAUSELLI, C., FEDERICO, C., and PIALLI, G., (2000), *Seismic expression of active extensional faults in northern Umbria (Central Italy)*, J. Geodyn. *29*, 309–321.

CZIRÓK, A., SOMFAI, E., and VICSEK, T., (1997), *Fractal scaling and power-law landslide distribution in a micromodel of geomorphological evolution*, Geol. Rundsch. *86*, 525–530.

DEL MONTE, M., FREDI, P., LUPIA PALMIERI, E., and SALVINI, F., (1999), *Fractal analysis to define the drainage network geometry*, Boll. Soc. Geol. It. *118*(1), 167–177.

DELLA SETA, M., DEL MONTE, M., FREDI P., and LUPIA PALMIERI, E., (2004), *Quantitative morphotectonic analysis as a tool for detecting deformation patterns in soft-rocks terrain: a case study from the southern Marches, Italy*, Géomorphologie *10*(4), 267–384.

DELLA SETA, M., DEL MONTE, M., FREDI P., LUPIA PALMIERI, E., and SBARRA, P., (2003), *Relations between morphodynamics and fractal dimension in some study areas of Italy*, Geogr. Fis. Dinam. Quat. *26*, 29–34.

GOKCEOGLU, C., and AKSOY, H., (1996), *Landslide susceptibility mapping of the slope in the residual soils of the Mengion region (Turkey) by deterministic stability analysis and image processing techniques*, Eng. Geol. *44*, 147–161.

GOLTZ, C., (1996), *Multifractal and Entropic Properties of Landslides in Japan*, Geol. Rundsch. *85*, 71–84.

GRASSBERGER, P., and PROCACCIA, I., (1983), *Characterization of Strange Attractors*, Phys. Rev. Lett. *50*(5), 346–349.

GREGORI, L., and MELELLI, L., Di fuoco e di acqua: forme e paesaggi delle "Città del Tufo", In L'ignimbrite di Orvieto-Bagnoregio (ed. Peccerillo A.) (Nuova Prhomos, Perugia 2012) pp. 113–134.

GRIFFITHS, J.S., (1999), *Proving the occurrence and cause of a landslide in a legal context*, B. Eng. Geol. Environ. *58*, 75–85.

GUZZETTI, F., MALAMUD, B.D., TURCOTTE, D.L., and REICHENBACH P., (2002), *Power-law correlations of landslide areas in central Italy*, Earth Planet. Sc. Lett. *195*, 169–183.

GUZZETTI, F., REICHENBACH P., ARDIZZONE, F., CARDINALI, M., and GALLI, M., (2006), *Estimating the quality of landslide susceptibility models*, Geomorphology 81, 166–184.

HERGARTEN, S., (2003), *Landslides, sandpiles, and self-organized criticality*, Nat. Hazards Earth Syst. Sci. *3*, 505–514.

ISPRA—Dipartimento Difesa del Suolo-Servizio Geologico d'Italia—Regione Umbria, (2006), Progetto IFFI (Inventario dei Fenomeni Franosi in Italia), http://www.progettoiffi. isprambiente.it/cartanetiffi/carto3.asp?cat=45&lang=IT.

IWAHASHI, J., WATANABE, S., and FURUJA, T., (2003), *Mean slope-angle frequency distribution and size frequency distribution of landslide masses in Higashikubiki area, Japan*, Geomorphology *50*, 349–364.

KAYASTHA, P., DHITAL, M.R., and DE SMEDT, F., (2013), *Evaluation and comparison of GIS based landslide susceptibility mapping procedures un Kulekhani watershed, Nepal*, J. Geol. Soc. India *81*(2), 219–231.

KOLMOGOROV, A.N., (1958), *A new invariant for transitive dynamical systems*, Dokl. Akad. Nauk. SSSR *119*, 861–864.

LI, C., MA, T., ZHU, X., and LI, W., (2011), *The power-law relationship between landslide occurrence and rainfall level*, Geomorphology *130*, 221–229.

LIEBOVITCH, L.S., Fractal and Chaos Simplified for the Life Sciences (Oxford University Press, New York 1998).

MALAMUD, B.D., TURCOTTE, D.L., GUZZETTI, F., and REICHENBACH, P., (2004), *Landslide inventories and their statistical properties*, Earth Surf. Process. Landforms *29*, 687–711.

MANDELBROT, B.B., (1967), *How long is the coast of Britain? Statistical self-similarity and fractional dimension*, Science *156*, 636–638.

MARK, D.M., and ARONSON, P.B., (1984), *Scale-Dependent Fractal Dimension of Topographic Surfaces; An Empirical Investigation, with Applications in Geomorphology and Computer Mapping*, Math. Geol. *16*(7), 671–683.

MEZUGHI, T.H., MAT HAHIR, J., GHANI RAFEK, A., and ABDULLAH, I., (2011), *Landslide Susceptibility Assessment using Frequency Ratio Model Applied to an Area along the E–W Highway (Gerik-Jeli)*, Am. J. Environ. Sci. *7*(1), 43–50.

NYKANEN, D.K., FOUFOULA-GEORGIOU, E., and SAPOZHNIKOV, V.B., (1998), *Study of spatial scaling in braided river patterns using synthetic aperture radar imagery*, Water Resour. Res. *34*(7), 1795–1807.

PECCERILLO, A., Plio-Quaternary Volcanism in Italy: Petrology, Geochemistry, Geodynamics (Springer, Heidelberg 2005).

PELLETIER, J.D., MALAMUD, B.D., BLODGETT, T., and TURCOTTE, D.L., (1997), *Scale-invariance of soil moisture variability and its implications for the frequency-size distribution of landslides*, Eng. Geol. *48*, 255–268.

PERUGINI, D., PETRELLI, M., and POLI, G., (2007), *Influence of landscape morphology and vegetation cover on the sampling of mixed plutonic bodies*, Miner. Petrol. *90*, 1–17.

PERUGINI, D., POLI, G., and MAZZUOLI, R., (2003), *Chaotic advection, fractals and diffusion during mixing of magmas: evidence from lava flows*, J Volcanol. Geoth. Res. *124*(3), 255–279.

PHILLIPS, J.D., (2006), *Deterministic chaos and historical geomorphology: A review and look forward*, Geomorphology 76, 109–121.

POURGHASEMI, H.R., MORADI, H.R., FATEMI AGHDA, S.M., SEZER E.A., GOLI JIRANDEH, A., and PRADHAN, B., (2013), *Assessment of fractal dimension and geometrical characteristics of the landslides identified in North of Tehran, Iran*, Environ. Earth Sci., doi:10.1007/s12665-013-2753-9.

PRADHAN, B., (2011), *An assessment of the use of an Advanced Neural Network Model with Five Different Training Strategies for the Preparation of Landslide Susceptibility Maps*, J. Data Sci. *9*, 65–81.

RÉNYI, A., (1959), *On the dimension and entropy of probability distribution*, Acta Math. Acad. Sci. Hungar. *10*(1–2), 193–215.

RICOTTA, C., ARIANOUTSOU, M., DÍAZ-DELGADO, R., DUGUY, B., LLORET, F., MAROUDI, E., MAZZOLENI, S., MORENO, J.M., RAMBAL, S., VALLEJO, R., and VÁZQUEZ, A., (2001), *Self-organized criticality of wildfires ecologically revisited*, Ecol. Model. *141*, 307–311.

RUFF, M., and CZURDA, K., (2008), *Landslide susceptibility analysis with a heuristic approach in the Eastern Alps (Vorarlberg, Austria)*, Geomorphology *94*(3–4), 314–324.

SUTEANU, C., Fractal-geometry-based analysis of microfabrics, In Proceedings of the Second European Workshop on the Analysis of Microfabrics in Geomaterials, (ed. Kruhl J.) (Technical University of Munich, Tectonics and Material Fabrics Section, Munich 2000) vol. 2(1).

SUTEANU, C., and IOANA, C., (2007), *Pattern identification in the dynamic fingerprint of seismically active zones*. Quatern. Int. *171–172*, 45–51.

SUTEANU, C., ZUGRAVESCU, D., and MUNTEANU F., The seismic activity in the Vrancea region in the light of Events Thread Analysis, In The Active Geodynamic Zone of Vrancea, Romania, Bucharest (eds. Zugravescu D., and Suteanu C.) (Publishing House of the Romanian Academy, Bucharest 2005) pp.103–110.

SÜZEN, M.L., and DOYURAN, V., (2004), *A comparison of the GIS based landslide susceptibility assessment methods: multivariate versus bivariate*, Environ. Geol. *45*(5), 665–679.

THEILER, J., (1990), *Estimating fractal dimension*, J. Opt. Soc. Am. A *7*(6), 1055–1073.

TURCOTTE, D.L., (1994), *Fractal Theory and the Estimation of Extreme Floods*, J. Res. Nati. Inst. Stand. Technol. *99*(4), 377–389.

TURCOTTE, D.L., (1999), *Self-organized criticality*, Rep. Prog. Phys. *62*, 1377–1429.

VAN WESTEN, C.J., CASTELLANOS, E., and KURIAKOSE, S.L., (2008), *Spatial data for landslide susceptibility, hazard and vulnerability assessment: An overview*, Eng. Geol. *102*, 112–131.

VAN WESTEN, C.J., RENGERS, N., and SOETERS, R., (2003), *Use of Geomorphological Information in Indirect Landslide Susceptibility Assessment*, Nat. Hazards 30, 399–419.

VERGARI, F., DELLA SETA, M., DEL MONTE, M., FREDI, P., and LUPIA PALMIERI, P., (2011), *Landslide susceptibility assessment in the Upper Orcia Valley (Southern Tuscany, Italy) through conditional analysis: a contribution to the unbiased selection of causal factors*, Nat. Hazards Earth Syst. Sci. *11*, 1475–1497.

YANG, Z.-J., and LEE, J.-H., (2006), *The fractal characteristics of landslides induced by earthquakes and rainfall in central Taiwan*, IAEG2006 48, 1–8.

YESILNACAR, E., and TOPAL, T, (2005), *Landslide susceptibility mapping: A comparison of logistic regression and neural networks methods in a medium scale study, Hendek region (Turkey)*, Eng. Geol. *79*(3–4), 251–266.

(Received March 29, 2014, accepted June 6, 2014, Published online July 3, 2014)

Reprinted from the journal

Pure Appl. Geophys. 172 (2015), 1975–1984
© 2014 Springer Basel
DOI 10.1007/s00024-014-0910-z

| Pure and Applied Geophysics

Fractal Dimension of the Hydrographic Pattern of Three Large Rivers in the Mediterranean Morphoclimatic System: Geomorphologic Interpretation of Russian (USA), Ebro (Spain) and Volturno (Italy) Fluvial Geometry

Carlo Donadio,[1] Fernando Magdaleno,[2,3] Adriano Mazzarella,[1] and G. Mathias Kondolf[4]

Abstract—By applying fractal geometry analysis to the drainage network of three large watercourses in America and Europe, we have calculated for the first time their fractal dimension. The aim is to interpret the geomorphologic characteristics to better understand the morphoevolutionary processes of these fluvial morphotypes; to identify and discriminate geomorphic phenomena responsible for any difference or convergence of a fractal dimension; to classify hydrographic patterns, and finally to compare the fractal degree with some geomorphic-quantitative indexes. The analyzed catchment of Russian (California, USA), Ebro (Spain), and Volturno (Italy) rivers are situated in Mediterranean-climate regions sensu Köppen, but with different geologic context and tectonic styles. Results show fractal dimensions ranging from 1.08 to 1.50. According to the geological setting and geomorphic indexes of these basins, the lower fractal degree indicates a prevailing tectonics, active or not, while the higher degree indicates the stronger erosion processes on inherited landscapes.

Key words: Fractal dimension, fluvial geomorphology, Mediterranean climate, USA, Europe.

1. Introduction

The evolution of a fluvial landscape and its current physiography are the result of mutual interaction between many factors and phenomena, which may be traced back to various Quaternary geomorphologic processes. The current waterscape conserves the stamp of the primary processes, both tectogenetic and morphoclimatic, which in the past contributed to its morphodynamic evolution, often superimposed on inherited morphologies (D'Alessandro *et al.* 2006).

In this paper, besides calculating for the first time the fractal dimension of three river drainage networks, we sought to compare their present-day hydrographic patterns to each other and to some significant geomorphic-quantitative indexes (Gardiner and Park 1978) showing a fractal behavior.

The selected river basins, located in California, Spain, and Italy, have well-studied but distinct geological history and geomorphologic settings, and all lie in Mediterranean climate regions. A Mediterranean climate is the climate typical of the Mediterranean Basin, and is a particular variety of subtropical climate. It also prevails in much of California, in parts of Western and South Australia, in southwestern South Africa, in sections of Central Asia, and in central Chile (Köppen and Geiger 1936). Areas with this climate receive almost all of their precipitation during their winter, autumn, and spring seasons, and have from four to six months during summer without significant precipitation. Thusly, holding climate constant amongst these basins, and knowing the specific lithological aspects of the catchments, it is possible to discriminate and quantify other factors, which have contributed in the past to, or currently support, the structuring of drainage networks. Among all, particularly the influence of tectonics and, consequently, the regional gradient are significant (Jones 2004; Burbank and Anderson

[1] DiSTAR, Department of Earth Sciences, Environment and Resources, University of Naples Federico II, Largo San Marcellino 10, 80138 Naples, Italy. E-mail: carlo.donadio@unina.it

[2] CEDEX, Centre for Studies and Experimentation on Public Works, Ministry of Public Works - Ministry of Agriculture, Food and Environment, Alfonso XII 3, 28014 Madrid, Spain. E-mail: fernando.magdaleno@cedex.es; fernando.magdaleno@upm.es

[3] Technical University of Madrid, Alfonso XII 3, 28014 Madrid, Spain.

[4] LAEP, Department of Landscape Architecture and Environmental Planning, University of California, Berkeley, 202 Wurster Hall 2000, 94720 Berkeley, CA, USA. E-mail: kondolf.berkeley@gmail.com

2011), considering that hydrographic patterns are ubiquitous and not strictly dependent only from lithology, but more often lithostructural control, fracturation, and morphoselection shape the fluvial geometry (Howard 1967; Kondolf *et al.* 2003; Perron *et al.* 2009). Hence, we applied fractal analysis (Mandelbrot 1967, 1975, 1983) in order to identify, through the degree of geometric irregularity of the drainage networks, the primary and secondary processes that contributed to their genesis (Klinkenberg 1992, 1994; Klinkenberg and Clarke 1992; Klinkenberg and Goodchild 1994; Gao and Xia 1996; Turcotte 1997), and to their morphological development (Xiao and Klinkenberg 1993).

2. Geomorphologic Features of the Rivers

The research was carried out on three main river basins of the Mediterranean morphoclimatic system (Fig. 1), which is characterized by a rainfall range of 500–1,200 mm/year and an average temperature of 10–16 °C/year. Spain and Italy in Europe have a climate subtype Csa, with average July temperatures of 20–27.5 °C, while California in the United States has a subtype Csb, which is more moderated by oceanic influences and has average July temperatures of 17.7–18.9 °C, sensu Köppen (Köppen and Geiger 1936; James 1966). The subtype Csa is the most common form of the Mediterranean climate, and it is known as a "typical Mediterranean climate". Regions with this form experience average monthly temperatures in excess of 22 °C during its warmest month and an average in the coldest month between 18 and −3 °C, with at least four months averaging above 10 °C, and wet winters. The subtype Csb is also termed "Cool-summer Mediterranean climate" (James 1966) and is a less common form of the Mediterranean climate. Regions with this subtype experience warm (but not hot) and dry summers, with no average monthly temperatures above 22 °C during its warmest month and an average in the coldest month between 18 and −3 °C and at least four - months averaging above 10 °C. Winters are rainy and can be mild to chilly but with a number of clear sunny days even during the wet season. In both sub-regions the climate is slightly continental sensu Ivanov (Pinna 1977), and is between humid and sub-humid in accordance with De Martonne (1941).

The Russian River (Fig. 2), 180 km long, is a third-order river located in Sonoma and Mendocino counties, debouching into the Pacific at Jenner, about 90 km north of San Francisco, California. The Russian River has a drainage area of 3,846 km², an average rainfall of 920 mm/year, a low sinuosity index $S < 1.5$ (slightly sinuous) and a low drainage

Figure 1
Location of Russian (R), Ebro (E) and Volturno (V) rivers in the Mediterranean morphoclimatic region, in USA (subtype Csb) and Europe (subtype Csa), respectively

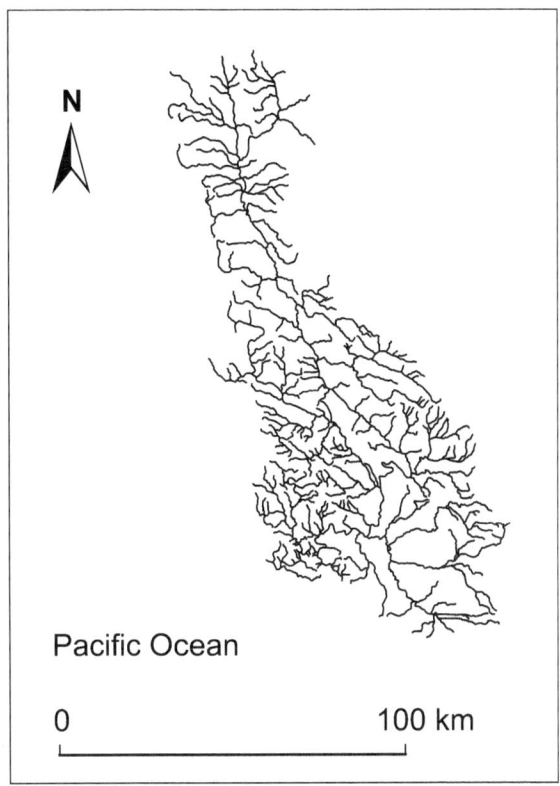

Depression, Castilian Plateau and Catalonian Coastal Range underlie most of the basin, with Holocene fluvial and marsh deposits near its mouth (MAGDALENO *et al.* 2012). Its drainage basin is underlain by NW–SE and NE–SW trending fault systems, and has a dendritic pattern.

The Volturno River (Fig. 4), 175 km long, drainage area of 5,550 km^2, average rainfall of 980 mm/year, is the main river of the Campania region in southwestern Italy. This fourth-order river flows along the Campania Plain *Graben*, west from the Apennines Chain, dissecting carbonatic units, volcanic rocks of the extinct Roccamonfina edifice and Phlegrean Fields pyroclastics. Holocene alluvial-marsh and aeolian-beach deposits form its mouth. The basin is affected by differential subsidence since the Plio-Quaternary and has an intermediate drainage density. The river crosses many cities and has high sinuosity $S \geq 1.5$ (from sinuous to meandering), thus, takes its name (*Volturnus*, turning in Latin). This urban river is regulated by two dams and is densely engineered since the 50's (DE PIPPO *et al.* 2008), discharging into the Tyrrhenian Sea, along a coastline that has been retreating at least since 1955 (PENNETTA *et al.* 2011).

Figure 2

Drainage network of Russian River, north California: the watercourse shows mostly a rectangular hydrographic pattern, locally intermediate between trellis and subdentric geometry

density. It is regulated by over 180 (mostly small) dams (KONDOLF AND BATALLA 2005; DEITCH and KONDOLF 2012). Sedimentary and volcano-metamorphic rocks of the Franciscan Complex, Great Valley Sequence, and Plio-Quaternary deposits underlie most of the basin, with Holocene alluvial and marine sediments near the mouth. The river crosses the active tectonic region of the San Andreas Fault. Its pattern is mostly rectangular, locally it is intermediate between trellis and subdendritic geometry (MAGDALENO *et al.* 2014 submitted).

The Ebro River (Fig. 3) is a 910-km long, fourth-order river, which drains 85,530 km^2, in northeastern Spain and discharges into the northwestern Mediterranean Sea. It has an average rainfall of 630 mm/year, a sinuosity index $S > 1.5$ (meandering) and low drainage density, and is regulated by over 187 dams (BATALLA *et al.* 2004). Sedimentary, metamorphic, and relic volcanic rocks of the Cantabrian Range, Vasque Mountains, Pyrenees, Iberian System, Ebro

3. Fractal Characterization of a Drainage Network

For our calculation we used satellite images and detailed topographic maps, at a scale of 1:25,000–1:50,000 with ETRS89 or WGS84 coordinate systems, from the US Geological Survey, CEDEX (Spain) and Campania Region (Italy), within a geographical information system (GIS) to isolate each drainage network (Figs. 2, 3, 4). The extracted georeferenced point coordinates were filtered to erase duplicates, then input in a specific software to compute 2D-fractal dimension through the correlation integral (GRASSBERGER and PROCACCIA 1983; LUONGO *et al.* 2000):

$$C(R) = 2[n(n-1)]^{-1} \sum_{i=1}^{n} \sum_{\substack{j=1 \\ j \neq 1}}^{n} \Theta\left(R - |X_i - X_j|\right). \quad (1)$$

The correlation integral counts the number of pairs with distance $|X_i - X_j|$ smaller than R. This is

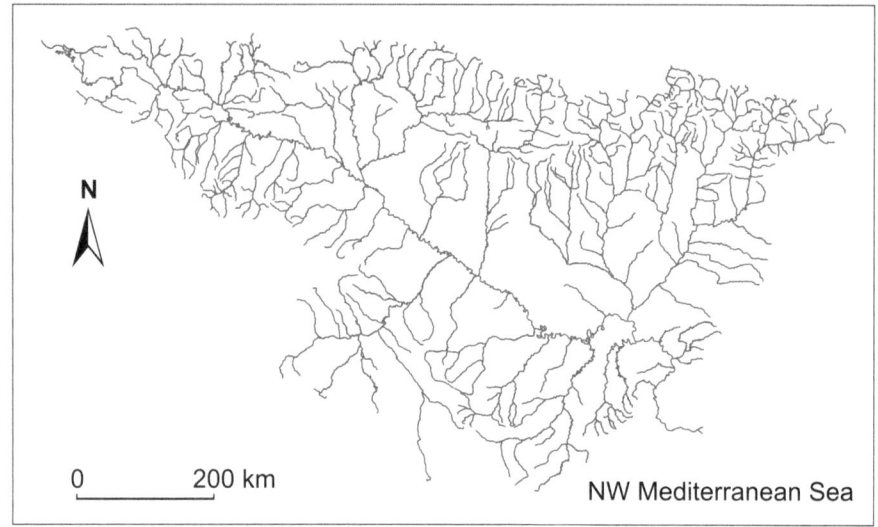

Figure 3
Drainage network of Ebro River, northeastern Spain: this large river shows a dense dendritic pattern, slightly asymmetric along the right bank

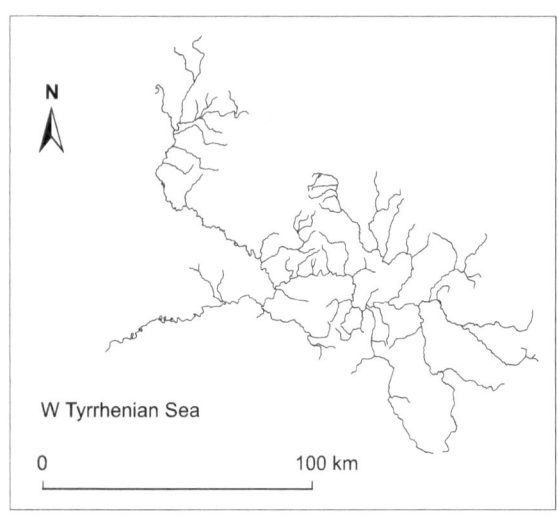

Figure 4
The drainage network of Volturno River, southwestern Italy, is a complex asymmetric pattern varying from dendritic (left bank) to subdentric (right bank)

done by taking each point in turn as a center and analyzing the distribution of the remaining points relative to it. The number of pairs provided by combinatorial calculus is equal to: $2\ [n(n-1)]^{-1}$.

In (1), $\Theta(x)$ is the Heaviside step function whose value is 0 for $x < 0$ and 1 for $x \geq 0$, n is the number of points available, X_i is the set of coordinates of the i-th point, and $2[n(n-1)]^{-1}$ is the normalization

factor representing the number of pairs such that $C(R)$ tends to 1 for R that tends to infinity.

If the n points of the aggregate have a fractal distribution, then:

$$C(R) = kR^D \tag{2}$$

or, equivalently, on a log–log scale graph, then

$$\log(C) = \log(k) + D\log(R), \tag{3}$$

where $C(R)$ is the cumulative frequency distribution of all the distance between the points, and k is a constant and D the fractal dimension.

We processed a minimum of 44,578 (Volturno River) to a maximum of 65,538 (Ebro River) xy georeferenced points composing the river networks, with a density of about 50 dots per linear kilometre. According to DE PIPPO et al. (2003) and D'ALES-SANDRO et al. (2006), we used the expression (1) rather than the box-counting algorithm (TURCOTTE 1997; RODRÍGUEZ-ITURBE and RINALDO 2001; DEL MONTE et al. 2007; SAA et al. 2007; SHEN et al. 2011). The latter algorithm counts the number of boxes of a grid occupied by at least one point with grid size doubled each time; instead, the correlation integral (1) counts the number of pairs of points whose distance falls within a set distance, each time doubled. Both methods calculate area clustering of the points analyzed and allow us to calculate a fractional dimension that represents the irregularity in the distribution of all the points, which compose the river network investigated. They also allow calculation of the so-called scaling region within which the property of scale invariance holds, i.e., the property that assimilates a small stretch of network, when extended, to a longer one. The advantage of applying algorithm (1) consists in the high level of confidence ($R^2 > 99\%$) reached in the scaling region computation (MAZZARELLA and TRANFAGLIA 2000), even if the correlation dimension (D_C) is lower than $\sim 5\%$ of the box-counting one (D_F).

To calculate the fractal dimension of the drainage networks and identify the range of distances within which the coordinates of the single points of the forms follow the fractal relation (3), we plotted on a log–log diagram the number C of pairs of points with a distance smaller than R, as a function of R. The smallest usable distance was set at the linear resolution of 30 m and gradually increased by a factor of 2. The best fit of the regression line of the least squares of $\log(C)$ on $\log(R)$ was obtained for distances between 1.6 and 200 km with a fractal dimension D equal to the topological dimension of a line. The results for the three Mediterranean rivers analyzed are shown in the log–log scale graphs (Fig. 5), considering the scaling region (3.2–5.3) within which the property of scale invariance exists (1.6–200 km).

Moreover, some geomorphic-quantitative indexes have been calculated, compared each other, and to the fractal degree of the drainage networks (Table 1). Particularly, considering that hydrographic patterns are similar to fractal trees (TOKUNAGA 1978; PECKHAM 1995), the sinuosity of a stream exhibits fractal behavior (SNOW 1989), and meander patterns are fractal (NAGATANI 1993), we computed the sinuosity through the simple relation:

$$S = l / L, \qquad (4)$$

where l is the curved river length and L the straight-line valley length calculated with a GIS, thus, their ratio $S < 1$ indicates a straight river, $1 < S < 1.5$ sinuous, $S > 1.5$ meandering; and the drainage density of the basin, through the expression:

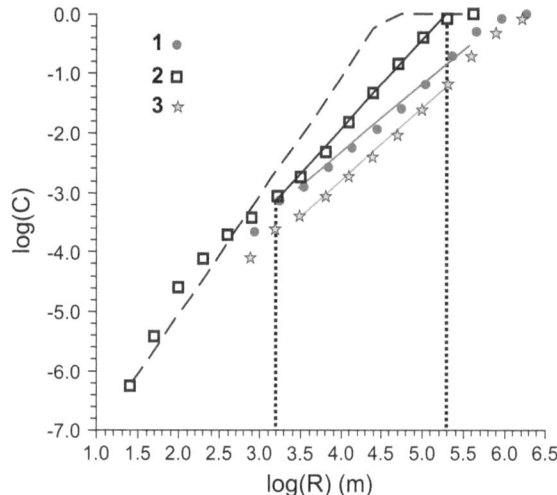

Figure 5
Log of number of pairs C of the areal coordinates of the three rivers (*1* Russian, *2* Ebro, *3* Volturno), with mutual distance smaller than R, as a function of $\log(R)$ (m). *Dashed line* is the random distribution with $D_C = 2$; *solid line* is D_C inside the scaling region (*dotted lines*)

$$D_k = \frac{\sum_{u=1}^{k} \sum_{i=1}^{Nu} L_u, i}{A_k}, \qquad (5)$$

where N_u is the number of channels of order u with length $L_{u,i}$ and A_k is the basin area of order k. All parameters are calculated with a GIS module. Another empirical expression allows correlation of the fractal dimension to Horton's laws (HORTON 1945):

$$D = \frac{\ln R_b}{\ln R_r},$$

where $R_b = N_i/N_{i+1}$ is the bifurcation ratio, N_i the number of streams of order i; and $R_r = r_{i+1}/r_i$ is the length-order ratio, where r_i represents the mean length of streams of order i. This relation suggests that standard stream-ordering parameters are directly related to the fractal degree of the drainage network. Differently from other authors (LA BARBERA and ROSSO 1989; TURCOTTE 1997), we have not applied this commonly used expression because, for the validity of Horton's laws, it requires that river networks are regular fractal trees, unlike the natural patterns of our case studies.

Reprinted from the journal

Table 1

Fractal dimension (D_C) of drainage network with confidence level (R^2), main hydrographic pattern, sinuosity (S), mainstem (Lr), total channel lengths (L_c), basin area (A), and drainage density (D_k) of the three rivers each compared to the other (1); some data from the literature of other Mediterranean (2) and non-Mediterranean climate rivers (3) are shown (see Fig. 7)

	Code	River	Pattern	D_c	R^2	S	L_r (km)	L_c (km)	A (km²)	D_k (km⁻¹)
1	R	Russian	Rectangular	1.08	1.00	<1.5	180	758.23	3,846	0.197
	E	Ebro	Dendritic	1.50	1.00	>1.5	930	12,073.57	85,530	0.141
	V	Volturno	Subdendritic	1.11	0.99	≥1.5	486.74	3,208.73	5,550	0.578
2	C	Volfe-Bell	Dendritic-pinnate	1.81	–					
	a	Trasubbie	Pinnate	1.73	0.99					
	b	Bretta	Pinnate	1.54	0.99					
	c	Rubbiattino	Dendritic	1.36	0.98					
	d	Fraginale	Pinnate	1.72	0.98					
	e	Vallelata	Parallel	1.77	0.99					
	f	Gravina di Matera	Rectangular	1.17	0.98					
	g	Cortilla	Dendritic-pinnate	1.50	0.90					
	h	Melacce	Dendritic-parallel	1.81	0.98					
	i	Lanzo	Dendritic-parallel	1.90	0.99					
	l	Gretano	Dendritic-parallel	1.88	0.99					
	m	Liri	Angulate	1.79	0.93					
	n	Ombrone	Dendritic-pinnate	1.54	0.99					
3	K	Kentucky	Pinnate	1.67	–					
	P	Powder	Parallel	1.77	–					
	M	Mississippi	Dendritic	1.83	–					

4. Geomorphologic Interpretation of the Fractal Dimension

The genesis of a river network is a gradual and complex phenomenon that depends on various factors and processes interacting during long and short time periods: tectonic structuring of the river basin and lithostratigraphic characteristics (BURBANK and ANDERSON 2011); areal erosion (FOURNIER 1960) and linear erosion (MAZZARELLA 1999) with development of channels (PERRON et al. 2009); and morphoselective erosion, edaphic factors and climate (BARTOLINI 2012).

Although they have a spatial geometry with globally constant curvature, natural forms show locally a geometry with variable curving, that is, with a degree of irregularity in space and with the property of self-similarity (scale invariance), termed dimension of internal homothety D insofar as it represents a geometric cascade (MANDELBROT 1983, 1987).

In general, the fractal geometry of river drainage networks, obtainable with the generalized method of Koch's curve, may be expressed by a fractional number between 1.1 and 1.8 (even though in nature it

is variable between about 1.16 and 1.66). In this case the former networks are less rugged than the latter, which tend to a Peano's curve with a fractal dimension $D = 2$, the dimension of a plane (MANDELBROT 1987). However, these sizes, under initial analysis, may not supply direct information on the main processes responsible for modeling the waterscape.

From the results of work carried out along three drainage networks of Mediterranean-type rivers it seems possible to derive, from the fractal degree, the primary and secondary processes behind modeling, distinguishing tectonic events from fluvial dynamics both responsible for river morphology, and identifying which is predominant, as stated by DE PIPPO et al. (2003) and D'ALESSANDRO et al. (2006). Actually, the fractal dimension $D = 1$, characterising the Euclidean dimension of a line, may be likened to the theoretical value obtained in the case of high geometric regularity. In nature, in the specific case of a river network where $D \sim 1$, this low value can only result from forms of tectonic genesis with monodirectional orientation (i.e., fault, faultline, overthrust). Instead, a fractal degree $D > 1$, characteristic of segmented curves or trees whose dimension tends

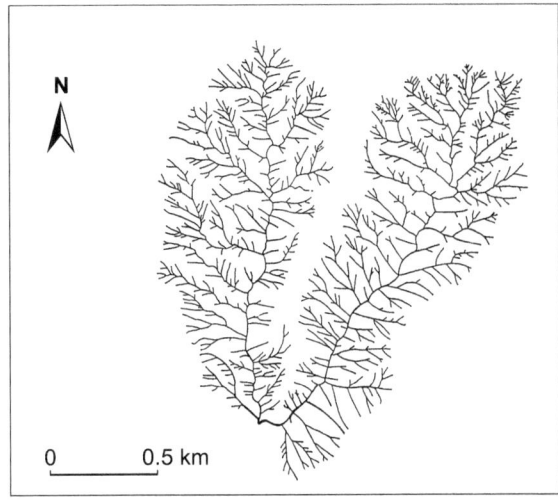

Figure 6

Drainage network in Volfe and Bell Canyons, San Gabriel Mountains in California: the third-order dichotomous stream shows a hydrographic pattern from dendritic (right branch) and subdendritic (left branch) to pinnate, developed by erosion processes on an inherited landscape, structured by tectonics; its fractal dimension is $D = 1.81$ (after TURCOTTE 1997)

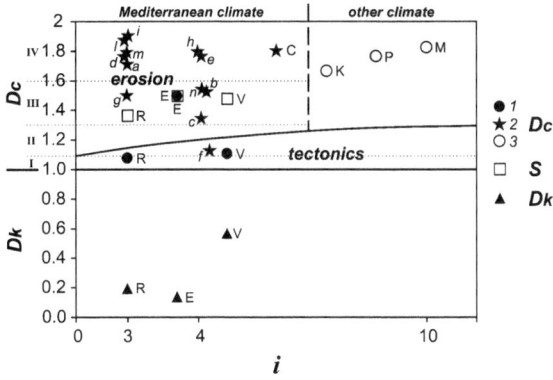

Figure 7

Diagram of a comparison between the values of fractal dimension (D_C) of fluvial drainage networks calculated with the correlation integral in this paper and some derived from literature, the river sinuosity (S, *open box*), the basin drainage density (D_k, *solid triangle*), and the stream order (i) sensu STRAHLER (1957). In the upper part of the diagram the 2D-fractal dimension refers both to Mediterranan climate rivers (*left*), i.e., the three ones analyzed (*1 solid circle*: R Russian, E Ebro, V Volturno) and those from the literature (*2 solid stars*: C, Volfe and Bell Canyons: TURCOTTE 1997; *a*, Trasubbie; *b*, Bretta; *c*, Rubbiatino; *d*, Fraginale; *e*, Vallelalata; *f*, Gravina di Matera; *g*, Cortilla; *h*, Melacce; *i*, Lanzo l, Gretano; *m*, Liri; *n*, Ombrone: DEL MONTE *et al.* 1999, 2007), and to rivers flowing in different morphoclimatic systems (right), derived from the literature (*3, open circle*: K, Kentucky: PECKHAM 1995; ANDREWS 2006; P, Powder: PECKHAM 1995; M, Mississippi: GARNETT 1986; TURCOTTE 1997), separated by a *dashed line*. The *curve* divides a region (below) with prevailing active or recent tectonics (low D_C) from the region (above) mainly characterized by erosion processes (high D_C), which have structured and modelled the current fluvial geometry. The lower part of the diagram shows particularly the relation between drainage density (D_k) of the three studied river basins and the other parameters. The *dotted lines* represent the limits of D_C intervals between the main Mediterranean-type drainage patterns identified: (*I*) rectangular, (*II*) subdendritic, (*III*) dendritic, (*IV*) pinnate, and parallel

to the Euclidean plane, is similar to the theoretical one obtained when geometric irregularity is greater. In this case the fractal degree results from shape and number of channels, such as sinuosity (SNOW 1989), meandering (NAGATANI 1993), drainage density of basin, and stream-ordering system (STRAHLER 1957) with a high hierarchy that is characterized by the modelling of fluvial erosion with tectonic phenomena absent or negligible.

Briefly, on this basis, the integer part of the fractal dimension value indicates the primary processes (i.e., straight shape due to tectonics), while the decimal part indicates the secondary ones (i.e., irregular shape due to channel erosion). Because tectonic activity tends to occur in pulses and to act over longer time scales than erosion (which is often spatially concentrated), if tectonics are not currently active, a low fractal dimension is likely related to the presence of an inherited landscape structured by tectonic events over Cenozoic time under conditions different from the present. Conversely, a high fractal dimension suggests intense erosion processes, which can also act on an inherited waterscape structured by past tectonics, as illustrated in small catchments of some California's canyons (Fig. 6).

5. Discussion and Conclusions

The fractal analysis is a valuable tool with which to examine and compare the drainage networks of rivers, independent of their size. Application of fractal geometry supplies a methodological and quantitative contribution to discriminate in the first analysis the relation between geomorphic processes and the degree of irregularity of fluvial networks.

In particular, the interpretation of fractal dimension in a geomorphologic key allows for discrimination of tectogenetic processes from morphogenetic ones, which act together and in coeval ways, but with different intensities, and prevailing over the other in certain periods, contributing to the overall development of the fluvial geometry in the catchment.

For this purpose, we analyzed three large rivers with similar Mediterranean climates, such that differences in the values of 2D-fractal dimension (calculated through the correlation integral) are attributable mainly to: (1) tectonics, when they result close to unity, i.e., to the Euclidean line; and (2) erosion, areal rather than linear, when the degree of geometric irregularity increases, and the values tend towards 2, i.e., to the Euclidean plane.

The fractal dimension of the three rivers ranged from ~ 1–1.5 (Table 1). This suggest that the geomorphology of these basins results from the overlay of recent geomorphic processes over a physiography of inherited waterscapes and structural control on hydrographic patterns (Fig. 7). Our interpretation partly disagrees with the model proposed by DEL MONTE et al. (1999, 2007), which imputes a high fractal degree mainly to strong tectonic control.

Tectogenesis was intense in the Plio-Pleistocene, although currently active in the Russian River basin ($D_C = 1.08 \pm 0.01$) with mostly rectangular pattern, low sinuosity and almost straight mainstem, and low drainage density; while morphogenesis that occurred in the Late Quaternary shows well preserved effects, as in the Ebro River basin ($D_C = 1.5 \pm 0.01$), with a regular dendritic network, high sinuosity, and low drainage density.

The different fractal degrees would indicate that fluvial basin evolution at certain periods was likely controlled by glacial fluctuations, at other times by effects related to Plio-Quaternary tectonics and locally to Pleisto-Holocene volcano-tectonic and bradyseismic activity, as in the Volturno River ($D_C = 1.11 \pm 0.01$) with complex subdendritic geometry, high sinuosity, and intermediate drainage density. Lastly, in historical times such phenomena seem to have been overlain by fluvial dynamics, changes due to climatic crises, and damming.

In particular, the diagram of Fig. 7 shows the results of fractal and geomorphic-quantitative analyses of these three Mediterranean-type rivers. The values of the fractal degree are compared to those from the literature of other Mediterranean rivers (DEL

MONTE et al. 1999, 2007), and non-Mediterranean morphoclimatic systems (Peckham 1995; TURCOTTE 1997). Overall, considering both the geological history of river basins and that a higher geometric irregularity is observed for $D_C \geq 1.25$ (MANDELBROT 1967, 1983, 1987), the values plotted suggest a boundary between regions with prevailing tectonic control (lower field) and erosion (upper field) for $D_C = 1.1$–1.25.

Furthermore, the diagram confirms a close relationship of direct proportionality between fractal dimension D_C and sinuosity S: when the latter increases, the geometric irregularity grows and, thus, also the fractal degree. Instead, fractal dimension seems independent from drainage density D_k, very likely due to the scaling property of drainage networks.

The fractal degree seems also correlable to the Horton-Strahler stream order number i, ranging from third-order to tenth-order, which depends on the presence of side tributaries differently by deterministic fractal trees, in accordance with TURCOTTE (1997). Indeed, in the theoretical models in which the empirical expression applies, the fractal dimension tends to 1 if, in the fractal tree the angle of divergence θ between the tributaries and the higher order stem is <60°, and tends to 2 for higher angular values ($D_C = 2$ for $\theta = 90°$), independently of stream order.

The gap between these models and drainage networks analyzed can be attributed to the fact that in the current waterscape, side tributaries of low-order often flow directly into mainstems of much higher order (e.g., first-order stream into third-order or fourth-order stem). In addition, the angles of convergence θ may vary considerably within the river basin for tectonic and morphostructural reasons, which together have controlled the evolution of its sub-basins.

Finally, our results suggest that the increase of the fractal dimension corresponds to a gradual transition from a drainage network with mainly rectangular pattern (Russian) to a subdendritic (Volturno) and then to the dendritic one (Ebro), corresponding to a gradient of decreasing tectonic control and to progressively greater influence of erosive processes and morphoselection that have increased the geometric irregularity of the drainage network. Thus, the more articulated networks have developed by superimposing current drainage patterns onto an inherited landscape that is shaped in a tectono-climatic regime different from the current Mediterranean one.

Synthesizing our case studies with those from the literature (totalling, among others, 16 Mediterranean climate rivers), we found that in the whole set of Mediterranean-type rivers analyzed (I) $Dc \leq 1.1$ is associated with rectangular hydrographic patterns, (II) $1.1 > Dc > 1.3$ corresponds to subdendritic types, (III) $1.3 > Dc > 1.6$ to dendritic models, and, finally, (IV) $Dc \geq 1.6$ to the pinnate and parallel ones (Fig. 7).

Intermediate cases of complex drainage networks falling along the limits of these subdivisions (e.g., dendritic-pinnate, parallel to angulate, subdendritic to trellis patterns, etc.) may reflect recent landscape rejuvenation due to local tectonic reactivation, exhumation of fault scarps by erosion, stream piracy, lithostructural elements, a high degree of fracturing, and high gradient, or combinations thereof.

In summation, although many models and geomorphic parameters have proven useful in describing drainage networks, they contain little information on their dynamical developing processes. The integration and comparison with fractal dimensions could contribute to a better understanding of fluvial morphostructural evolution and support quantitative classification of hydrographic patterns.

REFERENCES

ANDREWS, W.M. Jr. (2006), Geologic control on Plio-Pleistocene drainage evolution of the Kentucky River in central Kentucky, Kentucky Geological Survey, University of Kentucky (USA), Thesis 4, Series XII, pp. 216.

BARTOLINI, C. (2012), Is the morphogenetic role of tectonics overemphasized at times?, Boll. Geof. Teor. Appl. 53(4), 459–470.

BATALLA, R.J., GÓMEZ, C.M., and KONDOLF, G.M. (2004), Reservoir-induced hydrological changes in the Ebro River basin, J. Hydrol. 290, 117–136.

BURBANK, D.W., ANDERSON, R.S., Tectonic Geomorphology, 2nd ed. (Wiley-Blackwell., Oxford 2011).

D'ALESSANDRO, L., DE PIPPO, T., DONADIO, C., MAZZARELLA, A., and MICCADEI, E. (2006), Fractal dimension in Italy: a geomorphological key to interpretation, Zeit. Geom. N. F. 50(4), 479–499.

DE MARTONNE, E. (1941), Nouvelle carte mondiale de l'indice d'aridité, La meteorologie 1, 3–20.

DE PIPPO, T., DONADIO, C., MAZZARELLA, A., PAOLILLO, G., and PENNETTA, M. (2003), Fractal geometry applied to coastal and submarine features, Zeit. Geom. N. F. 48(2), 185–199.

DE PIPPO, T., DONADIO, C., PENNETTA, M., PETROSINO, C., TERLIZZI, F., and VALENTE, A. (2008), Coastal hazard assessment and mapping in Northern Campania, Italy, Geomorphology 97, 451–466.

DEITCH, M.J., and KONDOLF, G.M. (2012), Consequences of variations in magnitude and duration of an instream environmental flow threshold across a longitudinal gradient, J. Hydrol. 420–421, 17–24.

DEL MONTE, M., FREDI, P., LUPIA PALMIERI, E., and SALVINI, F. (1999), Fractal analysis to define drainage network geometry, Boll. Soc. Geol. It. 118, 167–177.

DEL MONTE, M., FREDI, P., LUPIA PALMIERI, E., and SBARRA, P. (2007), Some relations between fractal dimension of drainage network and geomorphology of drainage basins, Transactions Jap. Geom. Union 28(1), 1–21.

FOURNIER, F., Climat et Erosion (PUF, París 1960).

GAO, J., and XIA, Z. (1996), Fractals in physical geography, Progress Phys. Geogr. 20(2), 178–191.

GARDINER, V., and PARK, C. (1978), Drainage basin morphometry: review and assessment, Progress Phys. Geogr. 2, 1–35.

GARNETT, P.W. (1986), River meanders and channel size, J. Hydrol. 88, 147–164.

GRASSBERGER, P., and PROCACCIA, I. (1983), Characterization of strange attractors, Phys. Rev. Lett. 50, 346–349.

HORTON, R.E. (1945), Erosional development of streams and their drainage basins: hydrophysical approach to quantitative morphology, Geol. Soc. Am. Bull. 56, 275–370.

HOWARD, A.D. (1967), Drainage analysis in geologic interpretation: a summation, Amer. Ass. of Petroleum Geologist Bull. 51, 2246–2259.

JAMES, J.W. (1966), A modified Koeppen classification of California's climates according to recent data, California Geographer 7, 1–12 + map.

JONES, S.J. (2004), Tectonic controls on drainage evolution and development of terminal alluvial fans, southern Pyrenees, Spain, Terra Nova 16, 121–127.

KLINKENBERG, B. (1992), Fractal and morphometric measures: is there a relationship?, In Fractals in Geomorphology (eds. Snow R.S. and Mayer L.), Geomorphology 5, 5–20.

KLINKENBERG, B. (1994), A review of methods used to determine the fractal dimension of linear features, Math., Geol. 26, 23–46.

KLINKENBERG, B., and CLARKE K.C., Exploring the fractal mountains, In Automated Pattern Analysis in Petroleum Exploration (eds. Palaz I. and Sengupta S.) (Springer-Verlag, New York 1992), pp. 201–212.

KLINKENBERG, B., and GOODCHILD, M. (1994), The fractal properties of topography: a comparison of methods, Earth Proc. Landf. 17, 217–234.

KONDOLF, G.M., and BATALLA, R.J., Hydrological effects of dams and water diversions on rivers of Mediterranean-climate regions: examples from California, In Catchment Dynamics and River Processes: Mediterranean and Other Climate Regions (eds. Garcia C. and Batalla R.J.) (Elsevier, Amsterdam 2005) pp. 197–211.

KONDOLF, G.M., MONTGOMERY, D.R., PIÉGAY, H., and SCHMITT, L. (2003), Geomorphic classification of rivers and streams, In Tools in Fluvial Geomorphology (eds. Kondolf G.M. and Piégay H.) (John Wiley & Sons, Chichester 2003) pp. 171–204.

KÖPPEN, W., and GEIGER, R., Handbuch der Klimatologie. Vol. 1, Part C (Gerbrüder Borntraeger, Berlin 1936).

LA BARBERA, P., and ROSSO, R. (1989), *On the fractal dimension of stream networks*, Water Resources Res. *25*, 735–741.

LUONGO, G., MAZZARELLA, A., and DI DONNA, G. (2000*), Multifractal characterization of Vesuvio lava-flow margins and its implications*, J. Volc. Geotherm. Res. *101*, 307–311.

MAGDALENO, F., DONADIO, C., and KONDOLF, G.M. (2014), *30 year response of a Mediterranean river to damming in California, USA*, Hydrological Sciences Journal (submitted).

MAGDALENO, F., FERNÁNDEZ, J.A., and MERINO, S. (2012), *The Ebro River in the 20th century or the ecomorphological transformation of a large and dynamic Mediterranean channel*. Earth Surf. Proc. Landf. *37*(5), 486–498.

MANDELBROT, B.B. (1967), *How long is the coast of Britain? Statistical similarity and fractal dimension*, Science *155*, 636–638.

MANDELBROT, B.B. (1975), *Stochastic model is of the Earth's relief, the shape and the fractal dimension of the coastal lines, and the number area rule for the islands*, Proc. Nat. Acad. Sc. USA *72*, 3825–3828.

MANDELBROT, B.B., Gli oggetti frattali: forma, caso e dimensione (G. Einaudi ed., Torino 1987).

MANDELBROT, B.B., The Fractal Geometry of Nature (Freeman and Co., New York 1983).

MAZZARELLA, A. (1999), *Multifractal dynamic rainfall processes in Italy*, Theor. Appl. Climatol. *63*, 73–78.

MAZZARELLA, A., and TRANFAGLIA, G. (2000), *The fractal characterisation of geophysical measuring networks and its implications for an optimal location of additive stations: an application to a rain-gauge network*, Theor. Appl. Climatol. *65*, 157–163.

NAGATANI, T. (1993), *Dynamic scaling of river-size distribution in the Scheidegger's river network model*, Fractals *1*, 247–252.

PECKHAM, S.D. (1995), *New results for self-similar trees with applications to river networks*, Water Resources Res. *31*(4), 1023–1029.

PENNETTA, M., CORBELLI, V., ESPOSITO, P., GATTULLO, V., and NAPPI, R. (2011), *Environmental impact of coastal dunes in the area located to the left of the Garigliano river mouth (Campany, Italy)*. J. Coastal Res. SI *61*, 421–427.

PERRON, J.T., KIRCHNER, J.W., and DIETRICH, W.E. (2009), *Formation of evenly spaced ridges and valley*, Nature Letters *460*, 502–505.

PINNA, M., Climatologia (Utet, Torino 1977).

RODRÍGUEZ-ITURBE, I., and RINALDO, A., Fractal River Basins. Chance and Self-Organization (Cambridge University Press, UK 2001).

SAA, A., GASCÓ, G., GRAU, J. B., ANTÓN, J.M., and TARQUIS, A.M. (2007), *Comparison of gliding box and box-counting methods in river network analysis*, Nonlinear Processes in Geophysics *14*(5), 603–613.

SHEN, X.H., ZOU, L.J., ZHANG, G.F., SU, N., WU, W.Y., and YANG, S.F. (2011), *Fractal characteristics of the main channel of Yellow River and its relation to regional tectonic evolution*. Geomorphology *127*, 64–70.

SNOW, R.S. (1989), *Fractal sinuosity of stream channels*, Pure Appl. Geophys. *131*, 99–109.

STRAHLER, A.N. (1957), *Quantitative analysis of watershed geomorphology*, Trans. Am. Geophys. Un. *38*, 913–920.

TOKUNAGA, E. (1978), *Consideration on the composition of drainage networks and their evolution*, Geograph. Rep. Tokyo Metro. Univ. *13*, 1–27.

TURCOTTE, D.L., Fractals and Chaos in Geology and Geophysics (Cambridge University Press, UK 1997).

XIAO, Y., and KLINKENBERG, B. (1993), *Topographic characterization for geographic modeling*, Proc. GIS'93, Vancouver, 883–898.

(Received March 29, 2014, revised July 15, 2014, accepted July 18, 2014, Published online August 10, 2014)

Pure Appl. Geophys. 172 (2015), 1985–1997
© 2014 Springer Basel
DOI 10.1007/s00024-014-0937-1

Pure and Applied Geophysics

Erosion Triangular Facets as Markers of Order in an Open Dissipative System

GUIDO PALIAGA[1]

Abstract—The complexity and non-linearity of the morphogenetic system which is responsible for shaping the Earth's surface have been widely recognised by many authors who have documented the fractal nature of erosion. In this paper, two peculiar kinds of landforms are compared to point out ordered structures, i.e. triangular facets that arise in different geomorphic systems, due to the principle of morphologic convergence. Occurrence of triangular facets has been documented in mountainous areas in relation to base level changes and hydrographical network evolution; similarly shaped landforms are present even in recent tectonic uplift areas along faults. The spatial distribution of the two kinds of facets has been investigated in two river basins located in Liguria (northern Italy) and in a mountainous area in Oman. The results of this analysis document the different spatial features of the two kinds of facets.

Key words: Triangular facets, fractals, spatial distribution, morphogenetic system, self-organized system, landforms.

1. Introduction

According to the approach to modelling surface evolution introduced by LEOPOLD and LANGBEIN (1962) and SCHEIDEGGER and LANGBEIN (1966), the morphogenetic system is a complex system that requires a probabilistic rather than a deterministic methodology (SCHEIDEGGER 1961; SCHEIDEGGER and LANGBEIN 1966). Several evolution models followed this approach, including nonlinearity (PHILLIPS 1992) and many more processes, but the theoretical improvement was provided by SCHEIDEGGER (1987, 1992) in proposing a similitude between thermodynamic systems and geomorphologic systems, i.e. that the evolution is driven by entropy. Thus, in geomorphic systems, the roles of heat and temperature are assumed by mass and height, and the evolution is

guided by entropy growth until the maximum value consistent with the system is reached.

Deterministic chaos and complex system theory have been widely applied to geomorphic processes, pointing out the fractal nature of erosion (KORVIN 1992; TURCOTTE 1992; SIU-NGAM LAM and DE COLA 1993) and the scale invariance of processes. Geomorphic system evolution is ruled by deterministic chaos (SCHEIDEGGER 1994) and landforms do not correspond to a dynamic equilibrium condition, but to a self-organized order on the edge of chaos in an open dissipative system. Self-organization arises when a dissipative system is far from equilibrium (PRIGOGINE and STENGERS 1984): the evolution toward a new dynamic equilibrium condition causes fractal structures and patterns to arise. BAK *et al.* (1988) introduced this behaviour with the concept of self-organized criticality: the scale invariance that characterises this condition makes fractal analysis appropriate (TURCOTTE 1992). In such a situation, landforms with fractal features appear (SCHEIDEGGER 1994).

Within this theoretical framework, a peculiar kind of landform was analysed to search for markers of Scheidegger's theory, i.e. triangular facets.

2. Triangular Facets

Triangular facets are geomorphologic features observed in various contexts on the Earth's surface. According to morphologic convergence theory, different dominant processes may generate them: erosion or tectonic activity, and their mutual interaction.

2.1. Erosion Triangular Facets

Erosion triangular and trapezoidal facets are landforms generated by fluvial erosion and are related

[1] Professional Geomorphologist, via Livorno 1/15, 16146 Genova, Italy. E-mail: gpaliaga@gmail.com

to the hydrographical network evolution (BRANCUCCI and MARINI 1990a, b) induced by base level changes (Fig. 1). A base level drop, caused by sea-level change and/or tectonic uplift, induces acceleration of all erosion processes in a river basin. The increased linear incision may cause cutting of secondary divides, generating triangular and trapezoidal erosion facets, which are mostly characterised by facing the riverbed. They occur in mountainous areas where linear erosion processes are intense and when changing conditions in the geomorphic system (base level changes) trigger erosion cycles.

In the system theory approach, a base level change is regarded as a perturbation to the geomorphic system, introducing energy that will be successively dissipated by erosion, generating

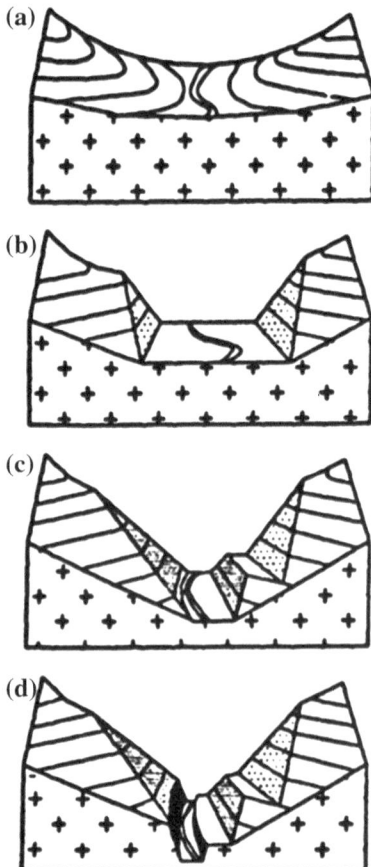

Figure 1
The genetic process of erosion triangular facets: **a** a base level drop, **b** causes activation of an erosion cycle in the catchment. The fluvial incision cuts the interfluves. Subsequent base level drops (**c**, **d**) generate other facets (from BRANCUCCI and MARINI 1990a)

landforms. Fractal features may appear as the system tends to a new dynamic equilibrium condition (SCHEIDEGGER 1994; PECKHAM 1995).

In Liguria, triangular and trapezoidal facets are present in the catchments facing the Ligurian Sea; all of them are characterised by a hierarchical structure (Fig. 2), probably related to several phases of global or local base level fluctuations that occurred in the Quaternary period as the region has been subjected to differential uplift movements and to the effects of sea-level changes (BRANCUCCI and MARINI 1990a, b).

2.2. Tectonic Triangular Facets

Tectonic triangular facets are frequently observed at normal fault scarps (COTTON 1950; SELBY 1985) as a result of relief cut by a direct fault system. Subsequently, erosion triggers the degradation process of the relief, causing the evolution of landforms (PETIT et al. 2009a), although the eventual reactivation of extensive movements may generate other triangular facets (Fig. 3), producing a hierarchical structure (WIWEGWIN et al. 2011). Imbrication structure has been recognised in relation to the evolution stage of the drainage and to the number of activation cycles of the fault system (PETIT et al. 2009b). Fault reactivation may even transform triangular into pentagonal facets (BRANCACCIO et al. 1978). BRANCACCIO et al. (1986) found such evidence in the central-southern Apennines (Italy).

Tectonic triangular facets are present in many regions of active extension.

3. The Study Areas

3.1. Liguria Region

This region facing the Mediterranean Sea is mountainous and characterised by small, steep catchments, mainly orthogonally oriented to the coastline. On the contrary, the north-oriented basins are less steep and are part of the large hydrographical network of the Po River.

Reliefs have been generated during the complex Alpine and Apennine orogeny and have been subjected to uplift movements related to the northward tilting that involved the territory in the Quaternary as

Hierarchical structure of erosional triangular facets in a catchment due to base level lowering - theoretical scheme

Hierarchical structure of erosional triangular facets in a catchment due to base level lowering

Real world situation: some facets are erased by subsequent erosional processes

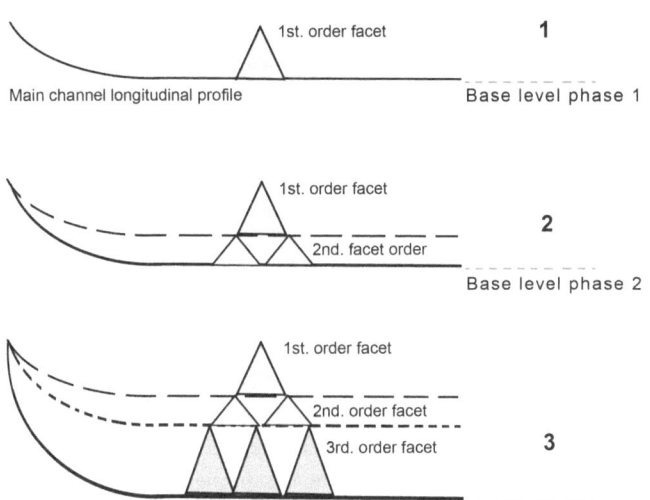

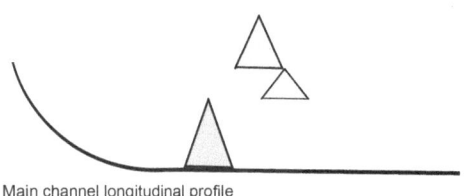

Figure 2

Left: the genetic process of erosion triangular facets; in the scheme, three base level drops generate three orders of facets along the main riverbed. *Right*: in real conditions, some facets may have been erased by subsequent erosion factors or may have not been generated due to unfavourable conditions (from PALIAGA 2004)

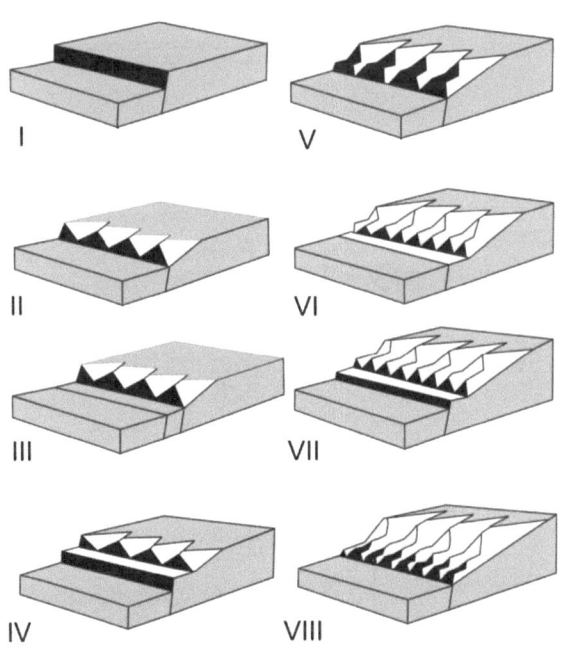

Figure 3

Different stages in the formation and evolution of tectonic facets: in IV and VII, reactivation of the fault generates new landforms (from WIWEGWIN *et al.* 2011)

a result of neo-tectonic activity (S.G.I. 1994). This movement caused the different geomorphologic settings of the Ligurian Sea and Po catchments.

Due to the tectonic and neo-tectonic activity, the area is characterised by fault systems generally oriented N–S and E–W; the river networks are largely set up on the fault systems.

The climate is typically Mediterranean along the coastline, but winter is colder in the hinterland; mean annual precipitation is between about 650 mm along the western coastline and 2,000 mm in the eastern hinterland (ARPAL 2013).

3.2. Bisagno Catchment

This catchment (Table 1) is characterised by strong steepness (Fig. 4) and a maximum altitude that exceeds 1,020 m a.s.l. The catchment orientation is partially parallel and partially orthogonal to the coastline according to the main riverbed setting, probably due to tectonic control. Lithology (Fig. 5) is mainly marl and subordinately slate and shale (S.G.I. 1994; PROVINCIA DI

Table 1

Main features of Bisagno and Varenna catchments (PALIAGA 2004)

Catchment	Surface (km²)	Strahler hierarchical order	Total stream number	Main river length (km)	Drainage density (km⁻¹)
Bisagno	96	5	974	26.6	3.58
Varenna	22	6	656	10.4	6.34

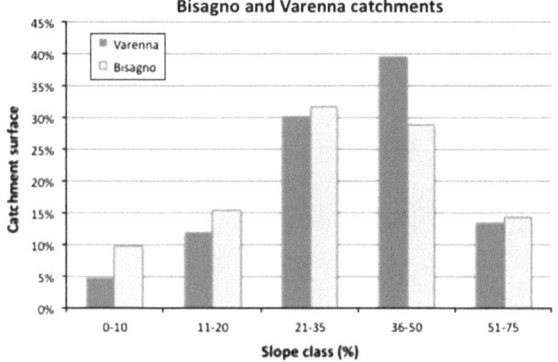

Figure 4
Slope distribution in Bisagno and Varenna catchments

GENOVA 2001a). Many landslides are present in the catchment, mainly in slate and shale substratum. The river network is characterised by a strong hierarchical anomaly (BRANCUCCI and PALIAGA 2005; PALIAGA 2004) mostly due to the network structure that is supposed to be affected by tectonic control; this network structure reflects the reactivation of erosion processes related to the last drop of the base level (BRANCUCCI and PALIAGA 2005; PALIAGA 2004). The main stream is in erosion condition until about 11 km from the basin mouth; in the area corresponding to the deposition regime, the catchment is characterised by strong urbanisation.

3.3. Varenna Catchment

This catchment is somewhat smaller than the Bisagno catchment, but has higher STRAHLER (1957) stream order (Table 1) and more regular shape; the orientation is orthogonal to the coastline, but the conformation shows a high level of asymmetry. The lithology (Fig. 6) is mainly ophiolite and subordinately sedimentary with great heterogeneity compared with the small extension of the basin

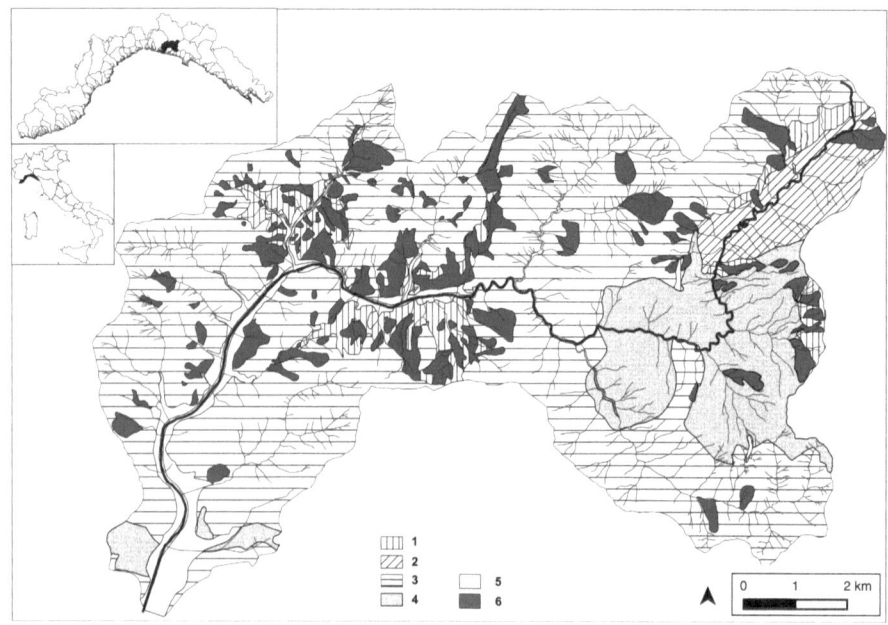

Figure 5
Geological map of Bisagno catchment: *1* shale (Late Cretaceous); *2* slate (Late Cretaceous); *3* marly limestone (Late Cretaceous–Paleocene); *4* marl (Pliocene); *5* alluvial deposits (Quaternary); *6* landslide deposits (modified from PROVINCIA DI GENOVA 2001a)

(S.G.I. 1991; PROVINCIA DI GENOVA 2001b). Two main direct fault systems oriented E–W and N–S have been recognised (MARINI 1987a, b). The river network is characterised by a low level of organization (BRANC-UCCI and PALIAGA 2005; PALIAGA 2004), and erosion processes are quite active; the drainage density value is significantly high.

3.4. Al Ansab Mountains, Oman

Oman's physiographic features (Fig. 7) mainly consist in a large desert area in the middle/west, divided from the coastline at the east by a high mountain belt (maximum altitude 3,004 m a.s.l.) that

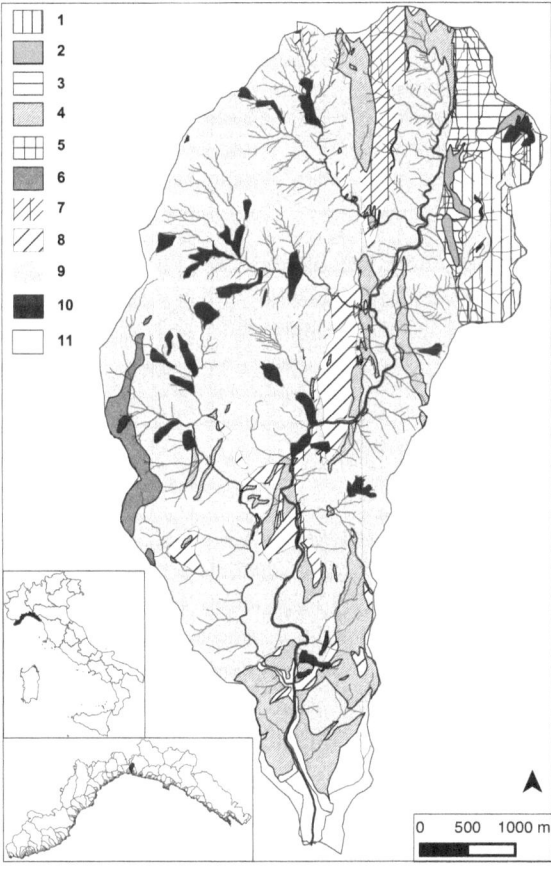

Figure 6
Geological map of Varenna catchment: *1* shale (Early Cretaceous); *2* bioclastic limestone (Late Triassic); *3* limestone (Triassic); *4* prasinite (Late Jurassic–Late Cretaceous); *5* dolomite (Early Triassic); *6* lherzolite (Early Jurassic); *7* metamorphic basalt (Malm); *8* metamorphic gabbro (Malm); *9* ophiolite (Dogger–Malm); *10* landslide deposits; *11* alluvial deposits (Quaternary); modified from PROVINCIA DI GENOVA (2001b)

corresponds to a vast obduction zone. Similarly to the Ligurian territory, Oman's mountainous region is characterised by a complex structural setting.

Three main tectonic events, which may be referred to as the two Hercynian and Alpine orogenic cycles, have generated the regional structures (VILLET *et al*. 1986a, b): a Hercynian event, an Eo-Alpine event and an Alpine (*s.s.*) event. The first one caused block-faulting and large-scale folding, the second took place in three main phases and the latter generated the uplift of large sectors of the studied area located in the Al Ansab catchment, close to Muscat (VILLET *et al*. 1986b). The uplift generated the peculiar morphology of the area with juxtaposition of a large plain (Fig. 7) and strong steep mountains. The lithology of the studied relief is dolomite, and the junction with the floodplain lies on a fault that is part of a system oriented NE–SW (Fig. 8).

The climate is arid and semi-arid with some differences from north to south and from the coastline to the internal desert regions. Precipitation ranges from 20 mm/year in the interior to 300 mm/year in the mountains (FAO 2009); summer is very hot with temperatures reaching 49 °C, while winter is mild.

4. Methods and Tools

Morphometry data in a geographic information systems (GIS) environment and direct survey were used to investigate triangular facet landforms in the studied areas. A 10-m digital elevation model (DEM) for the Bisagno and Varenna catchments was calculated starting from the available 5-m contour lines and spot height data (Regione Liguria, 1:5000–2003). For the Al Ansab Mountains, a 5-m DEM was used (Minister of Regional Municipalities and Water Resources, Sultanate of Oman).

In both cases, a contour map and a plan curvature map of the studied areas were calculated in SAGA GIS; the plan curvature map was filtered, enhancing the higher values that correspond to ridges and watersheds (HENGL and REUTER 2009; BLAGA 2012) (Fig. 9). In addition, a profile curvature map was calculated for all the areas. The two maps were used in connection with a shaded relief map and 5-m

Figure 7
The large floodplain and Al Ansab Mountains (Oman) in the background

contour lines. Erosion facets may be identified even in the DEM as an abrupt cut in main ridges descending to a stream (Fig. 10), while tectonic facets are all equally oriented to the fault system that generated them and can be identified at different heights along the relief (Fig. 11).

This approach enabled identification of triangular facet landforms in all areas, reducing the uncertainty level that would result from a purely subjective interpretation. Final control by high-resolution aerial photography (0.65-m pixel for Ligurian region, Studio SIT; 1-m pixel for Oman area, Minister of Regional Municipalities and Water Resources) was performed, and finally, a vector layer with triangular and trapezoidal facets was created for each studied area.

In addition, for the Ligurian case, a vector layer with a hydrographical Strahler order coded network was used. To perform the spatial distribution analysis of facets, a series of regularly spaced points were identified along the main stream starting from the head for erosion facets; parallel, regularly distanced lines from the main fault were used for tectonic facets.

The following triangular/trapezoidal facets were identified:

134 in Bisagno catchment (Fig. 12)
84 in Varenna catchment (Fig. 13)
85 in Al Ansab Mountains (Fig. 14)

5. Data Analysis and Discussion

The key consideration in analysing facets is their setting on the surface; the spatial distribution reflects the way processes act in generating both erosion and tectonic facets. Erosion facets may be created in different zones of a catchment and by subsequent and distinct erosion cycles; in the theoretical model, a drop of the base level implies acceleration of erosion that generates facets. The process starts from the main river mouth and, regressively, proceeds to the head of the river basin, generating facets as a response to the changed condition of the geomorphic system. The registered variation in the system produces its effects progressively along the whole river basin and is correlated to the hydrographical network structure.

In real conditions, the base level may have oscillated due to sea-level changes and to tectonic

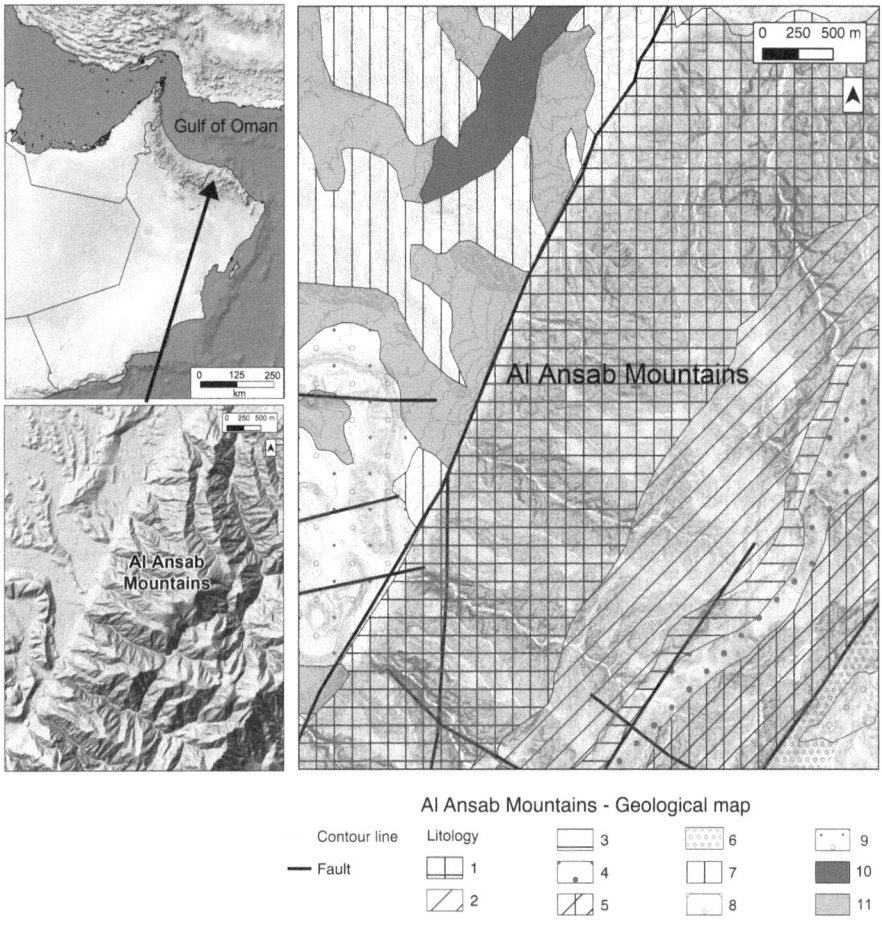

Figure 8

Geological map of Al Ansab Mountains area: *1* grey dolomite (Triassic); *2* black limestone (Late Permian); *3* volcanic rocks (Late Permian); *4* yellow dolomite (Late Permian); *5* dolomite–stromatolites (Late Proterozoic); *6* metasedimentary schist (Late Proterozoic); *7* conglomerate (Tertiary); *8* slope deposits; *9* marl/sandstone (Tertiary); *10* alluvial fans (modified from VILLET *et al.* 1986a, b)

uplift; therefore, landforms are, as normally occurs, the result of a superimposition of the effects of different processes. Moreover, a rise of the base level could even obliterate the lower-altitude facets previously created, while obviously the older the erosion cycle, the greater the possibility that all erosion processes will destroy the facets.

Following SCHEIDEGGER's (1994) theoretical approach, each drop of the base level can determine the rise of facets distributed in the catchment with some fractal features, as an effect of the self-organization of the geomorphic system (PECKHAM 1995; PERRON *et al.* 2009). Erosion is a typical dissipative process, and therefore finding fractal features could provide confirmation of the theory.

Tectonic facets are generated as a response to a sudden change in the geomorphic system that tends to a new dynamic equilibrium condition; erosion, in this case, tends to cause their evolution and, possibly, their destruction. Faulting, which is correlated to other dissipative processes, may occur with a main fault and a series of secondary ones, all of which may generate triangular facets. Moreover, reactivation of faults may cause an imbricate structure to arise (PETIT *et al.* 2009a).

These two facet-generating procedures are strongly different, in terms of both their genetic process and the role played by erosion. Thus, the spatial distribution analysis must take these differences into account; the following procedures described for the two cases were applied.

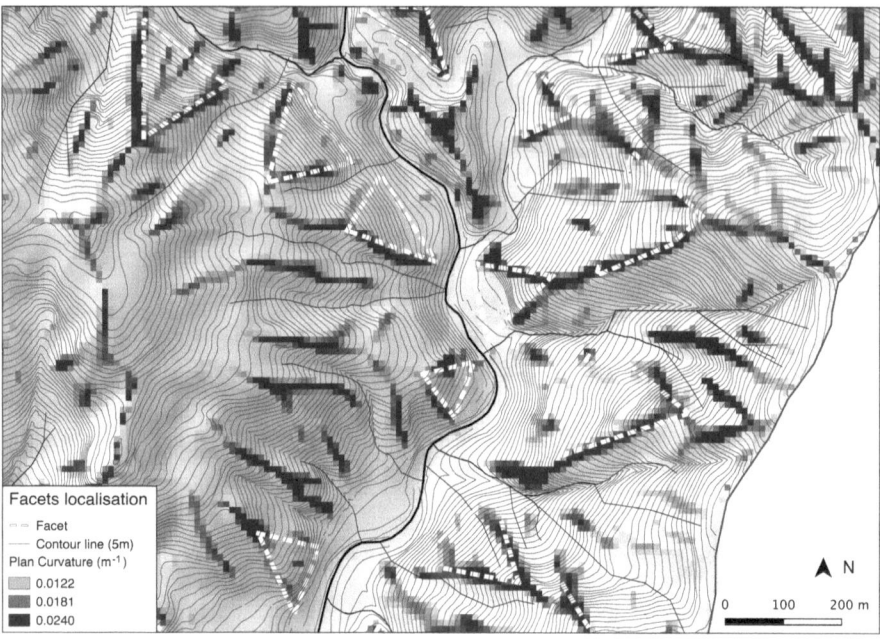

Figure 9
Facet localisation through plan curvature filtering analysis in the Varenna catchment. The higher values in plan curvature, in SAGA GIS calculation, correspond to ridges

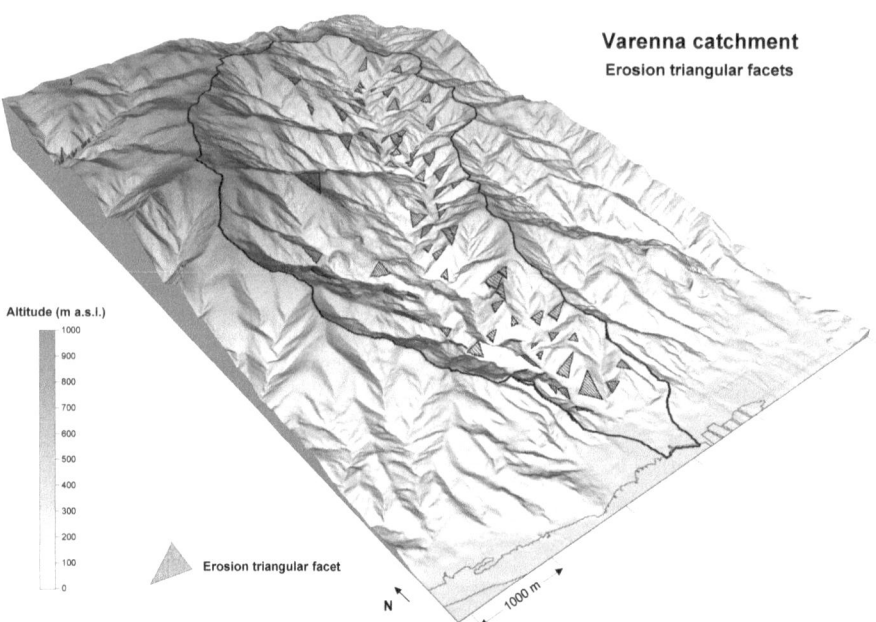

Figure 10
Erosion facet localisation in the 3D model, after analysis of plan and profile curvature

5.1. Erosion Facets

Some regularly spaced points were identified along the main riverbed starting from the head (Figs. 12, 13), and the numbers of facets in the partial catchments corresponding to each point were computed. The cumulative frequency was related to the distance from the riverhead and, according to Eq. (1), the fractal dimension D was computed as

Figure 11
Tectonic facet localisation in the 3D model, after analysis of plan and profile curvature

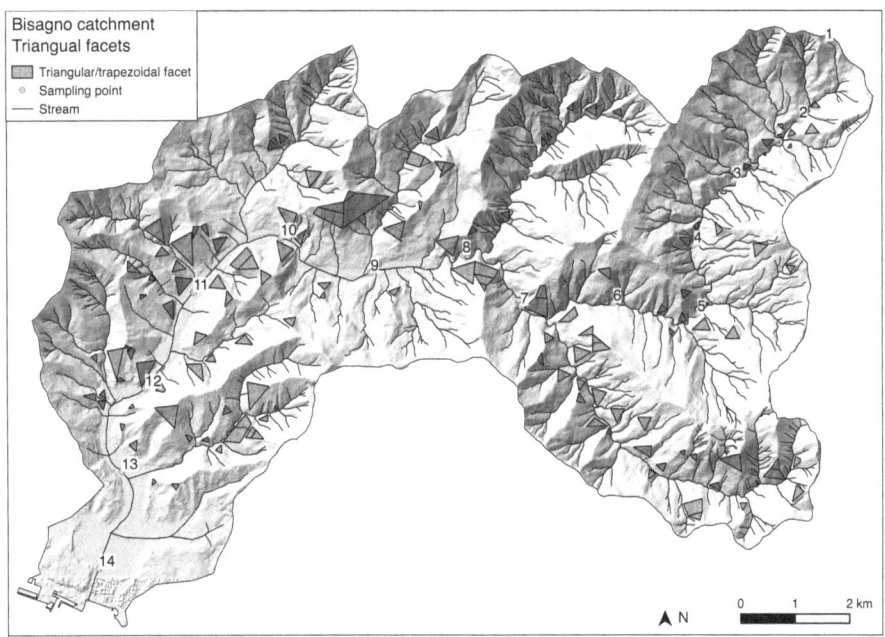

Figure 12
The erosion facets identified in the Bisagno catchment and the equally distanced numbered points (2 km) used to compute the facets in the relative sub-catchments

$$F_n = \frac{c}{r_n^D}, \qquad (1)$$

where F_n is the cumulative number of facets in the sub-catchment identified by point n, r_n is the distance of point n from the riverhead, D is the fractal dimension, and c is a constant. This calculation was adopted considering the facet's genetic mechanism, with the probable occurrence of more erosion cycles and observing their spatial distribution in the catchment.

The different extension of the two studied basins suggested using an interval of 2 km for the Bisagno one and 1 km for the Varenna one.

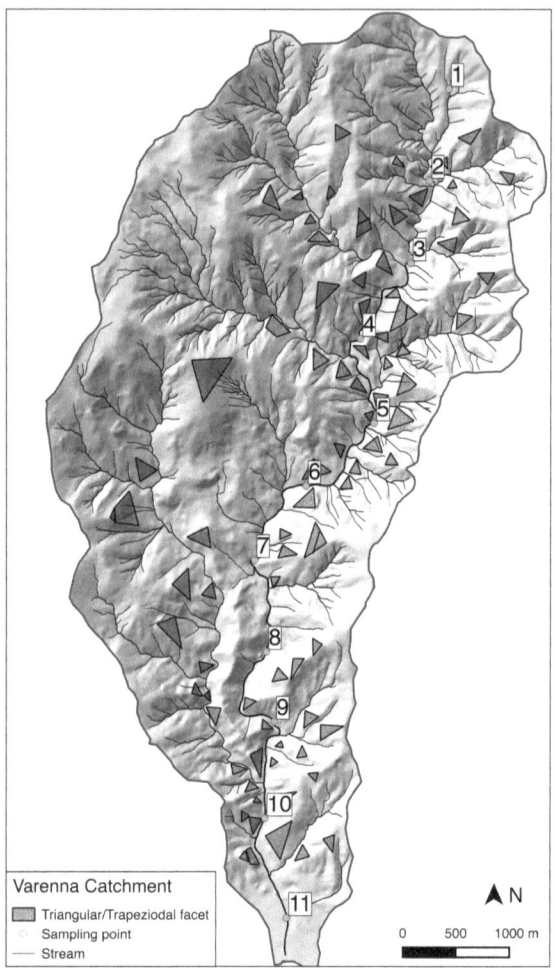

Figure 13
The erosion facets identified in the Varenna catchment and the equally distanced numbered points (1 km) used to compute the number of facets in the relative sub-catchments

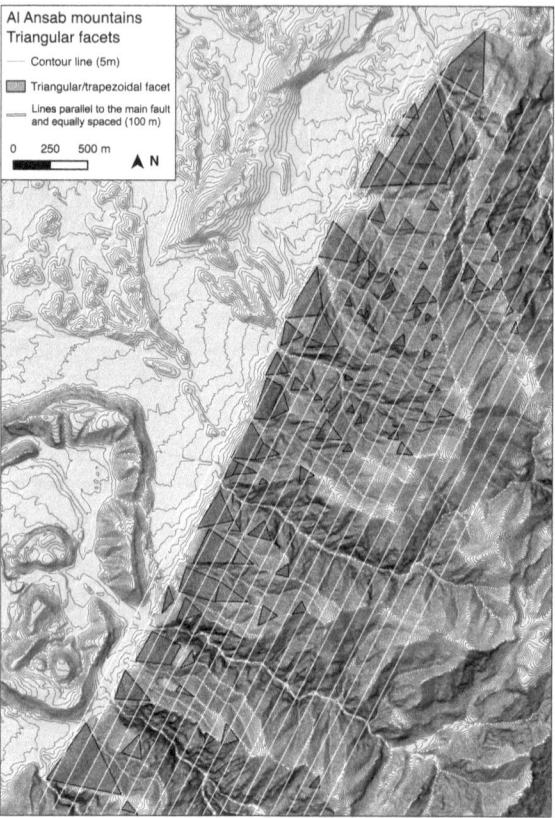

Figure 14
The tectonic facets identified in Al Ansab Mountains and the equally spaced lines parallel to the main fault used to compute their spatial disposition

The results of the spatial distribution analysis seemed to confirm a power-law distribution in the cumulative frequency of the erosion facets for both the Bisagno and Varenna catchments, albeit with some differences:

- Bisagno catchment (Fig. 15): the distribution appears to be characterised by two distinct linear sections. The sudden change could be related to the strong asymmetry of the catchment and to the peculiar hierarchical setting of the river network, i.e. Strahler order 5, and the strong hierarchical anomaly of the river network. Most of the facets lie on marl and, probably, those generated on shale have been erased due to the low resistance of this lithology. The following values were computed (D = fractal dimension):

$$D = 1.816 \quad R^2 = 0.97.$$

- Varenna catchment (Fig. 15): facets are mostly present on high-resistance lithology (ophiolite) and subordinately on others. The distribution appears to be more regular and characterised by the following values:

$$D = 1.182 \quad R^2 = 0.95.$$

5.2. Tectonic Facets

Starting from the trace of the main fault, some parallel lines at intervals of 100 m were drawn

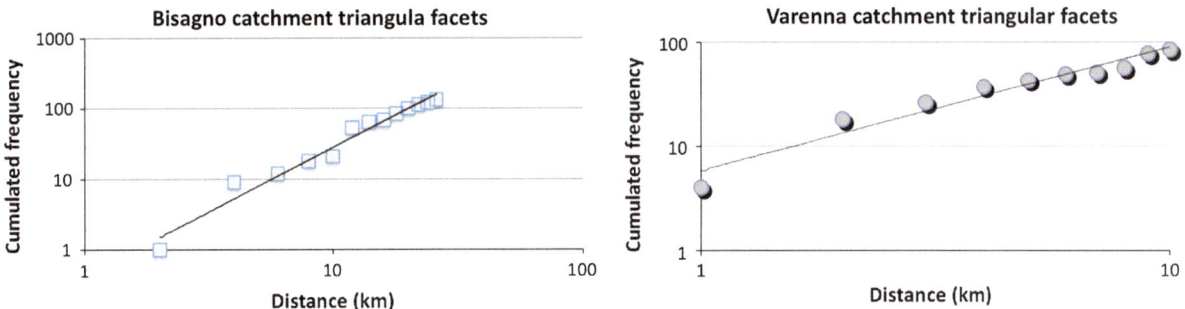

Figure 15
Cumulative number of erosion facets calculated in the sub-catchments identified at different distances from the main riverhead: Bisagno catchment on the *left* and Varenna catchment on the *right*

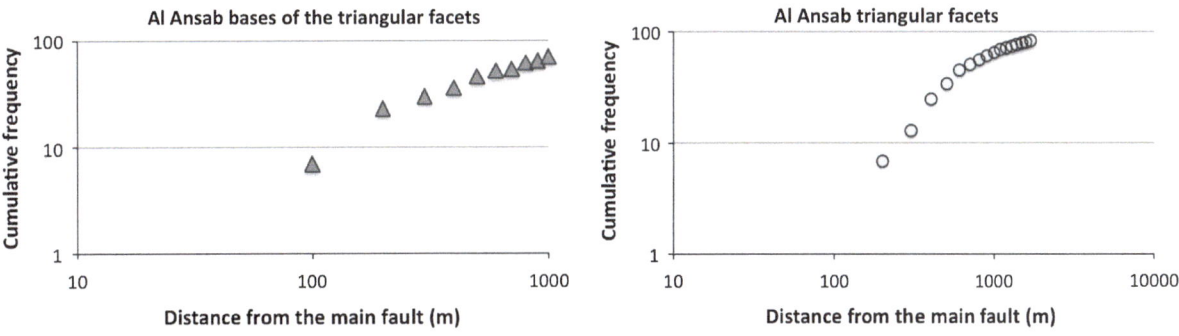

Figure 16
Cumulative number of facets calculated at fixed distances from the main fault: on the *left* computing only the base of the facets, and on the *right* computing the entire facet

(Fig. 14). The last line at a distance of 1.7 km corresponds to the last triangular facet identified. Facets within a fixed distance were computed according to two different methods, i.e. with the whole facet included within the distance, or only its base.

The results are uncertain and more complicated than in the case of erosion facets (Fig. 16). Computing the facets using their base alone, a power law seems to arise from the distribution. In this case, the following values were calculated:

$$D = 0.78 \quad R^2 = 0.933.$$

According to Mazzarella and Tranfaglia (2000), the low level of confidence ($R^2 < 0.95$) forbids assignment of the calculated D value to the fractal dimension of the set, which would be expected to be $D > 1$.

Counting the whole surface of facets comprised within the fixed and equally spaced distances from

the main fault seems to generate two linear sections in the cumulative distribution. For this reason the D value was not computed.

6. Conclusions

The analysis conducted on the identified facets supports a few preliminary conclusions. The spatial dispersion of the erosion facets in the sub-basins corresponding to equally distributed closure sections along the main riverbed appears to exhibit a power-law distribution. Their number is probably related to the hydrographical network structure but reflects the way in which erosion cycles are induced by base level changes affecting the catchment.

For erosion facets, the fluvial incision process is dominant, while tectonic activity acts as a possible triggering effect of base level changes, together with medium sea-level oscillations. Tectonic activity fulfils another effect: in the studied area, it represents a strong constraint on the hydrographical network evolution and thus, indirectly, contributes to the origination and spatial distribution of facets in the catchment. The two investigated catchments are different in dimension, shape, lithology and hydrographical network structure. All these differences may be reflected even in the spatial distribution of facets: the Bisagno catchment is strongly asymmetrical, has a complex setting in respect to the coastline, and its lithology is sedimentary, while the Varenna catchment is orthogonally oriented to the coastline, and its lithology is more various and prominently ophiolite. All these differences may have affected the way in which erosion cycles proceeded in the two catchments and, therefore, the different spatial distributions of facets as shown by the diverse fractal dimensions that emerged from the power-low distributions. Moreover, another difference may be the possible diverse intensity in basement tilting of the two basins.

As the dominant process in generating erosion facets is erosion, the observed power-low distribution may be interpreted following the dissipative system approach theory (SCHEIDEGGER 1994): the fractal feature of facets is a marker of self-organized order on the edge of the chaotic evolution of the geomorphic system. The observed fractal feature should be related to some erosion cycles. Further research efforts could investigate the hierarchical structure of facets, with the purpose of distinguishing them.

The dominant process in generating tectonic facets is faulting; erosion, in this case, contributes to the evolution and degradation of the landform once it has been created. In the studied area, no evidence of fractal features in the spatial distribution was recognised, but one cannot exclude that some other fractal property may arise. In this sense, further research is needed, possibly expanding the study areas.

These results are congruent with those of DE PIPPO et al. (2003) and D'ALESSANDRO et al. (2006); the highest value of the fractal dimension was found for a basin where the rocks are relatively more erodible, while in the Varenna basin the value is lower. Finally, the low value computed for the

tectonic facets is in accordance with the tectogenetic process.

REFERENCES

ARPAL (2013). *Centro funzionale meteoidrologico di Protezione Civile - Atlante climatico della Liguria*. Genova

BAK, P., TANG, C. and WIESENFELD, K. (1988). "*Self-organized criticality*". Phys. Rev. A *38* (1): 364–374.

BLAGA, L. (2012). Aspects regarding the significance of the curvature types and values in the studies of geomorphometry assisted by Gis. Analele UniversităŃii din Oradea–Seria Geografie Year XXII, no. 2/2012 (December), pp. 327–337 ISSN 1454-2749, E-ISSN 2065-1619.

BRANCACCIO, L., CINQUE, A. and SGROSSO, I. (1978), *L'analisi morfologica dei versanti come strumento per la ricostruzione degli eventi neotettonici*, 69° Congr. Soc. Geol. Ital., 3–4 ottobre 1978, Perugia Mem. Soc. Geol. It., *19*, 2–12.

BRANCACCIO, L., CINQUE, A. and SGROSSO, I. (1986), *Elementi morfostrutturali ereditati nel paesaggio dell'Appennino centro-meridionale*, Mem. Soc. Geol. Ital., *35*, 869–874.

BRANCUCCI, G., MARINI, M. (1990a). *Le faccette triangolari: un elemento utile nell'interpretazione morfogenetica delle valli liguri*. Mem. Soc. Geol. Ital., *45*: 893–898.

BRANCUCCI, G., MARINI, M. (1990b). *Stadi evolutivi della val Varenna (Riviera di Ponente, Liguria)*, Boll. Soc. Geol. Ital., *109*: 351–365.

BRANCUCCI, G., PALIAGA, G. (2005). Caratterizzazione *Geomorfica dei principali bacini idrografici della Liguria marittima–risultati preliminari*. Geografia Fisica e Dinamica Quaternaria, suppl. VII, 59–67.

COTTON, C. A. (1950), *Tectonic scarps and fault valleys*, Geol. Soc. Am. Bull., *61*, 717–758, doi:10.1130/0016-760661[717: TSAFV]2.0.CO;2.

D'ALESSANDRO, L., DE PIPPO, T., DONADIO, C., MAZZARELLA, A., and MICCADEI, E. (2006), *Fractal dimension in Italy: a geomorphological key to interpretation*, Zeit. Geom. N. F. *50*(4), 479–499.

DE PIPPO, T., DONADIO, C., MAZZARELLA, A., PAOLILLO, G., and PENNETTA, M. (2003), *Fractal geometry applied to coastal and submarine features*, Zeit. Geom. N. F. *48*(2), 185–199.

FAO AQUASTAT (2009), Water Report *34*.

HENGL, T., REUTER, H.I. (2009). Geomorphometry. Development in soil science, vol. *33*. Elsevier.

KORVIN, G. Fractal models in the earth sciences. Elsevier (1992).

LEOPOLD, L.B., LANGBEIN, W.B. (1962) The concept of entropy in landscape evolution. U.S. Geological Survey Professional Paper 500-A: 1–20.

MARINI, M. (1987a) *Deformazioni fragili e semifragili nella pietra di Finale (Liguria occidentale)*. Mem. Soc. Tosc. Sc. Nat., Ser. A, *93*: 21–55.

MARINI, M. (1987b) *Le deformazioni fragili del Pliocene ligure. Implicazioni nella geodinamica alpina*. Mem. Soc. Geol. Ital. *29*: 157–169.

MAZZARELLA, A., and TRANFAGLIA, G. (2000), *The fractal characterisation of geophysical measuring networks and its implications for an optimal location of additive stations: an application to a rain-gauge network*, Theor. Appl. Climatol. *65*, 157–163.

PALIAGA, G. (2004) Evoluzione morfologica dei principali bacini imbriferi del versante tirrenico ligure. PhD thesis, Università degli studi di Genova, Facoltà di scienze matematiche, fisiche e naturali BN 2005-2988T. Biblioteca Nazionale Centrale di Firenze.

PECKHAM, S.D. (1995), *New results for self-similar trees with applications to river networks*, Water Resour. Res. *31*, 1023–1029.

PERRON, J.T., KIRCHNER, J.W., and DIETRICH, W.E. (2009), *Formation of evenly spaced ridges and valley*, Nat. Lett. *460*, 502–505.

PETIT, C., GUNNELL, Y., GONGA-SAHOLIARILIVA, N., MEYER, B. and SEGUINOT, J. (2009a) *Faceted spurs at normal fault scarps: Insights from numerical modelling*, J. Geophys. Res., Vol. *114*, B05403, doi:10.1029/2008JB005955.

PETIT, C., MEYER, B., GUNNELL, Y., JOLIVET, M., SAN'KOV, V., STRAK, V. and GONGA-SAHOLIARILIVA, N. (2009b). *Height of faceted spurs, a proxy for determining long-term throw rates on normal faults: Evidence from the North Baikal Rift System, Siberia*. Tectonics, vol. *28*, TC6010, doi:10.1029/2009TC002555.

PHILLIPS, J.D. (1992). *The end of equilibrium?*. Geomorphology, *5*: 195–201.

PRIGOGINE, I., STENGERS, I., Order out of chaos (Bantam, New York 1984).

PROVINCIA DI GENOVA, Direzione pianificazione generale e di bacino. Bacino del torrente Bisagno-Piano di bacino stralcio (Genova 2001a).

PROVINCIA DI GENOVA, Direzione pianificazione generale e di bacino. Bacino del torrente Varenna-Piano di bacino stralcio (Genova 2001b).

SCHEIDEGGER, A.E. (1961) Theoretical geomorphology Springer-Verlag. Berlino, Gottinga, Heidelberg.

SCHEIDEGGER, A.E., LANGBEIN, W.B. (1966) Probability concepts in geomorphology Geological Survey Professional Paper 500-c: 1–14, Washington.

SCHEIDEGGER, A.E. (1987) The fundamental principles of landscape evolution. Catena supplement *10*: 199–210. Geomorphological models; theoretical and empirical aspects. Frank Ahnert (editor).

SCHEIDEGGER, A.E. (1992) *Limitations of the system approach in geomorphology*, Geomorphology, *5*: 213–217.

SCHEIDEGGER, A.E. (1994) *Hazards: singularities in geomorphic systems*. Geomorphology, *10*: 19–25.

SELBY, M.J. Earth's changing surface—An introduction to geomorphology. (Clarendon, Oxford, 1985).

SIU-NGAM LAM, N., DE COLA, L. Fractals in Geography. (Prentice-Hall, Englewood Cliffs, 1993).

S.G.I. SOCIETÀ GEOLOGICA ITALIANA ALPI LIGURI. Guide Geologiche Regionali. (BE-MA editrice 1991).

S.G.I. SOCIETÀ GEOLOGICA ITALIANA APPENNINO LIGURE-EMILIANO. Guide Geologiche Regionali. (BE-MA editrice 1994).

STRAHLER, A.N. (1957), *Quantitative analysis of watershed geomorphology*, Trans. Am. Geophys. Union *38*, 913–920.

TURCOTTE, D.L. Fractals and chaos in geology and geophysics. (Cambridge University Press 1992).

VILLET, M., LE METOUR, J., DE GRAMONT, X. (1986a) Explanatory notes of the geological map of Fanjah and map. Sultanate of Oman, Ministry of petroleum and minerals. BRGM.

VILLET, M., DE GRAMONT, X., LE METOUR, J. (1986b) Explanatory notes of the geological map of Sib and map. Sultanate of Oman, Ministry of petroleum and minerals. BRGM.

WIWEGWIN, W., SUGIYAMA, Y., HISADA, K. and CHARUSIRI, P. (2011) *Re-evaluation of the activity of the Thoen Fault in the Lampang Basin, northern Thailand, based on geomorphology and geochronology*. Earth Planets Space, *63*(9), 975–990.

(Received April 28, 2014, revised September 9, 2014, accepted September 18, 2014, Published online October 21, 2014)

Reprinted from the journal

Pure Appl. Geophys. 172 (2015), 1999–2008
© 2014 Springer Basel
DOI 10.1007/s00024-014-0906-8

Pure and Applied Geophysics

Fractal Dimension of Geologically Constrained Crater Populations of Mercury

Paolo Mancinelli,[1] Cristina Pauselli,[1] Diego Perugini,[1] Andrea Lupattelli,[1] and Costanzo Federico[1]

Abstract—Data gathered during the Mariner10 and MESSENGER missions are collated in this paper to classify craters into four geo-chronological units constrained to the geological map produced after MESSENGER's flybys. From the global catalogue, we classify craters, constraining them to the geological information derived from the map. We produce a size frequency distribution (SFD) finding that all crater classes show fractal behaviour: with the number of craters inversely proportional to their diameter, the exponent of the SFD (i.e., the fractal dimension of each class) shows a variation among classes. We discuss this observation as possibly being caused by endogenic and/or exogenic phenomena. Finally, we produce an interpretative scenario where, assuming a constant flux of impactors, the slope variation could be representative of rheological changes in the target materials.

Key words: Mercury's geology, impact processes, primary and secondary crust.

1. Introduction

Over the last few years, new data from the MESSENGER spacecraft (NASA mission) have shed new light on the geological history of Mercury. Endogenic phenomena such as contractional tectonics and volcanism extensively affected the surface of the planet, defining the morphologies observed by Mariner10 and MESSENGER. Exogenic processes such as asteroid impact-induced cratering also altered Mercury's surface, and may have contributed to volcanic activity on the planet, at least in the larger basins.

Data from both Mars and the Moon indicate that the inner solar system is characterised by two populations of impactors: one resulting from the Late Heavy Bombardment (LHB) (Gomes *et al.* 2005; Hartmann *et al.* 1981; Neukum *et al.* 2001; Marchi *et al.* 2009; Minton and Malhotra 2010) which ended about 3.8 Ga ago and was caused by migration of the giant planets (Morbidelli *et al.* 2001; Morbidelli *et al.* 2002; Gomes *et al.* 2005; Marchi *et al.* 2012), and another due to Near Earth Asteroids (NEA) (Bottke *et al.* 2002; Morbidelli *et al.* 2002; Strom *et al.* 2005; Strom *et al.* 2008; Strom *et al.* 2011). Considering that the analysis from Strom *et al.* (2005) obtained the same results for Mercury's highlands and Caloris plains and because of its location in the inner solar system, we expected Mercury's crater population to show the same behaviour on a global-scale analysis. The decay rate from LHB is still considered controversial, but most authors believe that the cratering projectile flux has been more or less constant for the last 3.5 Ga, at least for those impactors with diameters of less than 10 km (Hartmann *et al.* 1981; Neukum *et al.* 2001; Bottke *et al.* 2002; Marchi *et al.* 2009; Minton and Malhotra 2010). The effects of the LHB on a planetary body can also be constructive. In fact, LHB impactors have been hypothesized as providers of volatiles that could have contributed to the enrichment of the terrestrial planets (Trigo-Rodriguez and Martin Torres 2012; Trigo-Rodriguez 2013).

The crater represents the sum of the effects of three impact phases: coupling, excavation and modification (e.g., Melosh 1989; Xiao *et al.* 2014). The first two phases represents the transmission and propagation of the energy of the projectile to the target body, first producing melt and vaporization, then immediately after, the target material is deformed and ejected. Deformation is proportional to the energy of the projectile, is dependent on the target and projectile materials, and decreases with increasing distance from the impact. The result of these two

¹ Dipartimento di Fisica e Geologia, Università degli Studi di Perugia, Via A. Pascoli, 06123 Perugia, Italy. E-mail: pamancinelli@gmail.com; cristina.pauselli@unipg.it; diego.perugini@unipg.it; andrea-lupattelli@alice.it; costanzo.federico@unipg.it

phases is the transient crater. The last phase may last longer than the previous two, and involves the collapse of both the ejected material and the unstable portions of the crater rim, which will partially fill the crater basin. At the end of this phase, the final diameter of the crater is achieved.

Products and morphologies resulting from an impact on the surface of a planet can be used as a proxy to infer the structure and rheology of the planet's upper crust (MASSIRONI et al. 2009; MARCHI et al. 2011), and therefore to verify whether outcropping material is made of primary crust or of stratified igneous bodies and volcanic deposits.

Since the first data produced by the Mariner10, geological mapping of planet Mercury was one of the main goals to achieve in order to understand Mercury's surface evolution. Resolution of mapping increased with the quantity and resolution of data gathered by MESSENGER spacecraft (DENEVI et al. 2009). Crater statistics followed and constrained geological mapping, and were produced by many authors for different purposes (e.g. STROM and NEUKUM 1988; STROM et al. 2005; MARCHI et al. 2011). Despite crater statistics certainly being a valid method to gather information about a planetary surface, and its temporal evolution both in a relative and an absolute sense, global scale statistics were never produced for Mercury. The aim of this work is to present a first analysis of crater statistics on a global scale for the planet Mercury, and to test fractal behaviour of geologically constrained crater populations. In particular, we will present a size frequency distribution (SFD) constrained to geological observations, and discuss possible interpretations of the data, considering the variables that affect impact cratering.

2. Material, Methods and Preliminary Results

The basemap of this work (Fig. 1A) covers the whole planet and contains both new geological information acquired during MESSENGER's third flyby and orbital insertion, and more recently published data on outcropping geology (DENEVI et al. 2009). To achieve the maximum possible coverage, we started from that map and enlarged it in order to cover regions observed after MESSENGER's third

flyby and orbital insertion. To map these areas, we used the same parameters and unit classification as defined by DENEVI et al. 2009. The term 'geological units' will be used in the text to indicate outcropping formation grouped on the base of the reflectance that can be representative of a true mineralogical, and thus geological, change between units. The map was constructed in a GIS environment, by establishing a geological formation and a relative age value (geochronological information) for each deposit, as shown by DENEVI et al. 2009, who classified materials on the basis of reflectance and cratering, and attributed a relative timing of deposition to each unit.

The older unit, called Intermediate Terrains (IT), is representative of heavily cratered old regions. Above these terrains, Low Reflectance Material (LRM) encompasses low reflectance deposits, comprising of craters and basin ejecta. Within these LRM deposits there are a few darker outcrops, containing crater ejecta and which are defined as centres of LRM (LRM_C). The younger units, which are represented by smooth plains ranging from low to high reflectance, are subdivided into two deposits: 1) High Reflectance Plains (HRP) and Intermediate Plains (IP), with intermediate to high reflectance; and 2) Low Reflectance Blue Plains (LBP), which include later volcanic, low reflectance smooth plains. Within the HRP + IP unit, according to the new data from Flyby 3 and orbit insertion (HEAD et al. 2011), some fresh deposits are also introduced. These have been mapped around and inside the Rachmaninoff Basin (Fig. 1A). This new subunit (hereafter referred to as Late Volcanic Deposits; LVD) displays typical reflectance and cratering with respect to all the other units and is made up of smooth deposits. LVD fill topographic depressions, and are considered to be the products of volcanic activity that occurred in the Rachmaninoff region (PROCKTER et al. 2010), where LVD also registered in an anomalous secondary cratering (CHAPMAN et al. 2012). Other than in the South Rachmaninoff basin, other LVD have been mapped in the Firdousi Plains region. The recently observed volcanic floods that occurred in the northern polar region (HEAD et al. 2011) have also been mapped as LVD, in view of their morphological characteristics and of their spectral properties similar to the Firdousi Plains (Fig. 1A, B). However, on a

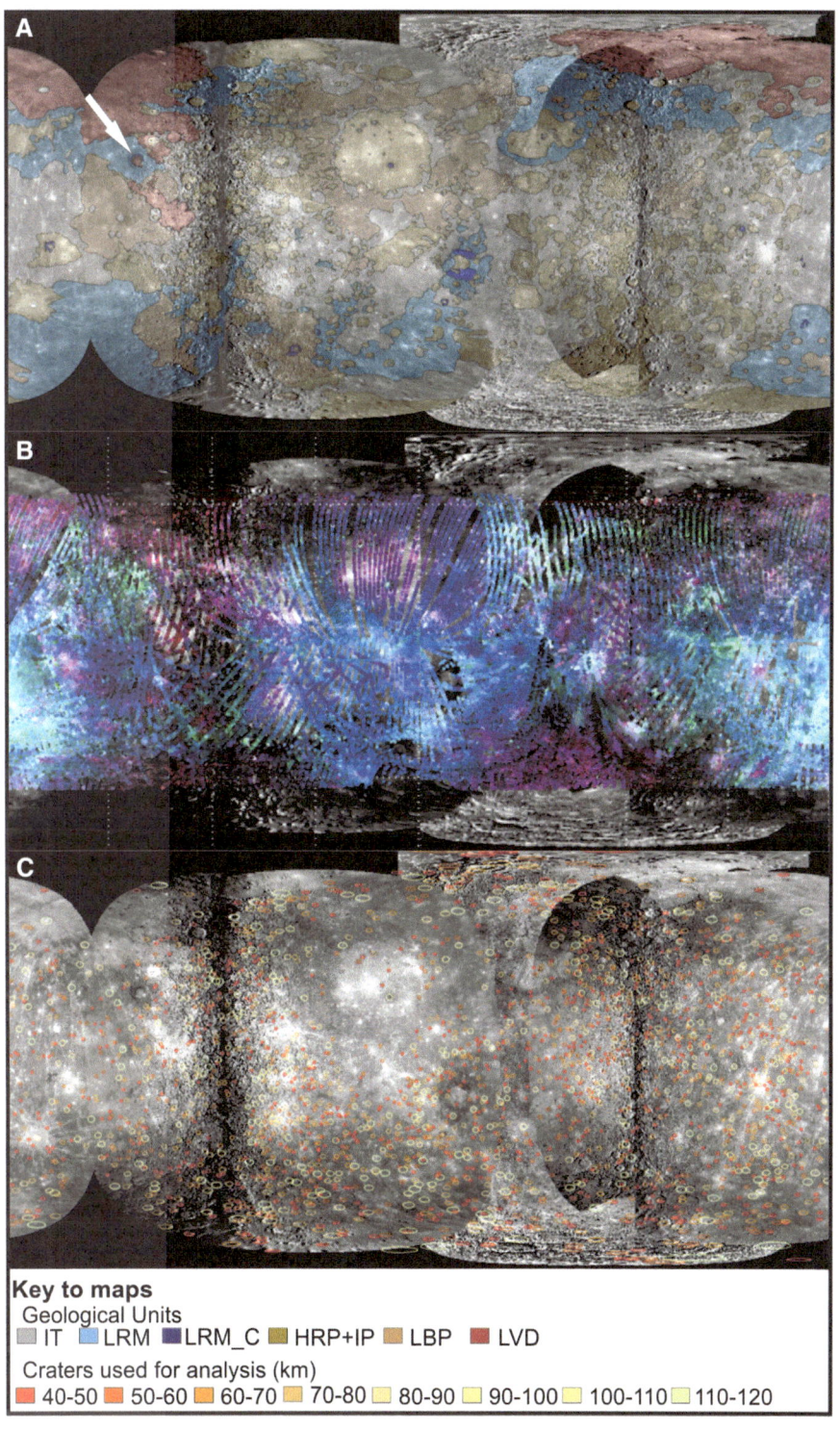

Key to maps

Geological Units

IT LRM LRM_C HRP+IP LBP LVD

Craters used for analysis (km)

40-50 50-60 60-70 70-80 80-90 90-100 100-110 110-120

Reprinted from the journal

◄

Figure 1

A Geological map, modified after DENEVI *et al.* 2009, and picture mosaic after Mariner10 and MESSENGER flybys. IT: Intermediate Terrains, relative Age: IV. LRM: Low Reflectance Material, relative Age: III. LRM_C: Low Reflectance Material Centre, relative Age: III. HRP + IP: Low Reflectance Plains and Intermediate Plains, relative Age: II. LBP: Low reflectance Blue Plains, relative Age: I. LVD: Late Volcanic Deposits, relative age: I [1–0.7 Ga for Rachmaninoff infill (MASSIRONI *et al.* 2009)]. White arrow: Rachmaninoff Basin. B Ultraviolet to near-infrared spectral variations mapped by Mercury Atmospheric and Surface Composition Spectrometer (MASCS) during the MESSENGER orbital phase modified from the file PIA14866 (link to the web page of this image is provided in the references). C Craters used for SFD analysis in Figure 4 and geo-chronological characterization of each crater. Colors show diameter classes from 40 to 120 km [crater data from FASSETT *et al.* (2011)]. Background image resolution used during mapping: 0.7 km/pixel. Equidistant cylindrical projection used on images and sinusoidal equal-area projection for area calculations. GIS software: QuantumGIS 1.7.1

reflectance data, this classification (i.e., units HRP, IP, LBP, LRM, LRM_C and IT) was made by DENEVI *et al.* 2009 according to colour and spectroscopic variations observed and described by previous works using images from the Mercury Dual Imaging System (MDIS) (ROBINSON *et al.* 2008) and data from the Mercury Atmospheric and Surface Composition Spectrometer (MASCS) (McCLINTOCK *et al.* 2008), respectively. This approach is further supported when the geologic map is compared to the mosaic of MASCS observations (Fig. 1A, B).

We assumed that terrains encompassed in the same unit were deposited in a brief time period (i.e., each unit is representative of a time step of the global-scale geological evolution of Mercury). This assumption is supported by the work of MARCHI *et al.* (2013), who found that the planet-scale resurfacing of Mercury was produced between 4.1 and 3.55 Ga ago and was related to LHB, which possibly triggered global-scale volcanic activity. This indicates that all global-scale units as mapped in Fig. 1A were "rapidly" deposited within this time interval, and thus that all deposits encompassed in the same unit were produced in a short time interval.

global scale, we consider LVD as part of the HRP + IP plains, according to previous work that classified the northern plains as similar to the HRP unit (HEAD *et al.* 2011).

Although the validity of this kind of geological mapping may be up for debate, particularly when terrains are classified primarily according to

Figure 2

Overlay of geological map, planetary mosaic and crater catalogue from the region west of Rembrandt basin. *T4* older fourth-order craters, *T3* third-order craters, *T2* second-order craters, *T1* younger first-order craters

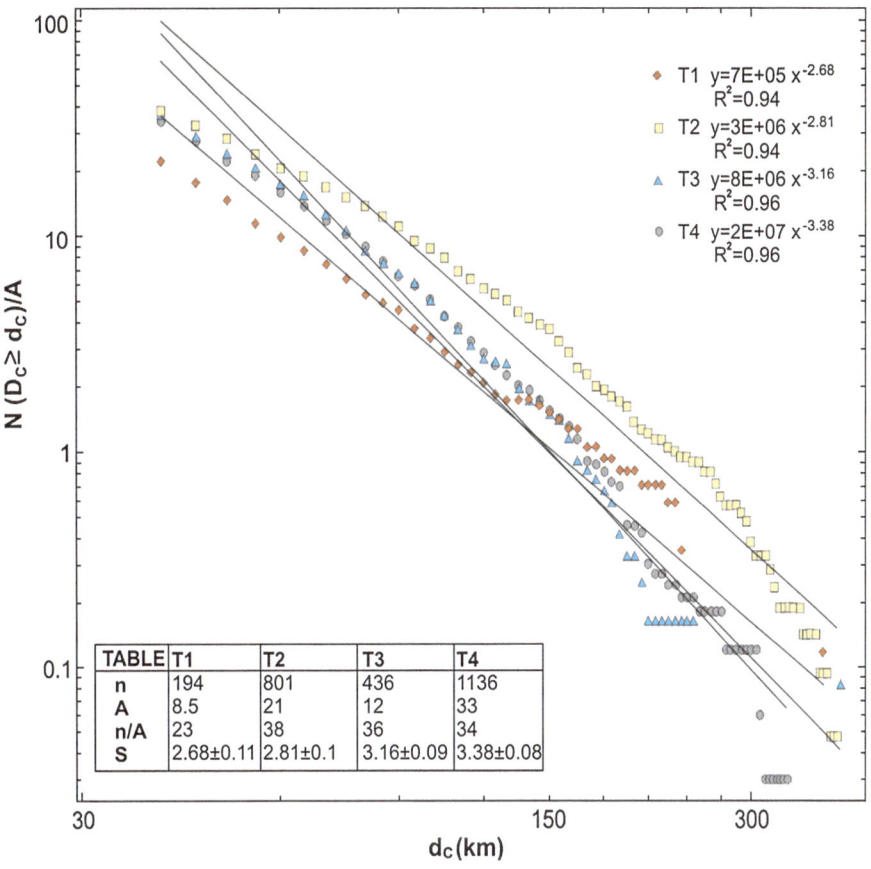

Figure 3

SFD analyses for crater classes defined with geo-chronological constraints (see text for details). Plot is built considering cumulative number (N) of craters with diameter D_C larger or equal than comparative diameter d_C for each class, normalized by area covered by geological unit on the whole planet (A). T1, T2, T3, T4 classes defined in text and in Fig. 2. Table: n/A ratio between number of craters per crater class (n) and A (units in 10E6 km^2). S: exponent of the cumulative SFD for each crater class (i.e., fractal dimension of each class). Classes encompass all observed craters with diameters between 40 and 412 km. Errors reported for the calculation of S are cumulative of errors resulting from linear fit and from data points

Stratigraphic data is more complex to gather from reflectance or spectral observations. Data produced by instruments on board the Mariner10 and MESSENGER missions did not allow for the easy building of a constrained model of the material in the immediate subsurface, also due to uncertainties over the local-scale mineralogical characterisation of outcropping terrains still needing to be clarified. This lack of stratigraphic constraints also has effects on formulating a definition of correct ages for outcropping material. However, the morphological features observed during GIS mapping, constrained by the degree of cratering of each deposit, do allow definition of a depositional order in time.

The above considerations allow us to assign the following relative ages, from the oldest to the youngest formations: Age IV for the IT, Age III for LRM and LRM_C, Age II for HRP, IP and LVD, and Age I for LBP deposits (Fig. 1). These ages are purely indicative of the depositional steps during which the mapped units were deposited from the older IT to the younger LBP. This order is established on the basis of morphological features and cratering. It is impossible to univocally define the absolute timing of the appearance of these units. However, it is clear that, from the beginning of the deposition of the IT unit to the end of the deposition of the LBP unit, a remarkable phase of the geological resurfacing of Mercury is

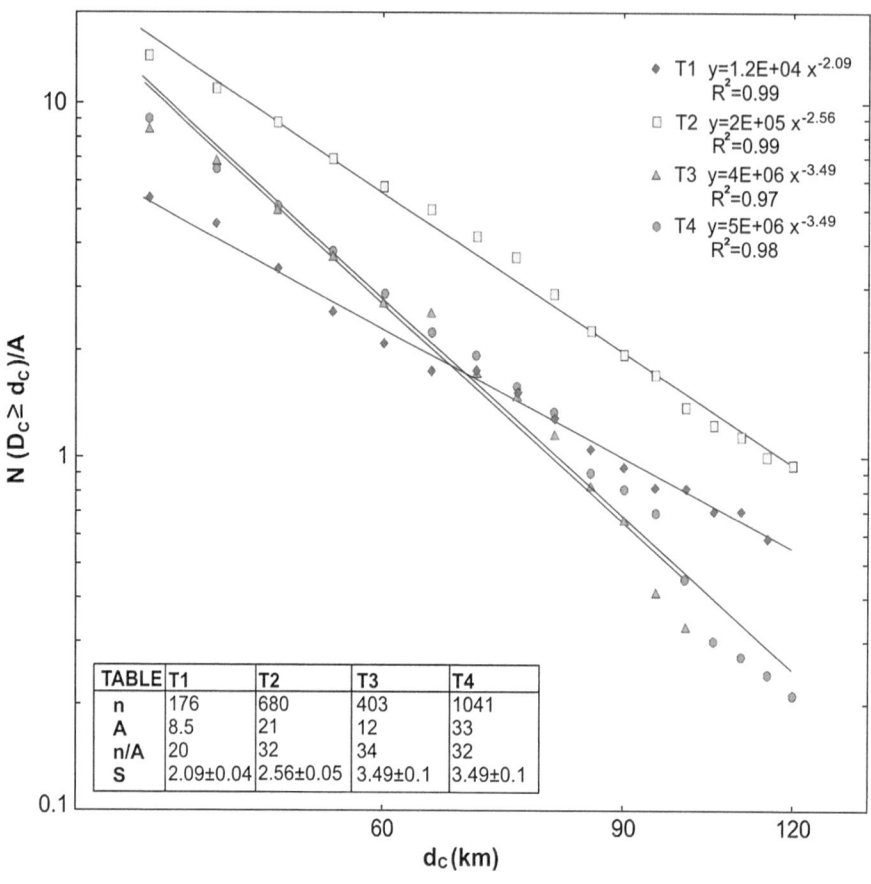

Figure 4
SFD analyses for crater classes defined with geo-chronological constraints and diameters between 40 and 120 km (see text for details). Plot is built considering cumulative number (N) of craters with diameter D_C larger or equal than comparative diameter d_C for each class, normalized by area covered by geological unit on the whole planet (A). T1, T2, T3, T4 classes defined in text and in Figure 2. Table: n/A ratio between number of craters per crater class (n) and A (units in 10E6 km^2). S: exponent of the cumulative SFD for each crater class (i.e., fractal dimension of each class). Errors reported for the calculation of S are cumulative of errors resulting from linear fit and from data points

shown (i.e., a time interval). Another limitation coming from the lack of constrained stratigraphic information is that we cannot locate or estimate units overlaps, either in global or in local-scale analysis.

Analysis of SFD was carried out starting from the recently published crater catalogue (FASSETT et al. 2011). From this catalogue, our analyses do not consider craters with diameter (D_C) smaller than 40 km, or the nine larger basins (i.e., craters with diameter larger than 412 km). The lower bound allows us to exclude the effects of secondary cratering that may produce uncertainties in the results; the upper bound is introduced to reduce redundancy of the major craters on the cumulative plot.

The geo-chronological data derived from the basemap is therefore combined with the position of the centre of each crater. This process organises the entire crater catalogue in four classes using the geo-chronological attributes of the basemap (i.e., out-cropping geology and relative ages). The GIS software automatically carries out this classification, but it is followed by a careful evaluation of each crater, to best characterise its relative age by considering its morphological features, the outcropping unit at its centre, and the morphological relations between this deposit and the surroundings. Based on these criteria, we identify four geo-chronological classes of craters, from the youngest to the oldest: T1, T2, T3 and T4 (Fig. 2).

Some assumptions were made in this process of crater classification: (1) if a crater is surrounded by a unit that is younger than the unit outcropping at its centre, the impact occurred prior to the deposition of the surrounding unit, which never filled the basin produced by the impact; (2) if a crater is filled by a unit younger than the surrounding unit, the impact is deemed to have occurred just before filling, and thus infill material deposition is "coeval" with the impact; (3) if an impact occurred in a region extensively covered by a single unit, the impact is deemed to have occurred soon after unit deposition and thus is "coeval" with the target unit.

From this classification, we obtain the number (n) of craters for each class, and by the re-projection of the map in Fig. 1A to sinusoidal equal area projection, we obtain the area covered by each unit (A). The ratio between n and A represents crater density of craters with diameters between 40 and 412 km. We found a relatively constant number of craters per 10E6 km^2 from T1 to T4; in particular, 34–38 for classes T4–T2 and 23 for T1.

The SFD analysis was performed for each geochronological class. The size-frequency plot $N(D_C \geq d_C)/A$ vs. d_C in Fig. 3 shows that crater classes T1–T4 have a fractal behaviour, with the number of craters inversely proportional to their diameter, and defines different linear trends following scale-invariant distributions, S being the exponent of the cumulative crater SFD (i.e., the fractal dimension of each crater class). The cumulative number of craters (N) is normalised to the area covered by each geological unit to which each crater class refers (A). It is interesting to note that the value of S decreases across classes from T4 (3.38 ± 0.08) to T3 (3.16 ± 0.09), T2 (2.81 ± 0.1) and T1 (2.68 ± 0.11), indicating a change in the fractal behaviour of the population. This change is likely driven by a variation of one or more of the properties affecting the impact (Eq. 1) and involves all the craters of the same class.

To further constrain our analysis and to avoid statistical redundancy of larger craters (in particular for classes T3 and T1), we removed craters with diameters larger than 120 km (Fig. 4). Reducing the upper limit allow us to analyse craters formed by the flux of asteroids of diameter (D_A) within the range from 1–3 to 3–10 km (i.e., $12 \leq D_C/D_A \leq 40$;

MELOSH, 1989; O'KEEFE and AHRENS 1993), and allows us to focus on asteroids considered to have a constant flux toward the terrestrial planet region over the last 3.5 Ga (BOTTKE et al. 2002; MINTON and MALHOTRA 2010). This further constraint permits us to highlight the effects of other parameters, which affect craterization aside from the variation of the flux of impactors (e.g. the variation of rheological properties of target material). Also, in this case, we found a relatively constant number of craters per 10E6 km^2 from T4 to T1; in particular, 32–34 for classes T4–T2 and 20 for T1. SFD still follows different linear trends with S, which is constant for classes T4 and T3 (3.49 ± 0.1) and decreases from classes T2 (2.56 ± 0.05) to T1 (2.09 ± 0.04) (Fig. 4).

3. Discussion

The results of the SFD analysis presented in Fig. 4 can be explained by one or more of the following hypotheses: (1) a change in the flux (number and/or size) of impactors through the Hermean surface, (2) the variable angle of incidence of impactors, (3) varying impact velocities, (4) a different rheological response of the planetary crust to impactors. These results can also represent effects of the overlap of the mapped units. The main limitations pertaining to this global scale approach came from the lack of chronological and stratigraphic constraints on unit deposition. These limitations are more evident and efficient in a global-scale approach than in a local scale approach, because of the wider distribution of deposits and the higher complexity of the geological scenario. Resurfacing events (producing unit overlaps), if present, are certainly a limitation to crater statistics (MICHAEL and NEUKUM 2010): increasing the area of interest for the analysis will increase the probability of including regions where units overlap. This, in turn, means that statistics of the underlying unit will be incomplete if calculated after the deposition of the mantling unit. This problem is certainly significant, and is directly related to the uncertainties concerning the sub-surface geometries and distribution of mapped units. The currently available data do not allow the exclusion of one or more of the hypotheses presented above, but the intent of this paper is to contribute to the debate with an approach to global scale analysis by providing relevant issues for discussion.

The first hypothesis cannot be excluded by the available data regarding the uncertainty of the definition of the age of deposition and the temporal relationships between one unit and the other. However, this analysis shows that the density of cratering [i.e., the ratio between the total number of craters (n) per each geological unit and the relative area coverage (A)] is more or less constant (Figs. 3, 4). In the absence of absolute age data about the deposition of the mapped units, and considering that all units have likely been deposited in a relatively brief time interval (MARCHI et al. 2013), the above observation suggests that during the observed time span or, at least, after the units were deposited, the flux of asteroids across Mercury's surface has undergone negligible variation.

Regarding the second and third hypotheses, the normal velocity component of an impact, depending on both velocity and angle of impact, is likely to directly affect the formation of craters (MELOSH, 1989; MARCHI et al. 2005; HOLSAPPLE and HOUSEN 2007; MASSIRONI et al. 2009: XIAO et al. 2014) in relation to the location of the impact at different latitudes and longitudes. However, the geological units to which classes T1–T4 refer are distributed across the whole planet (Fig. 1A). We believe that this outcome does not allow for the possibility that normal velocity variation could play a key role in generating the observed systematic variation of S from T4 to T1 crater populations, but currently, its effects cannot be precisely quantified.

The diameter of the crater can also be affected by varying only the angle of impact; however, this would not significantly affect the statistics of the whole population, because non-circular craters are only produced for very low impact angles (i.e., $< 10°$) (MELOSH, 1989).

The fourth hypothesis defines the variation in crater size as a function of the properties of the targeted geological unit (MASSIRONI et al. 2009; XIAO et al. 2014). The variation in S may, according to this hypothesis, reflect a change in the rheology of Mercury's upper crust (MARCHI et al. 2012).

The dependence of crater size on the rheology of the target material is given by:

$$R = f\{g, \rho_A, \rho_T, m, a, U, Y\} \tag{1}$$

in which R is the crater's final radius, g the gravitational acceleration, ρ_A and ρ_T the densities of asteroid and target, m the impactor's mass, a the impactor's radius, U the normal velocity component, and Y the strength of target material (HOLSAPPLE and HOUSEN 2007).

From the power-law relationship defining size-frequency distributions (Figs. 3 and 4), we obtain the result:

$$N(D_C \geq d_C)/A = Cd_C^{-S} \tag{2}$$

in which $d_C = 2R$. Thus, the variations of S could be representative of variations in the rheological parameters (i.e., strength and density) of the materials involved in the impact, possibly indicating the presence of a secondary crust and/or a compositional variation in the population of the LHB projectiles (MARCHI et al. 2012; BROŽ et al. 2013). However, we have not found any evidence to support an interpretation of the observed change in S as caused by a cometary origin of part of the LHB impactor population, and this is in agreement with the work of MARCHI et al. (2012).

Considering the results of the SFD for craters with diameters between 40 and 120 km (Fig. 4), if we assume that the SFD of asteroids with diameter between 3 and 10 km during the observed time interval related to the mapped geological units was constant (BOTTKE et al. 2002; MARCHI et al. 2009; MINTON and MALHOTRA 2010), and normal velocity component of the impacts to do not affect significantly the crater populations, the stratified crust model could be supported by some observations:

1. LRM and IT units (i.e., crater classes T3 and T4, respectively) were mapped as very similar deposits (DENEVI et al. 2009), the only difference lying in their colour properties. The slopes of their SFD are equal for craters with $40 \leq D_C \leq 120$ (Table in Fig. 4), suggesting (in a rheology-driven scenario) that the response to the impact was similar for both geological units, and therefore that the rheological properties of the units are similar. This observation is not supported when the upper limit in crater diameter is removed, i.e., considering craters with $40 \leq D_C \leq 412$ (Fig. 3), because

S changes from T4 to T3. However, in this case, *S* is certainly affected by the lack of craters larger than 250 km found in T3, and more generally, by the wider population of impactors.

2. The large number of volcanic plains, covering wide regions of the planet's surface (MALIN, 1976; STROM, 1977; SPUDIS and GUEST, 1988; DENEVI *et al.* 2009; HEAD *et al.* 2011; STOCKSTILL-CAHILL *et al.* 2012; Denevi *et al.* 2013), suggests that resurfacing involved planet-scale areas, and this observation is in agreement with recent works (MARCHI *et al.* 2013).

In this interpretation, the natural conclusion is that Mercury's surface is younger than 3.5 Ga, and therefore postdates the LHB. This observation contrasts with previous works [e.g., models proposing a differentiation in the origin and evolution of Mercury's impactor populations, in comparison with data from the Moon and Mars (STROM *et al.* 2005; STROM *et al.* 2008; STROM *et al.* 2011)]. The contrast might be resolved by considering the oldest unit IT (i.e., crater class T4) to be the target unit on which LHB acted. If this is so, LHB produced a greater number of impacts, and therefore an higher surface alteration within IT deposits, where volcanic activity was more intense (IVANOV *et al.* 2002; HEAD *et al.* 2011), contributing to the deposition of the succeeding younger units in a relatively brief phase after LHB. These units, filling basins previously created by LHB impacts, concealed large numbers of craters and IT deposits, leading to the present state in which, for population T4, the observed crater density is clear-cut. In this way, crater class T4 and therefore IT deposits become the key to definitive proof of the global-scale effects of LHB emplacement on Mercury's surface and crustal evolution (MARCHI *et al.* 2009). However, improvements in the development of the global geological map (with particular regard to older terrains) and of evolutionary models of resurfacing history will certainly contribute to refinement of our model.

4. Conclusions

We conclude that SFD analysis for geologically constrained global-scale crater classes, if properly constrained, will allow one to use crater statistics not only for in situ investigation (referring to a limited area and to a relative/absolute dating of deposits in this area), but also for wider applications. In fact, given the global coverage, these applications can help us to constrain asteroid flux models and global resurfacing models, and improve characterization of target and impactor properties (geological and rheological). New data from MESSENGER and the upcoming BepiColombo mission will help us to more precisely define the geological properties and the chronology of target materials' formation, while high-resolution images will improve the quality of the crater catalogue for low-resolution imaged areas.

Acknowledgments

We would like to dedicate this work to the memory of Angioletta Coradini, who played a leading role in many international scientific projects and was among the founders of space science in Italy. We all miss her enthusiasm and her contributions to planetary sciences. The authors also thank Caleb I. Fassett for making the crater catalogue available and for constructive comments on a preliminary version of this paper. We want to thank the two anonymous reviewers for their constructive and insightful comments, which improved the quality of the manuscript. This work was funded by the ASI-INAF BepiColombo agreement number I/022/10/0.

REFERENCES

BOTTKE Jr. W. F., *et al.* 2002. Debiased orbital and absolute magnitude distribution of Near-Earth Objects. Icarus *156*, 399.

BROŽ M., *et al.* 2013. Constraining the cometary flux through the asteroid belt during the late heavy bombardment. Astronomy and Astrophysics *551*, id. A117.

CHAPMAN C. R., *et al.* 2012. The young inner plains of Mercury's Rachmaninoff basin reconsidered (abstract #1607). 43rd Lunar and Planetary Science Conference.

DENEVI B. W., *et al.* 2009. The evolution of Mercury's crust: a global perspective from MESSENGER. Science *324*, 613.

DENEVI B. W., *et al.* 2013. The Distribution and Origin of Smooth Plains on Mercury. Journal of Geophysical Research - Planets *118*, doi:10.1002/jgre.20075.

FASSETT C. I., KADISH S. J., HEAD J. W., SOLOMON S. C. and STROM R. G. 2011. The global population of large craters on Mercury and comparison with the Moon. Geophysical Research Letters *38*, L10202.

GOMES R., LEVISON H. F., TSIGANIS K. & MORBIDELLI A. 2005. Origin of the cataclysmic Late Heavy Bombardment period of the terrestrial planets. Nature 435, 466-469.

GUEST J. E. and GAULT D. E. 1976. Crater populations in the early history of Mercury. Geophysical Research Letters 3, No. 3.

HARTMANN W. K., et al. 1981. Chronology of Planetary Volcanism by Comparative Studies of Planetary Craters, Basaltic Volcanism on the Terrestrial Planets, Pergamon Press, Elmsford, NY.

HEAD J. W., et al. 2011. Flood volcanism in the northern high latitudes of Mercury revealed by MESSENGER. Science 333, 1853.

HOLSAPPLE K. A. and HOUSEN K. R. 2007. A crater and its ejecta: An interpretation of Deep Impact. Icarus 187, 345.

IVANOV B. A., et al. 2002. Asteroids III. (University of Arizona Press, Tucson, AZ), volume 1, pp. 89-101.

MALIN M. C. 1976. Observations of intercrater plains on Mercury. Geophysical Research Letters 3, 581–584.

MARCHI S., MORBIDELLI A. and CREMONESE G. 2005. Flux of meteoroid impacts on Mercury. Astronomy and Astrophysics 431, 1123.

MARCHI S., MOTTOLA S., CREMONESE G., MASSIRONI M. and MARTELLATO E. 2009. A new chronology for the Moon and Mercury. The Astronomical Journal 137, 4936.

MARCHI S., et al. 2011. The effects of the target material properties and layering on the crater chronology: The case of Raditladi and Rachmaninoff basins on Mercury. Planetary and Space Science doi: 10.1016/j.pss.2011.06.007.

MARCHI S., BOTTKE W. F., KRING D. A. and MORBIDELLI A. 2012. The onset of the lunar cataclysm as recorded in its ancient crater populations. Earth and Planetary Science Letters 325, 27-38.

MARCHI S., et al. 2013. Global resurfacing of Mercury 4.0-4.1 billion years ago by heavy bombardment and volcanism. Nature 499, 59.

MASSIRONI M., et al. 2009. Mercury's geochronology revised by applying Model Production Function to Mariner 10 data: Geological implications. Geophysical Research Letters 36, L21204.

McCLINTOCK W. E., et al. 2008. Spectroscopic observations of Mercury's surface reflectance during MESSENGER's first Mercury flyby. Science 321, 62.

MICHAEL G. G. and NEUKUM G. 2010. Planetary surface dating from crater size-frequency distribution measurements: partial resurfacing events and statistical age uncertainty. Earth and Planetary Science Letters 294, 223-229.

MINTON D. A. and MALHOTRA R. 2010. Dynamical erosion of the asteroid belt and implications for large impacts in the inner Solar System. Icarus 207, 744.

MELOSH H. J. 1989. Impact cratering – A geologic process. Oxford monographs on geology and geophysics NO.11 (Oxford University Press, New York, NY).

MORBIDELLI A., PETIT J. M., GLADMAN B. and CHAMBERS J. 2001. A plausible cause of the late heavy bombardment. Meteoritics and Planetary Science 36, 371-380.

MORBIDELLI A., BOTTKE W. F., FROESCHLÉ C. and MICHEL P. 2002. Origin and evolution of Near-earth Objects. In Asteroids III, Bottke W. F., Cellino A., Paolicchi P. and Binzel R.P. (eds.), The University of Arizona Press, Tucson, AZ, USA, pp. 409-422.

NEUKUM G., IVANOV B. A., HARTMANN W. K. 2001. Cratering records in the inner solar system in relation to the lunar reference system. Chronology and Evolution of Mars, 96, 55–86, Kluwer Academic Publishers.

O'KEEFE J. D. and AHRENS T. J. 1993. Planetary cratering mechanics. Journal of Geophysical Research 98, 17,011.

PIA14866 is taken from the NASA/JHUAPL/Carnegie Institution MESSENGER photo journal, (http://photojournal.jpl.nasa.gov/catalog/PIA14866).

PROCKTER L. M., et al. 2010. Evidence for young volcanism on Mercury from the third MESSENGER flyby. Science 329, 668.

ROBINSON M. S., et al. 2008. Reflectance and Color Variations on Mercury: Regolith Processes and compositional heterogeneity. Science 321, 66.

SPUDIS P. D., and GUEST J. E. 1988. Stratigraphy and geologic history of Mercury, in Mercury, edited by F. Vilas, C. R. Chapman, and M. S. Matthews, pp. 118–164, University of Arizona Press, Tucson, Ariz.

STOCKSTILL-CAHILL K. R., et al. 2012. Magnesium-rich crustal compositions on Mercury: Implications for magmatism from petrologic modeling. Journal of Geophysical Research, 117, E00L15, doi:10.1029/2012JE004140.

STROM R. G. 1977. Origin and relative age of lunar and mercurian intercrater plains. Physics of the Earth and Planetary Interiors, 15, 156–172.

STROM R. G., and NEUKUM G. 1988. 'The Cratering Record on Mercury and the Origin of Impacting Objects', in F. Vilas, C.R. Chapman, and M.S. Matthews (eds.), Mercury, Univ. Arizona Press, Tucson, pp. 336–373.

STROM R. G., MALHOTRA R., ITO T., YOSHIDA F. and KRING D. A. 2005. The Origin of Planetary Impactors in the Inner Solar System. Science 309, 1847.

STROM R.G., CHAPMAN C. R., MERLINE W. J., SOLOMON S. C. and HEAD J. W. 2008. Mercury Cratering Record Viewed from MESSENGER's First Flyby. Science 321, 79.

STROM R. G. et al. 2011. Mercury crater statistics from MESSENGER flybys: Implications for stratigraphy and resurfacing history. Planetary and Space Science 59, 1960.

TRIGO-RODRIGUEZ J. M. and MARTÍN TORRES F. J. 2012. Clues on the importance of comets in the origin and evolution of the atmospheres of Titan and Earth. Planetary and Space Science 60, 3-9.

TRIGO-RODRIGUEZ J. M. 2013. Nitrogen in Solar System Minor Bodies: delivery pathways to primeval Earth. In The Early Evolution of the Atmospheres of Terrestrial Planets, Trigo-Rodríguez J. M. et al. (eds.), Springer, New York, 9-22.

XIAO Z., et al. 2014. Comparisons of fresh complex impact craters on Mercury and the Moon: Implications for controlling factors in impact excavation processes. Icarus 228, 260.

(Received February 7, 2014, revised May 8, 2014, accepted July 11, 2014, Published online August 7, 2014)

Pure Appl. Geophys. 172 (2015), 2009–2023
© 2014 Springer Basel
DOI 10.1007/s00024-014-0922-8

▌Pure and Applied Geophysics

Fractal Geometry-Based Quantification of Shock-Induced Rock Fragmentation in and around an Impact Crater

MD. Sakawat Hossain[1,2] and Jörn H. Kruhl[1]

Abstract—Shock-induced fragmentation structures of basement rocks and their limestone cover in and around the Ries impact crater (Germany) were recorded on outcrop, hand sample, and thin-section scale, and quantified mainly by fractal geometry methods. Quantification was performed by automated procedures and in areas of square-centimetres to square-decametres with a maximum resolution of micrometre scale. In 2D and on all scales, the fragmentation structures form complex, statistically self-similar patterns (fractals) with specific characteristics: (i) The pattern fractality is scale-dependent. (ii) Three different power-law relationships exist, which reflect the effect of three fragmentation processes. (iii) The fracture patterns are anisotropic and inhomogeneous over larger areas. (iv) Complexity and anisotropy of the fracture patterns vary systematically. Such systematic variation appears typical for impact-related fragmentation.

Key words: Impact, Fragmentation, Fracture pattern, Fractal geometry, Inhomogeneity, Anisotropy.

1. Introduction

Impact fragmentation is mainly caused by a primary shock and secondary refraction waves propagating through the rocks. Other causes include elastic rebound of the transient crater floor and gravitational collapse of the transient crater wall. Different fracture patterns have been recorded based on (i) field observations (Dalwigk 2003; Kumar and Kring 2008), (ii) cratering experiments (Buhl *et al.* 2013; Polanskey and Ahrens 1990), and (iii)

numerical modelling (Collins *et al.* 2004; Melosh and Ivanov 1999; Wünnemann and Ivanov 2003). These studies show that stress, strain, and strain rate are highest near the impact site and decrease with radial distance. Thus, shock-related shattering and fracturing of the target rocks rapidly decrease from the impact site outwards. The target rocks are intensively fractured from a micro- to a kilometre scale. Large-scale fractures with vertical to sub-vertical orientations and little or no displacement are dominant. Overall, the fractures form complex patterns, statistically self-similar (fractal) over a range of scales that can be best analysed by fractal-geometry techniques (Kaye 1993; Mandelbrot 1982). The quantitative aspects of such patterns are mostly represented by pattern anisotropy, inhomogeneity, and scaling behaviour (Kruhl 2013).

Quantification of fracture patterns and fragment size distributions (FSD) are increasingly important in investigations of stress fields (Chao and Kaneko 2004; Kaushik 2007; Leterrier *et al.* 2001; Quinta *et al.* 2012; Zhao *et al.* 2009), rock strength and deformability (Feng *et al.* 2009) and strain rate of deformation (Buhl *et al.* 2013). Fracture patterns may originate from different processes of fragmentation, for example, ballistic fragmentation or fragmentation in a slowly increasing stress field (Kaye 1989). FSD in fault rocks show variations in fractal dimensions, interpreted as resulting from different fragmentation processes, the number of events, energy input, confining pressure, selective fracturing of larger particles and, in certain situations, alteration of particles (Blenkinsop 1991). Shock experiments demonstrate that dynamic fragmentation influences the FSD and can be expressed as fractal (Buhl *et al.* 2013). Therefore, quantification of fracture patterns as well

Electronic supplementary material The online version of this article (doi:10.1007/s00024-014-0922-8) contains supplementary material, which is available to authorized users.

[1] Tectonics and Material Fabrics Section, Faculty of Civil, Geo and Environmental Engineering, Technical University Munich, 80333 Munich, Germany. E-mail: sakawat.hossain@tum.de

[2] Department of Geological Sciences, Jahangirnagar University, 1342 Dhaka, Bangladesh.

Reprinted from the journal

as fragment sizes are an important tools for getting deeper insight into impact-related fragmentation processes and their interaction.

The Ries meteorite crater (Nördlingen, Germany; Fig. 1) is one of the best preserved craters in the world and has been studied from various geological and geophysical perspectives. About 14.6 ± 0.1 million years ago (SCHWARZ and LIPPOLT 2013), an oblique impact of an approximately 1.5 km diameter achondritic meteorite (STÖFFLER et al. 2002) formed a crater of approximately 25 km diameter. The meteorite impacted a terrane that consisted of a 620–750 m thick sub-horizontally layered sequence of water-saturated sedimentary rocks of Triassic to Jurassic age, resting on basement rocks that experienced their last deformation and metamorphism about 320 million years ago (GRAUP 1977; HÜTTNER and SCHMIDT-KALER 1999). Prior to the impact, the sedimentary rocks did not undergo any deformation except for late-Eocene subsidence and mid-Miocene gentle uplift (PLANT et al. 2003; STRASSER et al. 2009).

The Ries meteorite crater is a complex impact crater with three distinct parts: centre, inner crater ring, and outer crater rim (STÖFFLER 1977). The inner ring boundary (Fig. 1) is mostly formed by up- and outward displacement of the crystalline basement. Between the inner ring and the outer crater rim is a megablock zone characterized by concentrically faulted separate blocks. The boundary of the outer rim is the structural margin of the Ries crater. Based on different fractal-geometry methods, our paper presents quantification of the inhomogeneity and anisotropy of shock-induced fracture patterns and their FSD. The results of FSD are discussed and compared with tectonically and experimentally produced FSD from other studies.

2. Materials, Methods, and Measurements

For fracture-pattern quantification, we photographed a vertical face in a quarry at the outer crater

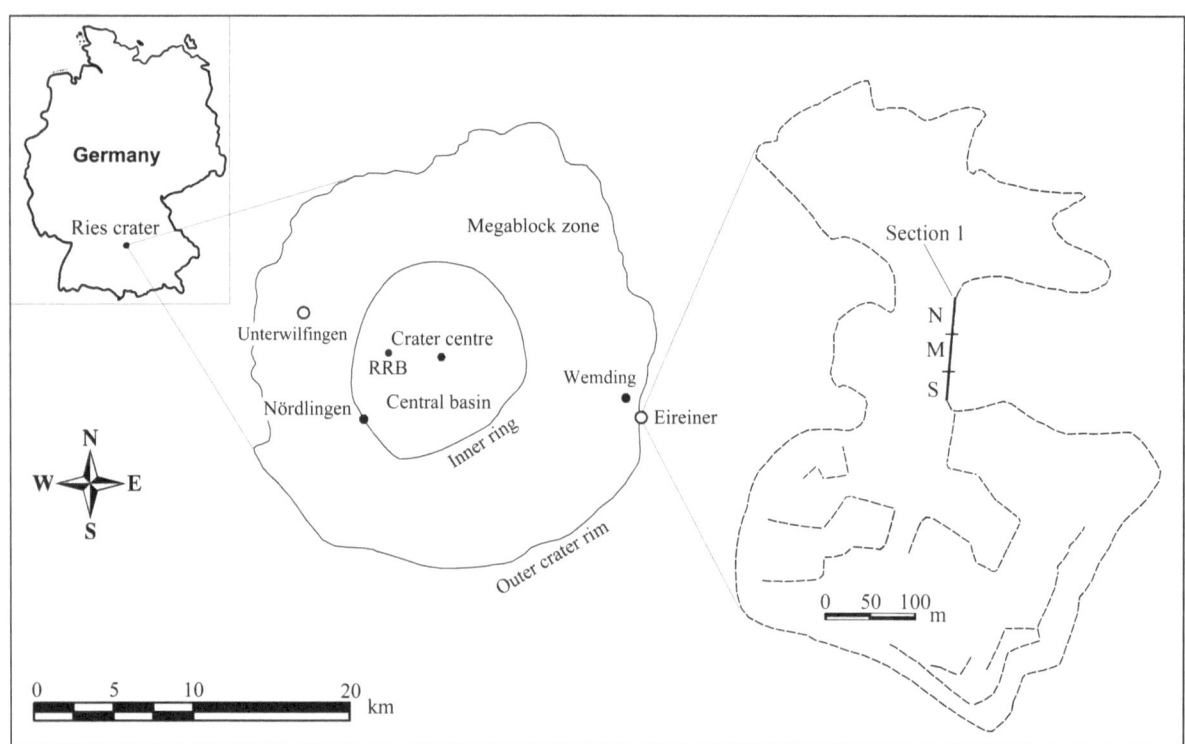

Figure 1
Ries impact area with crater outline, crater centre, location of Ries Research Borehole (RRB), boulder sample location (Unterwilfingen), and position of Eireiner quarry. The *dashed line* indicates the boundary of the quarry and quarry faces. Total length of section-1 (quarry wall), 114.03 m; average height, 9 m. *N*, *M*, and *S* are the north, middle, and south parts of the section, used for quantification

rim ESE of the crater centre (Fig. 1; the Eireiner quarry, section-1, approximately 45 m below the surface). The quarry is located in the 140–150 Ma old Upper Malm limestone (HÜTTNER and SCHMIDT-KALER 1999). We also quantified fracture patterns in core samples from the 1973 Ries Research Bore Hole (RRB) and patterns in two strongly fragmented boulders from the Unterwilfingen quarry (Fig. 1).

In the Eireiner quarry, 40 digital photographs of section-1 were taken from approximately the same distance and subsequently merged to an image representing a 117.35 m length of the quarry face. When taking the photographs, we used a medium focal length in order to minimize distortion. Other steps taken to minimize distortion, included overlapping the vertical sides of the photographs by approximately 35 % and cutting 5–10 % off the top and bottom of each photograph. Partly covered fractures at the base of the section and incomplete fractures close to the edges of the photographs were not digitized, except for a few (<15) large fractures that went out of the photographs. Some edge effects are present because of the uneven rock surface and fractures that were not fully exposed on the photographs. The full range of fractures in all orientations can only be measured in three orthogonal sections, with additional statistical corrections, which was not possible in this study owing to the limited rock exposures.

Individual fractures in the image were digitized manually using ArcMap10 and converted into a binary pattern representing a 114.03 m length (Fig. 2). Digitization and subsequent processing with Photoshop left a few isolated, randomly distributed pixels inside the pattern ('salt and pepper noise'), which were removed by a Matlab image processing algorithm (low pass and median filtering). Binary images of the boulder and core samples were processed in the same manner. Subsequently, the binary pattern from the quarry wall photographs was divided into 27 smaller parts for more convenient application of fractal-geometry methods. The minimum resolution of the analysis was about 5 mm. The steps undertaken from image acquisition to final processing are illustrated in supplementary Figure 1.

As mentioned, we also examined two boulders and six core samples. The two non-oriented boulders (limestone and granite) are each about 60 × 50 × 45 cm in size. Based on their lithology and location in the megablock zone, we infer that they are from that zone. The six core samples were obtained from the granitoid parts of the RRB core (GRAUP 1977; GUDDEN 1974), each 10 cm in diameters and 10–45 cm long. Two saw cuts parallel and perpendicular to the long-axes of the boulders and core samples were ground, polished and scanned with a

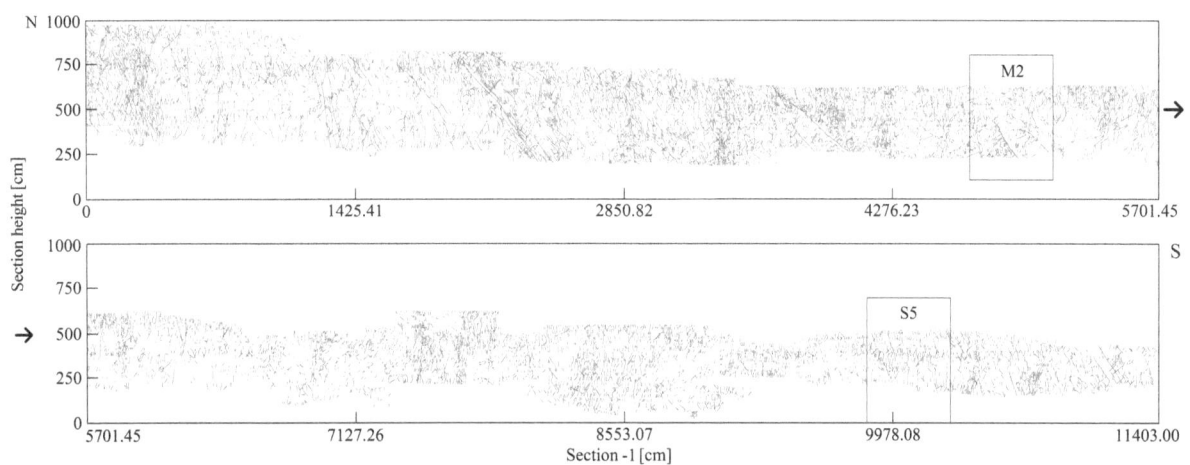

Figure 2

Binary image of fractures in section-1 of the Eireiner quarry after merging, digitizing and processing 40 high-resolution digital photographs. Each photograph imaged an area of roughly 4.6 × 6.9 m. One pixel in the binary image = 1.98 mm on the ground. The fractures were manually drawn on the basis of the photographs. The N, M, and S areas (Fig. 1) were divided into 10, 9, and 8 sub-sections, respectively. A total of 27 sub-sections (called N1, M2, S5, etc.) were produced. M2 and S5, locations of two sub-sections resolution of 1,200 dpi. In addition, thin sections from the core samples were scanned with 4,200 dpi resolution. The fracture patterns from all saw cuts and thin sections were manually digitized using Arc-Map10 and converted into binary images. The lower limit of the resolution is about 0.1 mm for the core samples and boulder rock samples and about 0.01 mm for the thin sections. In the boulder and core samples, fracture width is about 0.1 mm. In core thin sections, the fracture width is about 0.015 mm.

Fractality of the fracture patterns of all the 27 sub-sections was tested by 2D box counting (MANDELBROT 1982), using the Benoit® software. Because box counting is sensitive with respect to both the lower limit of resolution and the total area of analysis, larger and smaller box sides in a range of 150–0.6 cm were chosen, clearly far from the upper and lower limits of pattern resolution.

The possible inhomogeneity of the fracture pattern of section-1 was investigated by map counting (PETERNELL et al. 2010), where box counting is applied to a rectangular window that passes over the image and generates fractal box-counting dimensions for each window. These values are plotted as colour pixels in the centre of the windows. When counting is completed, a colour-coded map is produced on the basis of the distribution of all pixels. In this study, a window of 100×100 cm was used with a gliding distance of 25 cm in each step (i.e. with 75 % overlap).

The possible anisotropy of the fracture pattern and its spatial variation were explored by the modified Cantor dust method (MCDM) (VELDE et al. 1990; VOLLAND and KRUHL 2004) and by mapping of rock fabric anisotropy (MORFA) (PETERNELL et al. 2011). MCDM was performed with the AMOCADO software (GERIK and KRUHL 2009), an automated procedure that measures the 1D complexity of a pattern in different directions (i.e. the anisotropy of pattern complexity). MORFA combines MCDM with a gliding-window procedure, and thus quantifies the inhomogeneity of the anisotropy of the pattern complexity. Details of the method can be found in PETERNELL et al. (2011).

In boulders and core samples, fracture patterns analysed by box counting and FSD were determined and presented as cumulative frequencies on log–log plots. The slopes of the linear point arrangements are taken as fractal dimensions of the FSD. The computer program crack image analysis system (CIAS) (LIU et al. 2013) was used to calculate the geometric parameters and statistical analysis of the fragments in the binary image of boulders and core samples. The lower limit of the CIAS resolution is about five pixels, which includes about 98.5 % of the total number of fragments.

3. Results

3.1. Large-Scale Fracture Patterns

The larger fractures exposed in the Eireiner Quarry are dominantly ESE–WNW and NNE–SSW fractures, i.e. fractures oriented radially and tangentially to the centre of the Ries crater. These two orientations also determine the partly rectangular plan of the quarry (Fig. 1). Based on the outcrop conditions, the fractures could be recorded in sufficient numbers and with sufficient quality only on a roughly N–S oriented quarry face (section-1); therefore, the analysed fractures are generally radial to the crater centre. They are typically visible over distances of centimetre to decimetres. The few fractures of more than 1 m in length usually extend beyond the section-1 boundary. The fracture patterns were analysed by four fractal-geometry methods in order to identify pattern fractality and its inhomogeneity as well as inhomogeneity of anisotropy.

3.1.1 Fractality of Fracture Pattern

Application of the box-counting method resulted in two linear groups of data points whose regression lines clearly have different slopes (Fig. 3). For all 27 sub-sections, the switch from one regression line to the other occurs between the 10 and 20 cm box sizes. The slopes of the two regression lines, i.e. the fractal dimensions D_1 and D_2, of all sub-sections are 1.03–1.15 and 1.40–1.85, respectively. Along the section, D_2 varies strongly and systematically (Fig. 4). The same is true for D_1 (shown in supplementary Figure 2).

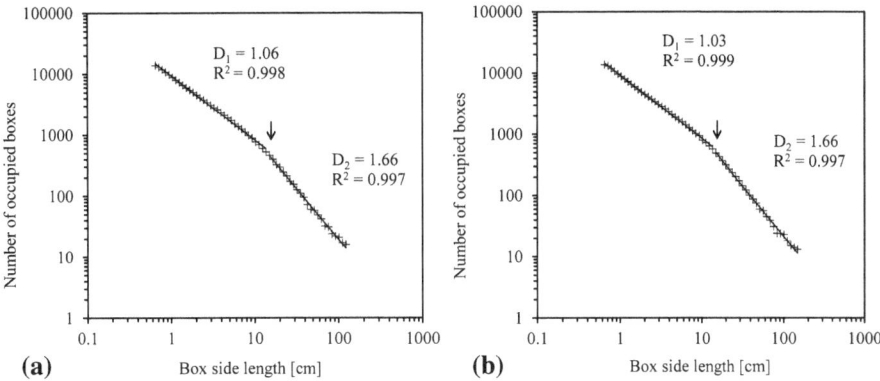

Figure 3

Box counting applied to sub-sections M2 (**a**) and S5 (**b**); locations shown in Fig. 2. For each sub-section, the cumulative frequency distribution shows two linear groups of points representing two fractal dimensions, D_1 and D_2. *Arrows* indicate the switch from one group to the other. R^2 correlation coefficient

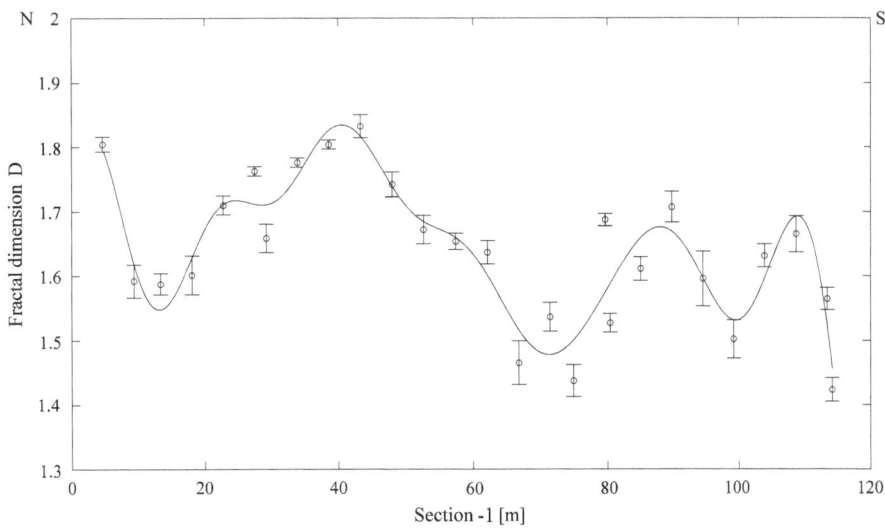

Figure 4

Variation of fractal box-counting dimension D along section-1 (summation of 27 sub-sections). The fractal dimension D_2 of each sub-section (Fig. 3) and its mean value is calculated by four initial and final box sizes. *Error bar* 95 % confidence interval of the mean value. *Solid curve* seventh order Fourier fit

3.1.2 Inhomogeneity of Fracture-Pattern Fractality

The binary fracture pattern, based on digital outcrop photographs of section-1, already reveals complexity and variation (Fig. 2). Map counting applied to this pattern produced a colour-coded map of box-counting dimensions D (Fig. 5). On the map, the D-values of 1.10–1.87 represent the variation in fracture pattern complexity over the analysed area, i.e. pattern inhomogeneity. The D-values do not differ randomly, but exhibit some regularity with several vertical to

sub-vertical elongate areas of high values of 1.7–1.87. The high-D areas are surrounded by more diffuse regions with low D-values of 1.15–1.55.

3.1.3 Inhomogeneity of Anisotropy of Pattern Complexity

Application of the MORFA method to the binary fracture pattern of section-1 resulted in a colour-coded map of anisotropy intensity (Fig. 6). The values 0.2–0.8 represent the variation in fracture

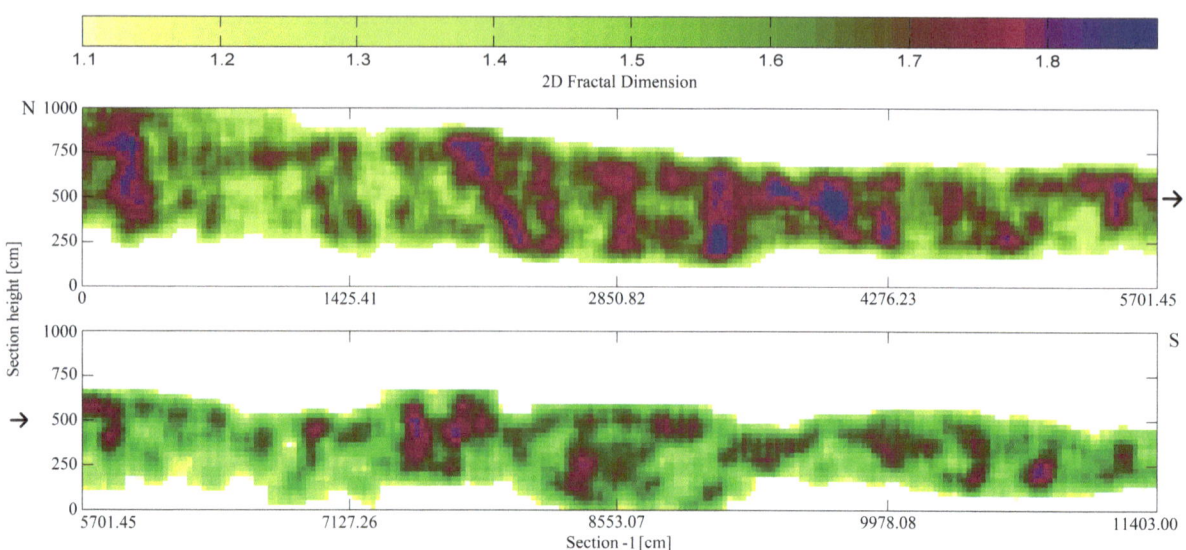

Figure 5
High-resolution map-counting method (PETERNELL *et al.* 2010) applied to the fracture pattern of section-1. See text for description of the method. *Rectangular* measurement window: 100 × 100 cm with 75 % overlap. The box-counting dimension is scaled by *colour index*. Each pixel is equivalent to an area of 25 × 25 cm. The colour index represents the intensity of inhomogeneity of the fracture pattern complexity, with 1.1 = low complexity (*light yellow*) and 1.87 = high complexity (*blue*)

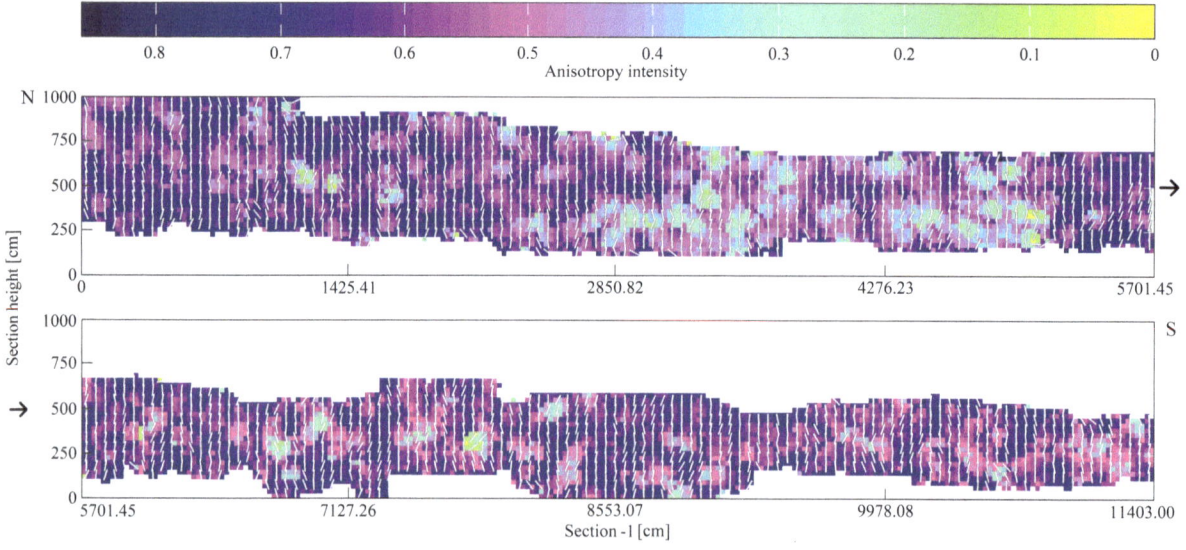

Figure 6
High-resolution mapping of rock fabric anisotropy (MORFA) method (PETERNELL *et al.* 2011) applied to the fracture pattern of section-1. Circular measurement window for MORFA: 100 cm diameter with 75 % overlap. The colour index represents the anisotropy intensity of pattern complexity, with 0 = isotropic (*yellow*) and 0.75 = high anisotropy (*blue*). Each colour pixel represents 25 × 25 cm. The *short white bars* represent the short axis of the fit ellipse, i.e., the general direction of the minimum pattern complexity

pattern complexity in different directions over the analysed area. High values of around 0.45–0.75 are dominant, while low values of 0.25–0.10 are observed only in diffuse areas. MORFA also performs an elliptical fit of the data points, whose short axis represents statistically the direction of lowest *D*-value or the short axis of anisotropy. This short axis is relatively constant and vertical to sub-vertical over the entire pattern, mostly in combination with relatively high anisotropy intensity (>0.45).

3.2. Small-Scale Fracture Patterns and Fragments

FSD and fracture patterns were also determined and quantified in boulders and core samples.

3.2.1 Boulders

Application of the box-counting method on fracture patterns of granitoid and limestone boulders (Figs. 7a, 8a) revealed that the fracture patterns are fractal in the range 0.9–20 mm (Figs. 7b, 8b). The

slopes of the regression lines, i.e. the fractal dimensions D, of 1.70 and 1.75 are in the range given for granite and carbonate rocks from fault zones (KEULEN et al. 2007).

FSD of the two boulders shows a power-law distribution in 20–100 mm^2 (Figs. 7c, 8c). Because the number of larger fragments is not statistically significant, these data points were excluded from the analysis. The smaller fragments were also excluded as the number of fragments is biased by low observability (because many fragments were too small to count). The slopes of the regression lines are 1.40 and 1.11 for the granitoid and limestone boulder, respectively, and those fractal dimensions

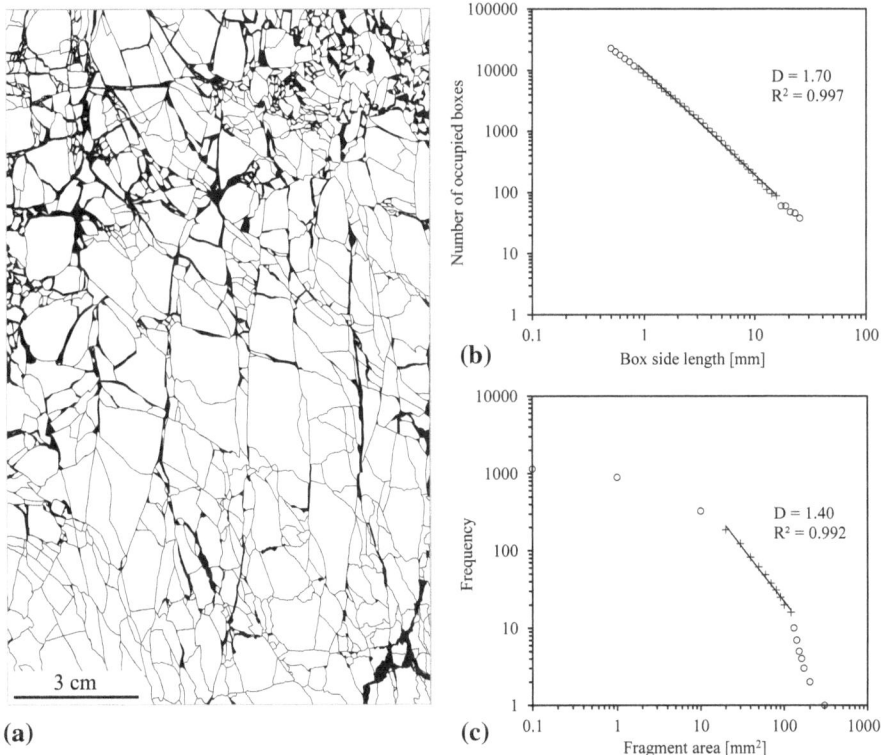

Figure 7

Analysis of fracture pattern from a granitic basement rock boulder collected in the Unterwilfingen quarry (Fig. 1). **a** Binary image of fractures (*black lines*) and regions of sub-microscopically fragmented material (*small black areas*). One pixel = 0.065 mm on the rock surface. **b** Box counting applied to the fracture pattern shown in (**a**). A power-law relationship (*solid line*) over more than one order of magnitude with a fractal dimension of 1.70 is indicated. *Open circles*: measurements excluded from analysis. R^2 correlation coefficient. **c** Cumulative size frequency distribution of fragment areas from (**a**). The areas are calculated by counting the number of pixels inside the fragments (LIU et al. 2011). A power-law relationship (*solid line*) over less than one order of magnitude with a fractal dimension of 1.40 is indicated. Measurements with low statistics (*large fragments*) or with truncation effects (BLENKINSOP 1991) because of small size and low observability (*open circles*) were excluded from the analysis. R^2 correlation coefficient

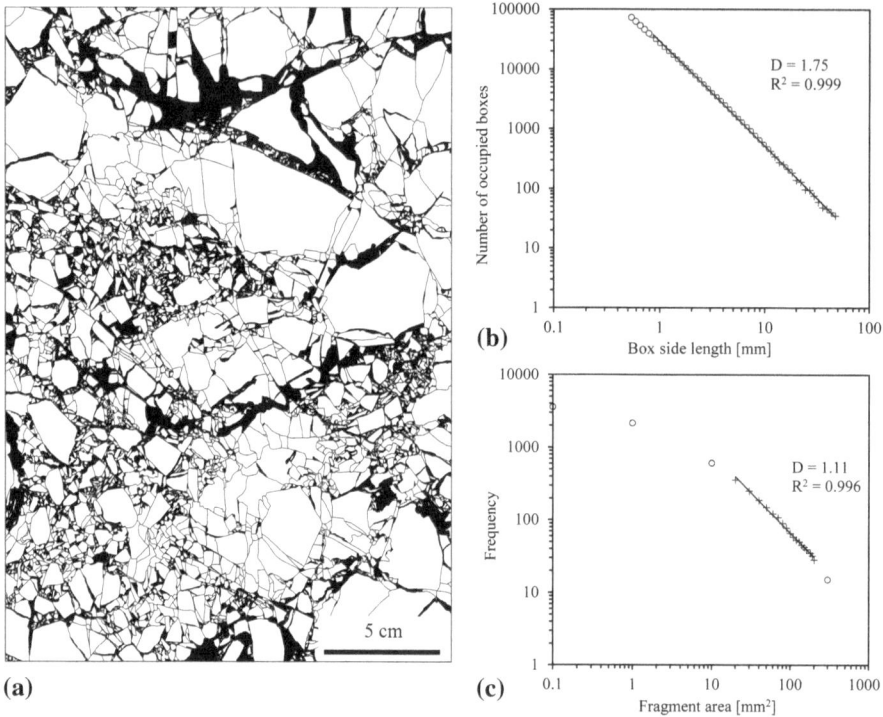

Figure 8

Analysis of fracture pattern from a limestone boulder collected from the Unterwilfingen quarry (location in Fig. 1). **a** Binary image of fractures (*black lines*) and regions of sub-microscopically fragmented material (*small black areas*). One pixel = 0.065 mm on the rock surface. **b** Box counting applied to the fracture pattern shown in (**a**). A power-law relationship (*solid line*) over more than one order of magnitude with a fractal dimension of 1.75 is indicated. Measurements with low statistics (*large boxes*) or with truncation effects (BLENKINSOP 1991) because of small size and low observability (*open circles*) were excluded from the analysis. R^2 correlation coefficient. **c** Cumulative size frequency distribution of fragment areas from (**a**). Area is calculated by counting the number of pixels inside the fragments (LIU *et al.* 2011). A power-law relationship (*solid line*) over less than one order of magnitude and a fractal dimension of 1.11 is indicated. Measurements with low statistics (*large fragments*) or with truncation effects (BLENKINSOP 1991) because of small size and low observability (*open circles*) were excluded from the analysis. R^2 correlation coefficient

are far below the range of values given in the literature (BLENKINSOP 1991).

3.2.2 Core Sample

Box counting applied to the fracture patterns of the six hand specimen scale core samples (Fig. 9a) shows two linear groups of data points with fractal dimensions D_1 and D_2 in the range 1.36–1.44 and 1.63–1.85, respectively (Fig. 9b). For all six analysed core samples, the switch from one interval to the other is between 1 and 2.5 mm. FSD of all core samples also shows two linear groups of data points with slopes 0.67–0.97 and 1.20–2.98, respectively (Fig. 9c). For all core samples, the switch from one

regression line to the other occurs between 45 and 65 mm² fragment area.

Application of box counting to the fracture pattern of the thin sections from the core samples (Fig. 10a) results in two linear groups of data points with fractal dimensions D_1 and D_2 1.29–1.35 and 1.80–1.88, respectively (Fig. 10b). For all thin sections, the switch from one interval to the other is between 0.4 and 0.5 mm. FSD of the core thin section also leads to two groups of points with clearly different regression line slopes (fractal dimensions D_1 and D_2) of 0.63–0.68 and 2.08–2.15 (Fig. 10c). For all core thin sections the switch from one regression line to the other occurs between 2.5 and 3.5 mm² fragment area.

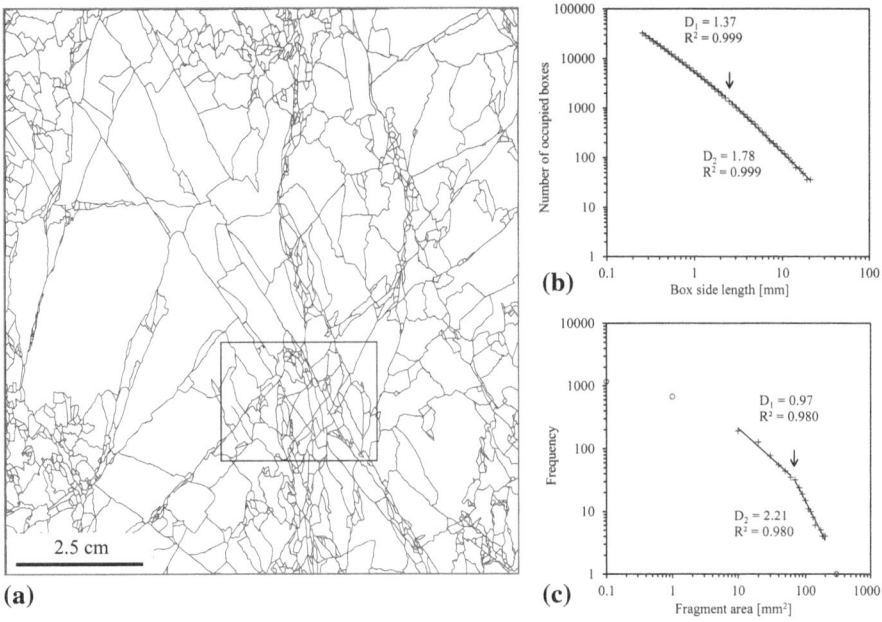

Figure 9
Analysis of hand sample scale fracture pattern from in the drill core from the Ries Research Borehole (RRB)-1973 (Centre for Ries Crater Impact Research, ZERIN); location shown in Fig. 1. **a** Binary image of fractures (*black lines*) from a vertical cut of a granitoid rock sample from a depth of 510.5 m. This image is an example of the six images analysed for this study. The *rectangle* outlines the area shown in thin section (Fig. 10a). One pixel = 0.025 mm on the rock surface. **b** Box counting applied to the fracture pattern shown in (**a**). The arrangement of measurement points shows two power-law relationships (*solid lines*) with fractal dimensions D_1 and D_2. The *arrow* indicates the switch from one linear point arrangement to the other. R^2 correlation coefficient. **c** Cumulative size frequency distribution of fragment areas from (**a**). The areas are calculated by counting the number of pixels inside the fragments (LIU *et al.* 2011). Two power-law relationships (*solid lines*) over less than one order of magnitude and fractal dimensions D_1 and D_2 are indicated. Because the number of larger fragments is not statistically significant and small fragments have low observability, those data points (*open circles*) were excluded from the analysis. The *arrow* indicates the switch from one linear point arrangement to the other. R^2 correlation coefficient

4. Discussion

Outcrop images in and around the Ries crater clarify that the country rocks are intensively fractured far beyond the range and style typical for regional brittle deformation (RAMSAY and HUBER 1987; SUPPE 1985). In particular, brittle structures are locally highly varied and form complex patterns. In order to characterize this type of deformation and understand its correlation with scale, pattern quantification is necessary. For this purpose, fractal geometry methods are useful, which can quantify the inhomogeneity and anisotropy of patterns as well as the combination of both (PETERNELL *et al.* 2010, 2011; VELDE *et al.* 1991; VOLLAND and KRUHL 2004). This study applies fractal methods that are designed for automated analysis, and thus can record patterns over large

areas with high precision (GERIK and KRUHL 2009; PETERNELL *et al.* 2010, 2011). In general, fragmentation largely follows power laws, but to different extents on different scales and at different locations. In addition, fragmentation patterns may be inhomogeneous and anisotropic.

Box counting applied to an approximately 114 m long, N–S oriented vertical section of Malm limestone, exposed about 13 km ESE of the crater centre (Eireiner quarry, Fig. 1), indicating fractality of the 2D fragmentation pattern in all parts of the section (Fig. 3), but with different fractal dimensions D_1 and D_2 on two different scales. The ranges of D_1 and D_2 values of 1.03–1.15 and 1.40–1.85 in the 27 subsections analysed are clearly differentiable (Fig. 4). Two different power-law relationships are usually interpreted as resulting from two different or two combined pattern-forming processes (KAYE 1989;

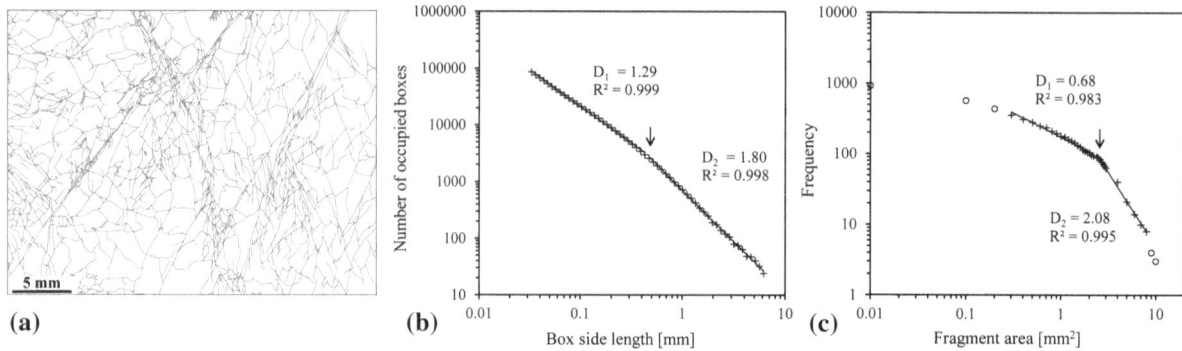

Figure 10
Analysis of thin section scale fracture pattern from drill core of the Ries Research Borehole (RRB)-1973; location shown in Fig. 1. **a** Binary image of fractures (*black lines*) from a thin section of a granitoid rock from a depth of 510.5 m. This image is an example of the three images analysed for this study. The area from which the thin section was taken is shown in Fig. 9a. One pixel = 0.006 mm in the thin section. **b** Box counting applied to the fracture pattern shown in (**a**). The arrangement of measurement points show two power-law relationships (*solid lines*) with fractal dimensions D_1 and D_2. The *arrow* indicates the switch from one linear point arrangement to the other. R^2 correlation coefficient. **c** Cumulative size frequency distribution of fragment areas from (**a**). The areas are calculated by counting the number of pixels inside the fragments (LIU *et al.* 2011). Two power-law relationships (*solid lines*) over less than one order of magnitude and fractal dimensions D_1 and D_2 are indicated. Because the number of larger fragments is not statistically significant and small fragments show low observability, those data points (*open circles*) were excluded from the analysis. The *arrow* indicates the switch from one linear point arrangement to the other. R^2 correlation coefficient

KRUHL 2013). We also consider our D_1 and D_2 values to have been produced by two processes.

We interpret one process to be the result of the contact and compression stage at the first phase of impact cratering. During this stage of cratering, strain rates are very high and small and closely spaced fractures develop (KENKMANN *et al.* 2014). Small, vertical to sub-vertical networks of parallel fractures, mainly restricted to single sedimentary beds (Fig. 2), are closely related to the contact and compression stage of deformation. We interpret the second process to be the result of the excavation stage during the intermediate phase of impact cratering. In this stage, individual fractures merge to form more complex fracture networks that lead to the second range of *D*-values.

The two different ranges of *D*-values can be interpreted as follows. The release wave related to the excavation stage fragments the rocks around the impact, which were earlier subjected to the shock wave. Consequently, the excavation stage increases the density of the fracture pattern within the early stage shock-wave-related, centimetre-scale fragmented limestone. Preferentially, in the more intensively fragmented regions, existing shock-related fractures were possibly partly reactivated, leading to a relatively smaller number of new fractures compared with less densely fractured regions. Thus, the fracture pattern is intensified at larger scales than at smaller scales, with increased *D*-values at the larger scale and decreased *D*-values at the smaller scale. The switch from one power-law relationship to the other at 10–20 cm (Fig. 3), which is the same for all of our sub-sections, represents the scale on which the impact-related fragmentations have transformed from early stage contact and compression to intermediate stage excavation. However, the third and final phase of impact cratering (crater modification) is not reflected in our box-counting analysis of the sub-sections. This is possibly due to scaling limitation. In the final stage of impact cratering, faults and fractures of hundreds of metres or even kilometres in length are formed (SPRAY 1997), and therefore, they are not visible on the scale of this study.

The seventh order Fourier fit with residual fractal dimensions ≤0.10 (Fig. 4) for the box-counting D_2 indicates a systematic variation of pattern fractality along the N–S section. The set of different D_2 fractal dimensions that occur at the scale larger than 20–30 cm is possibly related to the high deformation strain and inherent flaws in the rock. Inherent flaws are important as sites of weakness for the nucleation and coalescence of fractures (GRIFFITH 1920). These sites result in localization of deformation along specific fracture systems (FAULKNER *et al.* 2003; WILSON *et al.* 2003). Moreover, beyond a certain threshold limit of strain rate ($\sim 10^1$–10^3 s^{-1}) material inertia begins to affect the nucleation and propagation of the fractures (KIPP *et al.* 1980). Transition from narrow-spaced fracture networks to more localized large-scale fractures indicates the degree of strain localization shifting from a relatively homogeneous to a strongly heterogeneous distribution (KENKMANN *et al.* 2014).

More detailed quantification of the pattern variation with high-precession Map Counting (PETERNELL *et al.* 2010) revealed multiple vertical to sub-vertical elongate areas, dominated by high box-counting dimensions. The values of 1.7–1.85 correspond to more intensely fractured areas (Fig. 5). These elongated areas are regularly spaced at intervals of 2.5–3.5 m. They are also observed in other parts of the Eireiner quarry. Similar *D*-values are reported from fault zones and larger-scale tectonic regions (BOUR and DAVY 1997; HIRATA 1989; VOLLAND and KRUHL 2004). Even though the fractality of 2D patterns increases with increasing filling of the plane (MANDELBROT 1982), increasing fractality is also related to the geometric arrangement of the fracture pattern (PETERNELL *et al.* 2011). Hence, it can be argued that the high *D*-value of the presented fracture pattern indicates higher complexity as well as a higher degree of fragmentation. The overlay of the Map-counting *D*-value contours on the fracture map revealed the correlation between fracture density and fractal dimension values (Supplementary Fig. 3).

The regular variation in the fracture pattern complexity in section-1 seems to be typical for impact-related fragmentation. Moreover, based on fracture lengths, spacing, and location and depth of section-1, it can be assumed that these fractures are related to the transition zones between Grady-Kipp (below) and spall fragmentations (above) (MELOSH 1984, 1989).

Application of MORFA (PETERNELL *et al.* 2011) to the fracture pattern in section-1 leads to generally similar, but locally different results (Fig. 6). The directions of minimum pattern complexity are parallel to the vertical to sub-vertical fracture orientation in some parts of the section, whereas in other parts they differ. Mostly these differing orientations are not related to regions of low anisotropy, and therefore, are not an artefact of the method. They occur in numerous areas of up to 50 cm wide and 150 cm high, each of which shows a dominant direction of minimum pattern complexity with 15–20° angle to the vertical either to the north or to the south (Fig. 6). This indicates a strong and regularly oriented anisotropy of the pattern fractality, which is in agreement with the generally steep orientation of the single fractures. Because these directions roughly correlate with the main fracture orientation, a systematic change of fracture orientation is indicated. This might be correlated with shatter-cone style fractures on the decimetre to metre scale. Shatter cones of such size have been reported in the literature (OSINSKI and PIERAZZO 2013). In addition, these results show that MORFA is able to detect and quantify diffuse patterns of rock structures, patterns not measurable with conventional methods.

For each sample, fracture patterns and FSD of the boulders lead to a power-law relationship over one order of magnitude scale (Figs. 7, 8) and, therefore, indicate one single fragmentation process. This process is represented by material excavation through non-ballistic ejection during the impact (CHAO and MINKIN 1977; CHAO *et al.* 1978). The boulders we studied are assumed to be parts of larger highly fragmented blocks, which are transported over a short distance towards the crater centre. This conclusion is supported by their location inside the crater (Fig. 1). Because these blocks were only subjected to the shock-related fragmentation during non-ballistic ejection, only one single process is imprinted on the boulders, thus resulting in only one power-law relationship.

Analysis of the core samples leads to two distinct box-counting *D*-values (D_1 and D_2) of the fracture patterns and two values of FSD (Figs. 9, 10). The switch from D_2 to D_1 occurs at different scales for box counting and for FSD. As for the fracture pattern in the Malm limestone at the Eireiner quarry, the two different *D*-values are interpreted as resulting from two subsequent pattern-forming processes during the Ries impact. The first one is related to shock-wave fragmentation and the second to elastic rebound of the transient crater floor from a depth of 4.5–5 km to its current level of 500–600 m depth (GUDDEN 1974; WÜNNEMANN et al. 2005). In this stage, shear fracturing leads to additional diminution of the shock-wave fragments. Their re-fragmentation leads to a relatively higher number of small fragments (KEULEN et al. 2007) and, consequently, to two different D-values.

The box-counting D_1- and D_2-values of six analysed core samples are 1.36–1.44 and 1.63–1.85 (Fig. 9b), and those of the three analysed thin sections from the core samples are 1.29–1.35 and 1.80–1.88, respectively (Fig. 10b). These values are similar to the values of the fragmented Malm limestone of section-1 and of fragmentation elsewhere (AYUNOVA et al. 2007; CASTAING et al. 1996; CELLO 1997; HIRATA 1989; PARK et al. 2010). In addition, the D_1- and D_2-values are similar or show slight overlap. However, the values on the same scale in thin section and sample are different. Even though this is possibly related to the different resolutions of both patterns, the variation in D-values on different scales suggest scale-dependence of the fragmentation processes (KRUHL 2013).

The FSD of core samples also show two different fractal dimensions of 0.67–0.97 and 1.20–2.98, respectively (Fig. 9c). Specifically, the higher values fall into the wide range of values given for rock fragmentation (AN and SAMMIS 1994; BLENKINSOP 1991; CHESTER et al. 2005; STORTI et al. 2003). Impact

experiments show an increase of D-values with increasing impact energy (LANGE et al. 1984), and cratering and shock-recovery experiments suggest a correlation with strain rate (BUHL et al. 2013). Studies on natural impacts provide D-values within 1.2–1.8 (KEY and SCHULTZ 2011; ROUSELL et al. 2003). High values are reported to be related to increasing strain or strain localization (BILLI and STORTI 2004; DYER et al. 2012; MONZAWA and OTSUKI 2003). The wide range of fractal dimensions in the literature as well as in our investigations argues against shock-induced fragmentation being the only or dominant process during impact and suggests that different parameters, such as rock type and pre-existing rock structures, and their interaction have to be considered. In addition, the influence of the latter two stages of impact cratering (the excavation and modification stages) versus other parameters is expected to vary with time and distance from the point of impact and may produce a wide range of D-values.

In the core samples the D_2-values of the fracture patterns do not change with depth (Fig. 11a). The D_2-values of FSD, however, decrease from depths of 510.5 m to 773.3 m, but are higher at a depth of 1,169.5 m (Fig. 11b). The decrease in D with increasing depth below the crater floor has also been observed in impact experiments and is related to a zone of 'pervasive crushing and compaction' (BUHL et al. 2013). The same authors observed a zone of localized deformation, at a farther distance from the

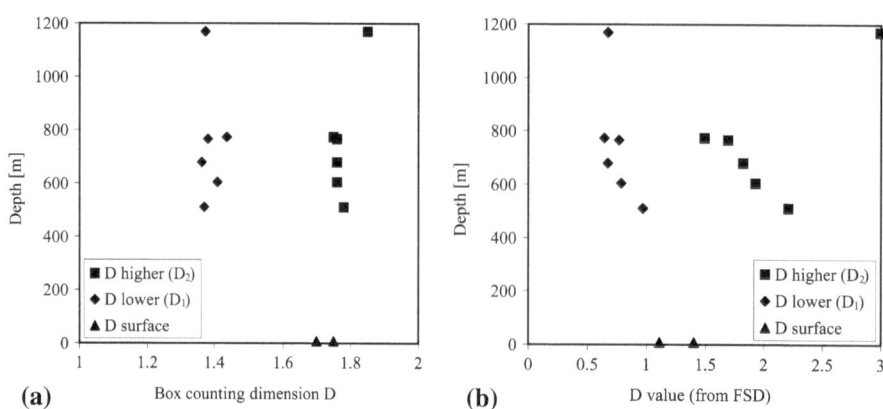

Figure 11
Variation of fracture pattern and fragment size distribution for boulders and core samples from the surface to a depth of 1,169.5 m. **a** Depth vs. D-value obtained from box-counting analysis of fracture patterns (Figs. 7b, 8b, 9b). **b** Depth vs. D-value obtained from fragment size distributions (Figs. 7c, 8c, 9c)

impact, with increased *D*-values. They related this increase to enhanced comminution due to stress localization, which leads to reduced fragment sizes. In our data, the high *D*-value at greater depth can possibly be ascribed to the same localized deformation. The zones of increased fragmentation, observed in limestone outcrops at the surface (Fig. 2), might correspond to zones of localized deformation. However, *D*-values from FSD are sensitive to truncation effects (KRUHL 2013) and should be interpreted carefully. FSD provides information about different fragmentation processes (KEULEN *et al.* 2007; KRUHL 2013), but not about the fragmentation pattern.

Summarizing, fractal geometry-based quantification of fragmentation structures in and around the Ries impact crater reveals (i) the general complexity, i.e. fractality, of these structures, (ii) the variation in complexity on different scales and in relation to distance from the impact centre, (iii) partly the effect of three different fragmentation processes, and (iv) the inhomogeneity of the fragmentation pattern over larger areas. A systematic variation in fracture orientation is indicated by the short axis of the anisotropy of the pattern, which might be correlated with shatter-cone style fractures on the decimetre- to metre-scale. The absence of a narrow range of fractal dimensions for FSD of the Ries core samples argues against shock-induced fragmentation being the only or dominant process during impact. The high *D*-value for FSD at greater depth can possibly be related to enhanced comminution due to stress localization, and might correspond to the zones of localized deformation. Fractal geometry, specifically when based on automated procedures, proves to be a powerful tool for quantifying, and thus analysing complex rock structures.

Acknowledgments

We are grateful to the colleagues from the Centre for Ries Crater Impact Research (ZERIN) in Nördlingen for providing core samples, to Mark Peternell for giving access to the map-counting and MORFA program system, and to Christian Stäb, Tim Yilmaz, Thomas Kenkmann, Alison Ord, and Tom Blenkinsop for helpful discussions. Reviews by Mark Peternell and an anonymous reviewer greatly improved the manuscript. We also express our gratitude to Namvar Jahanmehr and Klaus Mayer for sample and thin-section preparation. MD.S.H. gratefully acknowledges financial support by the German Academic Exchange Service DAAD (Grant A/11/75209).

REFERENCES

AN, L.-J., and SAMMIS, C. (1994), *Particle size distribution of cataclastic fault materials from southern California: A 3-D study*. Pure and Applied Geophysics, *143*, 203–227.

AYUNOVA, O.D., KALUSH, YU.A., and LOGINOV, V.M. (2007), *Relationship of the seismic activity of the Tuvinian and adjacent Mongolian areas with the fractal dimensionality of a fault system*. Russian Geology and Geophysics, *48*, 593–597.

BILLI, A., and STORTI, F. (2004), *Fractal distribution of particle size in carbonate cataclastic rocks from the core of a regional strike slip fault zone*. Tectonophysics, *384*, 115–128.

BLENKINSOP, T.G. (1991), *Cataclasis and processes of particle size reduction*. Pure and Applied Geophysics, *136*, 60–86.

BOUR, O., and DAVY, P. (1997), *Connectivity of random fault networks following a power law fault length distribution*. Water Resources Research, *33*, 1567–1583.

BUHL, E., KOWITZ, A., ELBESHAUSEN, D., SOMMER, F., DRESEN, G., POELCHAU, M.H., REIMOLD, W.U., SCHMITT, R.T., and KENKMANN, T. (2013), *Particle size distribution and strain rate attenuation in hypervelocity impact and shock recovery experiments*. Journal of Structural Geology, *56*, 20–33.

CASTAING, C., HALAWANI, M.A., GERVAIS, F., CHILÈS, J.P., GENTER, A., BOURGINE, B., OUILLON, G., BROSSE, J.M., MARTIN, P., GENNA, A., and JANJOU, D. (1996), *Scaling relationships in intraplate fracture systems related to Red Sea rifting*. Tectonophysics, *261*, 291–314.

CELLO, G. (1997), *Fractal analysis of a Quaternary fault array in the central Apennines, Italy*. Journal of Structural Geology, *19*, 945–953.

CHAO, E.C.T., and MINKIN, J.A. (1977), *Impact cratering phenomenon tor the Ries multiring structure based on constraints of geological, geophysical and petrological studies and the nature of the impacting body*. In: RODDY, D.J., PEPIN, R.O., and MERRILL, R.B. (eds.), Impact and Explosion Cratering: Planetary and terrestrial implications, Pergamon Press, New York, 405–424.

CHAO, E.C.T., SCHMIDT-KALER, H., and HÜTTNER, R. (1978), *Principal exposures of the Ries meteorite crater in southern Germany. Description, photographic documentation and interpretation*. Bayerisches Geologisches Landesamt, München, 84 pp.

CHAO, S.H., and KANEKO, K. (2004), *Rock Fragmentation Control in Blasting*. Materials Transactions, *45*(5), 1722–1730.

CHESTER, J., CHESTER, F., and KRONENBERG, A. (2005), *Fracture surface energy of the Punchbowl fault, San Andreas system*. Nature, *437*, 133–136.

COLLINS, G.S., MELOSH, H.J., and IVANOV, B.A. (2004), *Modeling damage and deformation in impact simulations*. Meteoritics and Planetary Science, *39*, 217–231.

DALWIGK, I., VON (2003), *Fracture pattern in a complex impact structure—what can it tell us about crater collapse? A new look at the Siljan impact structure.* EGS-AGU-EUG joint assembly, Nice (France), Abstract-10568.

DYER, H., AMITRANO, D., and BOULLIER, A.-M. (2012), *Scaling properties of fault rocks.* Journal of Structural Geology, *45*, 125–136.

FAULKNER, D., LEWIS, A., and RUTTER, E. (2003), *On the internal structure and mechanics of large strike-slip fault zones: field observations of the Carboneras fault in southeastern Spain.* Tectonophysics, *367*, 235–251.

FENG, Z.C., ZHAO, Y.S., and ZHAO, D. (2009), *Investigating the scale effects in strength of fractured rock mass.* Chaos, Solitons & Fractals, *41*, 2377–2386.

GERIK, A., and KRUHL, J.H. (2009), *Towards automated pattern quantification: time efficient assessment of anisotropy of 2D pattern with AMOCADO.* Computers & Geosciences, *35*(6), 1087–1097.

GRAUP, G. (1977), *Die Petrographie der kristallinen Gesteine der Forschungsbohrung Nördlingen 1973.* Geologica Bavarica, *75*, 219–229.

GRIFFITH, A.A. (1920), *Phenomena of rupture and flow in solids.* Philosophical Transaction Royal Society London, 221A, 163–198.

GUDDEN, H. (1974), *Die Forschungsbohrung Nördlingen 1973—Durchführung und erste Befunde.* Geologica Bavarica, *72*, 11–31.

HIRATA, T. (1989), *Fractal dimension of fault systems in Japan: fractal structure in rock fracture geometry at various scales.* Pure and Applied Geophysics, *131*, 157–169.

HÜTTNER, R., and SCHMIDT-KALER, H. (1999), *Geologische Karte 1:50000 Ries mit Kurzerläuterungen auf der Rückseite.* Tech. rep. Bayerisches Geologisches Landesamt, scale 1:50000, 1 sheet.

KAUSHIK, A.I. (2007), *Relationships between multiaxial stress states and internal fracture patterns in sphere-impacted silicon carbide.* International Journal of Fracture, *146*(1–2), 1–18.

KAYE, B.H. (1989), *A random walk through fractal dimensions.* Weinheim, VCH Publishers.

KAYE, B.H. (1993), *Chaos and Complexity.* VCH Publishers, Weinheim.

KENKMANN, T., POELCHAU, M.H., and WULF, G. (2014), *Structural geology of impact craters.* Journal of Structural Geology, *62*, 156–182.

KEULEN, N., HEILBRONNER, R., STÜNITZ, H., BOULLIER, A.-M., and ITO, H. (2007), *Grain size distributions of fault rocks: A comparison between experimentally and naturally deformed granitoids.* Journal of Structural Geology, *29*, 1282–1300.

KEY, W.R.O., and SCHULTZ, R.A. (2011), *Fault formation in porous sedimentary rocks at high strain rates: First results from the Upheaval Dome impact structure, Utah, USA.* GSA Bulletin, *123*, 1161–1170.

KIPP, M.E., GRADY, D.E., and CHEN, E.P. (1980), *Strain-rate dependent fracture initiation.* International Journal of Fracture, *16*(5), 471–478.

KRUHL, J.H. (2013), *Fractal-geometry techniques in the quantification of complex rock structures: A special view on scaling regimes, inhomogeneity and anisotropy.* Journal of Structural Geology, *46*, 2–21.

KUMAR, P.S., and KRING, D.A. (2008), *Impact fracturing and structural modification of sedimentary rocks at Meteor Crater, Arizona.* Journal of Geophysical Research, *113*, E09009.

LANGE, M.A., AHRENS, T.J., and BOSLOUGH, M.B. (1984), *Impact cratering and spall failure of gabbro.* Icarus, *58*, 383–395.

LETERRIER, Y., PELLATON, D., MENDELS, D., GLAUSER, R., ANDERSONS, J., and MÅNSON, J.-A. E. (2001), *Biaxial fragmentation of thin silicon oxide coatings on poly(ethylene terephthalate).* Journal of Materials Science, *36*(9), 2213–2225.

LIU, C., SHI, B., and TANG, C.-S. (2011), *Quantification and characterization of microporosity by image processing, geometric measurement and statistical methods: application on SEM images of clay materials.* Applied Clay Science, *54*(1), 97–106.

LIU, C., TANG, C.-S., SHI, B., and SUO, W.-B. (2013), *Automatic quantification of crack patterns by image processing.* Computers & Geosciences, *57*, 77–80.

MANDELBROT, B.B. (1982), *The Fractal Geometry of Nature.* Freeman & Co., New York.

MELOSH, H.J. (1984), *Impact ejection, spallation, and the origin of meteorites.* Icarus, *59*, 234–260.

MELOSH, H.J. (1989), *Impact Cratering: a Geological Process.* Oxford University Press, New York.

MELOSH, H.J., and IVANOV, B.A. (1999), *Impact Crater Collapse.* Annual Review of Earth and Planetary Sciences, *27*, 385–415.

MONZAWA, N., and OTSUKI, K. (2003), *Comminution and fluidization of granular fault materials: implications for fault slip behavior.* Tectonophysics, *367*, 127–143.

OSINSKI, G.R., and PIERAZZO, E. (2013), *Impact cratering: processes and products.* In: GORDON R. OSINSKI and ELISABETTA PIERAZZO (eds.), Impact cratering (1st edition), Blackwell Publishing Ltd., 45–64.

PARK, S.-I., KIM, Y.-S., RYOO, C.-R., and SANDERSON, D.J. (2010), *Fractal analysis of the evolution of a fracture network in a granite outcrop, SE Korea.* Geosciences Journal, *14*, 99–234.

PETERNELL, M., BITENCOURT, M.F., KRUHL, J.H., and STÄB, C. (2010), *Macro and microstructures as indicators of the development of syntectonic granitoids and host rocks in the Camboriú region, Santa Catarina, Brazil.* Journal of South American Earth Sciences, *29*, 738–750.

PETERNELL, M., BITENCOURT, M.F., and KRUHL, J.H. (2011), *Combined quantification of anisotropy and inhomogeneity of magmatic rock fabrics—an outcrop scale analysis recorded in high resolution.* Journal of Structural Geology, *33*, 609–623.

PLANT, J.A., WHITTAKER, A., DEMETRIADES, A., DE VIVO, B., and LEXA, J. (2003), *The geological and tectonic framework of Europe.* A contribution to IUGS/IAGC Global Geochemical Baselines, Geochemical Atlas of Europe, Part 1.

POLANSKEY, C.A., and AHRENS, T.J. (1990), *Impact spallation experiments: Fracture patterns and spall velocities.* Icarus, *87*, 140–155.

QUINTA, A., TAVANI, S., and ROCA, E. (2012), *Fracture pattern analysis as a tool for constraining the interaction between regional and diapir-related stress fields: Poza de la Sal Diapir (Basque Pyrenees, Spain).* Geological Society, London, Special Publications, *363*(1), 521–532.

RAMSAY, J.R. and HUBER, M. (1987). *The Techniques of Modern Structural Geology, Vol. 2: Folds and Fractures.* Academic Press, London.

ROUSELL, D.H., FEDOROWICH, J.S., and DRESSLER, B.O. (2003), *Sudbury Breccia (Canada): A product of the 1850 Ma Sudbury event and host to footwall Cu-Ni-PGE deposits.* Earth-Science Reviews, *60*, 147–174.

SCHWARZ, W.H., and LIPPOLT, H.J. (2013), ^{40}AR-^{39}AR Step heating of Nördlinger Ries Crater impact melts. Meteoritics and Planetary Science Supplement, id. 5191.

SPRAY, J.G. (1997), *Superfaults*. Geology, *25*(7), 579–582.

STÖFFLER, D. (1977), *Research drilling Nördlingen 1973: Polymict breccias, crater basement, and cratering model of the Ries impact structure*. Geologica Bavarica, *75*, 443–458.

STÖFFLER, D., ARTEMIEVA, N.A., and PIERAZZO, E. (2002), *Modeling the Ries-Steinheim impact event and the formation of the moldavite strewn field*. Meteoritics & Planetary Science, *37*(12), 1893–1907.

STORTI, F., BILLI, A., and SALVINI, F. (2003), *Particle size distributions in natural carbonate fault rocks; insights for non-self-similar cataclasis*. Earth and Planetary Science Letters, *206*, 173–186.

STRASSER, M., STRASSER, A., PELZ, K., and SEYFRIED, H. (2009), *A mid Miocene to early Pleistocene multi-level cave as a gauge for tectonic uplift of the Swabian Alb (Southwest Germany)*. Geomorphology, *106*, 130–141.

SUPPE, J. (1985), *Principles of Structural Geology*. Prentice-Hall Inc., Englewood Cliffs.

VELDE, B., DUBOIS, J., TOUCHARD, G., and BADRI, A. (1990), *Fractal analysis of fractures in rocks: the Cantor's Dust method*. Tectonophysics, *179*, 345–352.

VELDE, B., DUBOIS, J., MOORE, D., and TOUCHARD, G. (1991), *Fractal patterns of fractures in granites*. Earth and Planetary Science Letters, *104*(1), 25–35.

VOLLAND, S., and KRUHL, J.H. (2004), *Anisotropy quantification: the application of fractal geometry methods on tectonic fracture patterns of a Hercynian fault zone in NW-Sardinia*. Journal of Structural Geology, *26*, 1499–1510.

WILSON, J., CHESTER, J., and CHESTER, F. (2003), *Microfracture analysis of fault growth and wear processes, Punchbowl Fault, San Andreas system, California*. Journal of Structural Geology, *25*, 1855–1873.

WÜNNEMANN, K., and IVANOV, B.A. (2003), *Numerical modelling of impact crater depth-diameter dependence in an acoustically fluidized target*. Planetary and Space Science, *51*, 831–845.

WÜNNEMANN, K., MORGAN, J.V., and JÖDICKE, H. (2005), *Is Ries crater typical for its size? An analysis based upon old and new geophysical data and numerical modelling*. Geological Society of America Special Papers, *384*, 67–83.

ZHAO, Y.S., FENG, Z.C., LIANG, W.G., YANG, D., HU, Y.Q., and KANG, T.H. (2009), *Investigation of fractal distribution law for the trace number of random and grouped fractures in a geological mass*. Engineering Geology, *109*, 224–229.

(Received March 4, 2014, revised August 3, 2014, accepted August 7, 2014, Published online August 20, 2014)

Reprinted from the journal

Pure Appl. Geophys. 172 (2015), 2025–2043
© 2014 Springer Basel
DOI 10.1007/s00024-014-0874-z

▌Pure and Applied Geophysics

Complexity Phenomena and ROMA of the Earth's Magnetospheric Cusp, Hydrodynamic Turbulence, and the Cosmic Web

Tom Chang,[1] Cheng-chin Wu,[2] Marius Echim,[3,4] Hervé Lamy,[3] Mark Vogelsberger,[1] Lars Hernquist,[5] and Debora Sijacki[6]

Abstract—"Dynamic complexity" is a phenomenon observed for a nonlinearly interacting system within which multitudes of different sizes of large scale coherent structures emerge, resulting in a globally nonlinear stochastic behavior vastly different from that which could be surmised from the underlying equations of interaction. A characteristic of such nonlinear, complex phenomena is the appearance of intermittent fluctuating events with the mixing and distribution of correlated structures on all scales. We briefly review here a relatively recent method, ROMA (rank-ordered multifractal analysis), explicitly developed for analysis of the intricate details of the distribution and scaling of such types of intermittent structure. This method is then used for analysis of selected examples related to the dynamic plasmas of the cusp region of the Earth's magnetosphere, velocity fluctuations of classical hydrodynamic turbulence, and the distribution of the structures of the cosmic gas obtained by use of large-scale, moving mesh simulations. Differences and similarities of the analyzed results among these complex systems will be contrasted and highlighted. The first two examples have direct relevance to the Earth's environment (i.e., geoscience) and are summaries of previously reported findings. The third example, although involving phenomena with much larger spatiotemporal scales, with its highly compressible turbulent behavior and the unique simulation technique employed in generating the data, provides direct motivation for applying such analysis to studies of similar multifractal processes in extreme environments of near-Earth surroundings. These new results are both exciting and intriguing.

Key words: Fractals, ROMA, magnetospheric cusp, fluid turbulence, cosmic gas.

[1] Kavli Institute for Astrophysics and Space Research, Massachusetts Institute of Technology, Cambridge, MA 02139, USA. E-mail: tom.tschang@gmail.com; tsc@space.mit.edu

[2] Institute of Geophysics and Planetary Physics, University of California at Los Angeles, Los Angeles, CA 90095, USA.

[3] Belgian Institute for Space Aeronomy, 1180 Brussels, Belgium.

[4] Institute for Space Sciences, 077125 Bucharest, Romania.

[5] Harvard-Smithsonian Center for Astrophysics, Cambridge, MA 02138, USA.

[6] Institute of Astronomy, University of Cambridge, Cambridge CB3 OHA, UK.

1. Brief Description of ROMA

1.1. Preamble

Intermittent fluctuating events are popularly analyzed by use of the structure function and/or partition function methods. These methods investigate the multifractal characteristics of intermittency on the basis of the statistics of the full set of fluctuations. Because most of the observed or simulated intermittent fluctuations are dominated by fluctuations with small amplitudes, the subdominant fractal characteristics of the minority fluctuations—usually of larger amplitudes—are easily masked by those characterized by the dominant population. A new method of rank-ordered multifractal analysis (ROMA) was introduced to specifically address this concern (Chang and Wu 2008).

1.2. Monofractal Behavior

Consider, for example, a generic spatial series of a specific measure of physical turbulence: $\mu(x)$. To address its fluctuating characteristics, it is common to form the scale-dependent difference series $\delta\mu = \mu(x + \delta) - \mu(x)$ and consider the probability distribution functions (PDFs) $P(\delta\mu, \delta)$ for a range of spatial scales δ. Such PDFs for turbulent fluctuations are usually non-Gaussian with extended tails (Figs. 2, 4, 9, 12, 17, 18, 19, 20). If the phenomenon represented by the fluctuating measure is monofractal, i.e., self-similar, then the scale-dependent PDFs would map on to one scaling function P_s as follows (Chang et al. 2004):

$$P_s(\delta\mu/\delta^s) = \delta^s P(\delta\mu, \delta) \qquad (1)$$

where s is the scaling exponent.

Reprinted from the journal

To demonstrate this assertion, we note there is usually an irreducible basis of two independent power-law scale invariants for the variables (P, μ, δ): e.g., $\delta\mu/\delta^a = I$ and $P/\delta^b = J$, where (a, b) are the fractal exponents and (I, J) are constants—i.e. invariants in respect to the scale δ. If the functional form of $P(\delta\mu, \delta)$ is also invariant as the scale changes, it has been shown that a functional relationship exists between the two invariants (CHANG et al. 1973). Imposing the normalization condition for the PDFs, we obtain the one-variable scaling form, as shown in Eq. (1), where $s = a = -b$ is the lone fractal exponent (commonly known as the Hurst exponent). The PDFs are self-similar and monofractal because they map on to one master scaling function $P_s(Y)$ where $Y = \delta\mu/\delta^s$ is a global invariant and s is the only fractal exponent that enters the scaling expression. Such monofractal mapping was first used for analysis of solar wind turbulence by HNAT et al. (2002).

1.3. Structure Functions

A popular modus operandi designed to study the phenomenon of intermittency is based on the concept of "structure functions", S_q, defined by the moments of the PDFs:

$$S_q(\delta) = \int (\delta\mu)^q P(\delta\mu, \delta)\mathrm{d}(\delta\mu) \qquad (2)$$

The motivation here is that different moments emphasize different peaks in the fluctuating series.

In general, corresponding to each structure function of order q a fractal exponent ζ_q satisfying the power law relationship, $S_q = \delta^{\zeta_q}$, for some limited range of small values of δ may be defined. If $\zeta_q = \zeta_1 q$, then the fractal property of the fluctuating series in that range is characterized by the value of ζ_1. It may be easily demonstrated that PDFs satisfying the one-variable scaling form of Eq. (1) obey the monofractal property of $\zeta_q = \zeta_1 q$ with $s = \zeta_1$.

When the above linear relationship of ζ_q is violated, the fluctuating phenomenon is regarded as multifractal. In general, the nonlinear relationship between ζ_q and q is characterized by noticeable curvature for lower moment orders, q, and then becomes asymptotically a straight line for large

values of q. The reason for this linear asymptotic behavior is the unavoidable limitation of available sampling data.

Because the conventional structure function formalism is based on the moments of the full set of fluctuations (which are dominated by those of the small amplitudes), physical interpretation of the multifractal nature is not easily deciphered by merely examining the curvatures of the deviations from linearity, especially because of the generic linear asymptotic behavior. Furthermore, the structure function exponents are poorly defined because rarely do the actual data reveal truly power-law relationships between S_q and δ over the entire scaling range. In addition, even though structure function calculations may be performed conveniently for a fluctuating series for positive values of q, they invariably exhibit divergent characteristics for $q < 0$.

1.4. ROMA (Rank-Ordered Multifractal Analysis)

Thus, it seems reasonable to search for a procedure that enables investigation of the fractal, i.e., power-law, scaling behavior of the subdominant fluctuations by first appropriately isolating the minority populations and then performing the statistical investigation for each of the isolated populations. Such grouping of fluctuations must depend, somehow, on the sizes of the fluctuations. However, the groupings cannot depend merely on the raw values of the sizes of the fluctuations, because the ranges will be different for different scales. Therefore, we are led to proceed to rank-order the sizes of the fluctuations on the basis of the local invariant $Y = \delta\mu/\delta^s$ where s is the scaling exponent for each (local) grouping.

Consider a differential range of $\mathrm{d}Y$ in the vicinity of some scaled size $Y = \delta\mu/\delta^s$. We expect the fluctuations whose sizes fall within this differential range to have monofractal behavior characterized by the local scaling exponent, s, such that the differential structure function $\mathrm{d}S_q$ will vary with the scale as δ^{sq} according to:

$$\mathrm{d}S_q \triangleq (\delta\mu)^q P(\delta\mu, \delta)\mathrm{d}\delta\mu = \delta^{sq} Y^q P_s(Y)\mathrm{d}Y \qquad (3)$$

Given an ensemble of PDFs $P(\delta\mu, \delta)$, the corresponding multifractal spectrum $s(Y)$ may be obtained, approximately, (if the ansatz is valid) by integrating the functional differential expression Eq. (3) over small contiguous ranges of ΔY with the assumption that within each incremental range the scaling exponent s is essentially a constant. Thus, for a range of ΔY within (Y_1, Y_2), we form a range-limited structure function as follows:

$$\Delta S_q(\delta\mu, \delta) = \int_{a_1}^{a_2} (\delta\mu)^q P(\delta\mu, \delta) \mathrm{d}\delta\mu$$
$$\simeq \delta^{sq} \int_{Y_1}^{Y_2} Y^q P_s(Y) \mathrm{d}Y \qquad (4)$$

where $a_1 = Y_1 \delta^s$ and $a_2 = Y_2 \delta^s$. We may then search for the value of s such that the scaling property of the range-limited structure function that varies with s is $\Delta S_q(s) \sim \delta^{sq}$. If such a value of s exists, then we have found one region of the multifractal spectrum of the fluctuations such that the PDFs in the range of ΔY collapse on to one scaled PDF. Performing this procedure for all contiguous ranges of ΔY will produce the approximate rank-ordered multifractal spectrum $s(Y)$ we are seeking. The determined value of s for each grouping should be unaffected by the statistics of other subsets of fluctuations that are not within the chosen range ΔY and, therefore, should be quantitatively quite accurate. If this spectrum exists, the PDFs for all time lags collapse on to one master multifractal scaled PDF, $P_s(Y)$. The spectrum will be implicit, because Y is defined as a function of s (the local Hurst exponent).

The above procedure, commonly known as ROMA, was first introduced by CHANG and WU (2008). Since then, a flurry of activity in space plasma turbulent studies has utilized this procedure for analysis of multifractal and intermittent characteristics (CONSOLINI and DE MICHELIS 2011; TAM et al. 2010; reviewed by CHANG et al. 2010 and CHANG 2014).

We conclude this section by noting that definitions of self-similarity and intermittency of random functions are rather imprecise in the turbulence literature [as discussed by, e.g., FRISCH (1995)]. Because monofractal scaled PDFs are mathematically, by definition, self-similar, multifractal scaled PDFs are therefore non-self-similar and fluctuations represented by such PDFs may be defined as intermittent without ambiguity in the context of ROMA.

2. Turbulent Fluctuations in the Magnetospheric Cusp

The terrestrial magnetosphere is a bubble in the solar wind created by the magnetic field of the Earth. The cusp is characterized by the magnetospheric region through which plasma from the solar wind can have direct access to the upper ionized atmosphere of the Earth. The plasma in this region is highly turbulent and previous analysis on the basis of flatness, wavelet transforms, structure, and partition functions (YORDANOVA et al. 2004, 2005; ECHIM et al. 2007) indicated that the cusp magnetic field fluctuations were usually intermittent and multifractal in nature.

We review below some of the statistical properties of the magnetic energy fluctuations in the cusp region observed by Cluster (a constellation of four identical spacecraft in tetrahedral formation launched in 2000 by the European Space Agency with NASA participation traversing the vicinity and the interior of the Earth's magnetosphere). In space plasmas, physical variables, for example the magnetic field intensity, are sampled by satellite measurements along the orbit. Observed time (t) variations are assumed to be equivalent to spatial variations when the Taylor hypothesis (TAYLOR 1938) is valid, i.e. when a turbulent structure, for example an eddy, transits the spacecraft within a time period smaller than its own time of evolution. Figure 1 displays the time series of magnetic field measurements during a typical cusp passage.

An implicit partial removal of the dipole component of the magnetic field fluctuation data may be achieved by computing first the differences $\delta B^2(t, \tau)$ from the raw data with τ being the time scale and then the mean value of the fluctuations on each scale is subtracted yielding a new "ensemble" of fluctuations (ECHIM et al. 2007),

$$\delta b^2 \equiv \frac{\delta B^2(t, \tau) - \langle \delta B^2(t, \tau) \rangle}{\sigma^2} \qquad (5)$$

where the bracket indicates the ensemble average and σ is the variance. Typical PDFs were computed for the quantity, δb^2, where differences δB^2 have been calculated by moving an overlapping window of width $\tau = 2^j \delta t$ over the entire time interval with $\delta t = 0.0015$ s being the time resolution of the measurements and $j = 1, 2, \ldots, 15$ (Fig. 2).

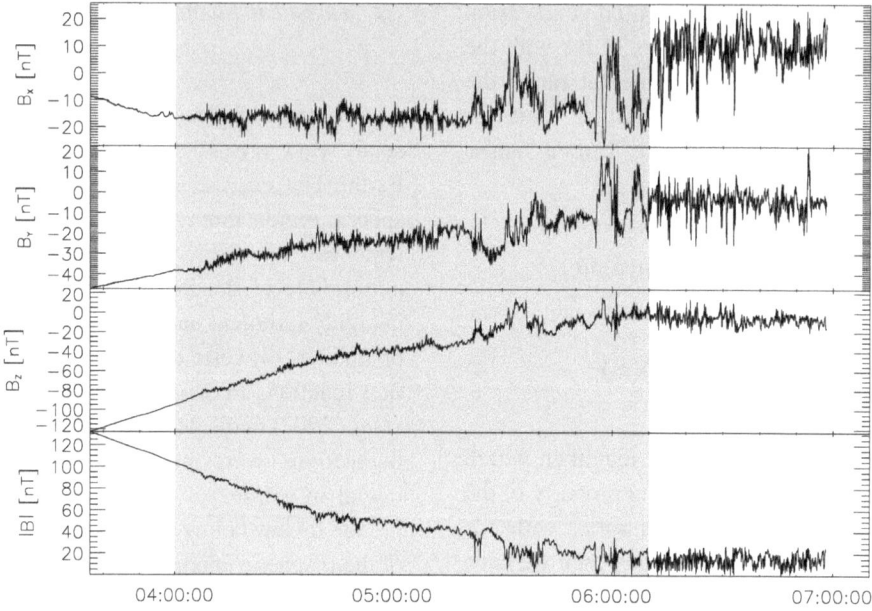

Figure 1
The components (*top* three panels) and intensity (*bottom* panel) of the magnetic field measured by Cluster-1. The interval corresponding to
traversal of the cusp region has been shaded in *gray* (Echim *et al.* 2007)

A ROMA analysis using the aforementioned
approximate integral technique was performed for the
chosen data set in the cusp region, (Lamy *et al.* 2008;
Echim and Lamy 2010, private communication). We note
from Fig. 3 that for small values of Y, the fluctuations
were persistent ($s > 0.5$), indicating the turbulence was
unstable and probably not yet completely fully developed. For larger values of Y, the fluctuations became antipersistent ($s < 0.5$) and the turbulence was probably
well-developed and became sparser and sparser as the
value of Y increased. (N.B.: The special situation for
$s = 0.5$ may be shown to correspond to fluctuations of
classical random diffusion.) The ROMA spectra for all
four spacecraft were very similar, indicating that,
although the magnetic field fluctuations were anisotropic, the magnetic energy density fluctuations in the
cusp were essentially statistically isotropic over the
distance covered by the cusp passage for spatial scales
of the separation distance between the Cluster spacecraft ($\sim 1,000$ km).

3. An Interlude: Refined Procedure for ROMA

The above discussion yields ROMA spectra that
are step-wise discontinuous. One might wish to

improve the calculated result by progressively
reducing the size of ΔY. However, this procedure may
be limited by the statistics within each ΔY because of
the statistically insufficient amount of available data.
Therefore, a refined method of obtaining a continuum
of the spectrum $s(Y)$ and the associated scaled PDF
$P_s(Y)$ was suggested by Wu and Chang (2011) as
described briefly below.

We can write the ROMA scaling relationships as:

$$P(X, \delta)\delta^{s(Y)} = P_s(Y) \text{ with } Y = X/\delta^{s(Y)} \qquad (6)$$

Thus, for a given value of s, we may plot $P(X, \delta)\delta^{s(Y)}$
against $Y = X/\delta^{s(Y)}$ for the various scales δ of the
PDFs. If the curves intersect at some point Y_1 then
ROMA is satisfied at $s = s(Y_1)$. On the other hand, if
only curves within some range of scales δ intersect,
then ROMA is satisfied at $s = s(Y_1)$ only for that
range of scales. (This, in fact, sets the stage for scenarios in which there are multiple ROMA scaling
ranges as we shall discover below.)

Continuing this procedure for the full range of
values of s leads us to a continuum ROMA $s(Y)$ and
the corresponding scaled PDF $P_s(Y)$. We may then
fine tune the result numerically by calculating the
PDFs from $s(Y)$ and $P_s(Y)$ and then compare them
with the original observed or numerically simulated

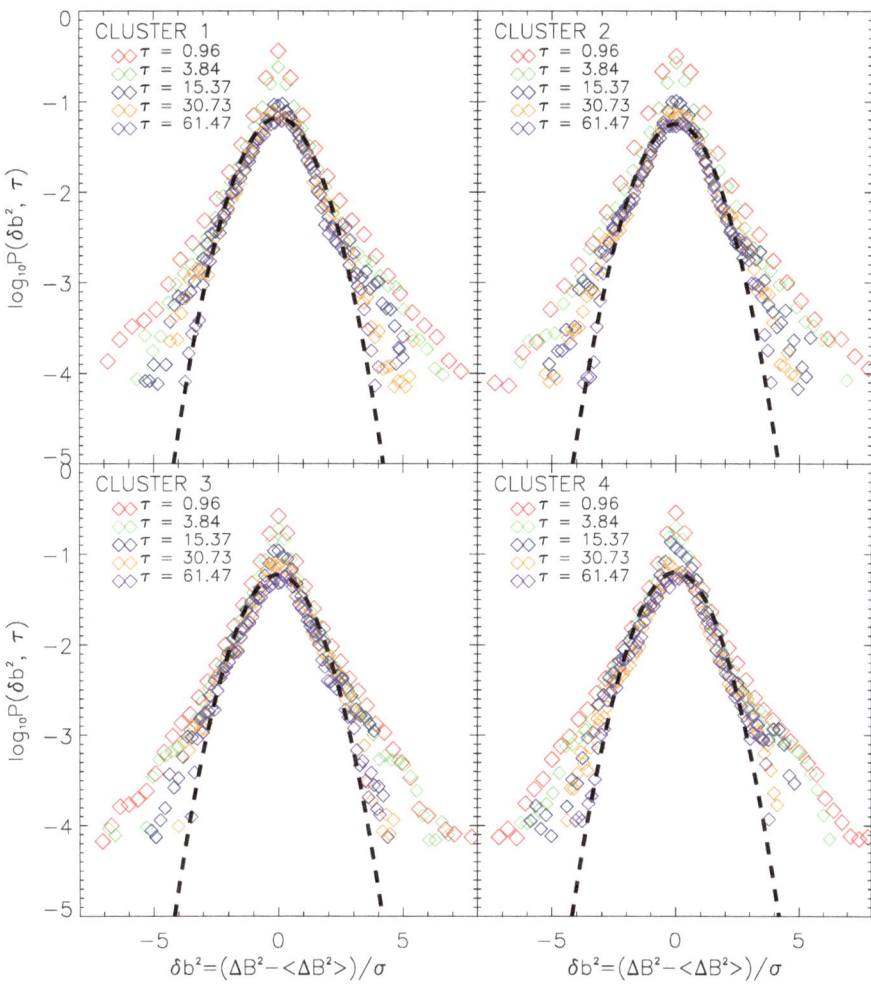

Figure 2

Typical PDFs of the magnetic energy density fluctuations measured by the four Cluster satellites in the cusp for the time period described in Fig. 1; the PDFs have been scaled with respect to their variance. The scales of τ are *color coded* and given in seconds (Echim *et al.* 2007)

PDFs. Such a procedure ensures that the results obtained are unique. In other words, the continua of $s(Y)$ and $P_s(Y)$ represent the full ensemble of the raw PDFs of the observational or simulated results.

In the sections below we shall apply this procedure to the velocity fluctuations of classical driven hydrodynamic turbulence and the complexity phenomenon of the cosmic gas.

4. Classical Driven Hydrodynamic Turbulence

It is well-known that fully developed turbulent fluid flows are intermittent and multifractal (Frisch 1995, and references therein). We have shown that

ROMA could be useful for analysis of fluid turbulence (Chang *et al.* 2010). We applied the technique to the Johns Hopkins University (JHU) large-scale direct numerical simulation turbulence database based on the Navier–Stokes equations (Perlman *et al.* 2007; Li *et al.* 2008).

Briefly, the data were obtained from a direct numerical simulation of forced isotropic turbulence of a periodic box of $(2\pi)^3$ on a $(1,024)^3$ grid using a pseudo-spectral parallel code. Energy was injected by keeping constant the total energy in modes such that their wave-number magnitude is less than or equal to 2. After the simulation reached a statistically stationary state, 1,024 frames for every 10 time steps of data, which included the 3 components of the velocity

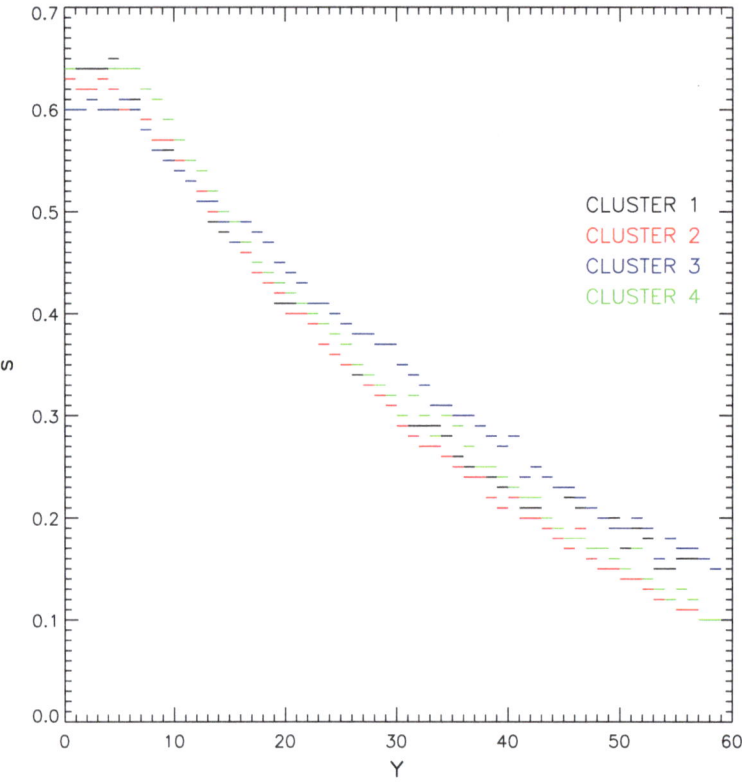

Figure 3
Rank-ordered multifractal spectra for the Cluster pass of Fig. 1 in the cusp. (Echim and Lamy 2010, private communication; Chang *et al.* 2010)

vector and the pressure, were generated and stored into the database. The duration of the stored data was approximately one large-eddy turnover time of 2.024. The radial energy spectrum averaged over this duration indicated the presence of an inertial range of wavenumbers between 8 and 60, approximately, corresponding to a spatial range from 17Δ to 128Δ where $\Delta = 2\pi/1,024$ is the grid spacing. Instead of the huge $1,024^4$ data points, 19 x-planes of data points were used in the analysis. They were arbitrarily selected at different x locations and different times. This set of 19 million data points provided sufficient statistics.

We considered the fluctuations of longitudinal velocity, $\delta v_{\parallel}$, defined by

$$\delta v_{\parallel}(\mathbf{r}, \delta) = (\mathbf{v}(\mathbf{r} + \delta\mathbf{i}) - \mathbf{v}(\mathbf{r})) \cdot \mathbf{i} \qquad (7)$$

where $\mathbf{i}$ is the unit vector and δ is the spatial scale. In the analysis $\mathbf{i}$ was either in the y or z-axis and δ

is in the range of $(16\Delta, 160\Delta)$. Figure 4 gives the PDF results for $\delta v_{\parallel}$ at scale $\delta = 64\Delta$. In computing the PDF, the range of $\delta v_{\parallel}$ was divided into 1,601 bins. For bin number i: $\left(i - \frac{1}{2}\right)\Delta_v < \delta v_{\parallel} \le \left(i + \frac{1}{2}\right)\Delta_v$. The bin size Δ_v was set as 8/1,601 because the maximum value of $|\delta v_{\parallel}|$ is slightly <4. The PDF is asymmetrical in $\delta v_{\parallel}$ and thus may be decomposed into a symmetrical and an antisymmetric part. The reason for the asymmetry is the 3D nature of the fluctuations.

A ROMA calculation for the PDFs was performed for the simulation results using the refined method. Interestingly, despite the asymmetric property of the PDFs, the ROMA spectrum was found to be symmetric as shown in Fig. 5a. The PDFs were mapped on to a scaled master curve $P_s(Y)$ which was asymmetric as shown in Fig. 5b, c.

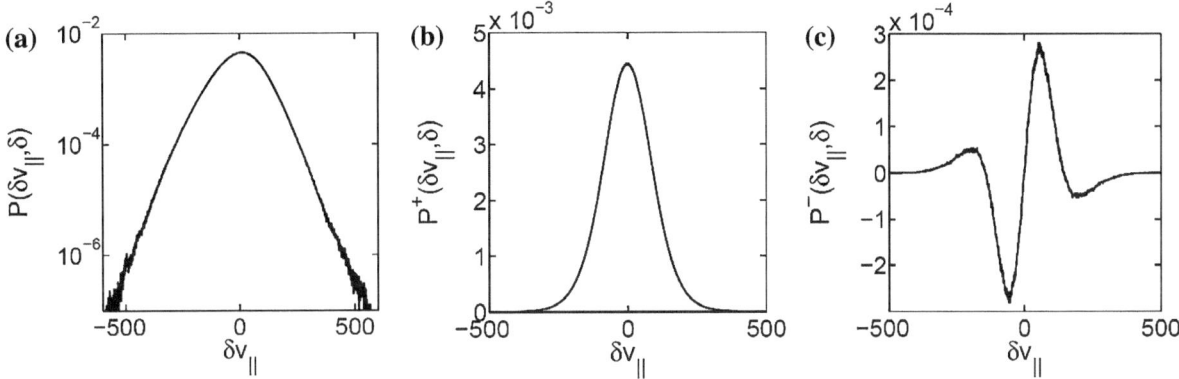

Figure 4

a PDF of $\delta v_\parallel$ for hydrodynamic turbulence at $\delta = 64\Delta$ in units of bin size. **b** and **c** Symmetric and antisymmetric plots (P^+, P^-) of the PDF (WU and CHANG 2011), respectively

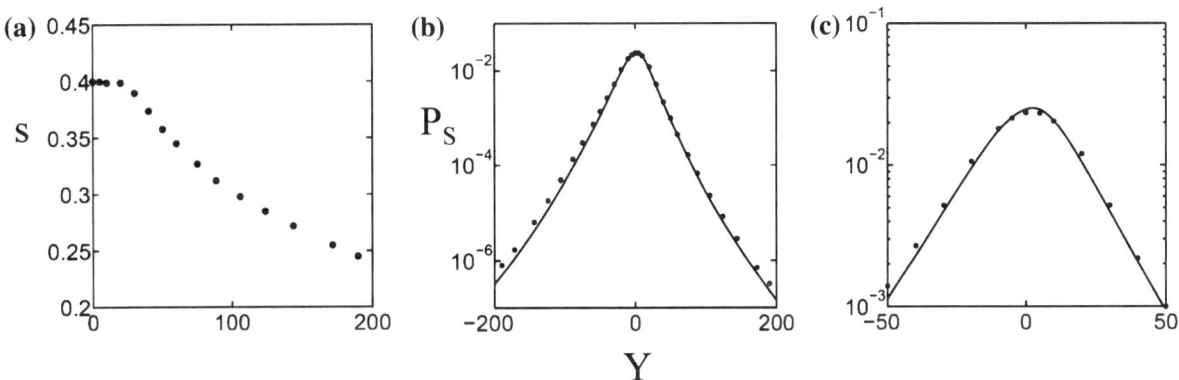

Figure 5

ROMA spectrum $s(Y)$ and scaled PDF $P_s(Y)$ for the fluctuations of longitudinal velocities for hydrodynamic turbulence. **a** $s(Y)$, which is symmetric with respect to $Y = 0$, has approximate monofractal behavior at $Y \leq 25$ and decreases monotonically at larger Y. **b** $P_s(Y)$ is asymmetric about $Y = 0$. **c** Magnification of **b** with smaller range of Y (Modified from WU and CHANG 2011)

Figure 6 shows the calculated $P(\delta v_\parallel, \delta)$ based on the ROMA scaling relationships of Eq. (6) and the results of $s(Y)$ and $P_s(Y)$ as shown in Fig. 5 for scales from $\delta = 24\Delta$ to 128Δ. The comparison with the PDFs from the data is also shown. The results demonstrate that in the inertial range the PDFs of the analyzed fluid turbulence exhibit multifractal scaling can be described using the ROMA decomposition analysis. Conversely, this also means that the two functions $s(Y)$ and $P_s(Y)$ faithfully reproduce all the PDFs in the inertial range. Thus, the results are unique.

In addition to the fluctuations of the longitudinal velocities, the PDFs of the fluctuations of the square of the velocity, v^2, which for incompressible flow are measures of the kinetic energy, also satisfied the ROMA scaling relationships. The characteristics of the ROMA spectrum here are very similar to those of the magnetospheric cusp magnetic energy fluctuations, Fig. 7. The difference is that turbulence analyzed here seems to be more well developed and the local Hurst exponent is everywhere antipersistent.

5. Unraveling the Complexity of the Cosmic Gas

5.1. Prologue

We are keen to include this third example of ROMA in this treatise both in the spirit of cross-discipline fertilization and exchange of scientific techniques and ideas, and in the demonstration of the importance of the role of multiple shock structures in

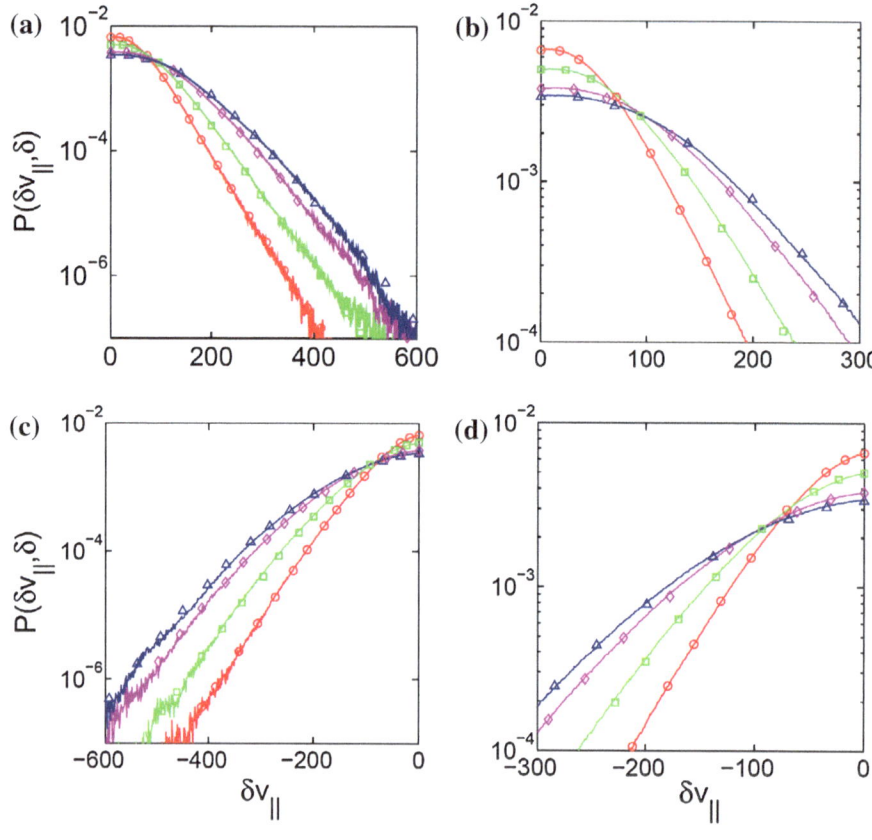

Figure 6
Plots of $P(\delta v_{\parallel}, \delta)$ for hydrodynamic turbulence: *solid curves* are from the simulation data and *markers* are from ROMA scaling relationships: *red* (*circles*) for $\delta = 24\Delta$; *green* (*squares*) for $\delta = 48\Delta$; *magenta* (*diamonds*) for $\delta = 96\Delta$; *blue* (*triangles*) for $\delta = 128\Delta$. **a** and **b** are for positive $\delta v_{\parallel}$; **c** and **d** are for negative $\delta v_{\parallel}$. **b** and **d** are enlarged plots

strongly turbulent and compressible gaseous media; situations that may arise in the Earth's upper atmosphere, ionosphere and magnetosphere, the heliosphere, and beyond. In addition, the unique moving mesh numerical simulation technique which provided the basic data for the ensuing analysis should prove to be especially useful for analyses of many realistic domains of the geo and space environments in which accurate multi-resolution dynamic studies are required. All the results reported below are new findings.

The commonly accepted theory of cosmic evolution that explains the clumpiness (filaments, pancakes, clusters, voids, etc.) of the baryonic matter content of the Universe is the ΛCDM (Lambda cold dark matter) model. It is based on the (FLRW) Friedmann–Lemaître equations and the Robertson–Walker expansion metric of Einstein's equations of

general relativity with inclusion of the cosmological constant term, Λ. In addition to the observable baryonic matter, the model includes a cold dark matter component in an effort to explain the observed anomalous rotational curves of the galaxies and gravitational lensing of light by the clusters of galaxies and to provide the gravitational backbone for cosmic evolution and structure formation. The cosmological constant (contributing to a constant negative pressure in some form of "dark energy") is included in the model to account for the accelerating expansion of the Universe (RIESS *et al.* 1999; PERMUTTER *et al.* 1999). An excellent up-to-date introduction to the subject is given by LIDDLE (2013).

Recent remarkable advances in supercomputing and numerical simulation based on the ΛCDM model have provided realistic results that give significant credence to the above theoretical modeling. The

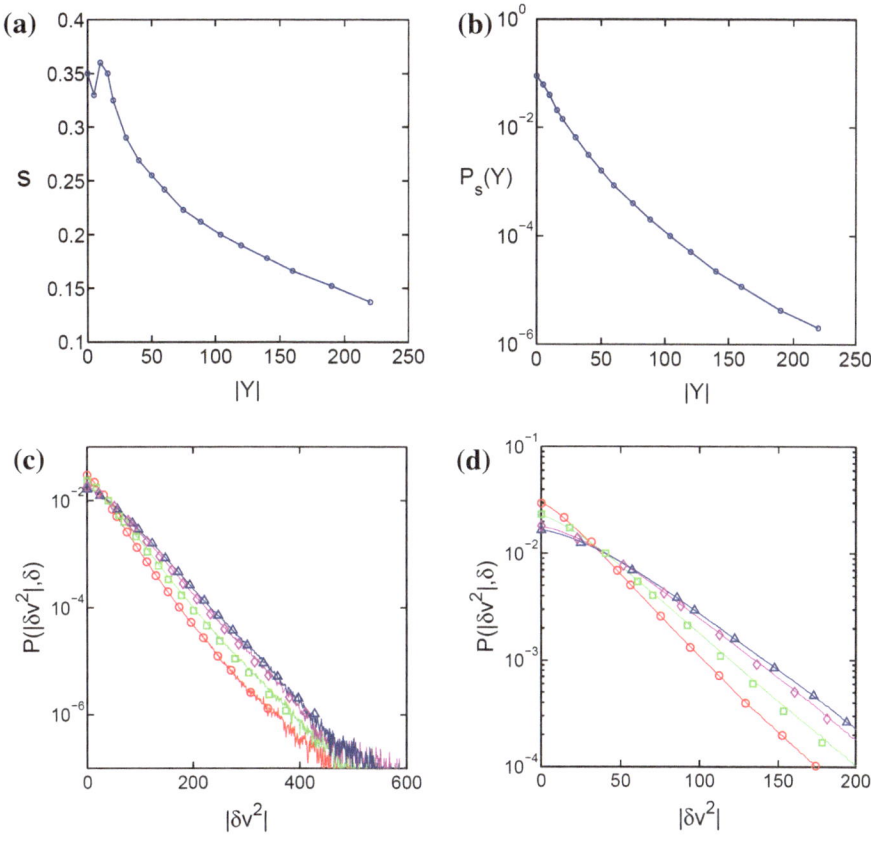

Figure 7

Fluctuations of v^2 for hydrodynamic turbulence. **a** and **b** ROMA spectrum $s(Y)$ and scaled PDF $P_s(Y)$. **c** plots of $P(\delta v^2, \delta)$: *solid curves* are from simulation data and *markers* are from ROMA scaling relationships: *red* (*circles*) for $\delta = 24\Delta$; *green* (*squares*) for $\delta = 48\Delta$; *magenta* (*diamonds*) for $\delta = 96\Delta$; *blue* (*triangles*) for $\delta = 128\Delta$. **d** is an enlarged plot for $\delta v^2 \leq 200$

simulations are generally based on the Newtonian approximation of the FLRW equations in terms of the expansion term and comoving coordinates (PEEBLES 1980). The physical parameters in the model are guided by observational inputs and constraints, for example those obtained by the WMAP (Wilkinson microwave anisotropy probe) survey and more recently by the Planck survey.

In particular, self-consistent ab-initio hydrodynamic simulations of the dark matter and baryonic gas based on the above formalism combined with reasonable feedback mechanisms, radiative cooling, UV ionization and heating, and other relevant physics have yielded reasonable comparisons of the simulated results with observations (VOGELSBERGER *et al.* 2013; TORREY *et al.* 2013, and references contained therein). For example, there have been studies of the galaxy stellar mass functions, star formation, and comparisons of

quasistellar absorption lines associated with the structure of the intergalactic medium, for example the Lyman-Alpha forest spectra (HERNQUIST *et al.* 1996; BIRD *et al.* 2013, and references contained therein).

There have been attempts at statistical studies of the overall density and velocity fields of the cosmic gas, notably those related to comparisons of observed and simulated power spectra and correlation functions. There is also, of course, the self-similar spectrum of condensates for the Friedmann cosmology of PRESS and SCHECHTER (1974) and its comparison with observational and simulation data. In addition, some studies of the intermittent (i.e., non-self-similar) fluctuations of the cosmic structure of the baryonic gas in terms of the traditional methods of structure and partition function analyses based on the simulated results have been reported (ZHU *et al.* 2011, and references contained therein).

In the following discussion we address the statistical properties of the density, kinetic energy, velocity, and linear momentum of the hierarchical baryonic gas distributions of simulation data by use of the ROMA technique. As indicated above, such studies will provide the quantitative tool to assess the similarities and differences of intermittent distributions on large, small, and intermediate scales. Because ROMA retains all the statistical information of the cosmic distributions, it not only contains the power spectra and correlation information that are usually reported in the observational and simulation literature, but also other quantitative information about the intermittent structures, which are useful for more in-depth comparisons with observations such as those considered by FANG (2006) and LOVEJOY et al. (2000).

Our ROMA discussed in this paper is based on the recent moving mesh AREPO simulation results of VOGELSBERGER et al. (2012). AREPO was developed by SPRINGEL (2010). It is a second-order accurate finite volume code that uses an unstructured moving mesh based on Voronoi tessellations using a set of moving mesh generating points. The physics implemented for the primordial helium–hydrogen mixture included optically thin radiative cooling, uniform but time-dependent ionizing UV background, and simple star formation and supernova feedback (SPRINGEL and HERNQUIST 2003). The simulation was performed in a periodic box of 20 h^{-1} Mpc (where h is the Hubble constant) and the input data were chosen to be consistent with the recent WMAP-7 measurements (KOMATSU et al. 2011) and other observational constraints. Initial conditions were generated at redshift $z = 99$ based on the spectrum fit of EISENSTEIN and HU (1999). The simulation evolved until $z = 0$. Other details may be found in the paper by VOGELSBERGER et al. (2012).

5.2. Scaling Properties of the Cosmic Density

The data analyzed in this subsection are taken from the snapshot at z(redshift) $= 0$ of the VOGELSBERGER et al. simulation (2012). The data given in terms of the moving mesh generating points and Voronoi tessellation were first appropriately redistributed to 512^3 grid of cubic cells and the fluctuations on different scales were then generated. We note from Fig. 8 that the distribution of the simulated cosmic gas involves densities that span many orders in magnitude. The figure indicates that even for just a two-dimensional slice of the simulation box, the hierarchical structure is already very complicated.

Our analysis, in a sense, is similar to the well-known PRESS and SCHECHTER (1974) (PS) idea of the mass distribution of the hierarchical condensates for the Friedmann cosmology. Through some heuristic arguments, the PS formalism predicted a power law scaling of the mass condensates and an exponential cutoff with characteristics equivalent to those of the monofractal scaling expression of Eq. (1) as discussed in the Sect. 1.2. The difference is that our analyses here are focused on the "cosmic gas" and their incremental changes of densities, etc., on different scales. In terms of the cosmic gas, our results contain more detailed description of its statistical distributions and scaling behavior, particularly those related to their intermittent characteristics. Figure 9 gives the PDFs for the density fluctuations of the gas on scales $\delta = 32\Delta, 64\Delta, 128\Delta, 256\Delta$ with $\Delta = 20/512\, h^{-1}$Mpc. The fluctuation size $|\delta\rho|$ is expressed in units of $(1/800)\, 10^{10} \odot /\Delta^3$ where $\odot$ is the solar mass. The distributions are distinctly non-Gaussian. We note that on these large scales, the PDFs are scale independent and, therefore, have asymptotically the same power-law behavior at large values of the fluctuation size $|\delta\rho|$. The scale-independent property indicates that the density fluctuations are self-similar and homogeneous for $\delta \geq 32\Delta$.

On smaller scales ($\delta = 4\Delta, 8\Delta, 16\Delta, 32\Delta$), the PDFs become non-self-similar and therefore the fluctuations are intermittent. The calculated ROMA spectrum $s(Y)$ and the scaled PDF $P_s(Y)$ for these scales are given in Fig. 10. The $s(Y)$ is persistent for small local scale invariant Y but soon becomes antipersistent as Y increases, indicating that the fluctuations are predominately fully developed on sufficiently large scaled sizes.

A comparison of the PDFs generated from ROMA scaling with those of the simulated data is shown in Fig. 11 using the same unit of the bin sizes as for

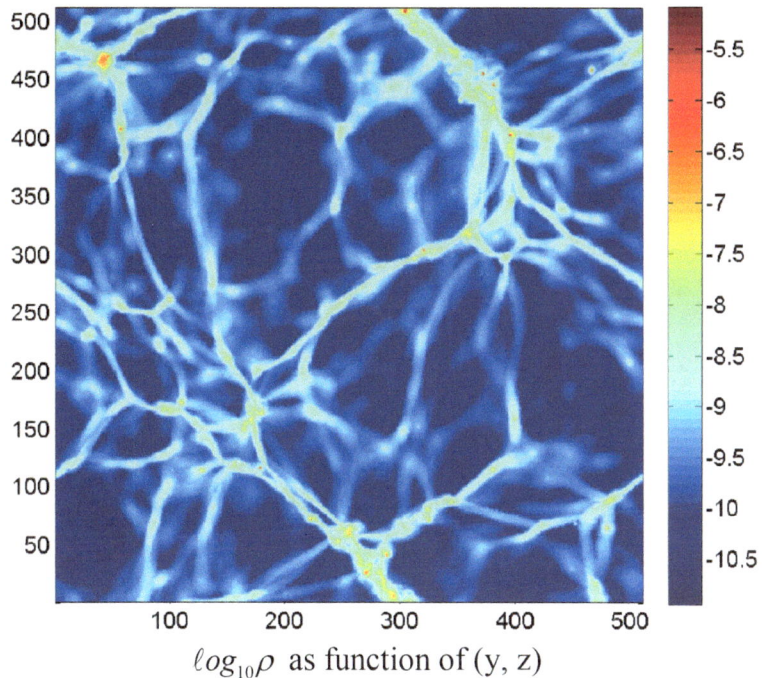

$$log_{10}\rho \text{ as function of (y, z)}$$

Figure 8

Density distribution ρ of the cosmic gas in units of $10^{10} \odot /(\mathrm{kpc}/h)^3$ for $0 < x < \Delta$ with $\Delta = 20/512\,h^{-1}\mathrm{Mpc}$, at the snapshot of $z(\text{redshift}) = 0$

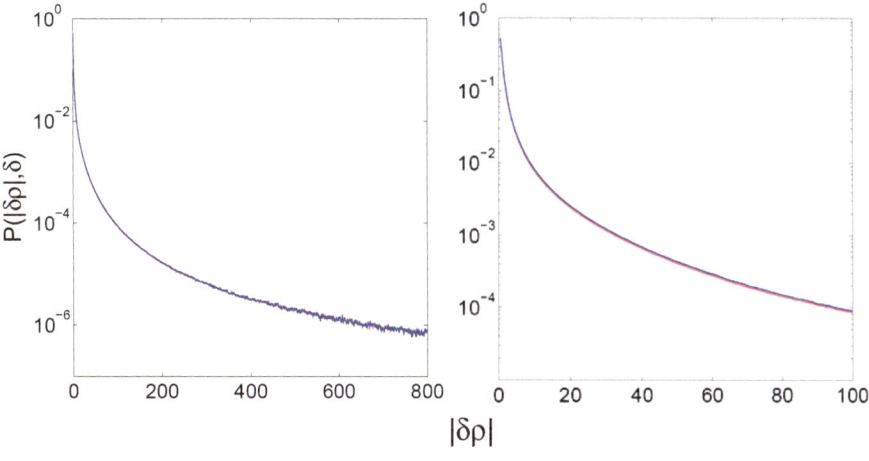

Figure 9

PDFs of density fluctuations $|\delta\rho|$ of the cosmic gas in units of $(1/800)$ times $10^{10} \odot /\Delta^3$ at $\delta = 32\Delta, 64\Delta, 128\Delta, 256\Delta$ with $\Delta = 20/512\,h^{-1}\mathrm{Mpc}$. *Curves* of *different colors* representing different scales lie on a *single curve* indicating scale independence. The figure on the *right* is a magnification of the figure on the *left*

Fig. 9. Except for the slight deviations for $\delta = 32\Delta$ the agreements on these small scales are quite striking. We note that the PDFs, although non-self-similar, nevertheless still have approximately the same power law behavior at large bin sizes.

5.3. Asymmetric Intermittency of the Longitudinal Velocity Fluctuations

Perhaps the most strikingly interesting intermittency scaling behavior of the cosmic gas is related to those characterized by the longitudinal fluctuations

Reprinted from the journal

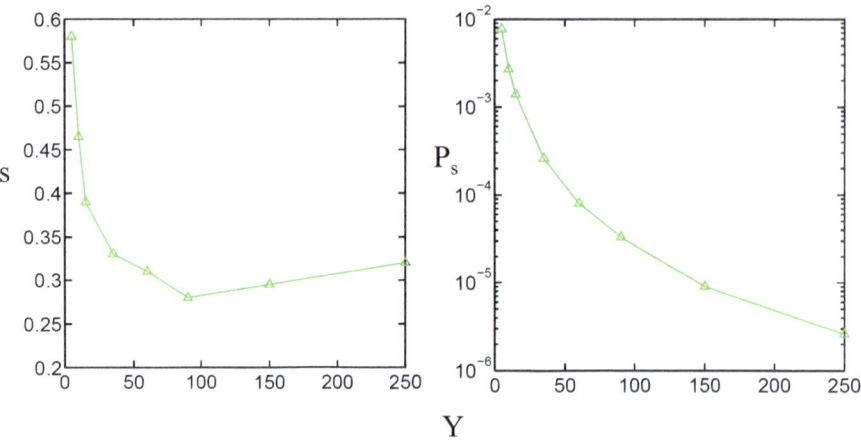

Figure 10
ROMA scaling for density fluctuations of the cosmic gas at $\delta = 4\Delta, 8\Delta, 16\Delta, 32\Delta$

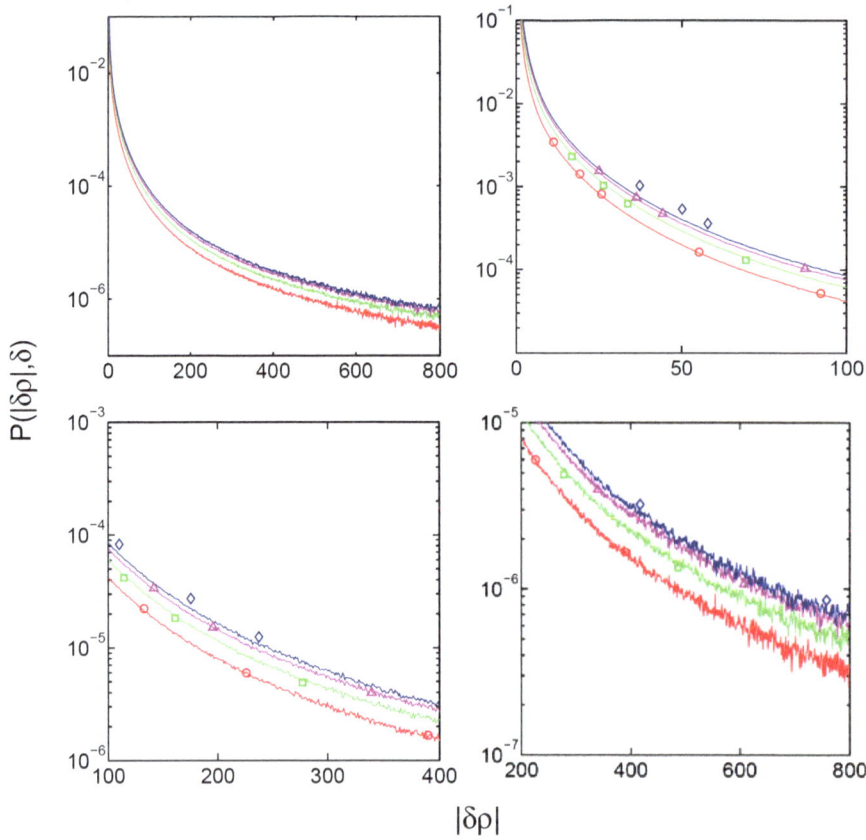

Figure 11
A sequence of magnifications of PDFs for density fluctuations of the cosmic gas generated by ROMA scaling (*markers*) and simulation data at $\delta = 4\Delta(red), 8\Delta(green), 16\Delta(magenta), 32\Delta(blue)$

$\delta v_\parallel$ as defined in Eq. (7). Figure 12 gives PDFs of $\delta v_\parallel$ for scales $\delta = 32\Delta, 64\Delta, 96\Delta, 128\Delta$ based on v_x, v_y, v_z, and the average PDFs over the three directions for the VOGELSBERGER *et al.* (2012) simulation data at z(redshift) ≈ 2.3. Because the values of $\delta v_\parallel$ lie within the range (944, −1115) km s^{-1}, in our

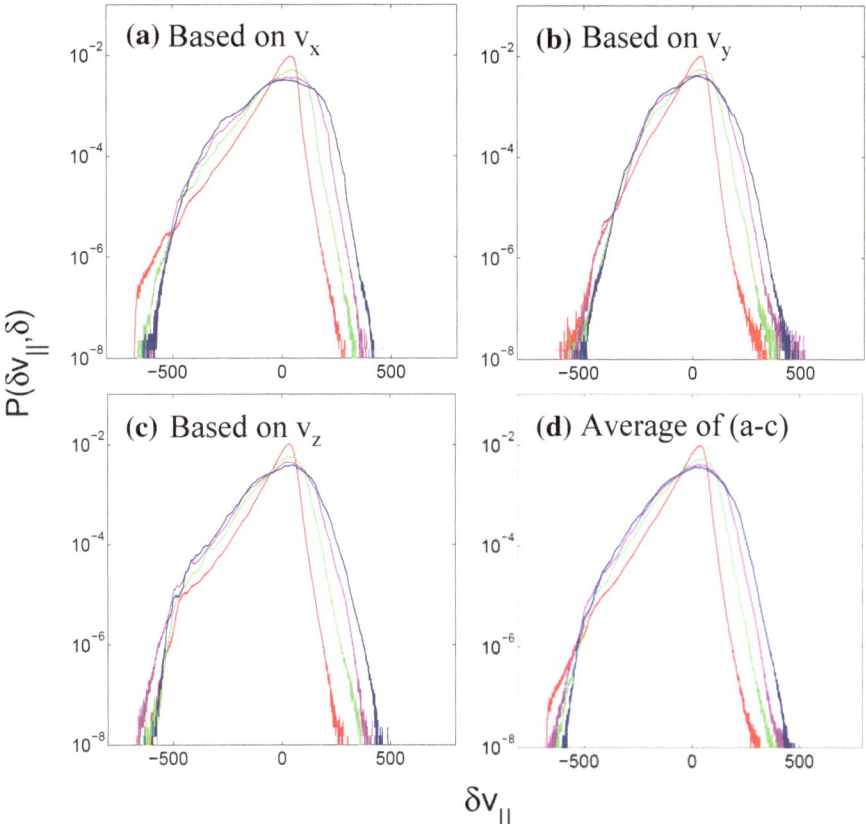

Figure 12

PDFs of $\delta v_\parallel$ of the cosmic gas in units of 1,100/800 km s^{-1}. $\delta = 32\Delta\,(red), 64\Delta\,(green), 96\Delta\,(magenta), 128\Delta\,(blue)$. **a** Based on v_x. **b** Based on v_y . **c** Based on v_z. **d** Average of (**a**–**c**)

analysis, we set the range to be $(-1,100,\ 1,100)$ km s^{-1}, and divide it into 1,600 bins, Thus, the units of $\delta v_\parallel$ are in 1,100/800 km s^{-1}.

Two properties are apparent from visual inspection of these plots. The first is that the fluctuations are isotropic on these scales. The second is the asymmetry in $\delta v_\parallel$. The asymmetry is much more pronounced than that of the PDFs for the classical driven hydrodynamic turbulence discussed in the Sect. 4. The reason may be the presence of shock and rarefaction waves in a compressible gas. It is easy to demonstrate that shock waves cause $-\delta v_\parallel$ fluctuations and rarefaction waves create $+\delta v_\parallel$ fluctuations. These asymmetric contributions to $\delta v_\parallel$ can become quite pronounced for highly compressible media with multiple shock events, for example the cosmic gas. Thus, we expect the intermittent fluctuating behavior of $\delta v_\parallel$ of the cosmic gas to be quite different from that for classical driven hydrodynamic turbulence

even if both of their traditional structure function spectra have similar nonlinear signatures.

ROMA spectra and scaled PDFs for the PDFs on these large scales are given in Figs. 13 and 14. For positive $\delta v_\parallel$, all PDFs collapse on to one scaled $P_s(Y)$. The corresponding $s(Y)$ is persistent for the small local scaling invariant Y but gradually becomes antipersistent as Y increases, indicating that the fluctuations are somewhat unstable for small sizes but become stable, sparsely distributed, and well developed for large size fluctuations. For negative $\delta v_\parallel$, all the PDFs for scales between $\delta = 32\Delta$ and 64Δ are describable by a scaled PDF with a ROMA spectrum for $-Y$ similar to that for the positive Y values. On larger scales, however, the PDFs could not be represented by a ROMA spectrum based on either the original or refined method of ROMA analysis (Fig. 14a). This extraordinary property will require further investigation.

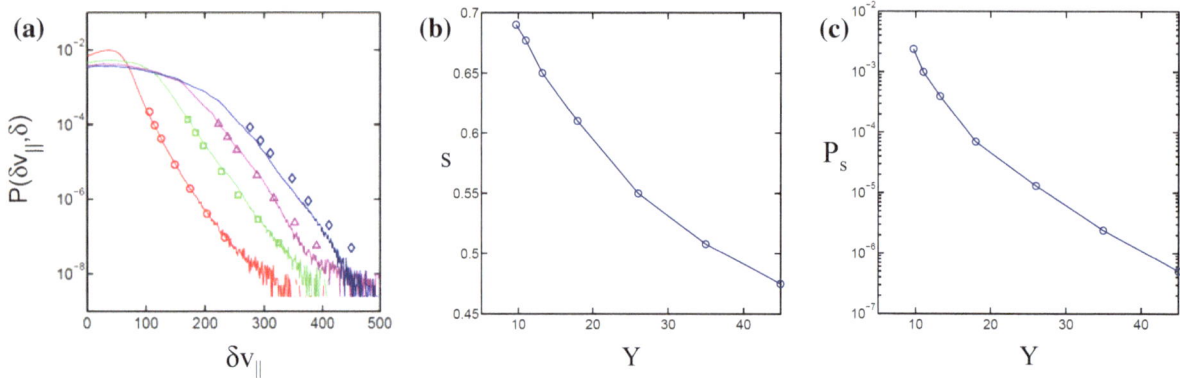

Figure 13

ROMA scaling and comparison with PDFs of simulation data for $+\delta v_\parallel$ at $\delta = 32\Delta(red), 64\Delta(green), 96\Delta(magenta), 128\Delta(blue)$. "$\delta v_\parallel$" of the cosmic gas in units of $1{,}100/800$ km s^{-1}. *Markers* show ROMA scaling using $s(Y)$ and $P_s(Y)$ **a** pdf, **b** s(Y), **c** P$_s$(Y)

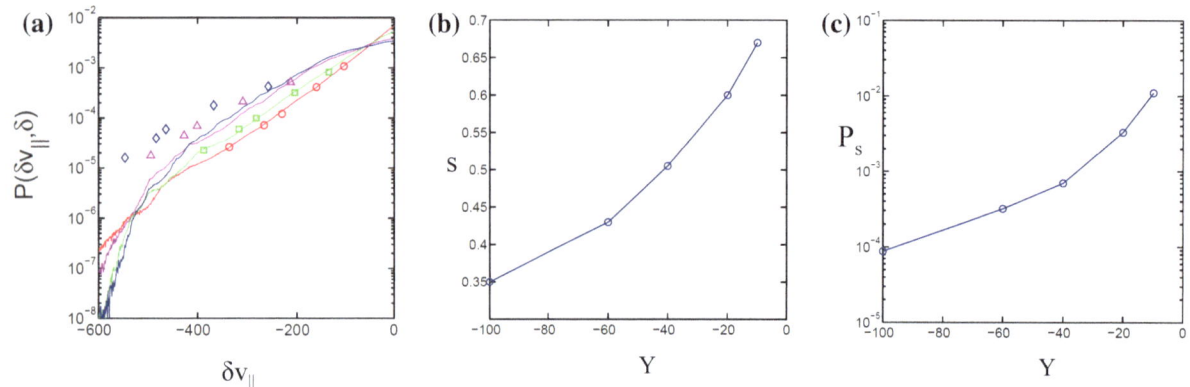

Figure 14

ROMA scaling and comparison with PDFs of simulation data for $-\delta v_\parallel$ at $\delta = 32\Delta(red), 64\Delta(green), 96\Delta(magenta), 128\Delta(blue)$. "$\delta v_\parallel$" of the cosmic gas in units of $1{,}100/800$ km s^{-1}. *Markers* show ROMA scaling using $s(Y)$ and $P_s(Y)$. **a** pdf. **b** s(Y). **c** P$_s$(Y)

For smaller scales, the PDFs map nicely on to separate ROMA scaling curves and the simulated results compare well with the ROMA scaling predictions. Skipping the intermediate details, we summarize the ROMA findings for all scales in Figs. 15 and 16. We note that aside from a restricted large positive Y region, the ROMA results for small scales and large scales fall on separate scaling curves indicating that the physics of the cosmological evolution process are different for scales larger and smaller than a demarcation region in the vicinity of 16Δ to 32Δ. AREPO simulation in terms of the mesh generating points and Voronoi cells has the capability of providing multi-resolutions for sparse and dense regions. It will be interesting to utilize this capability to analyze the dense regions on much smaller scales

with higher resolution to search for their special scaling properties.

5.4. *Fluctuations of Linear Momentum and Kinetic Energy*

In classically driven hydrodynamic turbulence, the longitudinal velocity is a measure for the corresponding linear momentum because the density is a constant. Because the density also fluctuates intermittently for the compressible cosmic gas, it would be interesting to investigate the behavior of the product of these two entities (i.e., the linear momentum density). Figure 17 presents the PDFs of the fluctuations of the longitudinal momentum density for $\delta = 32\Delta - 128\Delta$ with $\delta\rho v_\parallel$ in arbitrary

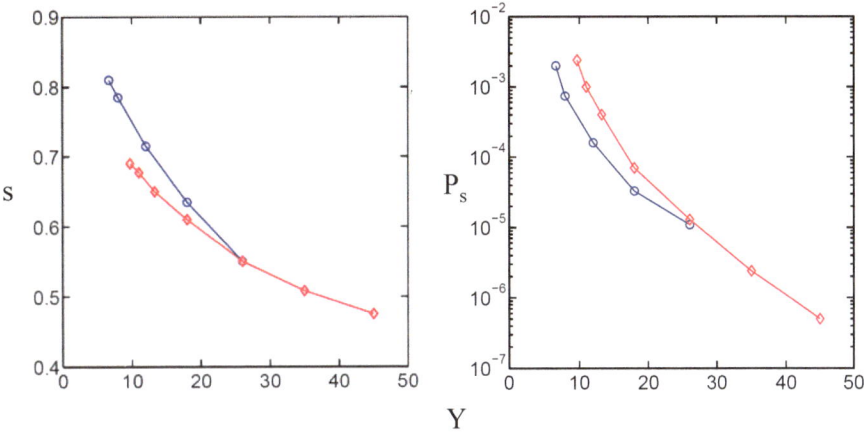

Figure 15

Combined ROMA $s(+Y)$ and $P_s(+Y)$ for longitudinal velocity fluctuations of the cosmic gas on small and large scales. *Blue circles* for δ between 4 and 32 and *red diamonds* for δ between 32 and 128 in units of Δ

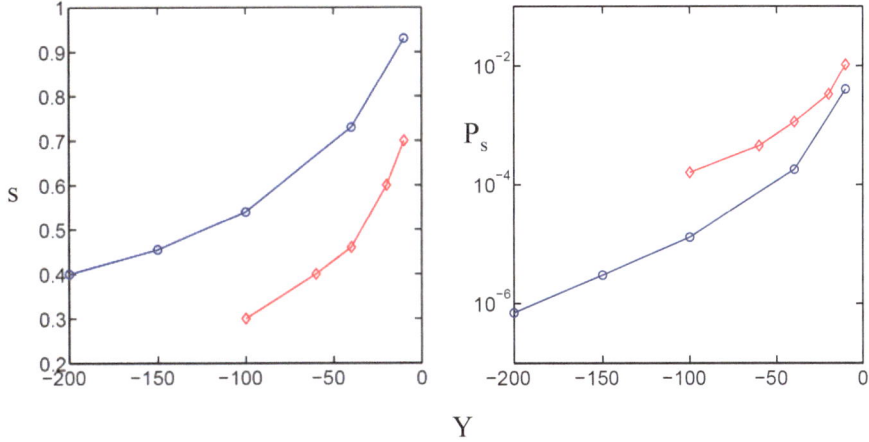

Figure 16

Combined ROMA $s(-Y)$ and $P_s(-Y)$ longitudinal velocity fluctuations of the cosmic gas on small and large scales. *Blue circles* for δ between 4 and 16 and *red diamonds* for δ between 32 and 56 in units of Δ

units. The PDFs are strongly non-Gaussian, but essentially scale independent and symmetrical. There appears only a slight scale dependence on smaller scales of $\delta = 4\Delta - 32\Delta$ (Fig. 18), and the asymmetry (if any) is noticeable only for small fluctuations for all scales.

The reason for this interesting phenomenon may be related to the fact that longitudinal momenta are continuous across both shock and rarefaction waves for compressible flow. On smaller scales, extra physics related to the real gas effects may affect this property for shock waves and therefore their scaling properties merit further investigation.

Also, for classical turbulence, v^2 is a measure of kinetic energy. It then provides us with motivation to analyze the scaling behavior of $E = \rho v^2/2$, i.e., the kinetic energy density of the cosmic gas. Figure 19 shows that the PDFs of energy density fluctuations for large scales of $\delta = 32\Delta - 128\Delta$ with δE in arbitrary units are essentially scale-independent and strongly non-Gaussian. In fact, it is only on very small scales that there is some scale dependence of the PDFs, Fig. 20.

The ROMA scaling relationships, $s(Y)$ and $P_s(Y)$, obtained from the PDFs for small scales (Fig. 21) have almost the same shapes as those for δv^2

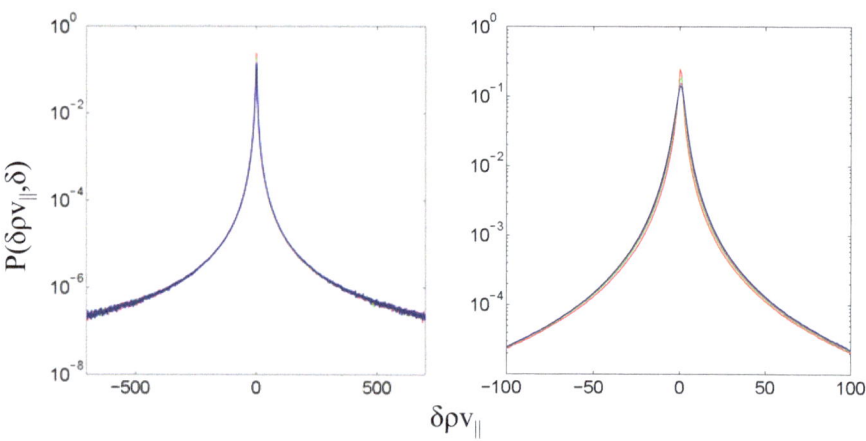

Figure 17
PDFs of $\delta\rho v_{\parallel}$ in arbitrary units of the cosmic gas for $\delta = 32\Delta(red), 64\Delta(green), 96\Delta(magenta), 128\Delta(blue)$. Maximum momentum density fluctuations $\sim 320\ 10^{10} \odot (km/s)/\Delta^3$

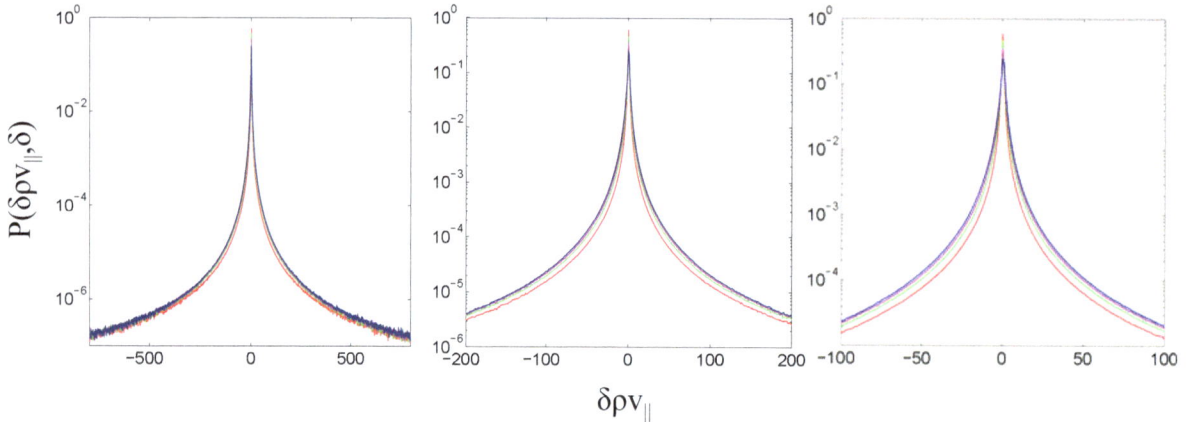

Figure 18
PDFs of $\delta\rho v_{\parallel}$ in arbitrary units of the cosmic gas for $\delta = 4\Delta(red), 8\Delta(green), 16\Delta(magenta), 32\Delta(blue)$

fluctuations of classically driven hydrodynamic turbulence (Fig. 7). The spectrum $s(Y)$ is everywhere antipersistent and it decreases as Y increases, indicating that the turbulence is fully developed.

6. Summary and Conclusions

We have provided a brief review of a relatively new method of multifractal analysis (ROMA) designed for in-depth studies of intermittent fluctuations arising from dynamic complexity. Results from such an implicit multifractal spectrum method have several advantages over the results obtainable

by use of traditional structure and partition functions. First, the utility of the spectrum is to fully collapse the unscaled PDFs. Second, the physical interpretation is clear. It indicates how intermittent the scaled fluctuations are once the spectrum is given. Third, determination of the values of the fractal nature of the grouped fluctuations is not affected by the statistics of other fluctuations that do not have the same fractal characteristics. Fourth, the method retains all the statistical information of the analyzed stochastic process and is unique and reversible.

As examples, the method was applied to analysis of the magnetic energy density fluctuations in the

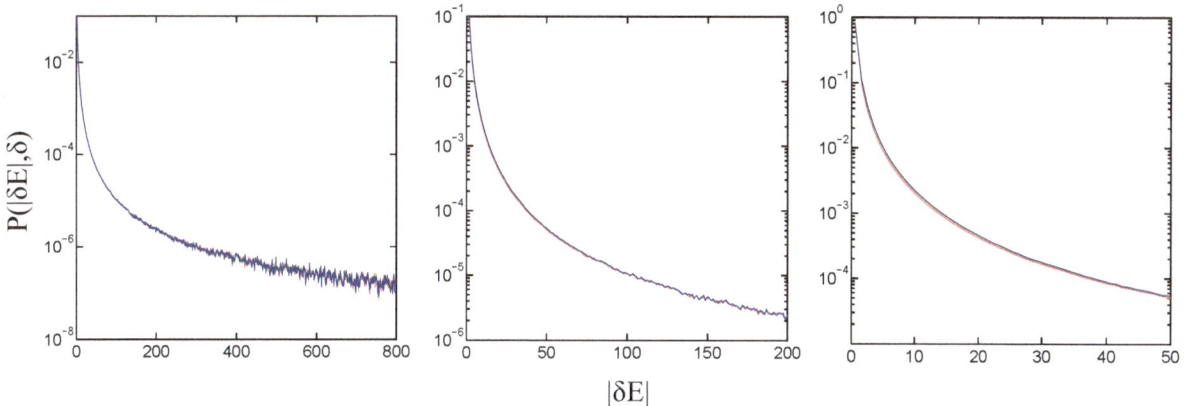

Figure 19
PDFs of $|\delta E|$ at $\delta = 32\Delta(red), 64\Delta(green), 96\Delta(magenta), 128\Delta(blue)$ for the cosmic gas. $|\delta E|$ in arbitrary units. Maximum energy density fluctuations $\sim 24{,}000 \; 10^{10} \odot (km/s)^2/\Delta^3$

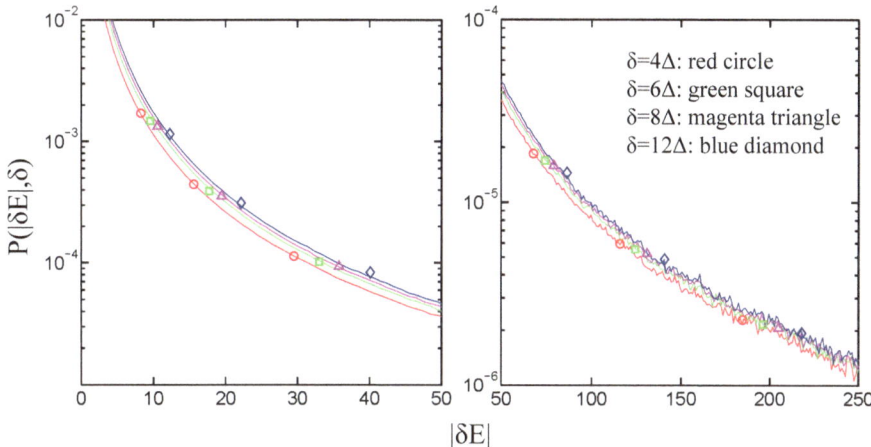

Figure 20
PDFs of $|\delta E|$ of the cosmic gas in arbitrary units on small scales. *Markers* show values obtained by the ROMA scaling relationships $s(Y)$, $P_s(Y)$ of Fig. 21

cusp region of the Earth's magnetosphere and to classically driven hydrodynamic turbulence.

In addition, for the purpose of relevant cross-discipline fertilization/exchange of scientific techniques and ideas, the paper concludes with a new and exciting analysis of results obtained from an AREPO moving mesh simulation of the cosmic gas. It was discovered that the intermittent characteristics of the longitudinal velocity fluctuations of the gas are vastly different from those generally observed for classically driven hydrodynamic turbulence, probably because of the robustly compressible and multiple shock nature of the cosmic baryonic medium. The fluctuations of the longitudinal momentum density of the gas, however, are nearly scale-independent on large scales. In fact, the PDFs for the mass, longitudinal momentum, and kinetic energy densities, though strongly non-Gaussian, are all scale-independent on large scales. On smaller scales, approximately 1–2 h^{-1}Mpc and below, the scale-dependence becomes important, indicating special physical effects are beginning to become important in affecting the statistics of the intermittent fluctuations of the cosmic gas.

All the above results indicate that ROMA will be a useful tool for studying and comparing the complexity effects of space plasmas, hydrodynamic turbulence, the climate and geo/space environment,

261

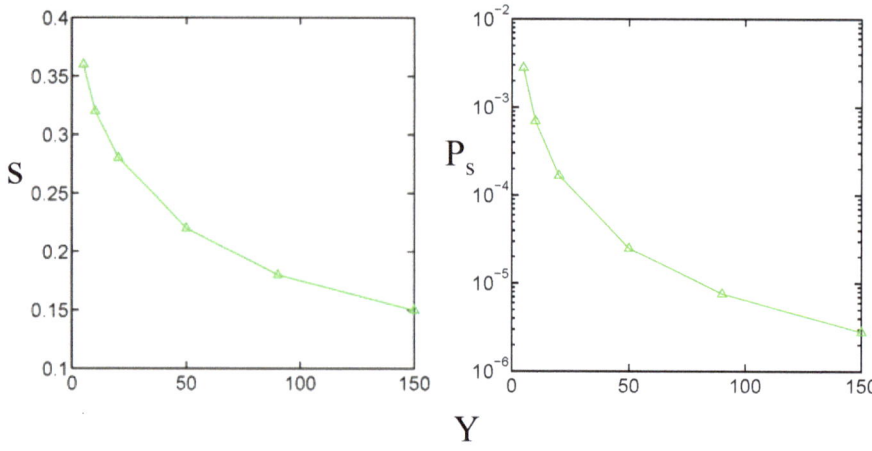

Figure 21
ROMA scaling relationships for the PDFs of $|\delta E|$ for $\delta = 4\Delta, 6\Delta, 8\Delta, 12\Delta$ for the cosmic gas

the cosmic web, and other fields of science. A distinct advantage of ROMA is its capability of spotting similarities and disparities among naturally occurring complexity processes.

Acknowledgments

This research is partially supported by the US National Science Foundation and the European Community's Seventh Framework Programme (FP7/ 2007–2013) under Grant agreement no. 313038/STORM. Tom Chang wishes to thank Dr. Diego Perigini for inviting him to present this combined review and report of new findings related to ROMA at the 6th International Conference on Fractals and Dynamic Systems in Geoscience in the spirit of providing cross-discipline fertilization/ exchange of scientific techniques and ideas in modern fractal analysis.

References

Bird, S., Vogelsberger, M., Sijacki, D., Zaldarriaga, M., Springel, V., and Hernquist, L. (2013), *Moving mesh cosmology: properties of neutral hydrogen in absorption*, Mon. Not. R. Astron., Soc., *429*, 3341.

Chang, T. (2014), *An Introduction to Space Plasma Complexity*, Cambridge University Press, New York, NY.

Chang, T., Wu, C. C., Podesta, Echim, M., Lamy, H., and Tam, S. W. Y. (2010), *ROMA (rank-ordered multifractal analyses) of intermittency in space plasmas – a brief tutorial review*, Nonlinear Processes in Geophysics, *17*, 545.

Chang, T., and Wu, C. C. (2008), *Rank-ordered multifractal spectrum for intermittent fluctuations*, Phys. Rev. E, *77*, 045401(R), doi:10.1103/Phys. Rev. E.77045401.

Chang, T., Tam, S. W. Y., and Wu, C. C. (2004), *Complexity induced anisotropic bimodal intermittent turbulence in space plasmas*, Phys. Plasmas, *11*, 1287.

Chang, T., Hankey, A., and Stanley, H. E. (1973) *Generalized scaling hypothesis in multicomponent systems. 1. Classification of critical points by order and scaling at tricriical points*, Phys. Rev. B, *8*, 346.

Consolini, G. and De Michelis, P. (2011), *Rank ordering multifractal analysis of the auroral electrojet index*, Nonlinear Processes in Geophysics, *18*, 277.

Echim, M. M., Lamy, H., and Chang, T. (2007), *Multipoint observations of intermittency in the cusp regions*, Nonlinear Processes in Geophysics, *14*, 325.

Echim, M, and Lamy, H. (2010), private communication

Eisenstein, D.J., and Hu, W. (1999), *Power Spectra for Cold Dark Matter and Its Variants*, Ap. J., *511*, 5, doi:10.1086/306640.

Fang, F. (2006), *Information of structures in galaxy distribution*, Ap. J., *644*, 678.

Frisch, U. (1995), *Turbulence*, Cambridge University Press, Cambridge, UK.

Hernquist, L., Katz, N., Weinberg, D. H., and Miralda-Escudé, J. (1996), *The Lyman-Alpha forest in the cold dark matter model*, Ap. J., *457*, L51.

Hnat, B., Chapman, S. C., Rowlands, G., Watkins, N. W., and Farrell, W. M. (2002), *Finite size scaling in the solar wind magnetic field energy density as seen by WIND*, Geophys. Res. Lett., *29*, 1446, doi:10.1029/2001GL014587.

Komatsu, E., et al. (2011), *Seven-year Wilkinson microwave anisotropy probe (WMAP) observations: cosmological interpretation*, Ap. JS, *192*, 18.

Lamy, H., M. Echim, and T. Chang (2008), Rank-ordered multifractal spectrum of intermittent fluctuations in the cusp: a case study with Cluster data, 37th COSPAR Scientific Assembly, Paper D31-0017-08, p. 1686.

LI, Y., PERLMAN, E., WAN, M., YANG, Y., BURNS, R., and MENEVEAU, C. (2008), *A public turbulence database cluster and applications to study Lagrangian evolution of velocity increments in turbulence*, J. Turbulence, *9*, 1, 2008.

LIDDLE, A. (2013), *An Introduction to Modern Cosmology*, Wiley, Chichester, West Sussex, UK

LOVEJOY, S., GARRIDO, P., and SCHERTZER, D. (2000), *Multifractal absolute galactic luminosity distributions and the multifractal Hubble 3/2 law*, Physica A, *287*, 49.

PEEBLES, P. (1980), *The large Scale Structure of the Universe*, Princeton University Press, Princeton, NJ.

PERLMAN, E., BURNS, R., LI, Y., and MENEVEAU, C. (2007), Data exploration of turbulence simulations using a database cluster, Supercomputing SC07, ACM, IEEE, doi:10.1145/1362622, 1362654.

PERMUTTER, S., *et al.*, (1999), *Measurements of Ω and Λ from 42 high-redshift supernovae*, Ap. J., *517*, 565.

PRESS, W. H., and SCHECHTER, P. (1974), *Formation of galaxies and clusters of galaxies by self-similar gravitational condensation*, Ap. J., *187*, 425.

RIESS, a., *et al.* (1999), *The rise time of nearby Type Ia supernovae*, Ap. J., *118*, 2268.

SPRINGEL, V. (2010), *E pur si muove: Galiliean-invariant cosmological hydrodynamical simulations on a moving mesh*, Mon. Not. R. Astron., Soc., *401*, 791.

SPRINGEL, V., and HERNQUIST, L. (2003), *The history of star formation in a Λ cold dark matter universe*, Mon. Not. R. Astron., Soc., *339*, 312.

TAM, S. W. Y., *et al.* (2010), *Rank ordered multifractal analysis for intermittent fluctuations with global crossover behavior*, Phys. Rev. E, *81*, 036404.

TAYLOR, G. I. (1938), *The spectrum of turbulence*, Proc. Royal Soc. London A, *164*, 476.

TORREY, P., VOGELSBERGER, M., GENEL, S., SIJACKI, D., SPRINGEL, V., and HERNQUIST, L. (2013), *A physical model for cosmological simulations of galaxy formation: multi-epoch*, arXiv:1305.4931v1.

VOGELSBERGER, M., SIJACKI, D., KEREŠ, D., SPRINGEL, V., HERNQUIST, L. (2012), *Moving mesh cosmology: numerical techniques and global statistics*, Mon. Not. R. Astron., Soc., *425*, 3024.

VOGELSBERGER, M., GENEL, S., SIJACKI, D., TORREY, P., SPRINGEL, V., and HERNQUIST, L. (2013), *A model for cosmological simulations of galaxy formation physics*, arXiv:1305.2913v3.

WU, C. C., and CHANG, T. (2011), *Rank-ordered multifractal analysis (ROMA) of probability distributions in fluid turbulence*, Nonlinear Processes in Geophysics, *18*, 261.

YORDANOVA, E., GREZESIAK, M, WERNIK, A.W., POPIELAWSKA, B. and STASIEWICZ, K. (2004), *Multifractal structure of turbulence in the magnetospheric cusp*, Ann. Geophys., *22*, 2431.

YORDANOVA, E., BERGMAN, J., CONSOLINI, G., KRETZSCHMAR, M., MATERASSI, M., POPIELAWSKA, B., ROCA-SOGORB, M., STASIEWICZ, K., and WERNIK, A.W., *Anisotropic scaling features and complexity in magnetospheric-cusp: a case study*, Nonlinear Processes in Geophysics, *12*, 817, 2005.

ZHU, W., FENG, L.-L., and FANG, L.Z. (2011), *Intermittence of the map of the kinetic Sunyaev-Zel'dovich effect and turbulence of the intergalactic medium*, Ap. J. Lett., *734*, L14.

(Received February 25, 2014, revised May 28, 2014, accepted June 4, 2014, Published online August 23, 2014)

Pure Appl. Geophys. 172 (2015), 2045–2056
© 2014 Springer Basel
DOI 10.1007/s00024-014-0909-5

❚ Pure and Applied Geophysics

Scaling Laws for the Distribution of Gold, Geothermal, and Gas Resources

THOMAS BLENKINSOP[1]

Abstract—Mass dimensions of natural resources have important implications for ore-forming processes and resource estimation and exploration. The mass dimension is established from a power law scaling relationship between numbers of resources and distance from an origin. The relation between the total quantity of resource and distance, measured by the mass-radius scaling exponent, may be even more useful. Lode gold deposits, geothermal wells and volcanoes, and conventional and unconventional gas wells are examined in this study. Mass dimensions and scaling exponents generally increase from the lode gold through geothermal wells to gas data sets, reflecting decreasing degrees of clustering. Mass dimensions are similar to or slightly less than the mass-radius scaling exponents, and could be used as estimates of the minimum scaling exponent in the common case that data are not available for the latter. All the resources in this study are formed by fluid fluxes in the crust, and, therefore, percolation theory is an appropriate unifying framework to understand their significance. The mass dimensions indicate that none of the percolation networks that formed the deposits reached the percolation threshold.

Key words: Mass dimension, fractal, resource, gold, percolation, unconventional gas resource.

1. Introduction

Scaling laws have been applied to many aspects of natural resources. MANDELBROT (1983) suggested that mineral distribution in the Earth might be a fractal dust, and this idea has been followed up for hydrothermal mineral deposits (e.g., CARLSON 1991; BLENKINSOP 1994, 1995; RAINES 2008; CARRANZA 2009) and petroleum deposits (BARTON and SCHOLZ 1995). Fractal relations between ore grade and tonnage were described by TURCOTTE (1986), and fractal aspects of structures in vein-hosted deposits have been described by SANDERSON *et al.* (1994), ROBERTS

et al. (1999), JOHNSTON and MCCAFFREY (1996) and NORTJE *et al.* (2006) among others. Fractal applications of geochemistry to natural resources have been well documented (e.g., AGTERBERG 1995; AGTERBERG *et al.* 1996; CHENG *et al.* 1994; CHENG 1999a, b, c, d). Describing the distribution of natural resources is useful in order to estimate total resources (e.g., BARTON and SCHOLZ 1995), and also has important implications for exploration strategies (e.g., FORD and BLENKINSOP 2008), and for processes by which natural resources form (e.g., ARIAS *et al.* 2011).

The box counting method has been widely applied to quantify the distribution of natural resources, for example, mineral deposits:

$$N(\delta) \sim \delta^{-D_b},$$

where $N(\delta)$ is the number of boxes of side length δ required to cover the deposits. D_b is the box-counting dimension, which is a measure of clustering (e.g., CARLSON 1991). Uniform or random distributions have $D_b = 2$; increasing degrees of clustering have smaller values of D_b. In the limit of a single point, the box counting dimension is 0. A more useful description of resource distribution may be given by the relation:

$$M(r) \sim r^{D_m},$$

where $M(r)$ is the mass of resource within a circle of radius r (e.g., LA POINTE 1995). If the mass of each resource occurrence is unity, this law describes the mass dimension D_m of the resource. D_m is also simply interpreted as a measure of the clustering of the resource distribution, and varies from 2 (uniform or random) to 0 (single point) with increasing clustering. The mass-radius relationship is sometimes expressed as the radial density function:

$$d(r) \sim r^{D_m - 2},$$

[1] School of Earth and Ocean Sciences, Cardiff University, Main Building, Park Place, Cardiff CF10 3AT, UK. E-mail: BlenkinsopT@Cardiff.ac.uk

where $d(r) = M(r)/\pi r^2$ is the density of deposits as a function of r (e.g., RAINES 2008).

However, the mass of resources typically varies at each location. This can be quantified by a general scaling law:

$$M(r) \sim r^{D_{mr}}.$$

D_{mr} is referred to here as the mass-radius scaling exponent. This exponent is potentially a more complete description of the distribution of natural resources because it measures variations in mass of resource at each resource location.

Mass dimensions have been investigated in diverse research fields, including astrophysics (e.g., DUVAL et al. 2010), neurobiology (e.g., CASERTA et al. 1995), particle science (LIAO et al. 2005), and texture analysis (e.g. BACKES and BRUNO 2013), but they have not been widely applied to natural resources. Box counting and mass dimensions have been determined for gold deposits (e.g., CARLSON 1991; BLENKINSOP 1994, 1995; CARRANZA 2009, 2010; CARRANZA et al. 2009; CARRANZA and SADEGHI 2010) and for petroleum deposits (BARTON and SCHOLZ 1995), but mass-radius scaling exponents are hardly reported in the literature. The primary aim of this paper is to investigate the applicability of mass dimensions and mass-radius scaling exponents for describing the distribution of some natural resources. Hydrothermal gold deposits, geothermal wells and volcanic vents, and gas wells are considered in this study. Each of the

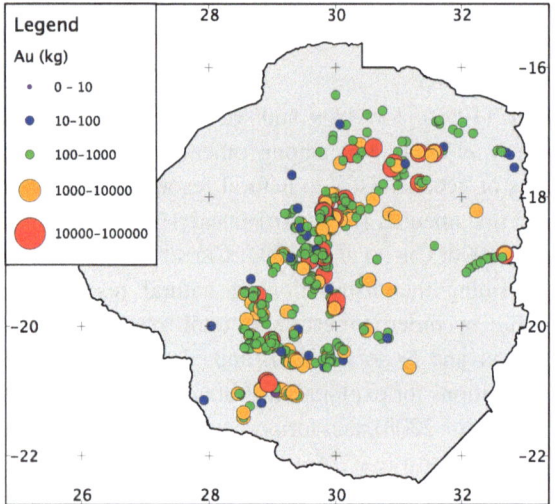

Figure 1
Gold mines in Zimbabwe, with symbols scaled by logarithm of gold production. From BARTHOLOMEW (1990). UTM coordinates, WGS84 Datum

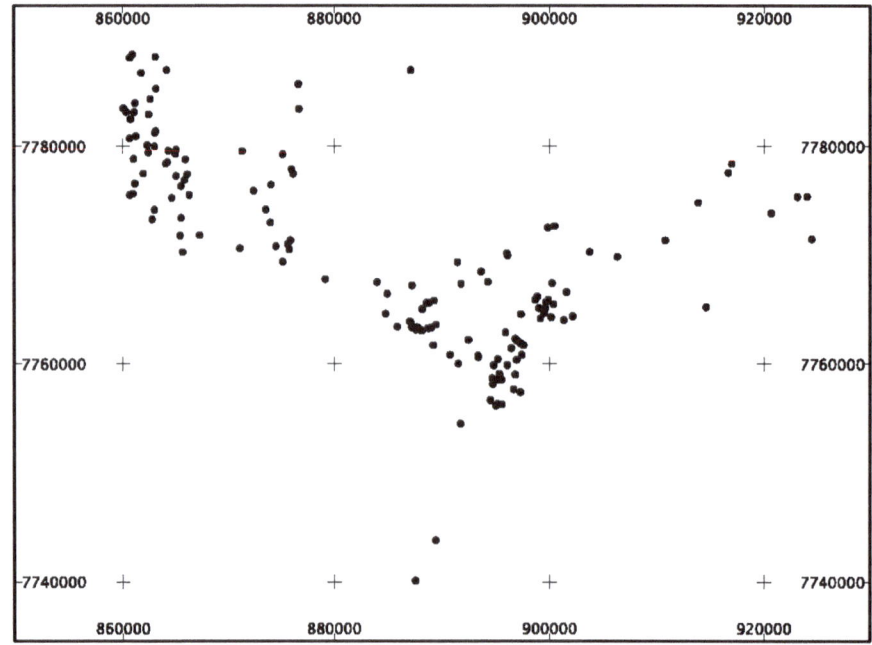

Figure 2
Gold occurrences in the Masvingo area, Zimbabwe. From WILSON (1964, 1968). UTM coordinates, WGS84 Datum

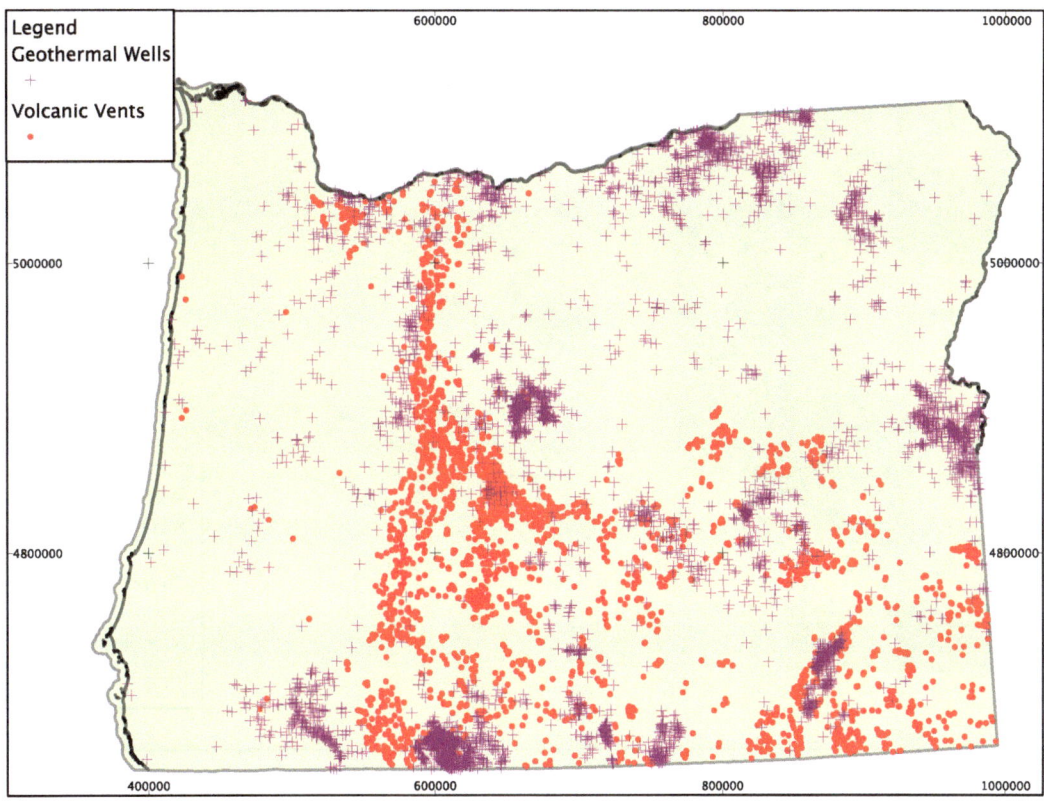

Figure 3
Geothermal wells and volcanoes, Oregon County. From http://www.oregongeology.com/sub/gtilo/index.htm. UTM coordinates, NAD83
Datum

data sets represents the product of fluid flow systems in the crust; hence, the relevance of percolation theory to the results is also considered.

2. Data and Methods

Mass dimensions and mass-radius scaling exponents have been determined in this study for Archean gold deposits in Zimbabwe (Figs. 1, 2), divided into a data set for the whole Craton and a more detailed data set from the Masvingo area. Geothermal wells and volcanoes in Oregon (Fig. 3), and conventional and unconventional gas wells in Pennsylvania (Figs. 4, 5, 6) were also analysed. Subsets of both types of gas wells could be identified that produced gas: these were distinguished as producing wells. Details of the data and sources are given in Table 1. In the Pennsylvania data, unconventional wells are considered as those drilled "for the purpose of or to be used for the

production of natural gas from an unconventional formation" (https://www.paoilandgasreporting.state. pa.us/publicreports/Modules/DataExports/ DataExports.aspx). All conventional wells are vertical, but most unconventional wells are horizontal. Virtually all unconventional wells, and by far the majority of conventional wells, produced gas only; there was some oil production from a few conventional wells.

Expanding circles used to count mass around a point were entirely constrained within the study area limits to avoid edge effects, and "mass" was normalized to the total value of the data sets ΣM, so that $M'(r) = M(r)/\Sigma M$. Two strategies were investigated for determining the exponents of the scaling laws:

1. A grid origin method, in which the mass was summed and averaged from expanding circles centred on 100 origins on grid nodes in the central part of the study area (cf. LA POINTE 1995).

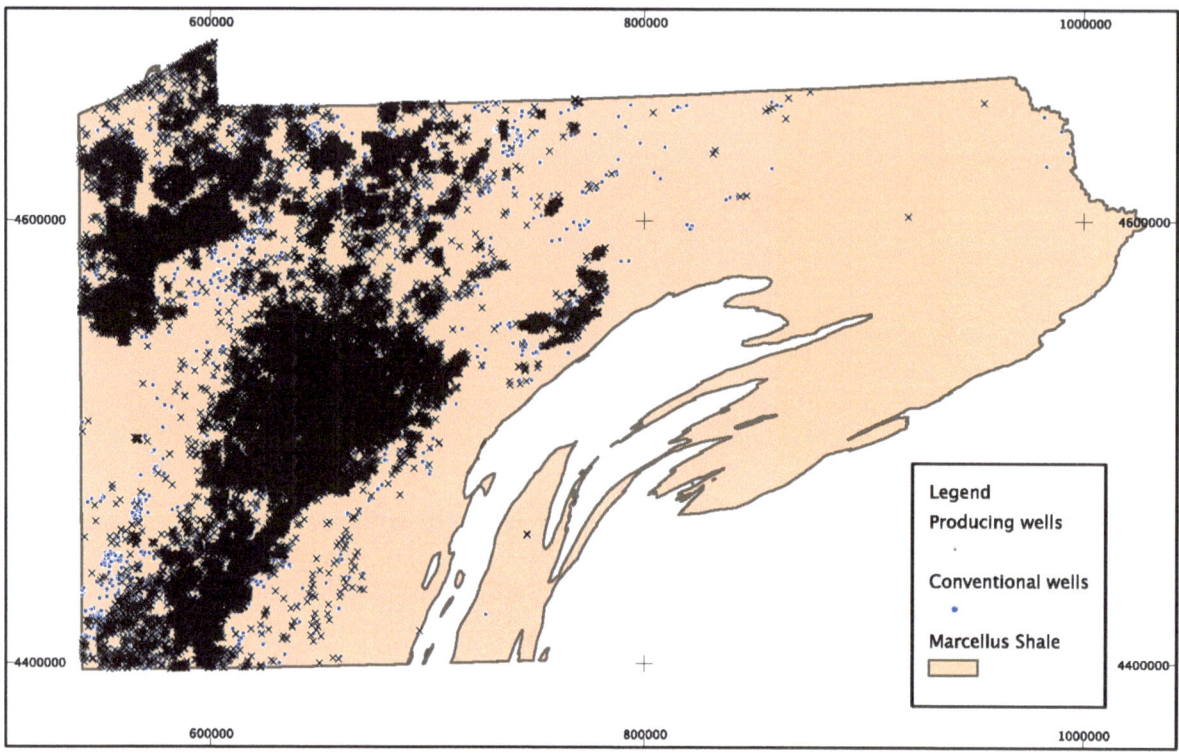

Figure 4
Conventional gas wells in Pennsylvania with producing wells distinguished. From https://www.paoilandgasreporting.state.pa.us/publicreports/
Modules/DataExports/DataExports.aspx. All Pennsylvanian maps are UTM coordinates with a WGS84 datum

2. A data point origin method, in which counting circles were centred on data points, and average values were taken from every circle used.

In order to evaluate these strategies, they were applied to the coordinates of a well-known fractal shape, the Koch curve, as well as to all data sets. The grid origin method produced exponents with values that were all near 2, including for the Koch curve, and showed little variation between data sets. By contrast, the data point origin method returned a value of 1.26 for the Koch curve (the curve has a fractal dimension of 1.26; e.g., PEITGEN *et al.* 2004), and discriminated sensitively between the data sets. Hence, it was used for all results shown in this study. The reasons for the differences in the two methods are not known.

Mass dimensions could be calculated for all data sets. Mass-radius scaling exponents could be calculated for gold production from the Zimbabwe Craton, and producing conventional and unconventional gas wells from Pennsylvania, because these data sets included resource figures. Exponents were obtained

by regression of log $M'(r)$ against log r over the linear part of the scaling relationship, for a range of r of 1 to 1.5 orders of magnitude. Lower and upper limits of regression were chosen by departure from a visual fit of a straight line.

3. Results

Linear parts of all data sets can be defined over at least an order of magnitude (Figs. 7, 8, 9, 10), justifying the above regression technique. The data sets showed two characteristic features. At both low and high values of r, the slopes of the log mass-radius relations were less than the central part of the data, where the regression was carried out (e.g., Fig. 7). Mass dimensions vary between 1.2 and 1.8 (Table 2). Standard errors of regression vary from 0.009 to 0.017, indicating that the range of mass dimensions measured reveals significantly different degrees of clustering between different data sets.

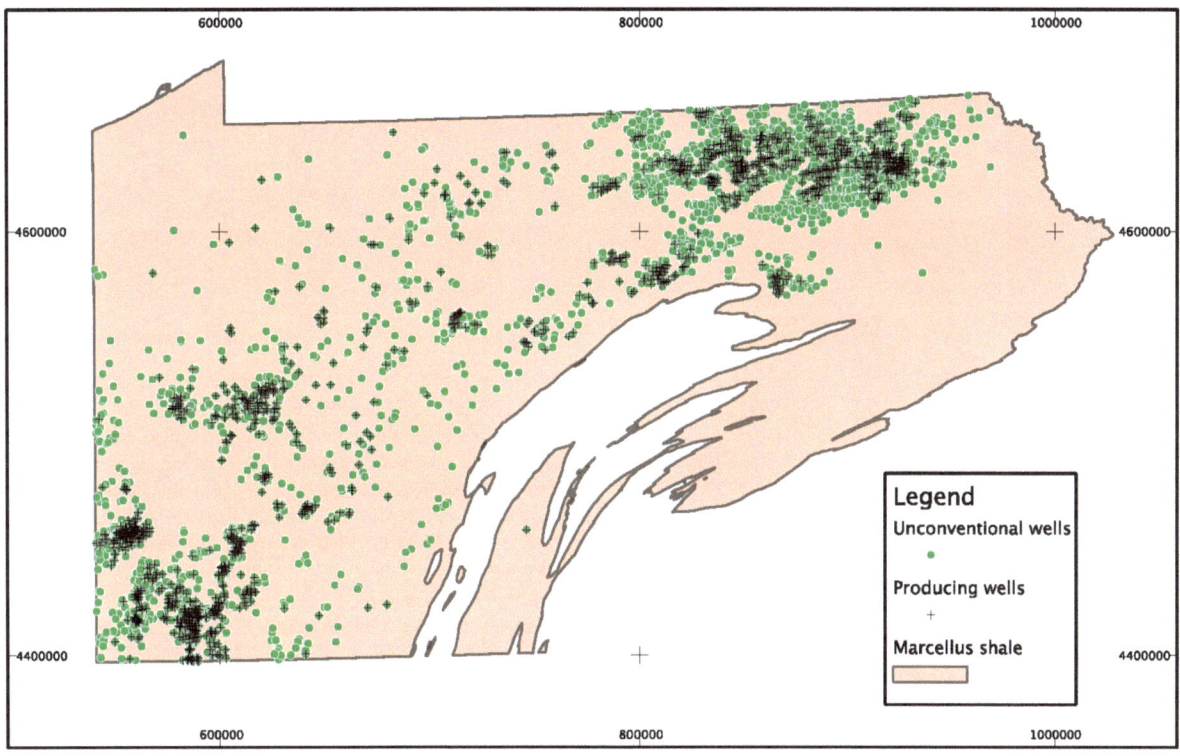

Figure 5
Unconventional gas wells in Pennsylvania, with producing wells distinguished. Source as in Fig. 4

The gold deposits of the Zimbabwe Craton have the lowest mass dimensions of all data sets considered, indicating the greatest degrees of clustering (Table 2). The mass-radius scaling exponent is within the regression error of the mass dimension for the Craton data set. The mass dimension for the Masvingo data set is significantly greater than for the Craton data.

The geothermal wells of Oregon have a stepped log mass-radius relation (Fig. 8) in which two segments of similar slope are offset from one another. The mass dimension of the larger part of the data is 1.23 (Table 2). The volcanic vents of Oregon have a mass dimension of 1.51 (Fig. 8).

Unconventional gas wells in Pennsylvania have mass dimensions of 1.26 (producing wells) and 1.45 (all wells) (Table 2). The mass-radius scaling exponent of the producing wells, 1.32, lies between these values. The highest values of mass dimension are from conventional gas production, (1.57 and 1.63 for all wells and producing wells, respectively). The

mass-radius exponent of the producing wells is the highest of any value measured, 1.72.

4. Discussion

4.1. Consistency With Previous Results

Mass dimensions of various types of gold deposits have been presented by BLENKINSOP (1994, 1995), CARRANZA (2009, 2010) and CARRANZA et al. (2009). In all these studies, different fractal dimensions are given at low and high r values, ranging from 0.54 for the low r values, to 1.52 (high r). Mass-radius scaling exponents were calculated for nine hydrocarbon plays by LA POINTE (1995) using area of hydrocarbon fields as a measure of mass, and reported as between 1 and 2. The mass dimensions and mass-radius scaling exponent obtained here are therefore broadly consistent with the few previous results reported in the literature from similar commodities.

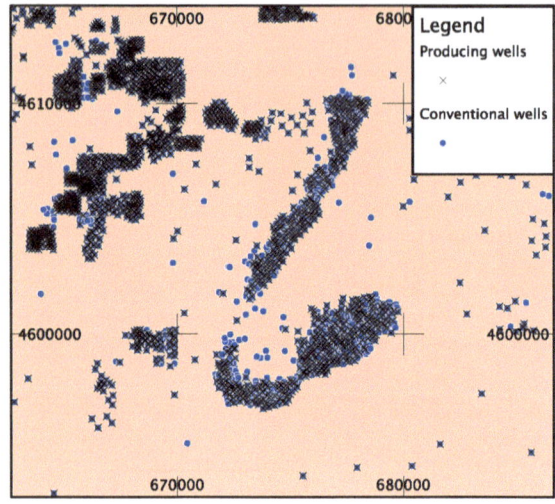

Figure 6

Detail of distribution of conventional gas wells, showing a structural control. Source as in Fig. 4

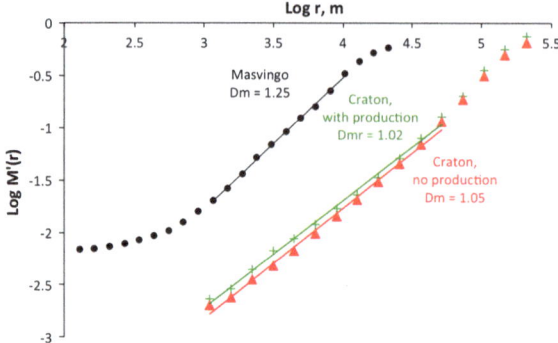

Figure 7

Variation of Logarithm of normalised mass M' with radius (logarithmic mass-radius function) for Masvingo and Craton gold deposit data sets, with regression lines and values of D_m and D_{mr} shown. All log–log plots use base 10 and radius is measured in m. D_m and D_{mr} are within the error for the Craton data set; the D_m value for the Masvingo data is larger. The Masvingo data show departures from a linear relationship at low values of r, possibly due to sampling effects, and also at high r, possibly due to some clustering near the periphery of the study area

4.2. Nonlinearity of Logarithmic Mass-Radius Scaling

The log mass-radius scaling relations examined are characteristically nonlinear at values of r generally less than 1,000 m, illustrated for the Masvingo data set in Fig. 7, and also seen at lower values of r than shown in Figs. 8, 9, 10 for the other data sets. This nonlinearity is similar to the "roll-off" observed in box-counting plots at low δ values (e.g. PICKERING et al. 1995). For gold deposits of the Zimbabwe Craton, this effect has been attributed to random sampling of a fractal data set (BLENKINSOP and SANDERSON 1999), and it seems likely that the same

explanation applies here, i.e., that the actual data sets represent samples of a true fractal distribution. The log mass-radius relations show less mass at high values of r than predicted by a linear relation. It is noticeable that the nonlinearity occurs at radii that are about $1/4$ of the maximum linear dimension of the study areas or greater. Counting circles with these large r values are only taken from the centre of the study areas; thus, concentrations of resources near the corners will not be included, possibly leading to a deficit in the case of clustering near the peripheries of the study areas.

Table 1

Data sources for this study

Commodity	Data	Location	N	Units	Source
Gold	Craton mine production	Zimbabwe	651	kg	1
Gold	Masvingo mines	Zimbabwe	147		2
Geothermal energy	Geothermal wells	Oregon	5,429		3
Volcanoes	Volcanic vents	Oregon	2,747		3
Gas	All conventional wells	Pennsylvania	62,931		4
Gas	All unconventional wells	Pennsylvania	8,686		4
Gas	Producing conventional wells	Pennsylvania	52,856	Mcf	4
Gas	Producing unconventional wells	Pennsylvania	2,878	Mcf	4

Mcf = 1,000 cubic feet. References: 1. BARTHOLOMEW (1990), 2. WILSON (1964, 1968), , 3. http://www.oregongeology.com/sub/gtilo/index.htm, 4. https://www.paoilandgasreporting.state.pa.us/publicreports/Modules/DataExports/DataExports.aspx

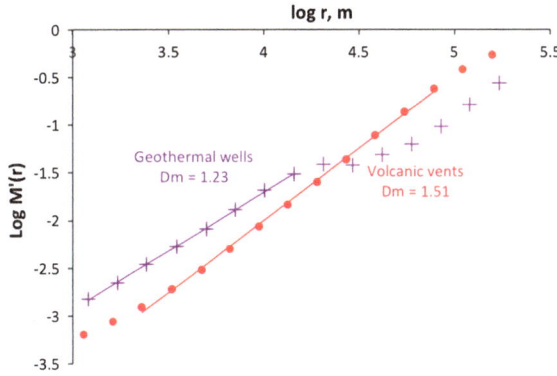

Figure 8

Logarithmic mass-radius function for Oregon data sets (volcanic vents and geothermal wells), with regression lines and values of D_m shown. The geothermal wells are more clustered (lower D_m) than the volcanic vents

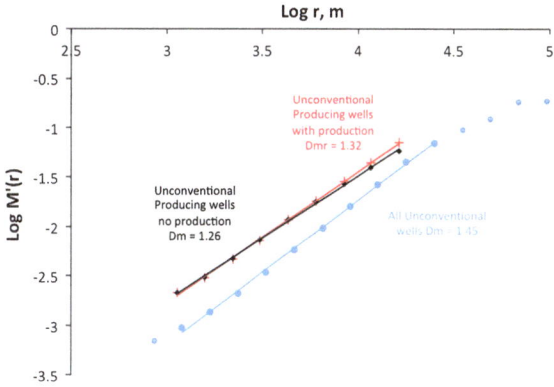

Figure 10

Logarithmic mass-radius function for Pennsylvania data sets (unconventional gas production), with regression lines and values of D_m and D_{mr} shown. D_{mr} for unconventional producing wells is greater than D_m. D_m for all wells is greater than the other values

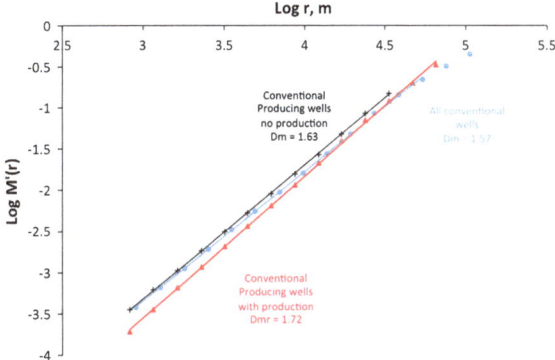

Figure 9

Logarithmic mass-radius function for Pennsylvania data sets (conventional gas production), with regression lines and values of D_m and D_{mr} shown. D_{mr} is greater than D_m for the producing conventional wells. D_m for all wells is less than either of the other values

4.3. Mass Dimensions of Data Sets vs. Natural Resources

True mass dimensions of natural resources should reflect the resource-forming processes. For hydrothermal mineral deposits and hydrocarbons, this may include elements of source distribution, fluid transport and deposition (trapping mechanisms). However, mass dimensions estimated from resource databases such as those used here will be influenced by the degree of exploration and other economic factors. The resource databases can be regarded as a sample of the true natural distributions, as discussed for the Zimbabwe data in BLENKINSOP and SANDERSON (1999).

How well the other data sets used here reflect the actual distribution of resources in the Earth is unknown. The production of gas from horizontal drilling (e.g., ARTHUR *et al.* 2008) could affect the distribution of wells on a one hundred m scale.

Despite the possible influence of nongeological factors, the results reported here make geological sense. Hydrothermal mineral deposits such as gold are strongly structurally controlled by specific deformation zones (e.g., GROVES *et al.* 1998; WITT and VANDERHOR 1998; COX 1999). This leads to strong clustering of gold deposits (BLENKINSOP 1994, 1995, CARRANZA 2009). Hydrocarbon resources, including gas, are also structurally controlled. The influence of structure can clearly be seen at a small scale (Fig. 6). Sources, seals, traps and burial history are also very important. The generally high values of mass dimensions for the gas resources of Pennsylvania reflect the presence of source rocks for both conventional and unconventional gas under much of the state.

The unconventional gas mass dimensions and mass-radius scaling exponents are less then the conventional values (Table 2) by more than the standard errors of regression. Shale gas is formed and trapped in situ in shales, so that the host rock is both source and reservoir. Thus, large parts of Pennsylvania that are underlain by the Marcellus shale, which is the main target for unconventional gas, may be productive (KARGBO *et al.* 2010). Since no hydrocarbon migration is involved in

Table 2

Mass dimensions (D_m) and mass-radius scaling exponents (D_{mr}) for data sets in this study

Resource	Gold	Geothermal	Gold	Unconventional gas	Unconventional gas	Geothermal	Conventional gas	Conventional gas
Data	Craton production	Wells	Mines	Producing wells	All wells	Volcanic vents	All wells	Producing wells
Region	Zimbabwe	OR	Masvingo	PA	PA	OR	PA	PA
D_m	1.05	1.23	1.25	1.26	1.45	1.51	1.57	1.63
E	0.022	0.009	0.014	0.012	0.017	0.017	0.006	0.005
R	0.997	1.000	0.999	1.000	0.999	0.999	1.000	1.000
L	1	1	1	1	1	2	1	1
U	52	14	10	16	25	78	54	33
D_{mr}	1.02			1.32				1.72
E	0.019			0.012				0.008
R	0.998			1.000				1.000
L	1			1				1
U	52			16				65

OR, PA Oregon, Pennsylvania. *E* standard error of regression, *R* correlation coefficient, *L, U* lower and upper limits of regression, km

unconventional gas formation, the unconventional wells might be expected to have a less clustered distribution, instead of the more clustered pattern suggested by the data. One possibility to account for the difference is that the unconventional resource is less thoroughly explored. It is notable that there is a decrease in density of unconventional wells in central Pennsylvania, but no corresponding decrease in conventional gas wells, and it is not clear that exploration has been less vigorous in this area. Depth and thickness of the Marcellus shale are variable, and the richest parts of the shale are in north central Pennsylvania (LEE *et al.* 2011), so the possibility remains that geological controls are a major influence on the distribution of unconventional gas wells shown in Fig. 5.

The distribution of geothermal wells is related to geothermal structure, which is a function of tectonics. The tectonics of Oregon are dominated by the Cascadia subduction zone, which creates the Cascade volcanic arc and determines the location of volcanoes (PRIEST 1990). Heat flow is thought to be influenced by the presence of partial melts in the mid crust at depths of 10 km (BLACKWELL *et al.* 1990). However, on a more local scale in North-Central Oregon, regional groundwater flow modifies the conductive flux by sweeping heat from young elevated rocks into adjacent older rocks at lower elevations (INGEBRITSEN *et al.* 1989; BLACKWELL *et al.* 1990). The lower mass dimensions of the geothermal wells compared to the volcanic vents

may be due to exploration being limited to areas of known enhanced geothermal gradient.

4.4. Percolation Theory: a Unifying Framework

The formation of all the georesources considered above is linked by fluid flow (Fig. 11). A possible unifying framework for considering the mass dimensions and mass-radius scaling exponents is, therefore, percolation theory. This concept has been applied to the formation of mineral deposits (COX 1999) in the context of fluid flow in fracture networks (e.g., RIVIER *et al.* 1985), and there is an extensive literature on applications of percolation theory to primary migration of hydrocarbons (e.g., CARRUTHERS and RINGROSEM 1998; CARRUTHERS 2003; CORRADI *et al.* 2009). General aspects of percolation theory may assist with interpretation of the results presented here (cf. COX 1999).

A percolation network consists of a lattice in which some sites are occupied, with a probability p of occupation (STAUFFER and AHARONY 1994). As the network evolves, p changes. Many aspects of percolation networks are fractal, for example the dimensions and numbers of clusters of occupied sites, and times for their evolution. As p increases, a critical stage is reached called the percolation threshold, defined as the point at which a continuous path of occupied nodes exists from one side of the network to the other, and the network changes from

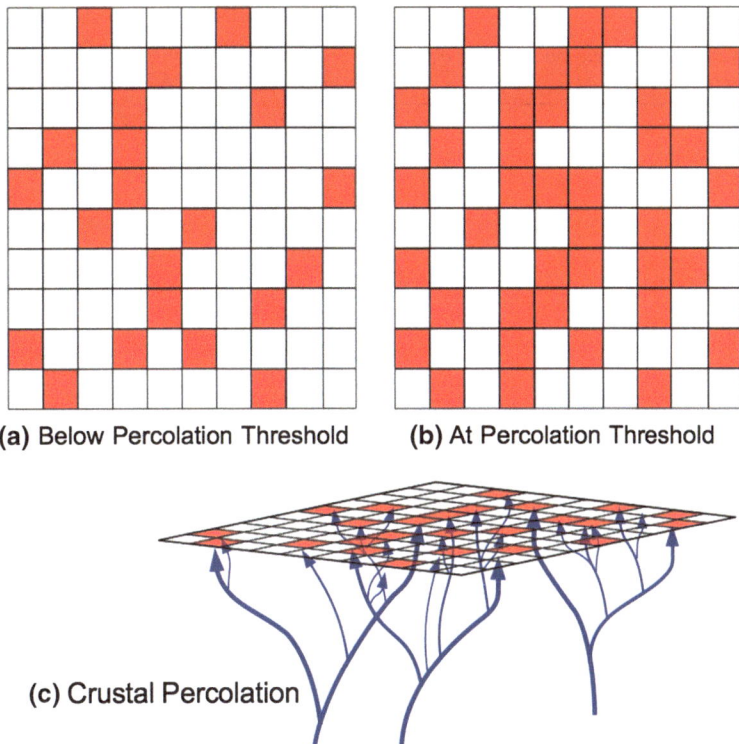

(a) Below Percolation Threshold **(b)** At Percolation Threshold

(c) Crustal Percolation

Figure 11
Schematic diagrams of the crust as a percolation network. **a, b** Maps of deposits (*red squares*) represent a section through the network, which is similar to a Bethe lattice. **a** Below the percolation threshold. **b** At the percolation threshold. **c** Percolation into a thin crust

closed to open (Fig. 11a, b). The percolation threshold occurs at a critical probability p_c. Fractal dimensions of percolation networks change over a considerable range as p increases, but can be simplified into three conditions: $p < p_c$, $p = p_c$ and $p > p_c$. Two-dimensional and three-dimensional mass dimensions for percolation networks consisting of a Bethe lattice (in which every site has the same number of neighbours and there are no closed loops) in these three stages are shown in Table 3.

The study areas of the Zimbabwe Craton, and for Oregon, and Pennsylvania, have linear dimensions of hundreds of km compared to crustal thicknesses of tens of km (Nelson 1992; Nguuri et al. 2000; Eagar et al. 2011). It may, therefore, be reasonable to compare the mass dimensions of this study to those of 2D percolation networks (Fig. 11c). All the mass dimensions measured here are below the mass dimensions of 2D Bethe lattices at the percolation threshold. In the case of gold deposits, this is intuitively reasonable. Once a backbone, network-

spanning cluster has formed in a hydrothermal system, the localization of fluid flow along this structure would preclude mineralization elsewhere. Mass dimensions of the gas wells are closer, but still less than, the 2D threshold value, which may be reflected in their more distributed pattern (Fig. 9). It is also reasonable that the gas has not attained a percolation threshold for the same reason as the gold deposits. Once a percolation threshold is reached, the reservoir would be breached and no resources would remain. The concept of fluid flow localisation at some threshold is vindicated by numerical modelling of flow in well-fractured rock masses by Sanderson and Zhang (1999), especially since these models do not start from percolation theory, but are based on fundamental mechanical principles. However conclusions about the relation between resource distributions and critical states must be tempered by the realisation that the measured patterns may partly reflect sampling issues discussed in the previous section.

Table 3

Fractal dimensions of 2 and 3D Bethe Lattices below, at and above the percolation threshold

	2D	3D
$p < p_c$	1.56	2
$p = p_c$	1.90	2.53
$p > p_c$	2	3

p is the probability of a lattice node being occupied, p_c is the probability at the percolation threshold (STAUFFER and AHARONY 1994)

4.5. The Relation Between Mass Dimension and Mass-Radius Scaling Exponent

Mass dimensions and mass-radius scaling exponents have obvious applicability to resource estimation (e.g., BARTON and SCHOLZ 1995; LA POINTE 1995). It is commonly hard to measure the mass-radius scaling exponent because accurate data for "mass" (resources) are difficult to obtain. The mass dimensions and mass-radius scaling exponents are within error for the gold data set, but the mass dimensions of conventional and unconventional gas production are similar but slightly smaller (outside the error limits) than the mass-radius scaling exponents. This may reflect the relatively uniform pattern of moderately large producing wells. These observations suggest that the mass dimension could be used as a minimum approximation for the mass-radius scaling exponent.

5. Conclusions

Mass dimensions of hydrothermal gold deposits, volcanic vents, geothermal wells and gas wells can be determined reliably from appropriate databases. How accurately these values reflect the true distribution of natural resources is not known, but the low mass dimensions of hydrothermal gold deposits compared to gas wells are consistent with a high degree of localization of the gold deposits due to strong structural controls. This contrasts with a relatively dispersed pattern of gas accumulations in Pennsylvania, for which the widespread presence of source rocks, and, for unconventional gas, of the Marcellus Shale as both a source and a host, is one of the most important factors

in determining their distribution. Mass dimensions of volcanic vents and geothermal wells are intermediate between the gold and gas values. The mass-radius scaling exponent (i.e., the variation of mass with distance including a measure of the resource) was estimated for gold and gas data sets. The mass dimension is similar or slightly less than the mass-radius scaling exponent, for which it could be used as a minimum estimate where resource estimates are not available. Percolation theory offers a framework for understanding the significance of the mass dimension and mass-radius exponents. The percolation threshold may not have been reached for the networks that generated the resources considered here.

Acknowledgments

Diego Perugini and the staff at the Universita degli Studi di Perugia are gratefully acknowledged for organizing the 6th International Conference on Fractals and Dynamic systems in Geosciences. Ian Merrick assisted with programming in C++. The journal reviewers, especially David Sanderson, are thanked for excellent comments, as is the Volume editor Jörn Kruhl.

REFERENCES

AGTERBERG F.P. (1995), *Multifractal modeling of the sizes and grades of giant and supergiant deposits*, Int. Geol. Rev. *37*, 1–8.

AGTERBERG, F.P., CHENG Q., and WRIGHT, D.F. (1996) *Fractal Modelling of Mineral Deposits*, In (eds) Proceedings of the International Symposium on the Application of Computers and Operations Research in the Minerals Industries (ed. Elbrond J, Tang X.) (Montreal) pp 43–53.

ARIAS, M., GUMIEL, P., SANDERSON, D.J., and MARTIN-IZARD, A. (2011), *A multifractal simulation model for the distribution of VMS deposits in the Spanish segment of the Iberian Pyrite Belt*, Comp. & Geosciences, 37, 1917–1927, doi:10.1016/j.cageo.2011.07.012.

ARTHUR, J.D., and BOHM, B., MARK LAYNE, M. (2008), *Hydraulic fracturing considerations for natural gas wells of the Marcellus Shale, Groundwater Protection Council Annual Forum. Cincinnati. 2008.*

BACKES, A.R., and BRUNO, O.M. (2013), *Texture analysis using volume-radius fractal dimension*, App. Math. Computation *219*, 5870–5875.

BARTHOLOMEW, D.S. 1990. *Gold Deposits in Zimbabwe*. Geological Survey of Zimbabwe Mineral Resources Series, *23*.

BARTON, C.C. and SCHOLZ, C.H. (1995), *The fractal size and spatial distribution of hydrocarbon accumulations; implications for* resource assessment and exploration strategy, In Fractals in petroleum geology and earth processes (ed. Barton, C.C. and La Pointe, P.R.) (Plenum Press, New York and London 1995), pp. 13–34.

BLACKWELL, D.D., STEELE, J.L., FROHME, M.K., MURPHEY, C.F., PRIEST, G.R., and BLACK, G. L. (1990), Heat Flow in the Oregon Cascade Range and its Correlation with regional gravity, Curie Point depths, and geology. J. Geophys. Res. 95, B12, 19475–19493.

BLENKINSOP, T. (1994), The Fractal Distribution of Gold Deposits: two examples from the Zimbabwe Archaean craton, in Fractals and Dynamic Systems in Geoscience (ed. Kruhl, J.) (Springer-Verlag) pp. 247–258.

BLENKINSOP, T. (1995), Fractal measures for size and spatial distributions of gold mines: economic applications, In Sub-Saharan Economic Geology (ed. Blenkinsop, T.G., Tromp, P.) (A.A. Balkema, Rotterdam) pp. 177–186.

BLENKINSOP, T.G., and SANDERSON, D.J. (1999), Are gold deposits in the crust fractals? A study of gold mines in the Zimbabwe craton, Geol. Soc. London, Spec. Publ. 155, 141–151.

CARLSON, C.A. (1991), Spatial distribution of ore deposits, Geology 19, 111–114.

CARRANZA, E.J.M., HALE, M., and MANGAOANG, J.C. (1999), Application of mineral exploration models and GIS to generate mineral potential maps as input for optimum land-use planning in the Philippines, Nat. Res. Research 8, 165–173.

CARRANZA, E.J.M. (2009), Controls on mineral deposit occurrence inferred from analysis of their spatial pattern and spatial association with geological features, Ore Geology Reviews 35, 383–400.

CARRANZA, E.J.M., Owusu, E.A., and HALE, M. (2009), Mapping of prospectivity and estimation of number of undiscovered prospects for lode gold, south western Ashanti Belt, Ghana, Min. Deposita 44, 915–938.

CARRANZA, E.J.M. (2010), From Predictive Mapping of Mineral Prospectivity to Quantitative Estimation of Number of Undiscovered Prospects, Resource Geology 61. 30–51. doi:10.1111/j.1751-3928.2010.00146.x.

CARRANZA, E.J.M., and SADEGHI, M. (2010), Predictive mapping of prospectivity and quantitative estimation of undiscovered VMS deposits in Skellefte district (Sweden), Ore Geol. Reviews 38, 219–241.

CARRUTHERS, D.J. (2003), Modelling of secondary petroleum migration using invasion percolation techniques, in: (Eds.), Multidimensional basin modeling, AAPG Discovery Series, (eds. Duppenbecker, S., Marzi, R.) (AAPG 2003) vol. 7, chapter 3.

CARRUTHERS, D., and RINGROSEM, P. (1998), Secondary oil migration: oil-rock contact volumes, flow behaviour and rates, Geological Society London Special Publications, 144, 205–220.

CASERTA, F., ELDRED, W.D., FERNANDEZ, E., HAUSAN, R.E., STANFORD, L.R., BULDEREV, S.V., SCHWARZER, S., and STANLEY, H.E. (1995), Determination of fractal dimension of physiologically characterized neurons in two and three dimensions. J. Neuroscience Methods 56, 133–144.

CHENG, Q. (1999), The gliding box method for multifractal modeling, Comput. Geosci. 25, 1073–1079.

CHENG, Q. (1999), Markov processes and discrete multifractals, Math. Geol. 31, 455–469.

CHENG, Q. (1999), Multifractality and spatial statistics, Comput. Geosci. 25, 949–961.

CHENG, Q. (1999), Spatial and scaling modelling for geochemical anomaly separation, J Geochem Explor 65, 175–194.

CHENG Q., AGTERBERG F.P., and BALLANTYNE S.B. (1994), The separation of geochemical anomalies from background by fractal methods, J. Geochem Explor. 51, 109–130.

CORRADI, A., RUFFO, P., CORRAO, A., VISENTIN, C. (2009), 3D hydrocarbon migration by percolation technique in an alternate sand–shale environment described by a seismic facies classified volume. Marine and Petroleum Geology 26, 495–503.

COX, S.F. (1999), Deformational controls on the dynamics of fluid flow in mesothermal gold systems, Geological Society London Special Publications 155, 123–140.

DUVAL, J., JACKSON, J.M., HEYER, M., RATHBONE, J., and SIMON, R. (2010), Physical Properties and Galactic Distribution of Molecular Clouds Identified in the Galactic Ring Survey, Astrophys. Journal, 723, 492–507.

EAGAR, K.C., FOUCH, M.J., JAMES, D.E. and CARLSON, R.W. (2011), Crustal structure beneath the High Lava Plains of eastern Oregon and surrounding regions from receiver function analysis, J. Geophys. Res., 116, B02313, doi:10.1029/2010JB007795.

FORD, A., and BLENKINSOP, T.G. (2008), Combining fractal analysis of mineral deposit clustering with weights of evidence to evaluate patterns of mineralization: Application to copper deposits of the Mount Isa Inlier, NW Queensland, Australia, Ore Geol. Rev. 33, 435–450.

GROVES, D.I., GOLDFARB, R.J., GEBRE-MARIAM, M., HAGEMANN, S.G., and ROBERT, F. (1998), Orogenic gold deposits: A proposed classification in the context of their crustal distribution and relationship to other gold deposit types, Ore Geol. Rev. 13, 7–27.

INGEBRITSEN, S.E., SHERROD, D.R., and MARINER, R.H. (1989), Heat Flow and Hydrothermal Circulation in the Cascade Range, North-Central Oregon, Science 243, 1458–1462.

JOHNSTON, J.D., and MCCAFFREY, K.J.W. (1996), Fractal geometries of vein systems and the variation of scaling relationships with mechanism, J. Struc. Geol. 18, 349–358.

KARGBO, D.M., WILHELM, R.G., and CAMPBELL, D.J. (2010), Natural gas plays in the Marcellus Shale: challenges and potential opportunities, Environ. Sci. Technol. 44, 5679–84.

LA POINTE, P.R. (1995), Estimation of Undiscovered Hydrocarbon Potential through Fractal Geometry, In Fractals in petroleum geology and earth processes (ed. Barton, C.C. and La Pointe, P.R.) (Plenum Press, New York and London 1995), pp. 35–57.

LEE, D., HERMAN, J., ELSWORTH, D., KIM, H., LEE, H., 2011. A critical evaluation of unconventional gas recovery from the Marcellus shale, northeastern United States. KSCE J. Civ. Eng. 15, 679–687.

LIAO, J.Y.H. SELOMULYA, C., BUSHELL, G., BICKERT, G., and AMAL, R. (2005), On different approaches to estimate the mass fractal dimension of coal aggregates, Part. Part. Syst. Charact. 22, 299–309. doi:10.1002/ppsc.200500978.

MANDELBROT, B.B., The Fractal Geometry of Nature, (W.H. Freeman and Company, New York 1983).

NELSON, K.D. (1992), Are crustal thickness variations in old mountain belts like the Appalachians a consequence of lithospheric delamination? Geology, 20, 498–502. doi:10.1130/0091-7613(1992)020<0498:ACTVIO>2.3.CO;2.

NGUURI, T. K., GORE, J., JAMES, D. E., WEBB, S. J., WRIGHT, C., ZENGENI, T. G., GWAVAVA, O., and SNOKE, J. A. (2000), Crustal structure beneath southern Africa and its implications for the formation and evolution of the Kaapvaal and Zimbabwe cratons,

Geophys. Res. Letters, *28*, 2501–2504. doi:10.1029/2000GL012587.

NORTJE, G.S., ROWLAND, J.V., SPÖRLI, K.B., BLENKINSOP, T.G., and RABONE, S.D.C. (2006), *Vein deflections and thickness variations of epithermal quartz veins as indicators of fracture coalescence*, J. Struct. Geol. *28*, 1396–1405.

PEITGEN, H-O, JURGENS, H., and SAUPE, D. (2004), *Chaos and Fractals: New Frontiers of Science*, (Springer 2004).

PICKERING, G., BULL, J. M., and SANDERSON, D. J. (1995), *Sampling Power-Law distributions*. Tectonophys, *248*, 1–20.

PRIEST, G.R. (1990), *Volcanic and tectonic evolution of the Cascade Volcanic Arc, central Oregon*, J. Geophys. Res., *95*(B12), 19583–19599, doi:10.1029/JB095iB12p19583.

RAINES, G.L. (2008), *Are fractal dimensions of the spatial distribution of mineral deposits meaningful?* Nat. Resour. Res., *17*, 87–97.

RIVIER, N., GUYON, E., and CHARLAIX, E. (1985), *A geometrical approach to percolation through random fractured rocks*, Geol. Mag. *122*, 157–162.

ROBERTS, S., SANDERSON, D.J., and GUMIEL, P. (1999), *Fractal analysis and percolation properties of veins*, Geological Society London Special Publication *155*. 7–16.

SANDERSON, D.J., and ZHANG, X. (1999), *Critical stress localization of flow associated with deformation of well fractured rock masses, with implications for mineral deposits*. In: McCaffrey, K.J.W., Lonergan, L. & Wilkinson, J.J. (eds) Fractures, Fluid Flow and Mineralization. Geological Society, London, Special Publications, *155*, 69–81.

SANDERSON, D.J., ROBERTS, S., and GUMIEL, P. (1994), *A Fractal Relationship between Vein Thickness and Gold Grade in Drill Core from La Codosera, Spain*, Econ. Geol. *89*, 168–173.

STAUFFER, D., and AHARONY, A., *Introduction to Percolation theory*, (CRC press, Florida 1994).

TURCOTTE D.L. (1986), *A Fractal Approach to the Relationship between Ore Grade and Tonnage*. Econ. Geol. *81*, 1528–1532.

WILSON, J.F. (1964), *The geology of the country around Fort Victoria*. Bulletin of the Rhodesian Geological Survey, *58*.

WILSON, J.F. (1968), *The geology of the country around Mashaba*. Bulletin of the Rhodesian Geological Survey, *68*.

WITT, W.K., and VANDERHOR, F. (1998), *Diversity within a unified model for Archaean gold mineralization in the Yilgarn Craton of Western Australia: An overview of the late-orogenic, structurally-controlled gold deposits*, Ore Geol. Rev. *13*, 29–64.

(Received February 25, 2014, revised May 30, 2014, accepted July 15, 2014, Published online August 5, 2014)

Pure Appl. Geophys. 172 (2015), 2057–2073
© 2014 Springer Basel
DOI 10.1007/s00024-014-0878-8

⎮ **Pure and Applied Geophysics**

Statistical Variability and Persistence Change in Daily Air Temperature Time Series from High Latitude Arctic Stations

CRISTIAN SUTEANU[1]

Abstract—In the last decades, Arctic communities have been reporting that weather conditions are becoming less predictable. Most scientific studies have not been able to consistently confirm such a trend. The question regarding the possible increase in weather variability was addressed here based on daily minimum and maximum surface air temperature time series from 15 high latitude Arctic stations from Canada, Norway, and the Russian Federation. A range of analysis methods were applied, distinguished mainly by the way in which they treat time scale. Statistical L-moments were determined for temporal windows of different lengths. While the picture provided by L-scale and L-kurtosis is not consistent with an increasing variability, L-skewness was found to change towards more positive values, reflecting an enhancement of warm spells. Haar wavelet analysis was applied both to the entire time series and to running windows. Persistence diagrams were generated based on running windows advancing through time and on local slopes of Haar analysis graphs; they offer a more nuanced view on variability by reflecting its change over time on a range of temporal scales. Local increases in variability could be identified in some cases, but no consistent change was detected in any of the stations over the studied temporal scales. The possibility for other intervals of temporal scale (e.g., days, hours, minutes) to potentially reveal a different situation cannot be ruled out. However, in the light of the results presented here, explanations for the discrepancy between variability perception and results of pattern analysis might have to be explored using an integrative approach to weather variables such as air temperature, cloud cover, precipitation, wind, etc.

Key words: Arctic climate, air temperature, pattern variability, L-moments, Haar wavelets, time scale.

1. Introduction: Aim and Scope

The sensitivity of the Arctic with respect to climate change, and the role played by this region in the framework of the climate system of the planet emphasize the importance of research on the climate of the Arctic, and in particular the high latitude Arctic (FRANCIS and VAVRUS 2012; OVERLAND and SERREZE 2012; PRZYBYLAK 2003; SERREZE and BARRY 2005; SERREZE and FRANCIS 2006; ZHANG *et al.* 2010). In this context, climatic variability on various scales represents a subject of particular interest. Pattern variability expressed, for instance, in surface temperature records, may have an impact on human health, as well as implications for ecosystems and agriculture (WEATHERHEAD *et al.* 2010). In the Arctic, changes in variability can have particularly important consequences for people's way of life, given their strong connection with the natural environment and their direct dependence on weather and on their ability to predict it based on traditional knowledge. In the last decades, especially after 1990, reports from Arctic communities emphasize the fact that weather conditions have become less predictable (GEARHEARD *et al.* 2010; KRUPNIK and JOLLY 2002; NAKASHIMA *et al.* 2012; WILLIAMS and HARDISON 2013). Scientific studies designed to identify and quantitatively describe the link between human perception and weather patterns have not led to consistent conclusions. WALSH *et al.* (2005) studied changes in temperature variance in Arctic regions throughout the second half of the twentieth century, without finding convincing evidence of increased variability. On the other hand, HOWE *et al.* (2013) documented a significant correspondence between human observation and the measured temperature change. Their research, however, concerns long-term average temperature values rather than temperature variability. KARL *et al.* (1995) found a moderate change in terms of cold and hot spell frequency and intensity, and a reduced variability in terms of day-to-day changes throughout the twentieth

¹ Geography Department and Environmental Science Department, Saint Mary's University, 923 Robie St., Halifax, NS B3H 3C3, Canada. E-mail: cristian.suteanu@smu.ca

century in the northern hemisphere. FRICH *et al.* (2002) found in the Arctic a decreasing number of frost days during the second half of the twentieth century. By looking at 5-day intervals of unusually high or low temperature, ALEXANDER *et al.* (2006) documented an increase in warm extremes and a decrease in cold extremes in most areas, including the Arctic. WEATHERHEAD *et al.* (2010) studied the relation between the perception of Inuit hunters regarding decreasing weather predictability and patterns in records from meteorological stations. They found a decrease in wind persistence—defined as similarity among wind conditions on successive days—over the last 15 years.

Most of the studies on this topic are mainly concerned with one particular temporal scale and usually involve monthly, seasonal, or annual temperature means (WALSH *et al.* 2011). A key question raised in this context was whether the lack of convergence between statistical evaluations and human perception was mainly caused by the ways in which we treat temporal scale (WEATHERHEAD *et al.* 2010). Statistical studies involve distinct, unique temporal scales, while humans perceive variability over an interrupted range of scales. The objective of this paper is to address the problem of variability change in high latitude Arctic stations: (a) by applying two approaches to variability evaluation, which differ from the point of view of the way in which they involve time scale, and (b) by considering variability change over a range of temporal scales.

2. Data: Stations Included in this Study

In spite of their limitations in terms of spatial distribution and temporal extent, instrumental surface air temperature (SAT) records acquired in meteorological stations are among the most valuable data sources in the Arctic. The number of meteorological stations in the northern regions is, in fact, relatively low, and their record lengths are more limited than in other areas. Such series of measurements for a decade offer, however, information that is not directly available from other sources.

We selected the locations for this study with the aim of analyzing relatively long daily records from a number of high latitude ($>70°N$) stations covering intervals of comparable length. Taking in consideration availability limitations and the fact that longer datasets are obtainable from fewer stations, the minimum record length was set at 60 years. Based on these criteria, daily minimum and maximum SAT data sets from 15 stations from Norway, the Russian Federation, and Canada were obtained for 15 stations (VINCENT *et al.* 2002; KLEIN TANK *et al.* 2002; KNMI 2013). The data for the Canadian Arctic stations were provided by the Climate Research Division of Environment Canada and had been subject to a homogenization process described by VINCENT *et al.* (2002). The data for the other regions were obtained from KNMI (2013). Only data from weather services were used in this study, since those obtained from the Global Communication System are considered to be less reliable (KNMI 2013). Station information is provided in Table 1. One can notice an exception from the established record length threshold in the case of Alert (57 years); this station was included, however, given its special status as the one with the highest latitude offering long-term records. As it is usually the case in the Arctic, the spatial distribution

Table 1

Stations representing data sources

Station name	Country	Latitude	Longitude	From	To
Alert	Canada	82.52	−62.28	1950	2006
Eureka	Canada	79.98	−85.93	1947	2008
Hopen	Norway	76.50	25.07	1939	2013
Mould Bay	Canada	76.23	−119.35	1948	2008
Ostrov Kotel'nyj	Russian Federation	76.00	137.87	1936	2011
Resolute	Canada	74.72	−94.98	1947	2008
Bjornoya	Norway	74.52	19.02	1937	2013
Dikson	Russian Federation	73.50	80.40	1936	2011
Maliye Karmakuly	Russian Federation	72.38	52.73	1943	2011
Fruholmen Fyr	Norway	71.09	24.00	1954	2013
Volochanka	Russian Federation	70.97	94.50	1949	2011
Chokurdah	Russian Federation	70.62	147.88	1944	2011
Clyde River	Canada	70.48	−68.52	1942	2008
Vardo	Norway	70.37	31.08	1951	2013
Dzalinda	Russian Federation	70.13	113.97	1942	2011

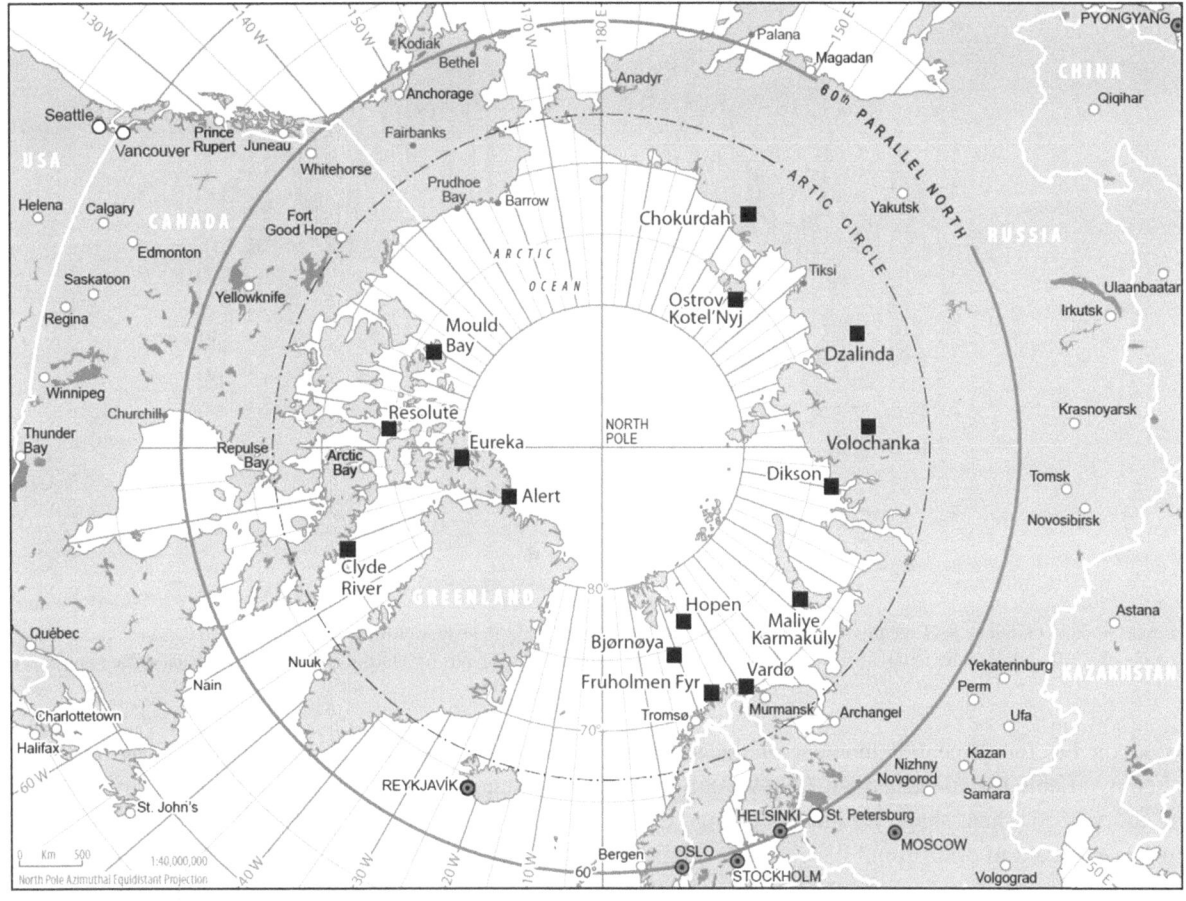

Figure 1
The analyzed stations (*solid squares*)

of stations is quite sparse with most stations having coastal positions and, thus, far from uniformly covering the variety of regional geographic conditions. Station locations are shown in Fig. 1. Samples from the daily minimum SAT records are presented in Fig. 2.

3. Methods

According to the objectives specified in Sect. 1, two distinct approaches to the evaluation of variability were considered. One focused on value distributions, specifically on the characterization of dispersion and of shape parameters. The other involved the identification and quantification of scaling properties over different ranges of time scale.

3.1. L-moments

Dispersion represents an important aspect of temperature time series and is, thus, often applied for their characterization. In many cases, variance or standard deviations are used (BOER 2010; ESAU *et al.* 2012; WALSH *et al.* 2005). In this study, we explored variability by using the statistic corresponding to standard deviation, calculated based on L-moments, as shown below. Shape parameters—skewness and kurtosis—offer, in their turn, effective means of assessing such distributions. Skewness reflects the symmetry of the distribution of a random variable and relies on its third central moment—when the distribution is unimodal, a positive skew is an indication of a longer right tail, while a longer left tail leads to a negative skew. Kurtosis represents a measure of the distribution "peakedness" and is

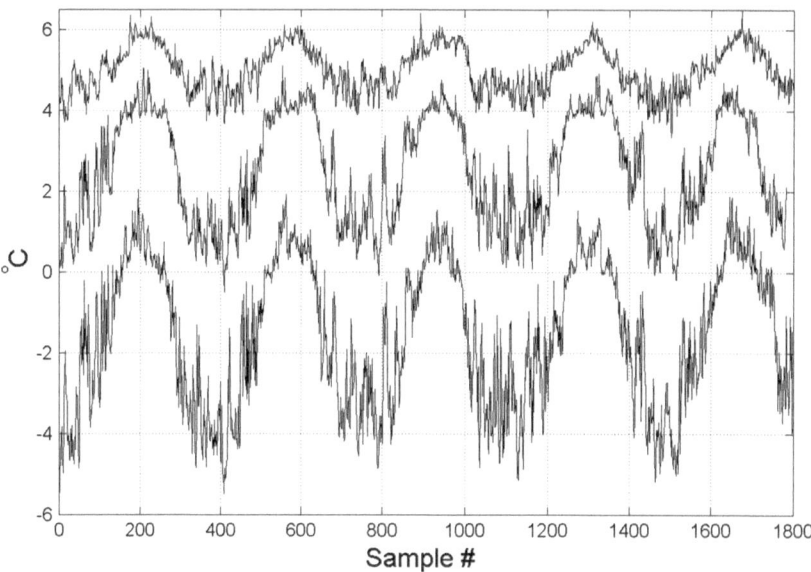

Figure 2
Samples of daily minimum SAT records illustrate differences in yearly amplitude and pattern irregularity. From *top* to *bottom*: Vardo, Dikson, Volochanka. All samples refer to the interval 1990–1994. For better legibility, the graphs for Dikson and Vardo have been shifted upwards by 4 and 5 units, respectively

based on the fourth central moment of the studied random variable (KENDALL and STUART 1983). Studying the way in which shape parameters change over time is expected to offer useful insights into temperature pattern change. In practice, the accurate estimation of shape parameters is difficult to accomplish with the help of central moments, and, therefore, another approach—L-moments—is recommended instead for this purpose (VON STORCH and ZWIERS 1999).

Introduced by HOSKING (1990), L-moments have similar meanings with central statistical moments, but they rely on linear combinations of order statistics. The rth L-moment is given by:

$$\lambda_r = r^{-1} \sum_{0}^{r-1} (-1)^k \binom{r-1}{k} EX_{r-k:r}, \qquad (1)$$

where $X_{k:r}$ is the kth smallest value in a sample of size r, and E is the expected value. The first moment is the expected lowest value in a sample of one element. It is, thus, equal to the first statistical moment, or the mean. The second moment, λ_2, or L-scale, plays the role of standard deviation; sometimes the ratio λ_2/λ_1 is used, which is equivalent to the coefficient of variation. Similarly with those using central statistical

moments, the shape parameters are based on higher L-moments, as follows: L-skewness is given by

$$\tau_3 = \frac{\lambda_3}{\lambda_2}, \qquad (2)$$

and L-kurtosis by:

$$\tau_4 = \frac{\lambda_4}{\lambda_2}. \qquad (3)$$

For the normal distribution, L-skewness is equal to zero (as is the case for the classical skewness definition), while L-kurtosis is equal to 0.1226; higher L-moments of odd values can also be computed and regarded as generalized measures of distribution symmetry (HOSKING 1990). A wide range of applications as well as theoretical studies have proven L-moments to be particularly robust and accurate and to outperform other methods (HOSKING 2006; HOSKING and WALLIS 1997; LOUCKS and VAN BEEK 2005; AKRAM and HAYAT 2014).

L-scale, L-skewness, and L-kurtosis were determined in this study for all the time series (daily minimum and maximum SAT) over intervals of various lengths, and their temporal change was recorded.

3.2. Scaling Aspects

The second approach consists of an evaluation of scaling properties of the studied time series. SAT time series are known to be characterized by auto-correlation functions that decay slowly, according to a power law, rather than an exponential (EICHNER et al. 2003; FRAEDRICH et al. 2009; KANTELHARDT et al. 2001). This pattern property can be identified over a wide range of scales, from days to years and decades. For these temporal scale ranges, a power law relationship is consistently found between the "size of the fluctuations" F and the time scale s over which they are evaluated. The meaning of such a power law relationship and its significance as a hallmark for "long-memory" processes has been a subject of extensive discussion (MARAUN et al. 2004; VAROTSOS et al. 2013). It has been pointed out that it is often tempting to fit log–log graphs with a straight line in a search for power laws, even when long-range corre-lations are not present (MARAUN et al. 2004): appropriate precautions must be taken before assign-ing exponents to an $F(s)$ relationship, and such steps were taken in this study (Sect. 4). On the other hand, the main focus of the present study concerns ways of identifying and evaluating aspects of variability, rather than an exploration into the nature and origin of long-memory processes.

This approach is substantially different from the one described in Sect. 3.1, in which the actual sequence of the values in the time series was not taken in consideration. The most important distinc-tion consists of the fact that in this case a virtually continuous range of scales is evaluated. The out-come of the analysis indicates whether and over what time scales the pattern is characterized by a consistent scaling property, and characterizes this property of the pattern. Different methods can be applied for this purpose. Detrended Fluctuation Analysis—DFA (KANTELHARDT et al. 2001; VAROT-SOS et al. 2008; EFSTATHIOU et al. 2011)—and wavelets analysis (HOLSCHNEIDER 1995) have proven to be among the most powerful and reliable ones. Both of these approaches are capable of accurately evaluating the scaling exponents associated with the time series, without being influenced by the potential presence of trends. DFA has been widely used in studies dedicated to climate variables (EICHNER et al. 2003; FRAEDRICH et al. 2009; LENNARTZ and BUNDE 2010; SUTEANU 2011; VAROT-SOS and KIRK-DAVIDOFF 2006; VAROTSOS et al. 2009). Wavelets have been typically applied for the reliable analysis of non-stationary time series and localized features, which has proven them to be particularly valuable in geophysics. In this study, we are not focusing on localized features. The singularities in the time series are invariant under translation, and our objective is to enhance the statistical relevance of the evaluation by taking averages over time series segments of different length. This goal is effectively addressed by Haar wavelet analysis (HAAR 1910; LOVEJOY et al. 2012). Haar wavelets analysis repre-sents a powerful tool for the study of climate variables on a wide range of temporal scales (LOVEJOY and SCHERTZER 2013). It is this method that we used here. A brief description of the method as applied in this study is given below.

The implementation of the method starts from the daily (minimum or maximum) temperature time series $t(i)$. As a first step, the seasonal variation is removed. The average value of every day of the year over the studied interval $\langle t(i) \rangle.$, is subtracted from the time series:

$$p(i) = t(i) - \langle t(i) \rangle. \tag{4}$$

The resulting time series is normalized:

$$w(i) = \frac{p(i) - \overline{p(i)}}{\text{std}(p(i))}. \tag{5}$$

The method can be applied directly to the time series $w(i)$; alternatively, it can be applied to its cumulative sum, or "profile" $q(i)$:

$$q(i) = \sum_{j=1}^{i} w(j). \tag{6}$$

Similar results are obtained with the two proce-dures, considering the relationship between the exponents H_p (associated with the profile) and H_o (corresponding to the original time series $w(i)$): $H_p = H_o + 1$. However, analysis experiments show that determining the exponent for the time series profile rather than directly for the time series $w(i)$ leads to more reliable results (SUTEANU 2013).

It is this procedure that we adopted for the present study.

The next step consists of establishing the relation between the "size of the fluctuation" F and the time scale s. To this end, we use the mother wavelet proposed by HAAR (1910), which is particularly simple and effective (Lovejoy et al. 2012):

$$\Psi(i) = \begin{cases} 1; & 0 \leq i \leq 1/2 \\ -1; & -1/2 \leq i < 0 \\ 0; & \text{otherwise} \end{cases} \tag{7}$$

The fluctuation size F is determined for every element i in the time series as the mean square difference between time series values of elements in the two intervals between i and $i - s/2$ and $i + s/2$, respectively. The $F(s)$ relationship is explored for a range of time scales s. If a power law is found:

$$F(s) \propto s^H, \tag{8}$$

then the exponent H characterizes the time series for the scale interval over which the power law (8) holds. The latter is often the case for long time scale intervals, from days to decades (SUTEANU 2013). Higher values of H obtained in this way (values closer to 1) indicate stronger persistence, i.e., a more pronounced tendency of the time series to preserve its growing (or decreasing) trace; intuitively, one may talk in this case about a "smoother" time series. Uncorrelated noise corresponds to an H value of 0.5. Lower values of H reflect an antipersistent behavior, which is dominated by stronger tendencies to revert the growing (decreasing) trace into a decreasing (growing) one.

In this study we applied the above methodology to the time series corresponding to minimum and maximum temperature values from all stations. Previous studies (SUTEANU and MANDEA 2012; SUTEANU 2013) have shown that the time series persistence not only varies with station location, but may also change over time to a significant extent. In order to investigate the temporal change in variability, according to the objectives of the current study, we also applied this methodology to successive temporal windows (the results are presented in Sect. 4).

The problems raised in Sect. 1 highlight apparent contradictions between the perception of change in weather variability and the results of various methods applied for the quantitative evaluation of patterns in measurement data. A question that might be raised is whether disagreements regarding the temporal change in variability can stem from the different temporal scales used to assess variability (WEATHERHEAD et al. 2010). To better address this question, we propose here a nuanced approach to scale-dependent pattern properties. Beyond the exponent determined for Eq. (8), we evaluate the $F(s)$ relation for every temporal scale interval used in the analysis, by determining the "local slopes" in these graphs (VAROTSOS et al. 2013). While the values of local slopes are still denoted by "H", in this case they have a different meaning: they are not power law exponents anymore, but rather characterize the $F(s)$ relation "locally", for each scale interval. This procedure is applied to running windows of length g shifted with a step z, defined in Sect. 4.1. A matrix is generated for each time series, with the elements H_{ij} representing the local slopes for all the applied temporal scales (i) and all the positions of the window in time (j). By representing these matrices in persistence diagrams, we may check if and in what ways pattern variability changes over time by distinguishing the time scales involved in the change, over various temporal intervals. In this context, the goal of the analysis is to distinguish aspects of variability and their change in time over different time scales, and not to determine a scaling exponent for the time series; therefore, persistence diagrams can contain scale intervals that would otherwise not be included in the detection and evaluation of a scaling exponent.

4. Results and Discussion

4.1. Results Concerning L-moments

The L-moments presented above were applied to the time series obtained according to Eq. (4) for successive temporal windows of different lengths. The window length g, shifted by steps z, was taken as: $g = 2 kz + 1$, $k = 0, 1, 2,\ldots$ and the value obtained for each window was assigned to its central element. In this study we took $z = 1$, while the choice of the window length proved not to make a substantial difference for the resulting pattern of change.

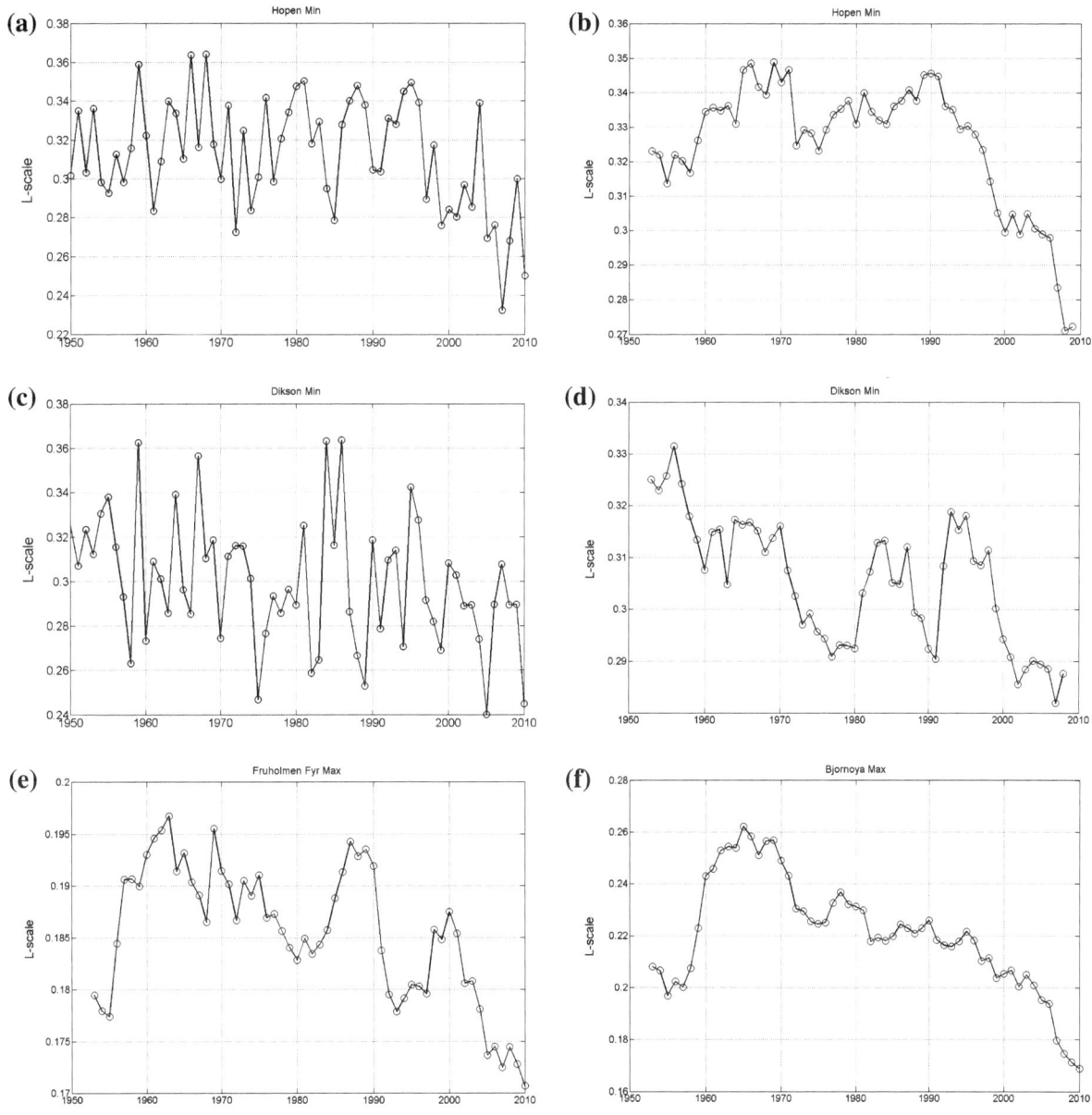

Figure 3

Examples of the change in L-scale over time; **a** and **c**: 1-year window width; **b**, **d**, **e**, **f**: 7-year window width, windows shifted by 1 year

Figure 3 shows examples of the temporal evolution of dispersion: L-scale for minimum temperatures in Hopen is evaluated with 1-year windows in Fig. 3a, and with 7-year windows in Fig. 3b. Results obtained with the same window parameters are shown for another station (Dikson) in Fig. 3c, d. The irregularity decreases with growing window length, but the overall tendency is the same, irrespective of the choice of the window size. This tendency is easier to

identify when longer windows are used (results for other two stations are presented in Fig. 3e, f). In all cases, one cannot find a general tendency for the L-scale (which corresponds to the standard deviation) to increase. If one concentrates on the time interval since 1990, for which the perception of an increasingly strong variability has been widely reported, one finds a lack of trend or a consistent decrease in L-scale. While the longer time scales can better

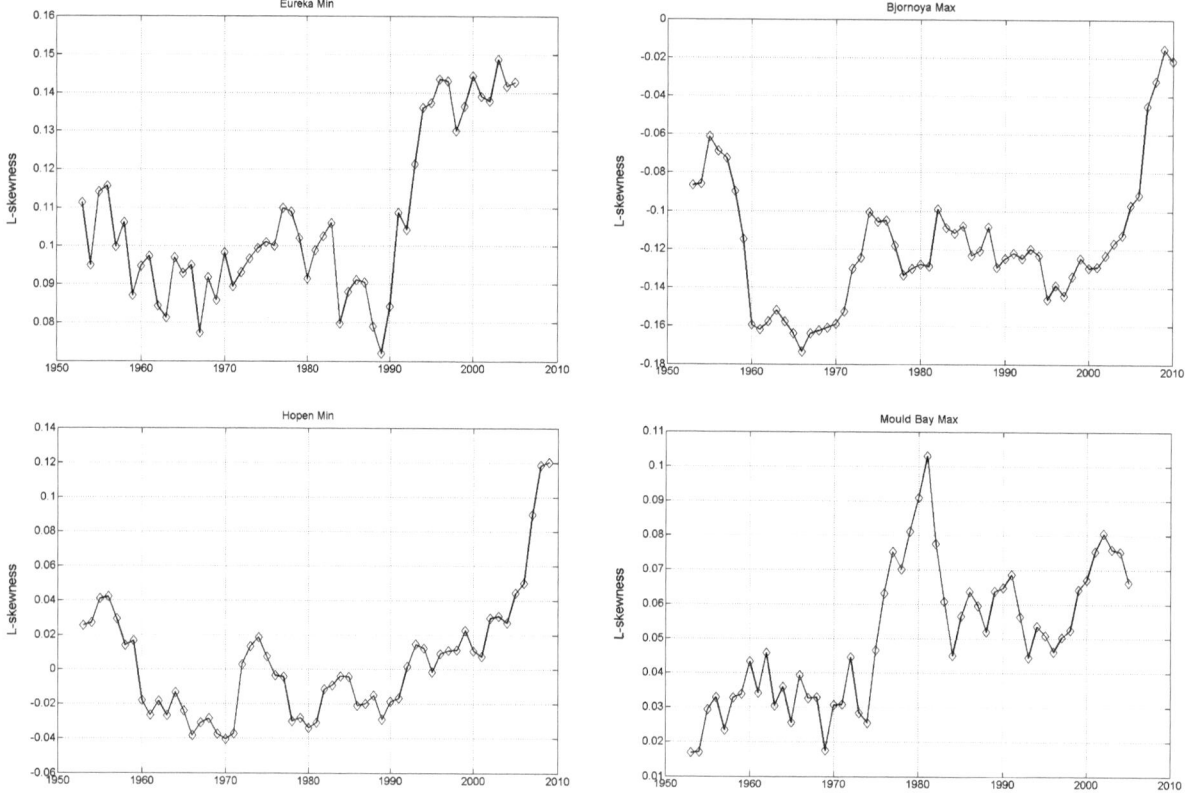

Figure 4
Examples of the change in L-skewness over time (7-year window width, windows shifted by 1 year)

highlight tendencies of change, the shorter time windows (of 1 year) have the merit of pointing out the fact that year-by-year changes involve relatively strong fluctuations (Fig. 3a, c), which are not captured by the picture of broader tendencies. By looking at the variation in L-scale in graphs such as those in Fig. 3a, c, one can see that it would be difficult to tell, based on the personal experience reflected in this variable alone, whether one faces an increasing variability from year to year or not.

The results of the evaluation of distribution symmetry, based on L-skewness, are different. We found in most cases a change towards more positive skewness (Fig. 4 shows examples from four stations). The positive and growing skewness can be related to warm spells reflected in longer tails on the right-hand side of the value distribution, which are not balanced out by correspondingly strong cold spells. This change in skewness can be seen especially after 1990, an interval which coincides with the one

reportedly becoming less and less predictable, as discussed in Sect. 1.

The results for L-kurtosis did not show similar tendencies with L-skewness. In fact, this characteristic of the distributions proved to lead to the most diverse scenarios of temporal evolution— several examples are shown in Fig. 5: for instance, L-kurtosis can emphasize a generally increasing trend (Fig. 5a), a generally decreasing trend (Fig. 5b), fluctuations without a clear trend (Fig. 5c), or even a slow increase followed by a sharp drop (Fig. 5d). No consistent picture could be identified for the studied datasets in terms of temporal change in L-kurtosis—neither for the whole available time intervals, nor for the interval from 1990 onwards.

Confronted with the question concerning a possible change in variability, the analysis based on L-moments provided mixed answers. Two of the determined parameters either lack a consistent

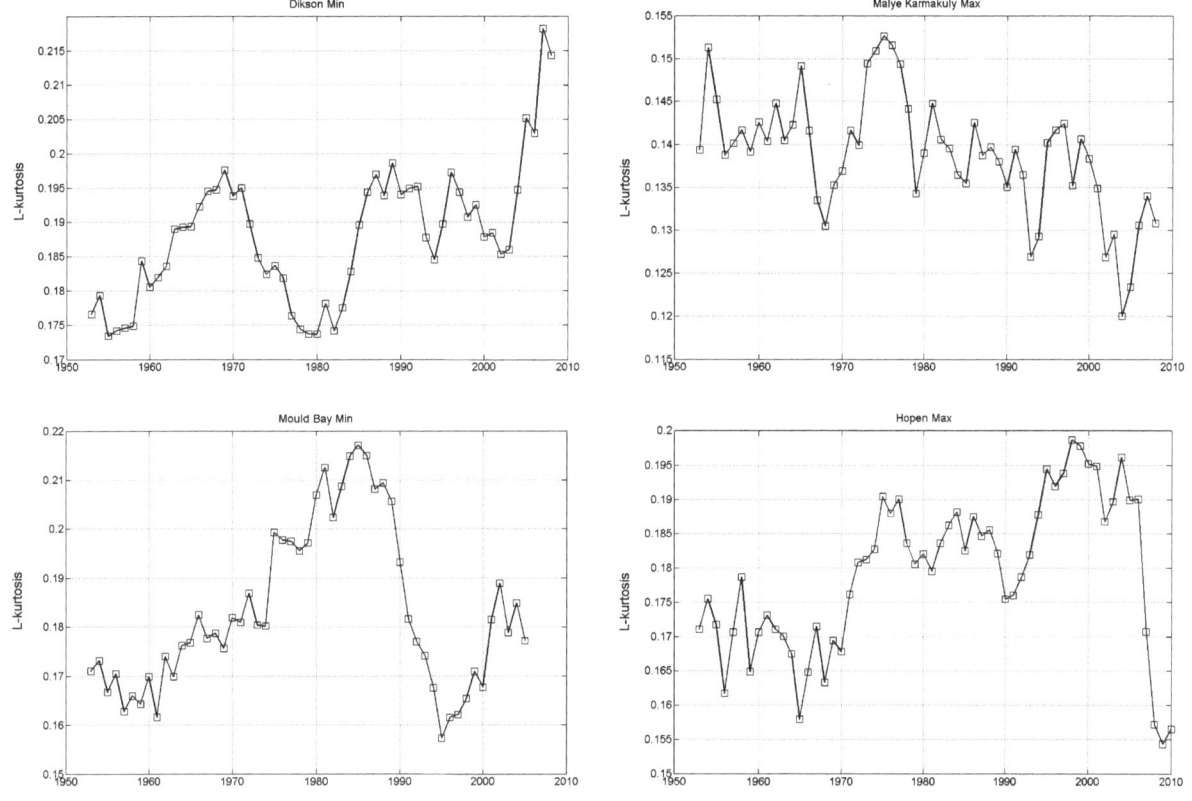

Figure 5
Examples of the change in L-kurtosis over time (7-year window width, windows shifted by 1 year)

tendency of change (L-kurtosis), or suggest decreasing variability (L-scale). L-skewness, however, points towards a change towards positive distribution skewness, which can be related to increasingly strong and/or frequent warm spells.

4.2. Results Concerning Scaling Aspects

The methodology presented in Sect. 3.2 was applied to all the time series corresponding to minimum and maximum SAT. As mentioned above, finding a power law relation by detecting straight lines in log–log graphs can be misleading (MARAUN et al. 2004), especially when the linear regression is not characterized by strong correlation. Even what is usually considered to be strong correlations may incorrectly suggest the presence of power laws, when gradual changes in slope take place. A careful analysis must be performed to confirm the consistent presence of an exponent that characterizes the pattern

over a certain scale range. Determining local slopes in $F(s)$ graphs in logarithmic coordinates can support this task; it also helps the identification of the temporal scale interval to which the scaling behavior can be assigned (MARAUN et al. 2004; VAROTSOS et al. 2013). The time series analyzed here proved to be characterized by strong scaling properties—Fig. 6a shows an example (Bjornoya). Local slopes were determined for different scale window lengths as recommended by VAROTSOS et al. (2013) and shown as a function of temporal scale. The representation in Fig. 6b refers to the graph in Fig. 6a: error bars correspond to 95 % confidence intervals for the slope values (CHATTERJEE and HADI 1986). One can notice in this example that the H-value fluctuates around 0.75 over a wide range of temporal scales; one can also notice scale range boundaries that should be taken in consideration when assigning the H exponent to the time series (approximately between 25 and 1,600 days). Fig. 6b illustrates the fact that an

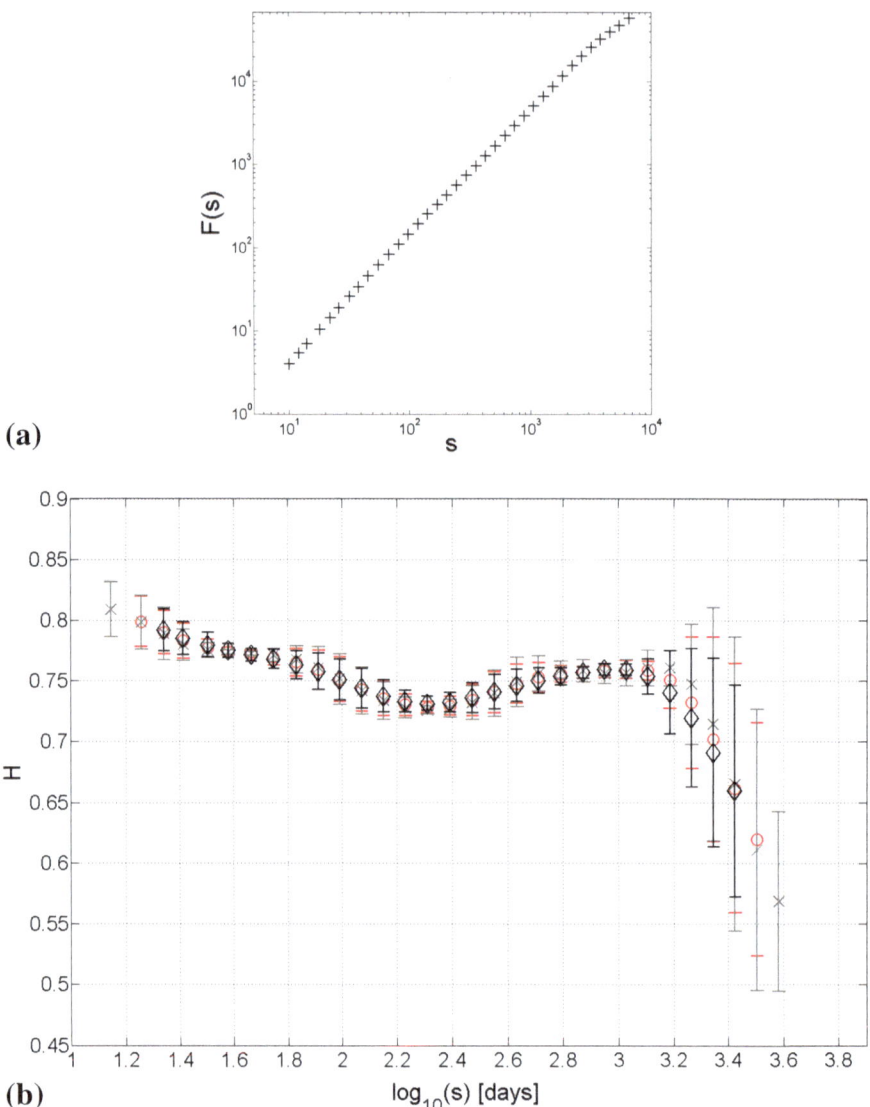

Figure 6

a Example of F(s) graph (Bjornoya, minimum temperature), *s* is shown in days; **b** local slopes for the graph shown above, for windows of 5 points (*x*), 7 points (*circles*), and 9 points (*diamonds*). *Error bars* show 95 % confidence intervals

incorrect choice of the scaling interval can lead to large departures in exponent value (MARAUN *et al.* 2004).

Well-defined power laws were found in the studied cases. Exponent values are presented in Fig. 7, along with their 95 % confidence intervals. The latter are typically narrower than those obtained with detrending fluctuation analysis (SUTEANU and MANDEA 2012). Differences in exponent values compared to SUTEANU and MANDEA (2012), however, seem to be mainly explained by the

selection of the scaling interval for the determination of the exponent. In most cases, minimum temperature time series have higher persistence than maximum temperature, in agreement with other studies (KIRALY *et al.* 2006; SUTEANU 2011; SUTEANU and MANDEA 2012).

Most values are clustered around a value of 0.7. Lower persistence can be noticed especially in the station located at the highest latitude (Alert). High latitude locations that are not close to the coast, Volochanka and Dzalinda, also stand out with their

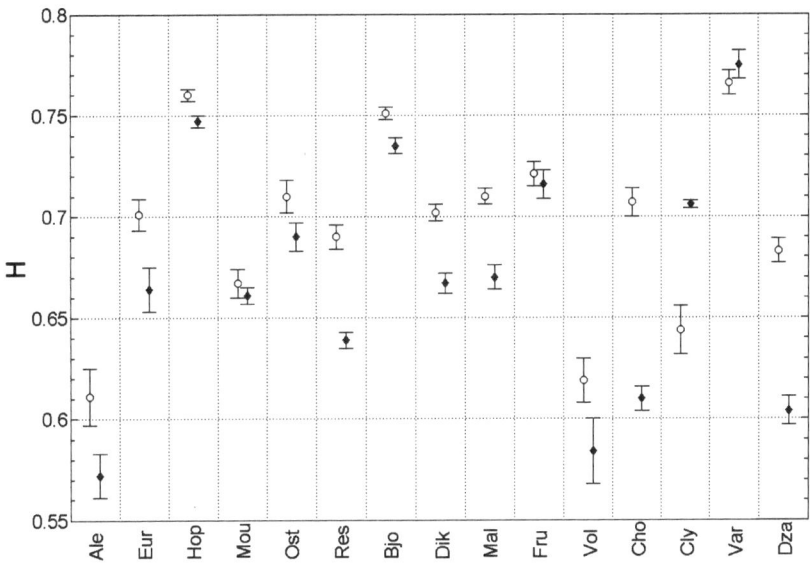

Figure 7
Scaling exponent H corresponding to the time series for minimum temperature (*open circles*) and maximum temperature (*solid circles*) for all the stations. *Error bars* show 95 % confidence intervals. Station *symbols* represent the first three letters of station names

lower persistence. This stronger variability, expressed on a wide range of temporal scales, is expected indeed in areas where temperature variation is not moderated by the effect of the ocean. In fact, some stations located in the immediate proximity of the coast or on small islands (Vardo, Hopen, Bjornoya, Fruholmen Fyr) enjoy the highest persistence; other studies have also found that oceanic locations are typically characterized by higher persistence than continental ones (EICHNER *et al.* 2003; KIRALY *et al.* 2006; SUTEANU 2011). For Chokurdah, which is not a coastal position either, only the maximum temperature pattern is characterized by lower persistence. Simple explanations for comparatively high or low persistence, such as those regarding the general effect of a large water mass or the thermal behavior of inland locations, may be helpful, to some extent, for our understanding of differences in persistence between locations. However, potential explanations regarding the persistence values found for temperature patterns in various locations should be provided with caution, given the nonlinear nature of the complex interacting processes that dominate different temporal scale ranges.

For the purpose of this study, it was important to investigate, beyond the overall time series, the extent to which and the ways in which pattern variability changes over time. To this end we applied the same methodology to running windows. The window length proved not to have a substantial impact on the outcome of this analysis, as was also the case in WALSH *et al.* (2005) and SUTEANU and MANDEA (2012). Results for 7-year long windows shifted by one year are presented here. This study showed that *H* values change over time, with certain stations emphasizing similar patterns of change. Examples are shown in Fig. 8 for two pairs of stations and in Fig. 9 for another group of three stations. It would be hazardous to advance possible explanations for the temporal variation of H, as is the case of statistical parameters reflecting changes in variability (WALSH *et al.* 2005). Although fluctuations in exponent values are strong, no station showed an indication of consistent increase or decrease in persistence. This analysis, which captures pattern properties over a range of scales (as opposed to one single scale as was the case in Sect. 4.1), does, therefore, not confirm the perception of increasing variability discussed above, neither over the available time interval of approximately 60 years, nor over the interval from 1990 onwards.

To further explore variability change and to address it as a function of scale, we applied the

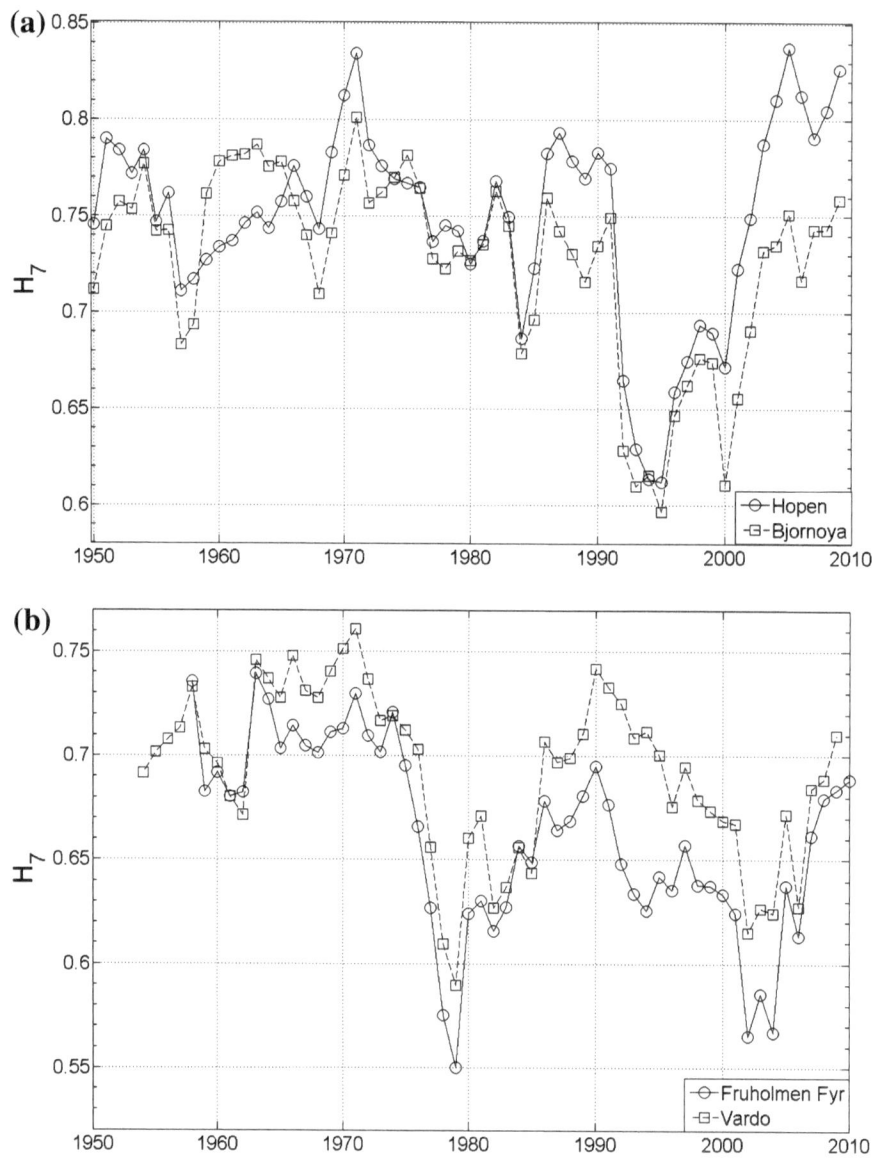

Figure 8

Changes in the scaling exponent over time, for 7-year wide windows shifted by one year (minimum temperatures). Pairs of stations characterized by similar temporal change: the two northern-most stations from Norway (**a**) and two other stations located at a lower latitude (**b**). In spite of their relative spatial proximity, the two pairs of stations are characterized by strong differences in terms of the temporal change in the scaling exponent

methodology proposed in Sect. 3.2 for the construction of persistence diagrams. Time windows of various length were explored—again, the window size does not have a significant influence on the outcomes of this procedure. Narrower window lengths offer a higher temporal resolution of diagrams (improving the information availability along the abscissa), but the shorter time series

resulting for each window decrease the scale range for the analysis (negatively influencing the information availability along the ordinate). Figures 10 and 11 show results obtained with a window length of 5 years and a shift step of one year. Colours (Fig. 10) or levels of grey (Fig. 11) represent the values of H for each scale interval. In contrast to the procedure performed and

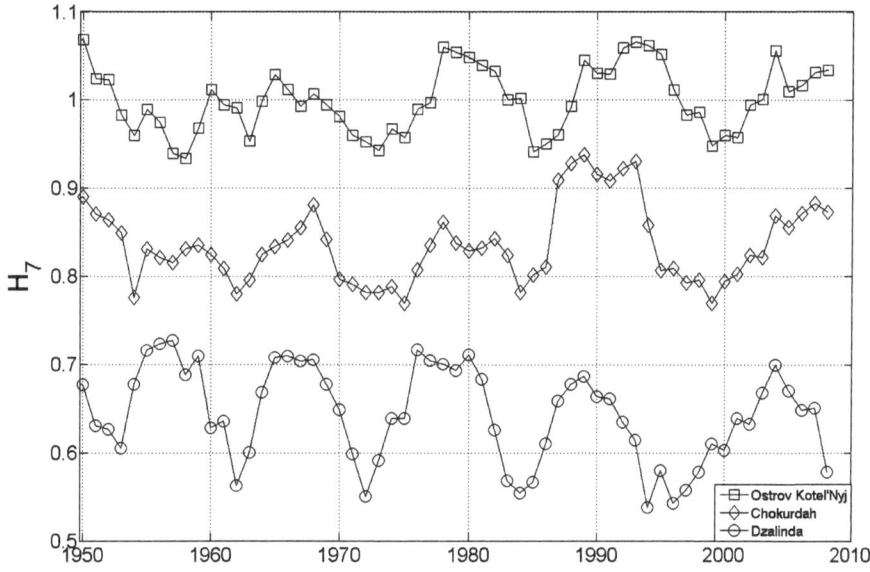

Figure 9

Same as in Fig. 7, for another group of three stations (graphs are vertically shifted by 0.1 units for better legibility). These stations are spatially more dispersed than those in Fig. 7

illustrated in Fig. 6b, here local slopes were always computed for each interval between two scale values, and not for groups of points. One can notice that the resulting diagrams offer a different view on pattern variability. Scale-related information can be followed across the recorded time interval. It is, thus, possible to notice scale ranges for which variability increased, as was the case, for instance, in Hopen right after 1990, or over which variability was relatively constant, which can be seen in Hopen and Bjornoya especially between 1960 and 1980 in the lower temporal scale range (Fig. 10). In most cases, alternative phases of increase and decrease in H can be observed, in some instances with a sharp drop to low values, as shown in Fig. 11 in the upper temporal scale range. Cross-sections through the persistence matrices—pertaining to a certain time scale and cutting through the succession of time windows—can, thus, be explored in order to study variability aspects of the time series that are otherwise not easy to detect. The study of the resulting diagrams for all stations did not reveal a consistent increase in variability over any broad scale interval among those investigated here. The diagrams shown in Figs. 10 and 11 represent

typical examples of outcomes of this type of analysis.

5. Conclusions

The question regarding the possible increase in weather variability, occurring especially during the last decades, was addressed based on daily minimum and maximum SAT time series from high latitude Arctic stations. Variability and its change over time were studied with a range of analysis methods, distinguished mainly by the way in which they treat time scale. Statistical L-moments (L-scale, L-skewness, and L-kurtosis) were determined for temporal windows of different lengths. Two of the moments—L-scale, which corresponds to the standard deviation, and L-kurtosis—did not provide a picture that would be consistent with an increasing variability. In contrast, L-skewness was found to change towards more positive values, which can be related to enhanced warm spells. This tendency is particularly noticeable after 1990. Haar wavelet analysis, which takes in consideration a spectrum of temporal scales, was applied both to the entire time series and to running windows. The results obtained for the entire time

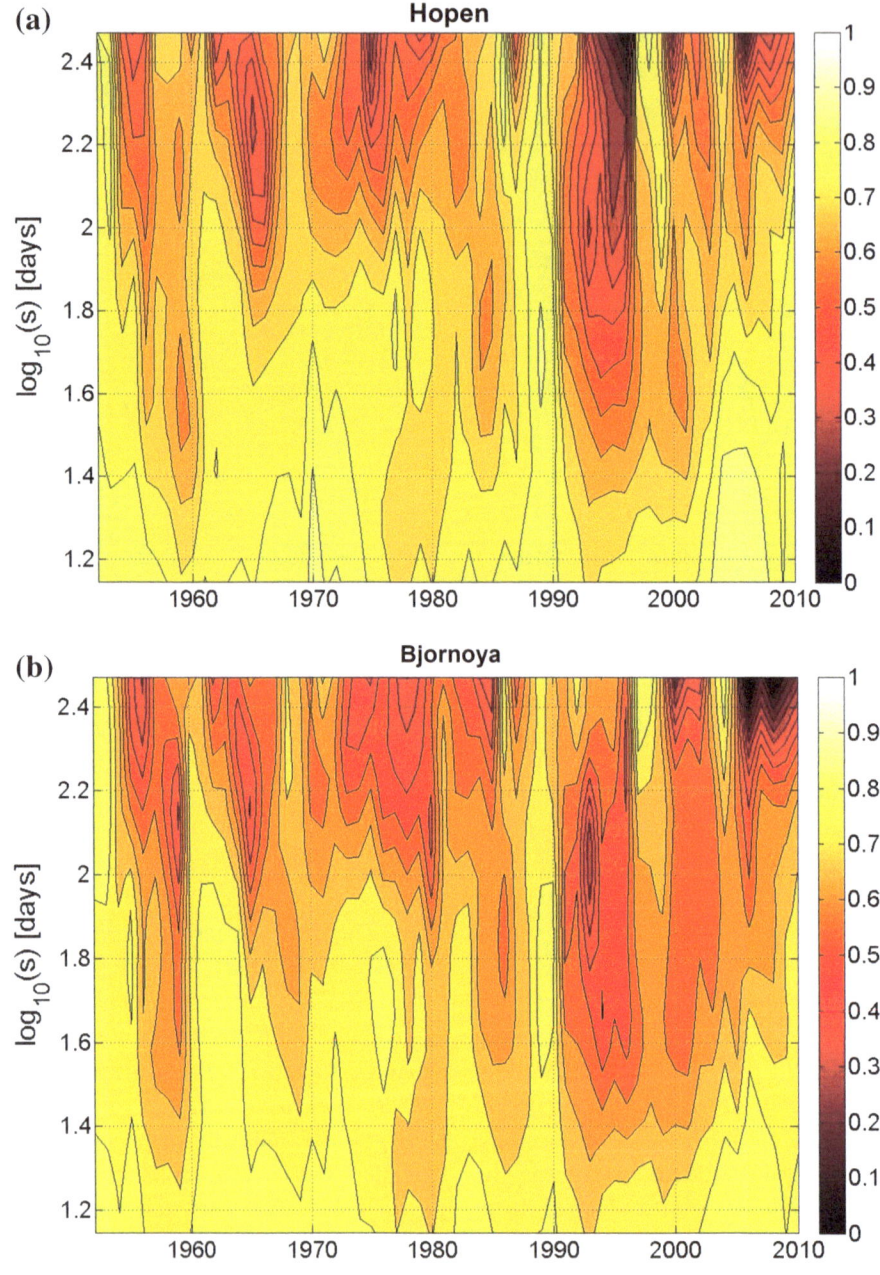

Figure 10

a, **b** Persistence diagrams (minimum temperature records) for Bjornoya amd Hopen. Colours correspond to the values of the H exponent as a function of time (5-year windows shifted by one year) and of the scale interval (indicated in the ordinate in logarithmic scale)

series show an H exponent clustering for most stations around a value of 0.7, with the lowest and highest values being obtained for continental and coastal/insular positions, respectively; moreover, the station located at the highest latitude, Alert, was found to be characterized by particularly low exponents. In general, minimum temperature records are associated with higher persistence than maximum temperatures. Persistence matrices were generated based on running windows advancing through time and on local slopes from Haar analysis graphs, which show the size of the fluctuation as a function of the

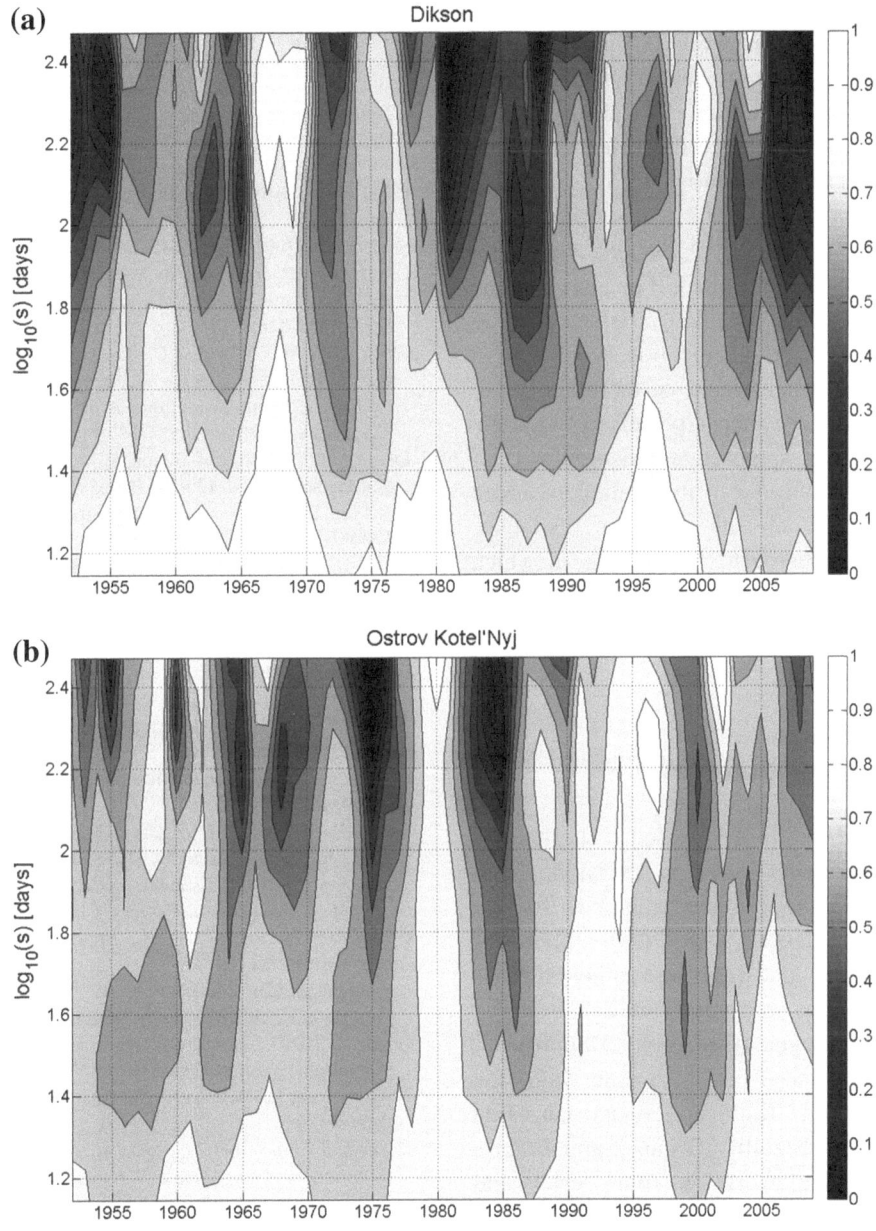

Figure 11

a, b Persistence diagrams (minimum temperature records) for Dikson and Ostrov Kotel'Nyj

time scale in logarithmic coordinates. These matrices were represented in persistence diagrams, which offer a nuanced access to aspects of pattern variability. They refer to time, on the one hand, and to temporal scale, on the other hand. Local increases in variability could, thus, be identified in some cases, but no consistent change in variability was detected in any of the stations over the studied temporal scales.

According to the results of this paper, the fact that the prevailing perception of decreasing weather predictability could not be reliably confirmed by statistical studies is not caused by the fact that the exploration did not include a range of temporal scales. The methods dedicated to the identification of pattern scaling properties, which were applied here, scan a wide range of temporal scales, and yet no consistent trend of increasing variability was found. The possibility for other intervals of temporal scale (e.g., days, hours, minutes) to reveal a different situation cannot be ruled out. However, in the light of the results presented here and of the statements made by members of Arctic communities (GEARHEARD *et al.* 2010; KRUPNIK and JOLLY 2002; NAKASHIMA *et al.* 2012; WEATHERHEAD *et al.* 2010), explanations might have to be searched for elsewhere. For instance, one may want to investigate weather variables such as temperature, cloud cover, precipitation, wind, etc., in an integrated way, which would be more similar to the manner in which weather is perceived by Arctic communities; such an approach would be, however, challenging, if only because of the limited access to weather information on all of these variables. Exploring such an approach in situations in which information on several weather aspects is available might offer premises to find a more effective way forward.

Acknowledgments

The author would like to acknowledge the help of Lucie Vincent (Climate Research Division, Environment Canada) who provided the data for the Canadian Arctic stations, and the support of KNMI, where the data from the other regions were obtained for this study. He would also like to thank William Flanagan (Geography Department, Saint Mary's University) for Fig. 1, and Katherine Dorey, Susan Murray, Katie Porter, and Christine Vincent (Climate Variability Group, Saint Mary's University) for their work on time series and metadata preparation and verification. This work was partly funded by the research grant "Space–Time Climate Variability Characterization" (Saint Mary's University) and the research grant 21/5.10.2011, Program TE, Romania.

REFERENCES

AKRAM, M., and HAYAT, A. (2014), *Comparison of Estimators of the Weibull Distribution*, Journal of Statistical Theory and Practice 8 2, 238–259, doi:10.1080/15598608.2014.847771.

ALEXANDER, L. V., ZHANG, X., PETERSON, T. C., CAESAR, J., GLEASON, B., KLEIN TANK, A. M. G., HAYLOCK, M., COLLINS, D., TREWIN, B.,

RAHIMZADEH, F., TAGIPOUR, A., RUPA KUMAR, K., REVADEKAR, J., GRIFFITHS, G., VINCENT, L., STEPHENSON, D. B., BURN, J., AGUILAR, E., BRUNET, M., TAYLOR, M., NEW, M., ZHAI, P., RUSTICUCCI, M., and VAZQUEZ-AGUIRRE, J. L. (2006), *Global Observed Changes in Daily Climate Extremes of Temperature and Precipitation*, Journal of Geophysical Research *111*, D05109, doi:10.1029/2005JD006290.

BOER, G. J. (2010), *Changes in Interannual Variability and Decadal Potential Predictability under Global Warming*, J. Climate 22, 3098–3109, doi:10.1175/2008JCLI2835.1.

CHATTERJEE, S., and HADI, A. S. (1986), *Influential Observations, High Leverage Points, and Outliers in Linear Regression*, Stat. Sci. *1*, 379–416.

EFSTATHIOU, M.N., TZANIS, C., CRACKNELL, A.P., and VAROTSOS, C.A. (2011), *New features of land and sea surface temperature anomalies*, International Journal of Remote Sensing, *32* (11), 3231–3238.

EICHNER, J.F., KOSCIELNY-BUNDE, E., BUNDE, A., HAVLIN, S., and SCHELLNHUBER, H.-J. (2003), *Power Law Persistence and Trends in the Atmosphere: a Detailed Study of Long Temperature Records*, Phys. Rev. E *68*, 046133.

ESAU, I., DAVY, R.., and OUTTEN, S. (2012), *Complementary Explanation of Temperature Response in the Lower Atmosphere*, Environ. Res. Lett. *7*, 044026, doi:10.1088/1748-9326/7/4/044026.

FRAEDRICH, K., BLENDER, R., and ZHU, X. (2009), *Continuum Climate Variability: Long-Term Memory, Extremes, and Predictability*, Int. J. Modern Physics B *23*, (28–29), 5403–5416.

FRANCIS, V.A., and VAVRUS, S.J/(2012), *Evidence Linking Arctic Amplification to Extreme Weather in Mid-Latitudes*, Geophys. Res. Lett. *39*, L06801.

FRICH, P., ALEXANDER, L.V., DELLA-MARTA, P., GLEASON, B., HAYLOCK, M., KLEIN-TANK, A.M.G., and PETERSON, T. (2002), *Observed Coherent Changes in Climatic Extremes during the Second Half of the 20th Century*, Clim. Res., *19*, 193–212.

GEARHEARD, S., POCERNICH, M., STEWART, R., SANGUYA, J. and HUNTINGTON, H.P (2010), *Linking Inuit Knowledge and Meteorological Station Observations to Understand Changing Wind Patterns at Clyde River, Nunavut*, Climatic Change *100*, 267–94.

HAAR, A. (1910), *Zur Theorie der orthogonalen Funktionensysteme*, Math. Ann. *69*, 331–371.

HOLSCHNEIDER, M., WAVELETS : An Analysis Tool (Clarendon Press, Oxford 1995).

HOSKING, J.R.M. (1990), *L-Moments: Analysis and Estimation of Distributions Using Linear Combinations of Order Statistics*, Journal of Royal Statistical Society B, *52*, 2, 105–24.

HOSKING, J.R.M. (2006), *On the Characterization of Distributions by Their L-Moments*, J. Stat. Planning and Inference *136*,193–198.

HOSKING, J.R.M., and WALLIS, J.R., Regional Frequency Analysis: an Approach Based on L-Moments (Cambridge University Press, Cambridge 1997).

HOWE, P.D., MARKOWITZ, E.M., LEE, T.M., KO, C.-Y., and LEISEROWITZ, A. (2013), *Global Perceptions of Local Temperature Change*, Nature Climate Change, *3*, 352–356, doi:10.1038/nclimate1768.

KANTELHARDT, J.W., KOSCIELNY-BUNDE, E., REGO, H.H.A., HAVLIN, S., BUNDE, A. (2001), *Detecting Long-Range Correlations with Detrended Fluctuation Analysis*, Phys. A *295*, 441–454.

KARL, T. R., KNIGHT, R. W. & PLUMMER, N (1995), *Trends in High-Frequency Climate Variability in the Twentieth Century*, Nature *377*, 217–220.

KENDALL, M.G., and STUART, A., The Advanced Theory of Statistics (Macmillan Publishers, London 1983).

KIRALY, A., BARTOS, I., JANOSI, I. M. (2006), *Correlation Properties of Daily Tmperature Anomalies over Land,* Tellus, *58A,* 593–600.

KLEIN TANK, A. M. G. *et al.* (2002), *Daily Dataset of 20th-Century Surface Air Temperature and Precipitation Series for the European Climate Assessment,* Int. J. of Climatol., *22,* 1441–1453.

KNMI (2013) Climate Explorer, http://climexp.knmi.nl/selectdailyseries.cgi. Accessed 21 January 2013.

KRUPNIK, I., and JOLLY, D. (eds.), The Earth is Faster Now: Indigenous Observations of Arctic Environment Change (Arctic Research Consortium of the United States, Fairbanks, Alaska 2002).

LENNARTZ, S., BUNDE, A. (2010), *Trend Evaluation in Records with Long Term Memory: Application to Global Warming,* Geophys Res Lett *36,* L16706.

LOUCKS, D. P., and VAN BEEK, E., Water Resources Systems Planning and Management: An Introduction to Methods, Models and Applications (UNESCO, Paris 2005).

LOVEJOY, S., SCHERTZER, D., and STANWAY, J.D. (2012), *Haar Wavelets, Fluctuations and Structure Functions: Convenient Choices for Geophysics,* Nonlin. Processes Geophys. *19,* 513–527.

LOVEJOY, S., and SCHERTZER, D., The Weather and Climate: Emergent Laws and Multifractal Cascades, (Cambridge University Press, Cambridge 2013).

MARAUN, D., RUST, H. W., and TIMMER, J. (2004), *Tempting Long-Memory – on the Interpretation of DFA Results,* Nonlinear Processes Geophys., *11,* 495–503.

NAKASHIMA, D.J., GALLOWAY, McLEAN, K., THULSTRUP, H.D., RAMOS CASTILLO, A., and RUBIS, J.T., Weathering Uncertainty: Traditional Knowledge for Climate Change Assessment and Adaptation, (UNESCO, and Darwin, UNU, 120 pp, Paris 2012).

OVERLAND, J.E., and SERREZE, M.C. (2012), Advances in Arctic Atmospheric Research, in Lemke, P., and Jacobi, H.-W. (eds.), Arctic Climate Change: The ACSYS Decade 11 and Beyond, Atmospheric and Oceanographic Sciences Library 43, Springer 11–26.

PRZYBYLAK, R., The Climate of the Arctic (Kluwer Academic Publishers, Dordrecht 2003).

Serreze, M.C., and Barry, R.G., The Arctic Climate System (Cambridge University Press, Cambridge 2005).

SERREZE, M. C., and FRANCIS, J. (2006), *The Arctic Amplification Debate,* Clim. Change *76,* 241–264.

SUTEANU, C. (2011): *Detrended Fluctuation Analysis of Daily Atmospheric Surface Temperature Records in Atlantic Canada,* Can. Geogr. *55,* 180–191, doi:10.1111/j.1541-0064.2010.00323. x.

SUTEANU, C., and MANDEA, M. (2012), *Surface Air Temperature in the Canadian Arctic: Scaling and Pattern Change,* Meteorology and Atmospheric Physics *118,* 3, 179–188.

SUTEANU, C. (2013), *Identifying Change in The Variability of Surface Air Temperature Patterns,* Pattern Recognition in Physics *1*(1),135–142.

VAROTSOS, C.A., EFSTATHIOU, M., and TZANIS, C. (2009), *Scaling Behaviour of the Global Tropopause,* Atmos. Chem. Phys. *9,* 677–683.

VAROTSOS, C. A., EFSTATHIOU, M. N., and CRACKNELL, A. P. (2013), *On the Scaling Effect in Global Surface Air Temperature Anomalies,* Atmos. Chem. Phys. *13,* 5243–5253, doi:10.5194/acp-13-5243-2013.

VAROTSOS, C.A., and KIRK-DAVIDOFF, D. (2006), *Long-memory Processes in Ozone and Temperature Variations at the Region 60 S–60N,* Atmos. Chem. Phys. *6,* 4093–4100, doi:10.5194/acp-6-4093-2006.

VAROTSOS, C.A., MILINEVSKY, G., GRYTSAI, A., EFSTATHIOU, M., and TZANIS, C. (2008), *Scaling effect in planetary waves over Antarctica,* International Journal of Remote Sensing, *29* (9), 2697–2704.

VINCENT, L.A., ZHANG, X., BONSAL, B.R., and HOGG, W.D. (2002), *Homogenization of Daily Temperatures over Canada.* J. Clim. *15,* 1322–1334.

VON STORCH, H., ZWIERS, F.W., Statistical Analysis in Climate Research (Cambridge University Press, Cambridge 1999).

WALSH, J.E., SHAPIRO, I., and SHY, T.L. (2005), *On the Variability and Predictability of Daily Temperatures in the Arctic,* Atmos Ocean *43*(3), 213–230.

WALSH, J.E., OVERLAND, J.E., GROISMAN, P.Y., and RUDOLF, B. (2011), *Ongoing Climate Change in the Arctic,* Ambio *40,* 6–16.

WEATHERHEAD, E., GEARHEARD, S., and BARRY, R. G. (2010), *Changes in Weather Persistence: Insight from Inuit Knowledge,* Global Environmental Change, *20*(3), 523–528, doi:10.1016/j.gloenvcha.2010.02.002.

WILLIAMS, T., and HARDISON, P. (2013), *Culture, law, Risk And Governance: Contexts of Traditional Knowledge in Climate Change Adaptation,* Climatic Change *120,* 531–544, doi:10.1007/s10584-013-0850-0.

ZHANG, Q., SUNDQVIST, H. S., MOBERG, A., KORNICH, H., NILSSON, J., and HOLMGREN, K. (2010), *Climate Change Between the Mid and Late Holocene in Northern High Latitudes—Part 2 Model-Data Comparisons,* Climate of the Past *6,* 609–626.

(Received April 1, 2014, revised June 4, 2014, accepted June 6, 2014, Published online July 3, 2014)

Reprinted from the journal

Pure Appl. Geophys. 172 (2015), 2075–2082
© 2014 The Author(s)
This article is published with open access at Springerlink.com
DOI 10.1007/s00024-014-0865-0

Pure and Applied Geophysics

On Microscopic Mechanisms Which Elongate the Tail of Cluster Size Distributions: An Example of Random Domino Automaton

Zbigniew Czechowski[1]

Abstract—On the basis of simple cellular automaton, the microscopic mechanisms, which can be responsible for elongation of tails of cluster size distributions, were analyzed. It was shown that only the appropriate forms of rebound function can lead to inverse power tails if densities of the grid are small or moderate. For big densities, correlations between clusters become significant and lead to elongation of tails and flattening of the distribution to a straight line in log–log scale. The microscopic mechanism, given by the rebound function, included in simple 1D RDA can be projected on the geometric mechanism, which favours larger clusters in 2D RDA.

Key words: Long tails, SOC, Cellular automata.

1. Introduction

Long tail distributions appear in many natural phenomena, e.g., in geophysics we can mention: size distribution of earthquakes, seismic faults, volcanic eruptions, and floods or hurricanes. We have only a limited understanding of the physical origins of long tails in these processes. Some models were introduced to explain this behaviour. In particular, SOC models, given by appropriate cellular automata, were proposed. They generate power laws; however, it is not exactly clear what mechanism is responsible for these long tail distributions in the models. However, it was frequently claimed that as a result of repetitive action of avalanches, the spatial correlations between different parts of the system appear, which lead to power distributions, like in critical phenomena. Interesting reviews of various approaches corresponding to this problem are presented in (Sornette 2006; Aschwanden 2013).

The goal of this paper is to show that microscopic rules of cellular automata should be considered as the main cause of long tail distributions. In order to prove this, we use the simple 1D cellular automaton, which is described by analytical equations. This feature enables us to make a clear analytical investigation of a role for microscopic rules in generating long tails. The automaton was introduced and analyzed in our previous papers (Czechowski and Białecki 2012a, b; Białecki and Czechowski 2013; Białecki 2013).

In Sect. 2 the rules and equations of the 1D model are briefly presented. The influence of some details of microscopic rules of the automaton and correlations on a shape of the cluster size distribution function is analyzed in Sect. 3. Section 4 introduces a similar 2D cellular automaton, in which inverse power tails have a geometric origin. In Sect. 5 we show that the simple 1D automaton can mimic and explain the behavior of a more complex 2D model.

2. The Model

The random domino automaton (RDA) is defined on a grid of N cells (e.g., 1D or 2D with different geometry), in which it is determined what is the neighbourhood of the cells. Cells can be in two states: empty or occupied by a single point. The rules of RDA on the grid are as follows: at each time step a point hits a randomly chosen cell; if this cell is empty, it becomes occupied with probability v or with probability $1 - v$ and the point is rebounded. If the cell belongs to a cluster of i neighbouring occupied cells then with probability $\mu(i)$ all cells of the cluster become empty (the avalanche of size i is triggered) or

[1] Institute of Geophysics Polish Academy of Sciences, Księcia Janusza 64, 01-452 Warsaw, Poland. E-mail: zczech@igf.edu.pl

with probability $1 - \mu(i)$ the point is rebounded from the grid. After an initial saturation process, the automaton reaches a statistically stationary state. The RDA may be treated as an extension (see for details CZECHOWSKI and BIAŁECKI 2012a; BIAŁECKI and CZECHOWSKI 2013) of the forest fire model studied in PACZUSKI and BAK (1993).

In the case of a 1D grid, in a statistically stationary state, the following set of equations for size distribution of occupied clusters n_i was derived (BIAŁECKI and CZECHOWSKI 2013):

$$n_1 = \frac{1}{\mu(1)/v + 2}\left((1 - \rho)N - 2n + n_1^0\right)$$

$$n_2 = \frac{1}{2\mu(2)/v + 2}\left(1 - \frac{n_1^0}{n}\right)$$

$$n_i = \frac{1}{i\mu(i)/v + 2}\left[2\left(1 - \frac{n_1^0}{n}\right)n_{i-1} + n_1^0 \sum_{k=1}^{i-2} \frac{n_k n_{i-1-k}}{n^2}\right]$$

$$i \geq 3, \tag{1}$$

where ρ (density of occupied cells), n (number of clusters), and n_1^0 (number of empty clusters of size 1) are expressed in terms of n_i, $\mu(i)$ and v:

$$\rho = \frac{1}{N}\sum_{i \geq 1} i n_i, \quad n = \sum_{i \geq 1} n_i,$$

$$n_1^0 = \frac{2n}{3 + \frac{2}{vn}\sum_{i \geq 1} i\mu(i)n_i}. \tag{2}$$

For given N, $\mu(i)$ and v, Eq. (1) with formulas (2) form a closed set of implicit equations for n_i.

In this paper we assume the rebound function $\mu(i)$ of the form $\mu(i) = \delta/i^\sigma$ with $\sigma = 0$, 1 and 2. Equation (1) may present a simple recurrence if we can find n, ρ, and n_1^0. This is possible for $\sigma = 1$ or $\sigma = 2$, because equations for moments

$$m_z = \frac{1}{N}\sum_{i \geq 1} i^z n_i, \quad m_z' = \frac{1}{N}\sum_{i \geq 1} \frac{\mu(i)}{v} i^z n_i \tag{3}$$

derived (BIAŁECKI and CZECHOWSKI 2013) in the following form

$$m_{z+1}' = 1 - m_1 - 2m_2 + 2\sum_{k=0}^{z-1} \binom{z}{k} m_k$$

$$+ \frac{2}{3m_0 + 2m_1'} \sum_{l,p=1}^{l+p \leq z} \binom{z}{l+p}\binom{l+p}{l} m_l m_p, \tag{4}$$

can be reduced to a closed set of two equations (for $z = 0$ and $z = 1$; in the case $\sigma = 1$),

$$\theta m_0 = 1 - m_1 - 2m_0$$
$$\theta m_1 = 1 - m_1 \tag{5}$$

or a closed set of three equations (for $z = 1$, $z = 2$, $z = 3$; in the case $\sigma = 2$).

$$\theta m_{-1} = 1 - m_1 - 2m_0$$
$$\theta m_0 = 1 - m_1$$
$$\theta m_1 = 1 + 3m_1 + 2m_1^2 \frac{2}{3m_0 + 2\theta m_{-1}}, \tag{6}$$

where $\theta = \delta/v$. Let us note that $m_0 = n/N$ and $m_1 = \rho$.

On the other hand, in the case of $\sigma = 0$ [i.e., $\mu(i) = \delta = \text{const}$] we have no closed set of moment equations, but n and n_1^0 can be expressed by ρ:

$$n(\rho) = \frac{N}{2}\left[1 - \left(\frac{\delta}{v} + 1\right)\rho\right], \tag{7}$$

$$n_1^0(\rho) = \frac{2n^2(\rho)}{3n(\rho) + 2N\frac{\delta}{v}\rho}, \tag{8}$$

where the first equation results from the balance for n (see Eq. 2 in BIAŁECKI and CZECHOWSKI 2013). Then, in order to find the stationary density, we put Eqs. (7) and (8) to the set (1) and solve numerically the following implicit equation (definition of density) for unknown variable ρ:

$$\rho = \frac{1}{N}\sum_{i \geq 1} i n_i(\rho). \tag{9}$$

3. Influence of Microscopic Rules of the Automaton and Correlations on the Length of the Tail of Cluster Size Distribution

We check how some details of microscopic rules of the automaton, such as a form of function $\mu(i)$, the parameter $\theta = \delta/v$, and possible correlations between clusters, can influence the shape of the cluster distribution function.

3.1. Role of a Form of Function $\mu(i)$

We assume $\mu(i) = \delta/i^\sigma$ and analyze the following three cases:

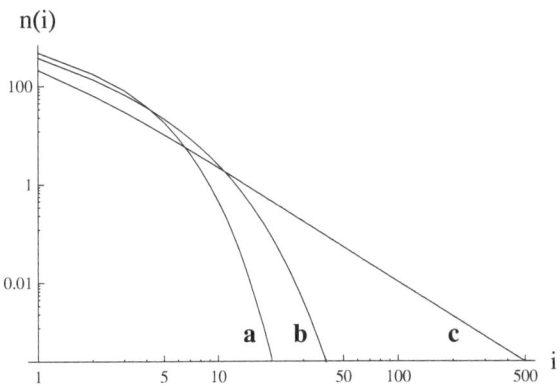

Figure 1
Cluster size distributions n_i calculated from Eq. (1) with $\mu(i) = \delta/i^{\sigma}$ for three cases: **a** $\sigma = 0$ and $\theta = 0.68$, **b** $\sigma = 1$ and $\theta = 1.81$, and **c** $\sigma = 2$ and $\theta = 7$. For each case the stationary density is $\rho \approx 0.3559$

(a) $\sigma = 0$
(b) $\sigma = 1$
(c) $\sigma = 2$.

Therefore, as it has been discussed in Sect. 2, the cluster size distribution n_i can be found from the recurrent Eq. (1). For each case we set for the stationary density the same moderate value $\rho = 0.3559$; this condition is realized for: $\theta = 7$ when $\sigma = 2$, $\theta = 1.81$ when $\sigma = 1$, and $\theta = 0.68$ when $\sigma = 0$. Figure 1 shows cluster size distributions n_i. The discrete distributions are presented by continuous lines for better clarity. For $\sigma = 0$ (constant μ) a short-tail exponential distribution is obtained. However, if the rebound function $\mu(i)$ is a decreasing function then a lengthening of the tail is observed—the faster the decrease the longer tail. For $\sigma = 2$ and $\theta = 7$ the tail is inverse-power with the exponent about -2.323.

Explanation of the mechanism is simple: the function $\mu(i)$ describes the probability that the particle, which hits a randomly chosen occupied cell belonging to a cluster of size i, triggers the avalanche (the cluster disappears). If $\mu(i)$ decreases with a cluster size then, in that trial, a probability of survival of larger clusters is greater than for smaller clusters. This mechanism leads to the bigger input of larger clusters in the distribution, i.e., to longer tails.

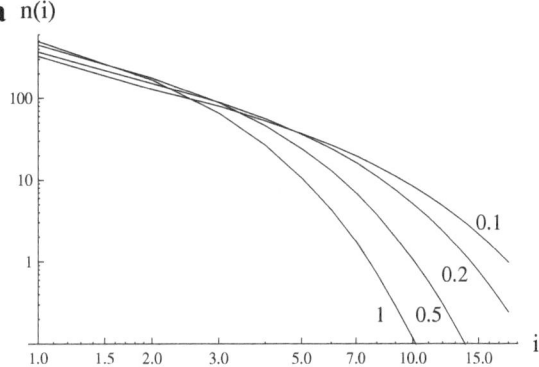

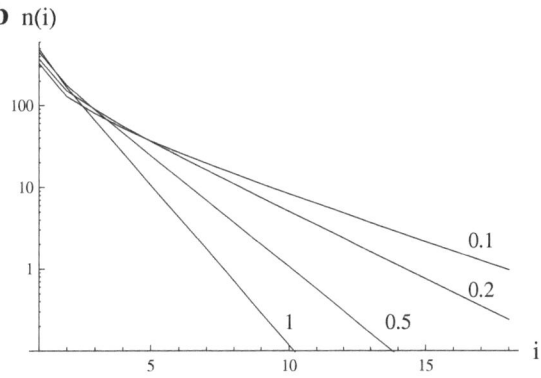

Figure 2
a Cluster size distributions n_i calculated from Eq. (1) with $\mu(i) = \delta$ for four assumed values of parameter $\theta = 1$, 0.5, 0.2 and 0.1 (in log–log scale). **b** Cluster size distributions n_i calculated from Eq. (1) with $\mu(i) = \delta$ for four assumed values of parameter $\theta = 1$, 0.5, 0.2 and 0.1 (in lin–log scale to show exponential tails)

3.2. Role of Parameter $\theta = \delta/v$

When the form of function $\mu(i) = \delta/i^{\sigma}$ and the exponent σ is established, then the value of parameter δ itself can have an influence on the shape of the cluster size distribution. In all formulas the parameter is appearing in the ratio $\theta = \delta/v$; therefore, we investigate an influence of changing θ separately for the three cases (a–c).

In case (a) decreasing parameter θ does not change an exponential dependence of the tail of cluster size distribution on the cluster size, but the exponential decrease is less and less steep (see Fig. 2a, b).

In case (b) a similar lengthening of tails is also observed (see Fig. 3a). Moreover, the widening central part of the distribution is achieving (in the

a n(i)

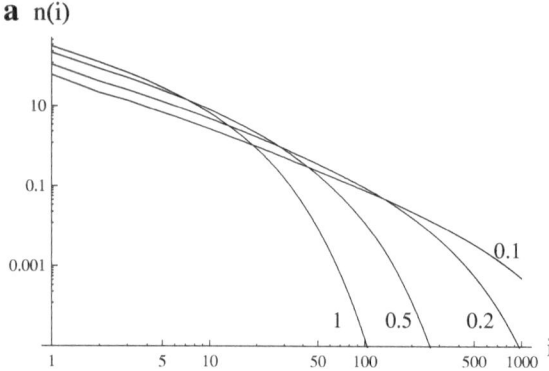

b n(i)

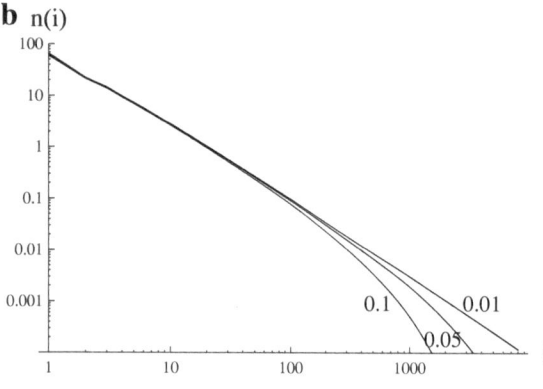

Figure 3
a Cluster size distributions n_i calculated from Eq. (1) with $\mu(i) = \delta/i$ for four assumed values of parameter $\theta = 1, 0.5, 0.2$ and 0.1. **b** Cluster size distributions n_i calculated from Eq. (1) with $\mu(i) = \delta/i$ for three assumed values of parameter $\theta = 0.1, 0.05$ and 0.01. Here the fixed product θN is maintained when θ decreases and N increases. For the case $\theta = 0.01$, the power exponent is $\alpha \approx -1.5$

limit $\theta \to 0$) the inverse-power slope $i^{-3/2}$. Figure 3b, in which the fixed product θN is maintained when θ decreases, clearly shows the behavior. This is in agreement with the asymptotic result of Motzkin numbers recurrence (BIAŁECKI 2012a) applied to RDA.

In case (c) decreasing parameter θ causes decreasing inverse-power exponents of the tail, e.g., if θ drops from 8.0 to 7.0, the exponent drops from 2.83283 to 2.32307 (see Fig. 4). Therefore, in this case we also observe lengthening of tails with decreasing θ. In Fig. 4 two graphs of n_i are presented for each θ, because the closed set (6) of three moment equations has two solutions: $(m_1^{(1)}, m_2^{(1)}, m_3^{(1)})$ and $(m_1^{(2)}, m_2^{(2)}, m_3^{(2)})$ if $\theta > 6.936236$.

In order to explain the common (in the three cases under investigation) influence of the parameter θ on

n(i)

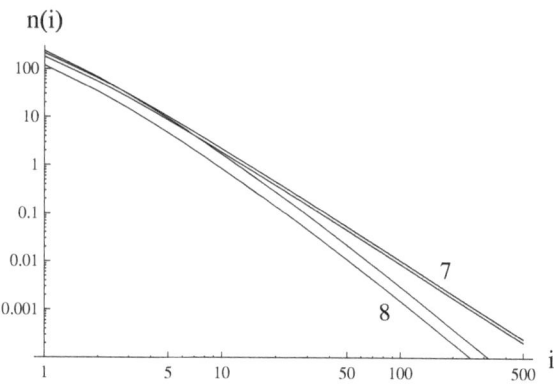

Figure 4
Cluster size distributions n_i calculated from Eq. (1) with $\mu(i) = \delta/i^2$ for two assumed values of parameter $\theta = 8$ and 7. For each θ *two lines* are presented because Eq. (6) has two solutions. Power exponent for the pair $\theta = 8$ is $\alpha \approx -2.83$ and for the pair $\theta = 7$ is $\alpha \approx -2.32$

tails of cluster size distributions, we should note that decreasing θ causes increasing stationary density ρ (when $\theta \to 0$ then $\rho \to 1$). The increase of density means reducing the number of empty cells on the grid and, as a result, it reduces a possibility of creation of clusters of size 1 (and their next growth to small clusters of size 2, 3, ...). On the other hand, the probability of growth of clusters, due to hitting an empty cell on a cluster perimeter or due to joining two neighbor clusters if the particle hits the empty cell between them, increases. These mechanisms lead to the growth of the number of greater clusters at the expense of the number of small clusters (because the source of n_1 is being depleted). This is observed as the lengthening of tails.

3.3. Role of Correlations Between Clusters

Equation (1) was derived under the assumption that clusters are distributed independently on the grid (i.e., the length of the next cluster on the grid does not depend on the length of the previous one). The assumption can be justified only for small and moderate densities. However, for big densities the correlations can be significant. Here, we analyze the influence of correlation on a shape of the cluster size distribution function. For the test we use 1D RDA with constant $\mu(i) = \delta$, and we compare analytically computed n_i with the results of simulations. When we

a n(i)

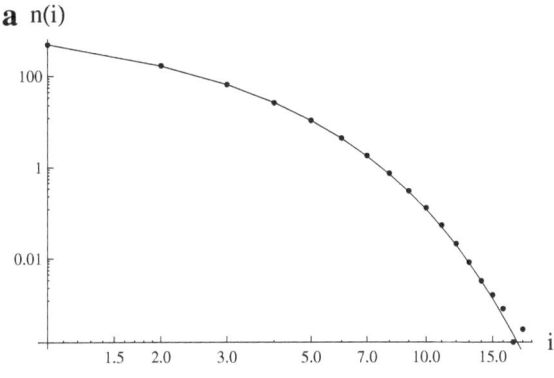

b n(i)

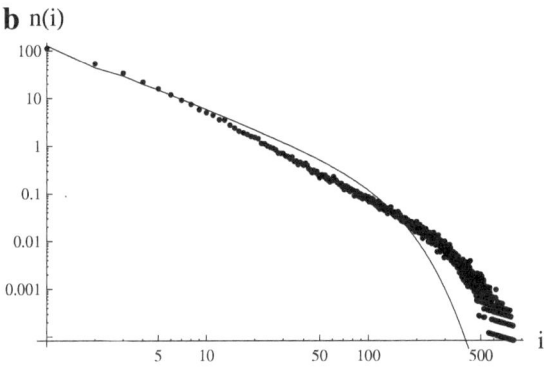

Figure 5

a Comparison of cluster size distributions n_i calculated from Eq. (1) (*line*) with generated by simulation of RDA (*points*) for the case $\mu(i) = \delta$ and $\theta = 1$. The stationary density is $\rho \approx 0.3076$. The influence of correlations is negligible. **b** Comparison of cluster size distributions n_i calculated from Eqs. (1) (*line*) with generated by simulation of RDA (*points*) for the case $\mu(i) = \delta$ and $\theta = 0.002$. The stationary density is $\rho \approx 0.8099$. Influence of correlations is significant. There is a lengthening of the tail, and a remarkable flattening of the simulated distribution to a straight line ($\alpha \approx -1.78$) in the central part of the graph is visible

assume $\theta = 1$, the moderate value of stationary density $\rho = 0.3076$ is obtained. Figure 5a shows a good agreement between the analytical curve and the numerical result represented by points. This means that correlations are too weak to deform the cluster size distribution. However, for a very small $\delta = 0.002$, the stationary density increases to $\rho = 0.8099$ and the influence of correlations is significant. First of all, on Fig. 5b we observe a lengthening of the tail of a simulated distribution when $i > 200$, but, what is more important, a remarkable flattening of the distribution to a straight line in the range $i \in (4, 300)$ is evident. In log–log

scale this means power-law behaviour. The exponent is about -1.78.

Spatial correlations between different parts of the system were often taken into account as a main factor leading to power laws in SOC models (see SORNETTE 2006 Chapter 15; CHEN and WU 2010; CORRAL *et al.* 2008). In this section correlation between clusters were discussed.

4. Two-Dimensional Random Domino Automaton with Constant μ

We investigate RDA with constant rebound parameter μ on a 2D square grid. Cells are called neighbours, if they have one side in common, but not if they only touch at one corner. The 2D case is much more complex than 1D RDA because i-clusters have perimeter lengths $t(i, r_i)$ depending on the cluster size i and, moreover, on different geometric configurations r_i of i-clusters. Because of these complexities, instead of deriving full equations for n_i we are presenting only a draft of the balance equation. We use the mean perimeter $t(i)$ averaged over all possible configurations r_i.

A number n_i of clusters of size i can increase if a new point hits an empty cell on the perimeter of a $(i - 1)$-cluster and the cell becomes occupied. The probability is $Cvt(i - 1)n_{i-1}$. The parameter C denotes the probability that the empty cell does not belong to a perimeter of any other cluster (to avoid the linking process). On the other hand, if the empty cell belongs to the overlap of perimeters of two, three or four clusters, whose common size is $i - 1$, then these clusters join in a cluster of size i. Estimation of the probability of this process is too complex for 2D RDA, so we denote it as the "fusion term".

Losses of n_i can be due to hitting an empty cell (and changing its state) on a perimeter of i-cluster. The probability is $vt(i)n_i$ or if the point hits an occupied cell of the i-cluster and triggers the avalanche then the probability is $\mu i n_i$. Therefore, the draft of the balance equation has the form:

$$\mu i n_i + vt(i)n_i = Cvt(i-1)n_{i-1} + v \cdot \textit{(fusion term)}. \tag{10}$$

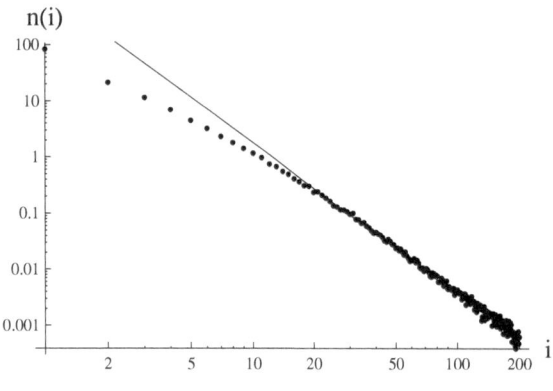

Figure 6

Cluster size distribution n_i generated by 2D RDA with constant $\mu(i) = \delta$ and $\theta = 1$ (*points*). *Line* presents the inverse power function with $\alpha = -2.7$

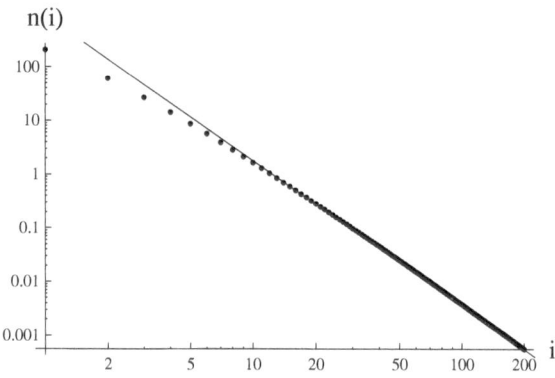

Figure 7

Cluster size distribution n_i generated by 1D RDA with $\mu(i) = \delta/i^2$ and $\theta = 7.74$ (*points*). *Line* presents the inverse power function with $\alpha = -2.7$

From the percolation theory (STAUFFER and AHARONY 1992) we use the relation $t(i) \sim i$, valid for large clusters. If we put $t(i) = ai + b$ to Eq. (10) then the equation for the number of i-clusters in stationary state is:

$$n_i = \frac{1}{i(a + \mu/v) + b}[(ai + b - a)Cn_{i-1} + (fusion\ term)].$$

(11)

5. Correspondence Between Privilege Mechanisms in 1D RDA and in 2D RDA with Constant μ

The comparison of the draft Eq. (11) with the respective equation for 1D RDA with $\mu(i) = \delta/i^2$ [i.e., the third of the Eq. (1) set, in which $c = 1 - n_1^0/n$, $d = n_1^0/n^2$ and numerator and denominator were multiplied by i]:

$$n_i = \frac{1}{2i + \delta/v}\left[2icn_{i-1} + id\sum_{k=1}^{i-2} n_k n_{i-1-k}\right] \quad (12)$$

shows a close resemblance of both equations. The "*fusion term*" in Eq. (11) contains sums of products of two, three and four cluster distributions and strongly depends on geometric configurations of these clusters, which are associated with their perimeters. Therefore, the factor i should appear in the term (like in Eq. 12).

Let us check if the similar Eqs. (11) and (12) can lead to similar distributions for n_i. In the case of 2D

RDA we base it on numerical simulations. The distribution function of clusters calculated from the generated data is shown in Fig. 6. The distribution has an inverse power tail with the fitted exponent 2.7. For 1D RDA we can use Eq. (1). Then, assuming $\mu(i) = \delta/i^2$ and $\delta/v = 7.74$, the distribution of n_i has also the inverse power tail with exponent 2.7 (see Fig. 7).

Moreover, in the case of 1D RDA we can also apply another approach. The 1D automaton has an important feature in that it provides one-to-one correspondence between rebound parameters and size distribution of clusters (BIAŁECKI and CZECHOWSKI 2013). Therefore, we are going to reconstruct such a function $\mu(i)$, which can give the pure inverse power distribution $n_i \sim i^{-2.7}$ in the finite range $i = 1, \ldots, K$ (for $i > K$ we put $n_i = 0$). We use the following relations:

$$\frac{\mu(1)}{v} = \frac{1}{n_1}\left[(1 - \rho)N - 2n + n_1^0\right] - 2$$

$$\frac{\mu(2)}{v} = \frac{n_1}{n_2}\left(1 - \frac{n_1^0}{n}\right) - 1$$

$$\frac{\mu(i)}{v} = \frac{1}{i}\left[2\left((1 - \frac{n_1^0}{n})\frac{n_{i-1}}{n_i} - 1\right) + \frac{n_1^0}{n_i}\sum_{k=1}^{i-2}\frac{n_k n_{i-1-k}}{n^2}\right]$$

$$i \geq 3.$$

(13)

In these equations the parameter n_1^0 is found numerically from the implicit equation

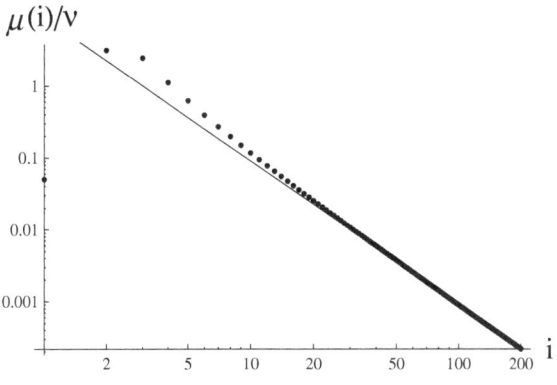

Figure 8

The rebound function $\mu(i)/\nu$ (*points*) reconstructed from Eq. (13), where $n_i = 650/i^{2.7}$ for $i = 1, \ldots, 200$, $\rho \approx 0.3281$. *Line* presents the inverse-power function with $\alpha = -2$

$$n_1^0 = \frac{2n}{3 + \frac{2}{\nu n} \sum_{i \geq 1} i\mu(i, \ n_1^0) \, n_i} \qquad (14)$$

in which for $\mu(i)$ we put Eqs. (13) and the summation is over the finite range 1, 2, ... , K. The reconstructed $\mu(i)$ is shown in Fig. 8. For $i > 20$ the function has an inverse power shape with the exponent close to two (continuous line).

The origin of long tails in this 1D RDA has been discussed in Sect. 3.1. This is a mechanism of privilege of greater clusters related to the survival rate given by the assumed rebound function $\mu(i) \sim 1/i^2$. In the case of 2D RDA with constant μ, the privilege has a geometric nature. The greater clusters have longer perimeters $t(i)$, which means greater probability of growth of these clusters due to linking of neighbouring clusters, when a site on the intersection of their perimeters becomes occupied. As a result, the number of greater clusters increases and long tail cluster size distribution appears. The mechanism was analyzed in (CZECHOWSKI 2003) for the case of the percolation process on Bethe and the 2D lattice.

In 1D RDA we have a constant perimeter equal to two for every cluster, but the survival rate (which increases with cluster size) can fulfill the role of the geometric privilege. In this meaning the simple 1D RDA mimics and explains behavior of the more complex 2D model.

In seismology both our automata reflect two main properties of earthquake processes: slow accumulation of energy at a constant rate to the fault, caused by the relative movement of tectonic plates, and an abrupt release of energy in earthquakes (avalanches). The motivations for using such simple automata are as follows: full physical modeling of earthquake process is an extremely complex task; therefore, many details must be averaged or simplified in some way. Moreover, the rules of the automata can be very simple, but they can lead to complex evolution of the system and produce rich patterns. The description of models by handy mathematical equations enables us to understand relations between microscopic mechanisms [e.g., given by the rebound function $\mu(i)$] and macroscopic characteristics of seismic phenomena, as, for example, the energy frequency power distribution or recurrence interval statistics (BIAŁECKI 2012b).

6. Conclusions

The tasks put forward in the introduction could be achieved only because we had at our disposal 1D RDA introduced earlier. This cellular automaton has simple rules, but this feature let us derive analytical equations describing the statistically stationary state. In particular, due to the analytical description, the influence of correlations on a form of cluster size distribution can be analyzed.

First of all, we have shown how the choice of rebound function $\mu(i)$ generates the form of n_i. The faster $\mu(i)$ is decreasing, the longer the tail of cluster size distribution. For constant $\mu(i) = \delta$, exponential tails are generated; however, for $\mu(i) = \delta/i^2$ the tail is described by a power law. We explain this behaviour by the privilege mechanism. The probability of survival of larger clusters is greater than that of smaller clusters (larger clusters are privileged).

The role of the parameter δ in the rebound function $\mu(i)$ was also analyzed. It describes the probability of triggering an avalanche irrespectively of the cluster size, if a point hits a randomly chosen occupied cell. Decreasing of the parameter causes a lengthening of tails; however, this regularity should be referred rather to the rise of density on the grid.

When the density is big enough, the role of correlations between clusters on the grid becomes essential and the analytical Eq. (1) is not valid. Comparison between analytical solutions and simulation results explains the role of correlations. Correlations induce a lengthening of the tail and a remarkable flattening of the distribution to a straight line. Even for constant $\mu(i) = \delta = 0.002$ (then density $\rho \approx 0.81$), the cluster size distribution obtained from simulations resembles a power law in some range of i.

For small and moderate densities; however, the correlations are weak and can be omitted. Then the analytical description is valid. We have shown that equations for 1D RDA with $\mu(i) = \delta/i^2$ are similar to appropriate equations for 2D RDA with constant $\mu(i) = \delta$. In this way we can state that the simple 1D RDA mimics and explains the behavior of a more complex 2D model. The probability of survival of clusters [given by $1/\mu(i)$] in 1D case refers to the probability of growth of clusters [given by the cluster perimeter $t(i)$] in the 2D case, but both describe some privilege of larger clusters.

An inspiration to the concept of privilege, which can be treated as a generalization of Simon's (SIMON 1955, 1960) idea of preferential attachment, was an analysis of the coagulation equation in the crack fusion problem (CZECHOWSKI 1993). In our papers (CZECHOWSKI 2001, 2003, 2005; CZECHOWSKI and ROZMARYNOWSKA 2008), it was shown that in many cases (i.e., percolation processes, cellular automata, Cantor set, resource redistribution model, return-to-the-origin problem in random walk, multiplicative processes, and multiplication of probabilities) the hidden privilege can be extracted and explained. Moreover, a relevance between the privilege concept and the nonlinear approach was found. The present paper is a continuation of those investigations.

Acknowledgments

This work was partially financed by the project (Contract No. DEC-2012/05/B/ST10/00598) carried out by the Institute of Geophysics, Polish Academy of Sciences on the order of the National Science Centre.

Open Access This article is distributed under the terms of the Creative Commons Attribution License which permits any use, distribution, and reproduction in any medium, provided the original author(s) and the source are credited.

REFERENCES

ASCHWANDEN, M., ed., Self-Organized Criticality Systems (Open Academic Press, Berlin Warsaw 2013).

BIAŁECKI, M. (2012a), *Motzkin numbers out of Random Domino Automaton*, Physics Letters A *376*, 3098–3100.

BIAŁECKI, M. (2012b), An explanation of the shape of the universal curve of the Scaling Law for the Earthquake Recurrence Time Distributions, arXiv:1210.7142 [physics.geo-ph].

BIAŁECKI, M. (2013), *From statistics of avalanches to microscopic dynamic parameters in a toy model of earthquakes*, Acta Geophysica *61*, no 6, 1677–1689.

BIAŁECKI, M., and CZECHOWSKI, Z. (2013), *On one-to-one dependence of rebound parameters on statistics of clusters: exponential and inverse-power distribution out of Random Domino Automaton*, J. Phys. Soc. Jpn. *82*, 014003 (9 pp).

CHEN, H.S., and WU, G.Y. (2010), *Effects of pair correlation on mean-field theory of BTW sand pile model*, Physica A *389*, 2339–2350.

CORRAL, A., TELESCA L., and LASAPONARA, R. (2008), *Scaling and correlations in the dynamics of forest-fire occurrence*, Phys. Rev. E *77*, 016101, doi:10.1103/PhysRevE.77.016101.

CZECHOWSKI, Z., (1993), *A kinetic model of nucleation, propagation and fusion of cracks*, J. Phys. Earth *41*, 127–137.

CZECHOWSKI, Z., (2001), *Transformation of random distributions into power-like distributions due to non-linearities: application to geophysical phenomena*, Geophys. J. Int. *144*, 197–205.

CZECHOWSKI, Z., (2003), *The privilege as the cause of the power distributions in geophysics*, Geophys. J. Int. *154*, 754–766.

CZECHOWSKI, Z., (2005), *The importance of the privilege in resource redistribution models for appearance of inverse-power solutions*, Physica A *345*, 92–106.

CZECHOWSKI, Z., and ROZMARYNOWSKA, A. (2008), *The importance of the privilege for appearance of inverse-power solutions in Ito equations*, Physica A *387*, 5403–5416.

CZECHOWSKI, Z., and BIAŁECKI, M. (2012a), *Ito equations out of domino cellular automaton with efficiency parameters*, Acta Geophysica *60*, no 3, 846–857.

CZECHOWSKI, Z., and BIAŁECKI, M. (2012b), *Three-level description of the domino cellular automaton*, Journal of Physics A: Math. Theor. *45*, 155101 (19 pp).

PACZUSKI, M., and BAK, P. (1993), *Theory of the one-dimensional forest-fire model*, Phys. Rev. E *48*, R3214–R3216.

SIMON, H.A. (1955). *On a class of skew distribution functions*. Biometrika *42*, 425–440.

SIMON, H.A. (1960). *Some further notes on a class of skew distribution functions*. Information and Control *3*, 80–88.

SORNETTE, D., Critical Phenomena in Natural Sciences (Springer-Verlag, Berlin Heidelberg 2006).

STAUFFER, D., and AHARONY, A., Introduction to Percolation Theory (Taylor and Francis, London 1992).

(Received March 25, 2014, revised May 21, 2014, accepted May 23, 2014, Published online June 15, 2014)